B. G. Jansen van Rensburg

Oral Biology

B. G. Jansen van Rensburg, B.D.S.,

H. Dip. Dent, B. Sc., M. Sc. (Dent. Sci)
Professor of Oral Biology
University of Stellenbosch
South Africa

Quintessence Publishing Co, Inc

Chicago, Berlin, London, Tokyo, São Paulo, Moscow, Prague, Sofia, and Warsaw

Originally published in Afrikaans as *Mondbiologie* by Juta en Kie, Bpk 1981, Posbus 123, Kenwyn 7790.

Library of Congress Cataloging-in-Publication Data
Van Rensburg, B. G. Jansen.
  (Mondbiologie. English)
  Oral biology/B. G. Jansen van Rensburg.
  p. cm.
  Includes bibliographical references.

  ISBN 0-86715-271-0
  1. Mouth. 2. Teeth. I. Title.
  (DNLM: 1. Mouth–anatomy & histology. 2. Mouth–physiology. 3. Tooth–anatomy & histology.
  4. Tooth–physiology.
  WU 101 V217m 1995a)
  RK280. V3613 1995
  612.3' 1–dc20
  DNLM/DLC
  for Library of Congress

© 1995 by Quintessence Publishing Co, Inc

Published by Quintessence Publishing Co, Inc
551 N. Kimberly Dr.
Carol Stream, IL 60188

All rights reserved
This book or any part thereof may not be reproduced,
stored in a retrieval system, or transmitted in any form or
by any means, electronic, mechanical, photocopying,
recording, or otherwise, without prior written permission
of the publisher

Typesetting: Type Design Fotosatz- und Layoutservice GmbH, Berlin, Germany
Printing and binding: Loibl Druck und Gestaltung, Neuburg, Germany

Printed in Germany

Preface

It is unfortunate that a schism exists between the basic sciences, such as general embryology, anatomy, histology, physiology, and biochemistry, and clinical dentistry in many undergraduate dental training programs. A need is apparent for a core curriculum in oral biology in which selected aspects of the basic sciences are elaborated and targeted at the oral and related environment to enable the dental clinician to effectively diagnose problems in this field. A general dilemma facing student and teacher alike is the relative absence of a modern textbook giving an authoritative overview of the field generally accepted as constituting basic oral biological sciences. A large number of highly specialized works have to be consulted for information on often very limited aspects of the subject. It is hoped that this book provides a satisfactory answer to both these needs.

The contents of Part I are intended as an introduction to oral biology and deal with general aspects of embryology, gross anatomy, histology, and physiology of the oral cavity, its contents, and related systems. In addition, attention is given to genetics, blood groups, hemostasis, evolution, and relevant aspects of anthropology and comparative anatomy. The author regards a knowledge of these subjects as a prerequisite for the more detailed study of the contents of Part II.

Part II deals with the teeth and surrounding structures, tooth eruption, the temporomandibular joint, salivary glands and saliva, fluorides, mastication, taste, and deglutition.

Although also intended as a reference work for dental clinicians, this text is of particular value to undergraduate and postgraduate students. The bibliography at the end of each chapter has been carefully selected to provide further information on specific topics, if required. Moreover, review questions are included to help focus the reader's attention on fundamental principles contained within each chapter.

I am deeply grateful to Dr Helmut Heydt, Executive Director of the Dental Association of South Africa, for continued encouragement and unceasing efforts to make this publication possible, and to Mr Horst-Wolfgang Haase of Quintessenz Verlags-GmbH for most helpful co-operation in the production of this book. I am indebted to my Dean, Professor Wynand Dreyer, for inspiration continually received and to faculty members for never-ending encouragement.

My sincere thanks go to Elizabeth Lückhoff for her meticulous preparation of the manuscript and to my friend and colleague, Professor Vincent Phillips, for his careful perusal of the text and helpful suggestions.

B G J Van Rensburg

This work is dedicated to dental
students all over the world

Contents

Preface		5
PART I		17
1	**The cell**	19
	General remarks	19
	The cell membrane	20
	The cytoplasm	22
	The nucleus	25
	Arrangement of cells	27
2	**A historical review of embryology**	29
3	**Fertilization, establishment of germ layers, and fate of the primary embryonic tissues**	33
	Fertilization	33
	Establishment of the germ layers	35
	Fate of the primary embryonic tissues	41
4	**Epithelium**	45
	General remarks	45
	Types of epithelium	46
5	**Connective tissue, including bone and cartilage**	53
	General remarks	53
	Connective tissue	53
	Bone	58
	Cartilage	61

6	**Blood**	**65**
	General remarks	65
	Red blood corpuscles (erythrocytes)	65
	White blood cells (leukocytes)	66
	Blood platelets	69
	Blood plasma	70
	Lymph	70
	Blood values	71
7	**Muscle**	**73**
	General remarks	73
	Smooth muscle	73
	Striated skeletal muscle	74
	Cardiac muscle	78
8	**Transition from embryo to fetus**	**79**
9	**Endocrinology**	**81**
	The thyroid gland	81
	The parathyroid glands	82
	The pituitary gland	82
	The thymus gland	83
	The adrenal glands	83
	The pancreas	84
	The gonads	84
10	**Growth and development**	**87**
	General differences between the sexes	87
	Changes in physiologic function	88
	Skeletal age	89
	Dental age	89
	Factors which may influence growth	90
11	**Development of the nervous system**	**95**
	Development of the head fold	95
	Development of the central nervous system	97
	The spinal cord and peripheral nerves	103

12	**General functional and histologic aspects of the nervous system**	**105**
	General remarks	105
	Functions of the nervous system	106
	Histology of nervous tissue	109
	The trigeminal nerve	112
13	**Development of the face**	**117**
14	**Development of the palate and the nasal cavity**	**127**
15	**Development of the septomaxillary complex**	**133**
	The nasal capsule	133
	The maxilla	135
	The premaxilla	137
	The lacrimal bone	137
	The palatine bone	138
	The vomer	138
	The maxillary sinus	138
16	**Development of the mandible, the temporomandibular joint, and muscle**	**141**
	The mandible	141
	The temporomandibular joint	145
	Muscle	147
17	**Development of the pharyngeal arches and the tongue**	**149**
	The pharyngeal arches	149
	The tongue	154
18	**The mouth, pharynx, and nose**	**159**
	The mouth	159
	The pharynx	163
	The nose	163
19	**The digestive system**	**165**
	The esophagus	165
	The stomach	165
	The small intestine	166
	The large intestine	166
	Functions of the alimentary tract	166

20	**The respiratory system**	**169**
	The trachea and the extrapulmonary bronchi	169
	The lungs	169
21	**Introduction to genetics**	**173**
	Chromosomes and genes	173
	The human karyotype (chromosome set)	173
	Gene loci and alleles	175
	A character or trait	176
	Homozygotes and heterozygotes	176
	Dominance, recessivity and co-dominance	177
	Human sex chromosomes and sex determination	177
	Sex-linked inheritance	178
22	**Blood groups**	**181**
	General remarks	181
	The Rh (Rhesus) factor	183
	Some aspects of the genetics of blood	185
	Distribution of the ABO groups	186
23	**Hemostasis**	**187**
	General remarks	187
	Blood clotting	187
	Bleeding time	188
	Clotting time	190
	Blood clotting factors	190
24	**The cardiovascular system**	**193**
	The heart and major blood vessels	193
	The blood circulation after birth and the functions of the heart	194
	The fetal blood circulation	196
	A fainting attack	196
25	**Development of the arterial circulation of the head and neck**	**199**
26	**Lymphatic drainage of the scalp, face, oral cavity, and associated structures**	**209**

27	**Charles Darwin (1809-1882)**	**215**
28	**Man's position in the animal kingdom**	**219**
	General remarks on classification	219
	Man's position	220
	Mastication and heterodontism	222
29	**Teeth through the ages**	**227**
	General review	227
	The masticatory apparatus of man with reference to evolution and the infuence of diet	235

PART II 237

1	**A tooth and its surroundings**	**239**
	The development of a tooth	239
	Tooth eruption	240
	The components of a tooth	241
	The periodontium	242
2	**Development of the teeth – a general review**	**247**
	The origin of the dental lamina	247
	The dental papilla	248
	Changes in the dental lamina	248
	The origin of the enamel organ	248
	The epithelial sheath of Hertwig	252
	Changes in the papilla	252
	The dental follicle	254
	The tooth germ	254
	Enamel, dentine, and cementum	254
3	**Epithelium-ectomesenchyme interaction in tooth development**	**257**
	General remarks	257
	The origin of ectomesenchyme	257
	The mechanism of epithelium-ectomesenchyme interaction in tooth development	258

4	**Development of dentine**	**263**
	General remarks	263
	Differentiation of odontoblasts	263
	Formation of the matrix and mineralization	264
	Intratubular (peritubular) and intertubular dentine	268
5	**Dentine**	**271**
	General remarks	271
	Chemical composition	271
	General structure	271
	Odontoblasts and odontoblast processes	272
	Predentine	273
	Dentinal tubules	273
	Intratubular (peritubular) dentine	275
	Interglobular dentine	275
	Incremental lines	275
	Granular dentine (granular layer of Tomes)	277
	Amelodentinal junction (enamel-dentine junction)	277
	Dentinocemental junction	279
6	**Development of enamel**	**281**
	General remarks	281
	Determination of crown form	281
	Differentiation of the ameloblasts	282
	Formation of the matrix and early mineralization	285
	The nature of enamel matrix	285
	Mineralization and maturation of enamel	286
	The reduced enamel epithelium and enamel cuticle	286
7	**Enamel**	**289**
	General remarks	289
	Chemical composition	289
	The structure of enamel	290
	Enamel crystals and rod sheaths	290
	Cross striations	292
	Lines of Retzius	292
	Bands of Hunter-Schreger	295
	The amelodentinal junction	296
	Enamel tufts	296
	Enamel spindles	296
	Enamel lamellae	299

	Enamel permeability	299
8	**Development of cementum**	**301**
	General remarks	301
	Changes in the root sheath	301
	Formation of cementum	302
	Types of cementum	305
	Disturbed development	307
9	**Cementum**	**309**
	General remarks and chemical composition	309
	Structure and distribution	309
	Cementocytes	310
	Sharpey's fibers	310
	The amelocemental (enamel-cementum) junction	313
	Functions and clinical behavior	313
10	**The dental pulp**	**317**
	General remarks	317
	General appearance and structure	318
	Cells of the pulp	320
	Fibers of the pulp	320
	Blood vessels of the pulp	320
	Lymphatics of the pulp	321
	Nerve supply	321
11	**Form, arrangement and chronology of teeth**	**323**
	General remarks	323
	The primary dentition	329
	The secondary dentition	339
	Differences between primary and secondary teeth	363
12	**Composition of teeth**	**367**
	Methods to obtain samples of tooth material	367
	Hardness	368
	General remarks on chemical composition	368
	Enamel	369
	Dentine	369
	Cementum	370
	Trace elements	370

13	**Permeability and age changes of teeth**	**373**
	General remarks	373
	Permeability of enamel	373
	Permeability of dentine	374
	Permeability of cementum	374
	Withdrawal of minerals from enamel and dentine	375
	Age changes in teeth and their reaction to irritation	375
14	**Oral mucosa**	**379**
	General remarks	379
	Classification and nature of oral mucosa	380
15	**Calcium metabolism and bone mineralization**	**389**
	General remarks	389
	Calcium absorption	390
	Hormonal influences	390
	Osteogenesis (chemical aspects)	391
	Osteogenesis (morphological aspects)	393
	Important information	397
16	**The jaws**	**401**
	General remarks	401
	The maxilla	401
	The mandible	405
17	**The alveolar process**	**409**
	General remarks	409
	Structure of the alveolar process and the bone of the jaws	409
	The facial buttresses	411
18	**The periodontium**	**415**
	General remarks	415
	The gingiva — a general description	416
	The gingival collar	420
	Gingival fibers (gingival ligament)	421
	The periodontal ligament — a general description	423
	The periodontal fibers	424
	Blood and nerve supply of the periodontium	426
	Metabolism of the periodontal ligament	429
	Development of the periodontium	430

19	**Nerve supply and sensitivity of teeth**	**435**
	General remarks	435
	Nerve supply to the dentine and pulp	435
	Dental pain	437
	Theories of pain perception in dentine and tooth pulp	439
20	**Tooth eruption**	**443**
	General remarks	443
	The role of cell proliferation	444
	The role of tissue fluid pressure	444
	The role of the periodontal ligament	445
21	**The temporomandibular joint**	**449**
	The articular surface of the temporal bone	449
	The condyle of the mandible	451
	The capsule of the joint	452
	The articular disc (meniscus)	454
	The joint cavities (joint compartments)	454
	Mandibular movements	455
22	**Salivary glands**	**459**
	Embryological development	459
	Classification of salivary glands	460
	Components of salivary glands	460
	General remarks on mucous cells, serous cells, and the arrangement of cells in a mixed gland	462
	Description of the salivary glands with reference to the nature of the saliva	466
23	**Saliva**	**469**
	General remarks	469
	General factors influencing the secretion of saliva	470
	Nervous control of secretion of saliva	471
	Functions of saliva	474
	Composition of saliva	477
	Halitosis (Fetor oris)	478
24	**Fluoride**	**481**
	A historical review	481
	Metabolism of fluoride	482

	Toxicity of fluoride	485
	Fluoride in teeth	487
	The protective effect of fluoride on tooth enamel	489
	The detrimental effect of fluoride on teeth	490
25	**Mastication**	**493**
	The mechanics of mastication	493
	The dynamics of mastication	496
26	**Taste**	**501**
	General remarks	501
	Primary (basic) tastes	501
	Taste and chemical structure	502
	Other stimuli associated with taste	503
	Adaptation to taste	503
27	**Deglutition**	**505**
	General remarks	505
	The first phase (oral phase)	505
	The second phase (pharyngeal phase)	507
	The third phase (esophageal phase)	508
	Nervous control of deglutition	508

Part I

I. The cell

General remarks

The existence of two major cell types is recognized, namely prokariotic cells, i. e. cells without a definitive nucleus, such as algae and bacteria, and eukariotic cells, i. e. cells with a functional nucleus. Eukariotic cells are characterized by well developed nuclei and a complex internal cytoplasmic organization composed of specialized organelles (Fig 1).

The body is made up of virtually countless cells which are connected by means of a variable quantity of intercellular material. Each cell is enveloped by a cell membrane (plasma membrane) and contains a nucleus which is surrounded by a nuclear membrane.

Two events take place when a zygote undergoes further development, namely

1. Cell division (mitosis), and
2. Maturation or differentiation in which cells assume specific properties to enable them to perform different functions, e. g. contraction as in muscle. Some cells, such as neurons, lose the capacity for mitosis when fully differentiated. Other cells, such as liver cells, retain this ability.

The cells of the body display a very large structural diversity and are in most instances not dependent on one another for survival.

Each cell derives a supply of oxygen and nutrients from the bloodstream from which it forms its own components through internal metabolism and from which energy is derived for a diversity of functions such as chemical, mechanical, or electrical activity.

The number of chemical reactions which take place within an active cell is so staggering that it is difficult to comprehend how everything can take place within a structure that appears to be of such a simple nature through a light microscope. Advanced research has, however, shown that a simple cell can simulate a large industrialized city with factories, storehouses, streets, power stations, and the ability to neutralize intruders and to dispose of waste products.

The cell membrane

The cell membrane, as viewed through an electron microscope, is a trilaminar structure, often termed a unit membrane, which consists of a specific arrangement of certain molecules (Fig 2). One of the models to illustrate the cell membrane is the Singer–Nicholson fluid mosaic model. This consists of a bimolecular lipid layer associated with protein. The lipids are mainly phospholipids, although most membranes also contain glycolipids and cholesterol. The phospholipids are represented as possessing a polar end which contains the phosphate and is soluble in water (hydrophilic) and two non-polar tail ends composed of lipids which are not soluble in water (hydrophobic). The non-polar ends of the double layer of phospholipid molecules are opposed to one another in the middle of the membrane and make up the lighter interspace of the membrane. The outer polar ends of the phospolipids, and the associated proteins which are found outside as well as within the phospholipid layer, account for the external dark laminae of the trilaminar structure of the membrane as seen on the electron micrograph. Although other models for illustrating the structure of the cell membrane exist, it is believed that all membranes conform basically to the above description.

Functions of the cell membrane

Motility

The membrane does not possess any inherent motility. As a result of contraction or relaxation of ectoplasmic contractile microfilaments, however, the membrane can move, and this results in certain cells being able to migrate through some tissues. White blood cells (leukocytes), for example, possess a highly developed motility which can lead to phagocytosis, in which finger-like projections (pseudopodia) of the cell membrane engulf particles such as bacteria which are taken into the cell and dealt with as described later in this chapter.

Cell recognition

The cell membrane contains specific substances (antigens) whereby the body is able to recognize and tolerate its own cells. Cells of another person with different antigens are consequently regarded as intruders and attacked.

Cell adhesion

The cell membrane is concerned with adhesion between identical cells of the same organ. Not all cells display this ability, e. g. blood cells in which such an ability would not be compatible with life. Cancer cells also possess a lesser degree of intercellular adhesion which accounts for the fact that they may metastasize and invade surrounding tissues.

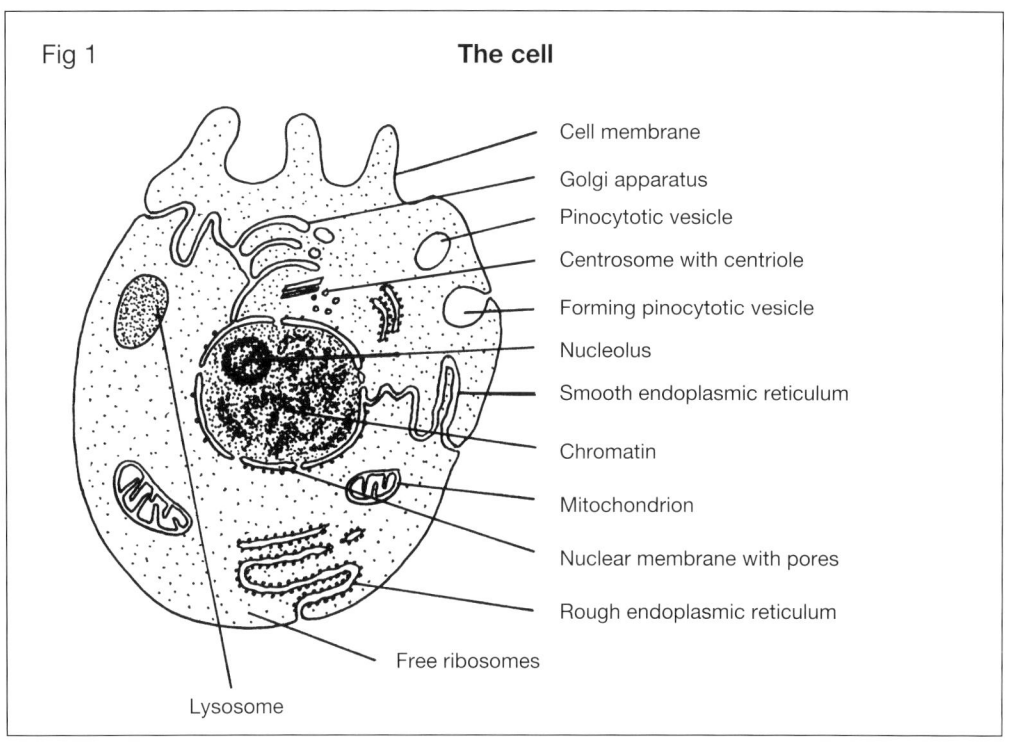

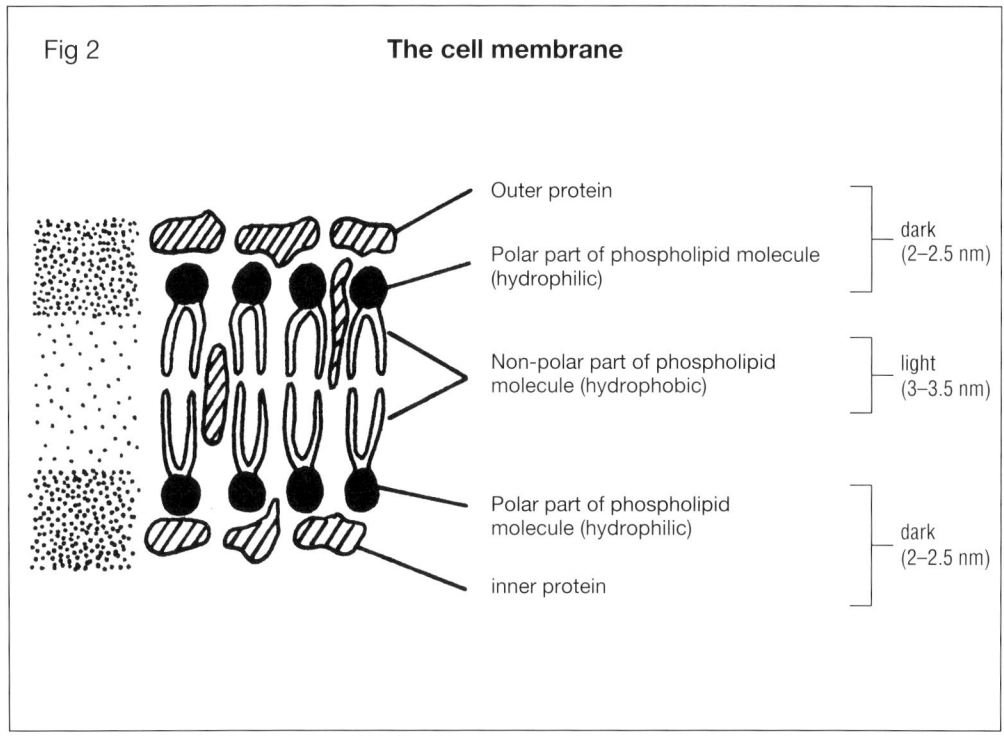

Cell growth

Mitotic activity in the epidermal cells adjacent to a wound is much increased and this increased activity continues until the wound edges meet. Contact between similar cells inhibits mitotic activity in such a situation, a phenomenon known as contact inhibition. It appears that this is a further function of the cell membrane.

Exchange activity

The cell membrane is a surface through which exchange of substances between the intracellular and extracellular compartment takes place. The surface of some cells possesses short finger-like processes (microvilli) which increase the surface area of the membrane and facilitate absorption.

The cytoplasm

The cytoplasm is that part of the protoplasm outside the nucleus and has a complex structure. It is divided into numerous compartments by membranes which are continuous with the cell membrane and with the nuclear membrane.

Endoplasmic reticulum

The endoplasmic reticulum (ER) is responsible for most of the subdivision of the cytoplasm which contains numerous other membrane-bound structures or organelles such as mitochondria, lysosomes etc. The endoplasmic reticulum consists of a series of membranes arranged to form continuous tube-like structures and cisternae. In these spaces the secretions of some glands appear first.

Ribosomes

Granules (ribosomes) which are rich in ribonucleic acid (RNA) may be found on the membranes of the endoplasmic reticulum and give it a granular (rough) appearance. This rough endoplasmic reticulum manufactures protein for "export" by the cell. Similar granules lie free in the cytoplasm and are not associated with the ER. These free ribosomes manufacture protein for "own consumption". Endoplasmic reticulum without associated ribosomes is termed smooth endoplasmic reticulum and plays an important role in the synthesis of non-protein substances. The rough and smooth endoplasmic reticulum are continuous with one another, as well as with the outer lamina of the nuclear membrane and with the cell membrane. In a sense the fluid in the lumen of the endoplasmic reticulum is therefore extracellular.

Golgi apparatus

The Golgi apparatus consists of closely packed stacks of membrane-bound cisternae with associated vesicles. These cisternae, which are not associated with ribosomes and are smaller than the endoplasmic reticulum, surround the centrosome which contains one or more centrioles. The Golgi apparatus is best developed in secretory cells of glands and is found in a supranuclear position. Secretory products formed by the rough endoplasmic reticulum are transported to the Golgi apparatus by means of small vesicles which are pinched off the ER and merge with the membrane of the Golgi apparatus. The product then leaves the Golgi as vesicles (secretory "granules") which are pinched off and pass to the secretory surface of the cell where mergence with the cell membrane enables the contents to exit the cell and to enter the lumen of the duct (Fig 3).

Centrioles

The centrioles are found as a pair in a specialized zone of the cytoplasm, the centrosome, which lies close to the nucleus. In glandular secretory cells it lies on the luminal side of the nucleus. The two centrioles are together called the diplosome and lie perpendicular to each other. Each is an elongated cylindrical organelle, in the wall of which nine evenly spaced elongated parallel tubular units are found. In cell division the centrioles are concerned with the organization of the spindle microtubules. The centrioles are self-replicating and the diplosome is reformed in each daughter cell.

Mitochondria

Most cells contain mitochondria, red blood corpuscles (erythrocytes) being an exception. They usually have an elongated rod-like shape and are bounded by a double unit membrane. The inner membrane extends into the interior of the mitochondrion for variable distances in the form of folds (cristae) which greatly enlarge the membrane surface area of the organelle (Fig 4). Mitochondria are extremely active in metabolism, an outstanding example being the internal enzymatic activity resulting in the Krebs citric acid cycle and energy production in the form of adenosine triphosphate (ATP), the most important source of cellular energy. The mitochondria are rightly regarded as the power plants of the cell.

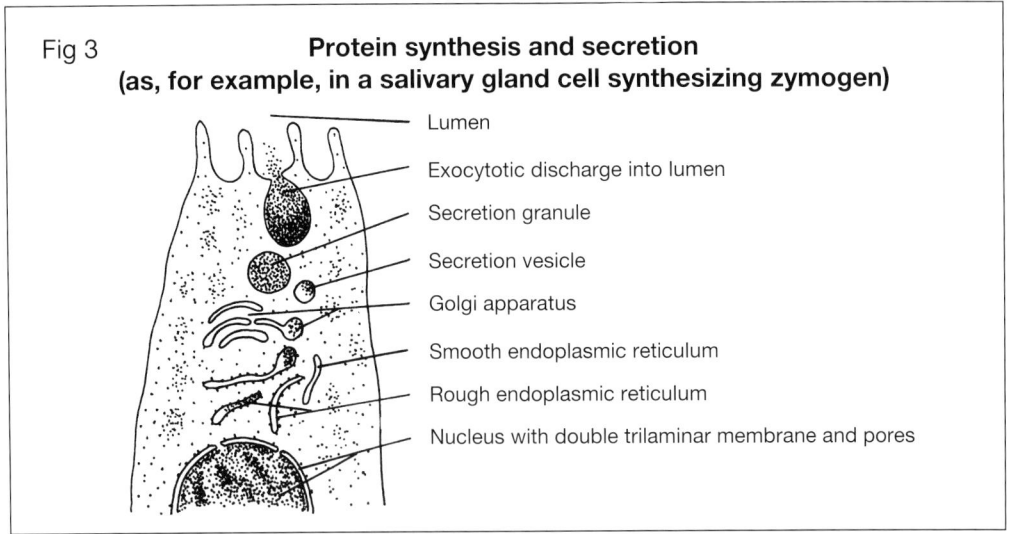

Summary of events:

1. The synthesis of proteins takes place in association with ribosomes on the endoplasmic reticulum. The proteins then reach the lumen of the endoplasmic reticulum.
2. The protein is transported through the cisterna to the Golgi apparatus via the smooth endoplasmic reticulum.
3. The secretion vesicles are pinched off from the Golgi apparatus. They are changed into secretion granules by concentration of the protein which is then discharged into the lumen of the tubule by exocytosis.

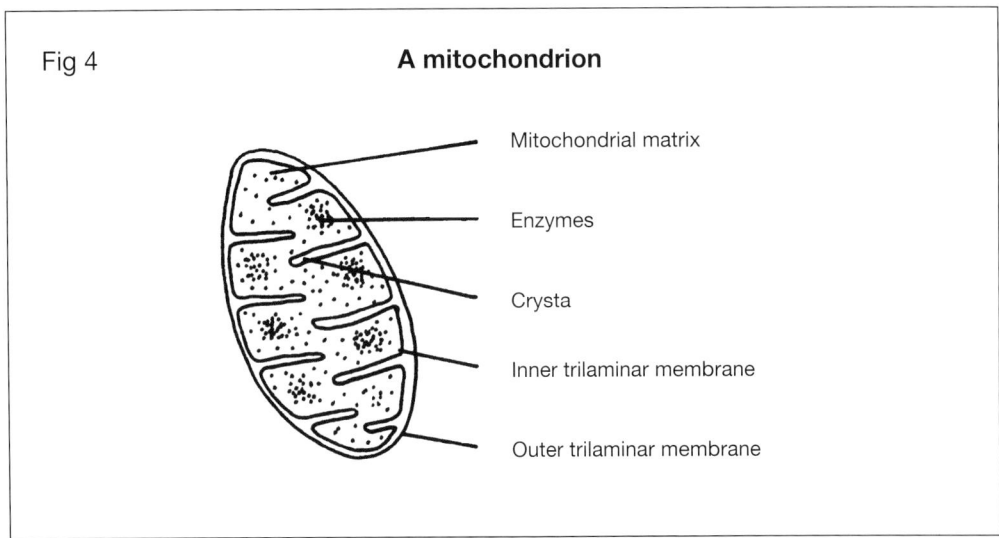

Microtubules and microfilaments

Electronmicroscopy reveals that most cells contain microtubules and microfilaments. The role of these structures will be explained where appropriate.

Lysosomes

Lysosomes are round membrane-bound bodies containing enzymes (hydrolases) which break down proteins, carbohydrates and nucleic acids at an acid pH. They therefore possess the ability to destroy bacteria and foreign substances which enter the cell. The granules of polymorphonuclear leukocytes are in fact lysosomes, which is indicative of a phagocytic action.

Pinocytosis and phagocytosis

These terms are used to describe the ability of a cell to ingest fluid particles (pinocytosis) or solid material (phagocytosis). When a droplet of fluid or a solid particle surrounded by fluid contacts the surface of the cell, the cell membrane forms an invagination and engulfs the droplet or solid particle which is subsequently found within a membrane-bound vacuole inside the cytoplasm. The term endocytosis, broadly speaking, encompasses both pinocytosis and phagocytosis.

The membrane-bound vacuole containing the ingested material is termed a phagosome. When a phagosome becomes incorporated into a lysosome, for example for purposes of enzymatic digestion, the resulting body is termed a phagolysosome. When the process is completed the phagolysosome fuses with the cell membrane to release the products of digestion, a process termed exocytosis.

Autophagy refers to the process whereby a cell digests its own organelles to provide reserve nutrient material for cell survival in states of starvation, or as part of a normal process of removal of damaged organelles.

The nucleus

The nucleus is surrounded by a nuclear envelope consisting of a double unit membrane. Spaced on the circumference of the nuclear envelope are the nuclear pores, which have a complex structure. The pores, with a diameter of 50-70 nm, enable the transfer of large molecules (for example RNA) between the cytoplasm and the nucleus. The outer unit membrane of the nuclear envelope is continuous with other cytoplasmic membrane systems.

The nucleus is a most important structure, since it contains codified information which is transmitted to daughter cells and from one generation to the next. The chemical substance responsible for this is a nucleic acid, deoxyribonucleic acid (DNA) which is able to replicate and make the above possible. DNA is acidic, stains blue with a basic dye such as hematoxylin, and is termed chromatin. The term gene denotes the hypothetical, or supposed unit of inheritance of a characteristic by a next generation. DNA and RNA are responsible for directing the synthesis of proteins by the cell and specifying the unique sequence of amino acids in the product. The DNA is confined to the nucleus but is able to transmit instructions to the cytoplasm via messenger RNA which is an RNA transcript of the gene or genes. This mRNA attaches to ribosomes which then translate the coded instructions and thereby direct the desired sequence of amino acids in the forming of a polypeptide.

Chromosomes

The DNA molecules do not lie free in the nucleoplasm but are arranged in the form of chromatin threads termed chromosomes. Each resting somatic cell contains a specific number of chromosomes. This is the diploid number. In man the diploid number is 46, of which 23 were inherited from one parent and 23 from the other. Two chromosomes are concerned with sex and are termed the sex chromosomes, while the other 44 are autosomes.

Part of the RNA content of the nucleus is found in the nucleolus. A cell may have more than one nucleolus. The nucleoli are round, dense bodies but do not have a limiting membrane. They display variable basophilia and acidophilia in different cells.

The Barr bodies

A separate small mass of chromatin is found in the interphase nuclei of human and other females. This Barr body (or sex chromatin) was first described by Murray Barr and Bertram in 1949 in nerve cells of female cats and represents one of the two female sex chromosomes.

It is usually seen as a small stained body with a diameter of approximately 1 µm next to the nucleolus or adjacent to the nuclear envelope and is easily detected in smears, e. g. from the oral mucosa. A cell from a normal female is said to be sex chromatin positive with a singe Barr body.

Cell division

Cell division occurs in two ways, by means of mitosis and meiosis. In mitotic division the cell and all its constituents, including the nuclear DNA, divides to form two identical daughter cells, each with a diploid number of chromosomes (46). This type of division is typical of somatic cells. Meiosis (or reduction division) occurs in the development of germ cells in the ovaries or the testes. In this more comples division the daughter cells receive one half the number of chromosomes (the haploid number), totalling 23. At fertilization the diploid number of chromosomes is restored.

Arrangement of cells

Cells are grouped together to form tissues. Four types of tissue are traditionally described, namely

1. epithelia,
2. connective tissue,
3. muscle, and
4. nervous tissue.

Epithelia

Epithelial cells are closely packed together and cover surfaces or line hollows such as the mouth. In some situations, for example in glands, the arrangement of epithelial cells is more complex and specialized.

Connective tissue

Connective tissue cells are usually separated from one another by a jelly-like material (or ground substance) in which fibres, such as collagen, are found. Connective tissue can also contain primitive multipotential cells which can differentiate to form a large variety of cells such as fibroblasts, fat cells, and hemopoietic cells.

Bone, cartilage, tendons, and fibrous tissue are alle forms of connective tissue and share a supportive function.

Muscle

Muscular tissue is divided into skeletal muscle, smooth muscle, and cardiac muscle. Muscle cells (or muscle fibers) all share a contractile function.

Nervous tissue

Nervous tissue consists of nerve cells (neurons) and supporting cells, the neuroglia. The properties of excitation and conduction of a nerve impulse are highly developed in nervous tissue.

Selected bibliography

1. Cole, A. S. and Eastoe, J. E. (1988) *Biochemistry and Oral Biology*, 2nd edition. London: Wright.
2. Fawcett, D. W. (1986) *A Textbook of Histology*, 11th edition. Philadelphia: W. B. Saunders Company.
3. Kelly, D., Wood, R. L. and Enders, A. C. (1984) *Bailey's Textbook of Microscopic Anatomy*, 18th edition. Baltimore: Williams and Wilkins.
4. Krause, W. J. and Cutts, J. H. (1986) *Concise Text of Histology*, 2nd edition. Baltimore: Williams and Wilkins.
5. Leeson, T. S., Leeson, C. R. and Paparo, A. A. (1988) *Text/Atlas of Histology*. Philadelphia: W. B. Saunders Company.
6. Osborn, J. W. and Ten Cate, A. R. (1983) *Advanced Dental Histology*, 4th edition. Bristol: Wright.

Review questions

1. Discuss the morphology and functions of the cell membrane.
2. Discuss the cytoplasm.
3. Discuss the nucleus.

2. A historical review of embryology

It is part of the nature of man to be interested in his own origin and mode of development. In the case of medical and dental students a knowledge of embryology is of the utmost importance, since it explains the development of the human body. Each part of the body develops according to a complex plan and when this is disturbed a maldevelopment may result.

Embryology is the study of the growth and differentiation which an organism undergoes during its development from a single fertilized cell to a complex independent living being. For our purposes, however, embryology implies the phase of human development before birth and covers the first period of a person's life history or ontogeny.

Every animal starts life in the form of a simple cell, a fertilized egg or zygote. A zygote is formed by two cells, namely the germ cells (gametes) of the parents. When the gamete of the father (spermatozoon) unites with the gamete of the mother (egg cell or ovum) a zygote results. The gametes are prepared for fertilization by meiosis, a reduction division of the chromosomes. The number of chromosomes in a normal somatic cell is 46, consisting of 22 pairs of autosomes and one pair of sex chromosomes. Together these form the diploid number. During meiosis in the gonads of the two parents the members of each pair of chromosomes are separated and each gamete therefore receives one half of the chromosome set (the haploid number). When fertilization occurs the two halved sets unite, resulting in a restoration of the diploid number in the zygote. This is then followed by mitosis which leads to growth and differentiation.

Man has since the earliest times speculated about his origin. The earliest idea was that living beings had their origin in spontaneous generation or development from no clearly definable precursor. Aristotle (384-322 B. C.) was the first person to consider the matter seriously and was of the dual opinion that an embryo either exists in a preformed state (the theory of preformation) or originates in some unorganized state (spontaneous generation). He finally favored the latter and considered that an embryo originates in organized menstrual blood of the mother, a variation of the theory of spontaneous generation. Even during the 17th century this was a popular theory, but the development of the compound light microscope in about 1590 by Hans and Zacharias Janssen inevitably had a drastic influence on scientific thinking in this matter.

The human spermatozoon was first seen by Hamm and von Leeuwenhoek in 1677, shortly after the first description of female ovarian follicles by De Graaf in 1672. The human ovum had not, however, been discovered at this time. Even after these developments the contribution of the two sexes to a new life was uncertain and scientists were divided into two opposing camps. The animalculists believed that the head of a spermatozoon contains a miniature fully formed individual, a theory based on a "discovery" and drawing by Hartsoeker in 1694, and that this miniature being only received nutrients inside the mother. On the other hand, the ovists believed that the female possessed a miniature being which was stimulated to grow by the seminal fluid of the father. The standpoint of the ovists was strengthened by the discovery in 1745 by Bonnet that the eggs of some insects could develop parthenogenetically without the intervention of the male. Spermatozoa were even regarded by some as parasites in seminal fluid. The above thoughts were part of the theory of preformation which stated that a miniature being was present in either the father or the mother and only required the correct stimulation to develop further. This theory was popular until approximately the year 1800.

Spallanzani (1729-1799) and Kaspar Friedrich Wolff (1733-1794) were largely responsible for a diminished interest in the theory of preformation. Spallanzani showed experimentally that both sexes were necessary for reproduction while Wolff suggested the theory of epigenesis. This theory implies that the material from which an animal will later develop exists in the egg in the form of unorganized granular material and that an organizing factor is responsible for meaningful continued growth and development. Wolff's conclusions were based on his study of chicken eggs, and his "granules" probably refer to nuclei. Wolff was also responsible for drawing attention to the significance of the germ layers and described how formless "granules" initially became organized into germ layers which subsequently resulted in organ formation. Obviously the above findings did not provide all the answers to the controversy, but they clearly proved that no miniature animal was initially present.

Although the eggs of fish, birds, and reptiles were, of course, familiar since ancient times, the eggs of mammals were first discovered by Karl Ernst von Baer in 1827. Von Baer was also the first scientist to emphasize that the general basic characteristics of a large group of animals, for example the notochord of vertebrates, appeared earlier in ontogenetic development than specialized characteristics of members of such a group, for example the hair of most mammals or the feathers of birds. This observation is often called von Baer's law.

Cleavage of a zygote to form cells which, through growth and differentiation, lead to the formation of an embryo was first described by Prévost and Dumas in 1824 but it was only in 1839 that Schleiden and Schwann recognized the significance of the cell as the structural and functional unit of the different organs. In 1875 Hertwig was the first person to observe and describe the fertilization of an egg by a spermatozoon, while von Beneden in 1883 proved that the male and female gametes contributed the same number of chromosomes to a zygote.

Driesch (1900), possibly one of the first experimental embryologists, and others were responsible for eliminating many misconceptions which still existed. It became apparent that in many living forms the daugther cells of a zygote each possess the potential to develop into a complete embryo when experimentally seperated from each other. The significance of this finding is that each cell contains in its genetic make-up the coded characteristics of an entire organism.

In the light of present knowledge it is not unreasonable to state that in a certain sense growth is epigenetic while development and differentiation is preformative, being dictated by genetic factors.

Selected bibliography

1. Arey, L. B. (1974) *Developmental Anatomy*, revised 7th edition. Philadelphia: W. B. Saunders Company.
2. Hamilton, W. J., Boyd, J. D. and Mossman, H. W. (1972) *Human Embryology*, 4th edition, revised by Hamilton, W. J. and Mossman, H. W. Cambridge: W. Heffer and Sons Ltd.
3. Patten, B. M. and Carlson, B. M. (1974) *Foundations of Embryology*, 3rd edition. New York: McGraw - Hill Book Company.

Review questions

1. What is embryology?
2. Discuss briefly the points of view of the animalculists and the ovists.
3. Discuss the contribution of Wolff (1733-1794) to the history of embryology.
4. Discuss the significance of the findings of Driesch (1900) and other experimental embryologists.

3. Fertilization, establishment of the germ layers, and fate of the primary embryonic tissues

Fertilization

Fertilization occurs when the male and female gametes (the spermatozoon and the ovum) unite to form a zygote.

A mature spermatozoon possesses a head, neck, middle-piece and tail and is approximately 60 µm long (Fig 5). The head contains the condensed nucleus of the cell. Behind the head, the spermatozoon consists of a very short neck, the middle section, and finally the very long tail or flagellum. Detailed descriptions of spermatozoa may be found in the bilbiography. Spermatozoa are formed in the testes but are not functionally active before they leave the body in the seminal fluid during ejaculation. A normal male will emit some 200 million sperms in a single ejaculation, after which they become motile, the tail performing an undulating movement which propels the sperm forward. Spermatozoa have the ability to fertilize an ovum for up to 48 hours after ejaculation.

The ovum is apparently not as complex in structure as a spermatozoon. After ovulation it is a round cell with a diameter, including the surrounding zona pellucida, of 110-150 µm (Fig 6). It is surrounded by cellular remains (corona radiata) of the follicle within which it formed in the ovary. After ovulation, and when it enters the abdominal (cephalic) opening of the uterine tube on its way to the uterus, the ovum is still involved in the final stages of meiosis which only terminate in consolidation of the nucleus of the ovum after the sperm head enters the ovum during fertilization. This probably occurs in the abdominal one third of the uterine tube.

The spermatozoon may penetrate the ovum at any point on its circumference. The head swells and its nuclear material becomes visible, the nuclear membranes of both nuclei disappear and the chromosomes unite (Fig 7). The chromosomes of the newly formed zygote almost immediately commence the first cleavage division (Fig 8).

The immediate results of fertilization may be summarized as follows:
1. Resoration of the diploid chromosome number;
2. Determination of sex;
3. Initiation of mitotic cell division.

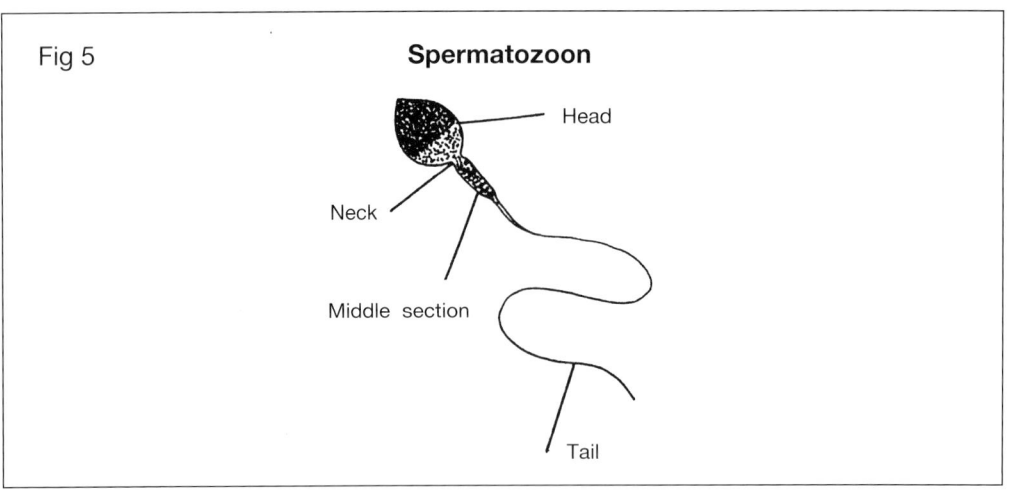

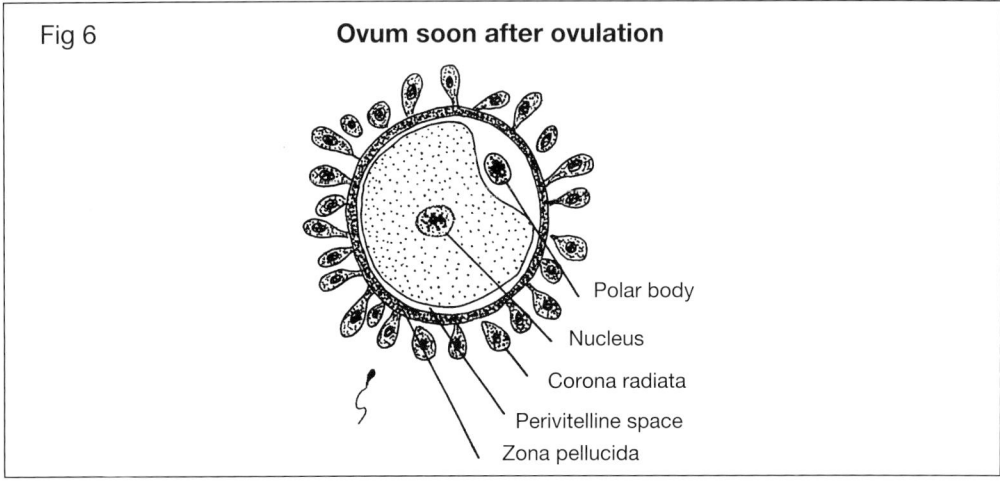

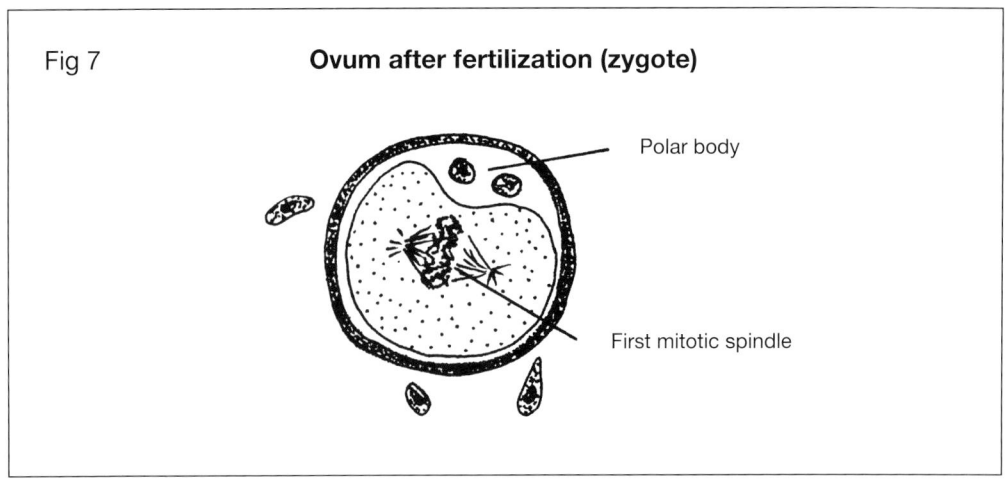

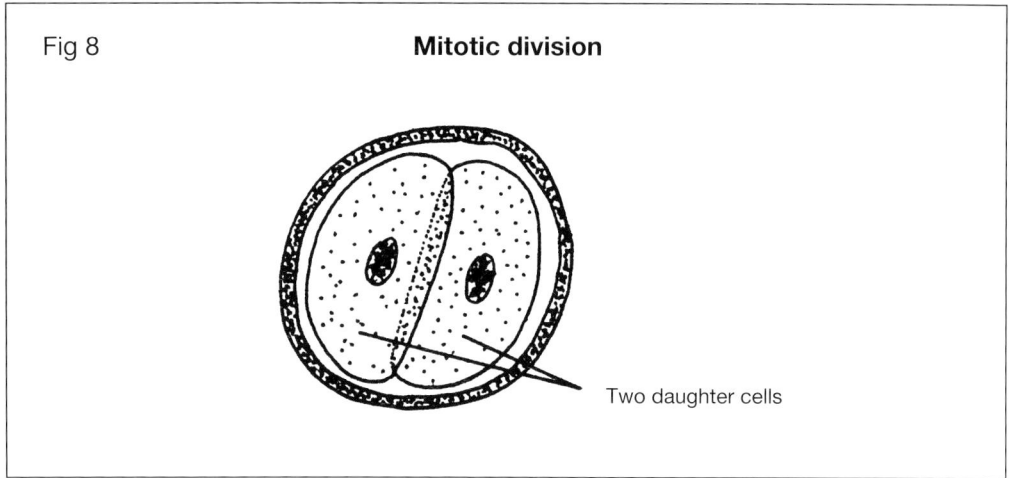

Fig 8 Mitotic division

Two daughter cells

The somatic cell of a male has 44 autosomes ("ordinary" chromosomes), arranged in 22 pairs, plus two sex chromosomes denoted X and Y (the male sex chromosomes), a total of 46. As a result of meiotic division in the gonads to produce a male gamete (spermatozoon), both the number of autosomes and the pair of sex chromosomes are halved. The resulting two gametes contain the same complement of autosomes (22) but either an X or a Y sex chromosome.

In the female the same division occurs but since the sex chromosome pair consist of two X chromosomes, the resulting ovum can only possess an X chromosome. At fertilization the following combinations can therefore occur: the spermatozoon contains either an X or a Y chromosome while the ovum can only be X. If a spermatozoon possessing an X chromosome fertilizes an ovum the embryo will be female (44 autosomes plus XX sex chromosomes). On the other hand, if a Y spermatozoon fertilizes an ovum the resulting embryo will be male (44 autosomes plus XY sex chromosomes). In this way the sex of an individual is determined.

Establishment of the germ layers

Fertilization is followed by rapid mitotic cell division, resulting in daughter cells which are termed blastomeres (Fig 9). The "golf-ball" like little mass of cells is known as a morula (mulberry) and consists of a group of centrally placed cells (inner cell mass) surrounded by a peripheral layer of cells (the future trophoblast).

At the time of entry of the morula into the uterine cavity, or shortly afterwards, fluid enters it and separates the inner cell mass from the peripheral cells on one side. This gives rise to a blastocyst in which the inner cell mass is eccentrically situated. At this stage, 7–8 days after fertilization, the blastocyst attaches to the uterine wall by means of the trophoblast (implantation) (Fig 10). The placenta will later develop in this area of implantation. The *period of the ovum* has now been completed and is followed by the *period of the embryo* lasting to approximately 8 weeks after fertilization. The *period of the fetus* follows and is continued throughout pregnancy, which normally ends in birth after approximately 280 days.

The first sign of differentiation in the inner cell mass is the appearance of somewhat flattened cells on its inner surface. These cells, the primary endoderm, spread along the periphery of the blastocyst cavity and from the inner lining of the trophoblast (Fig 11, I). The remaining cells of the inner cell mass which are still intimately related to the overlying trophoblast become columnar and so form the primary ectoderm, which is subsequently separated from the trophoblast by a space (the amniotic cavity). The blastocyst cavity which is lined by primary endoderm is now termed the primary yolk sac of the embryo. The primary endoderm is at this stage intimately attached to the primary ectoderm and the two cell layers form the bilaminar embryonic disc.

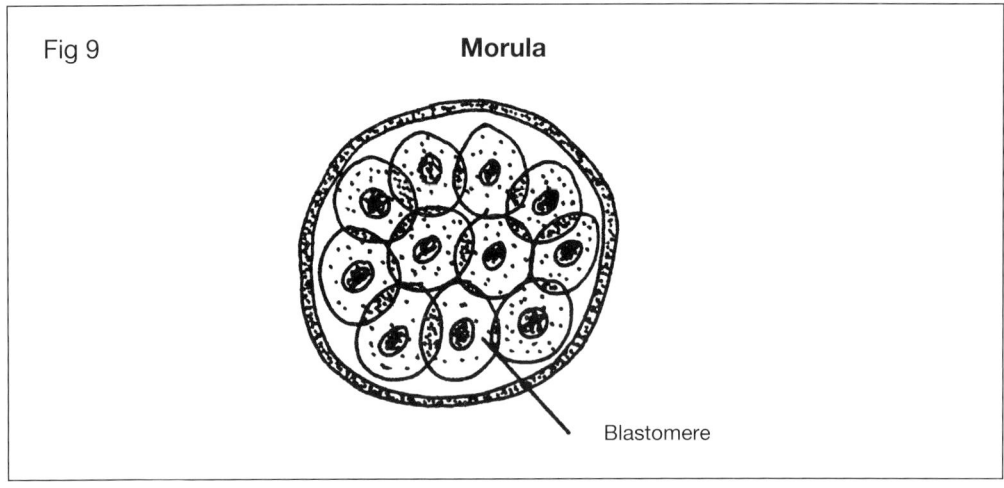

Fig 9 — Morula — Blastomere

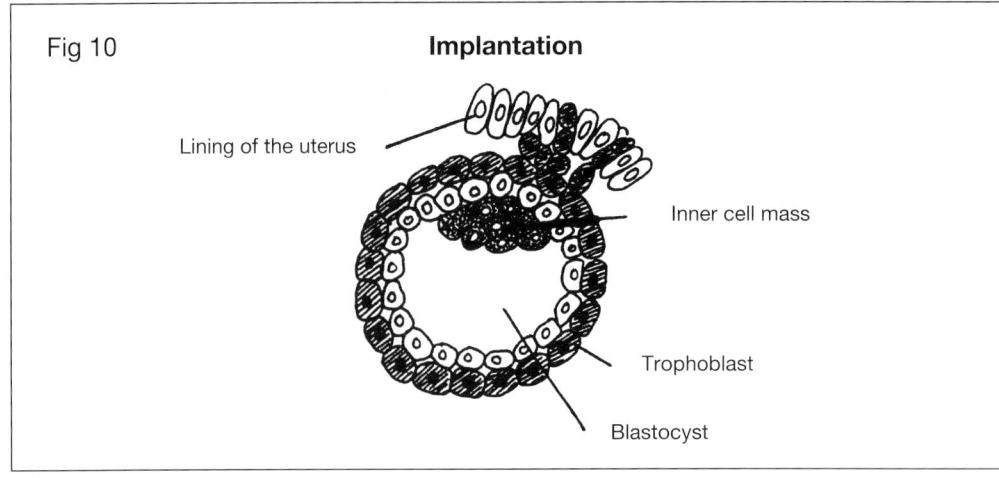

Fig 10 — Implantation — Lining of the uterus, Inner cell mass, Trophoblast, Blastocyst

Fig 11   **Further developments in zygote (sagittal section)**

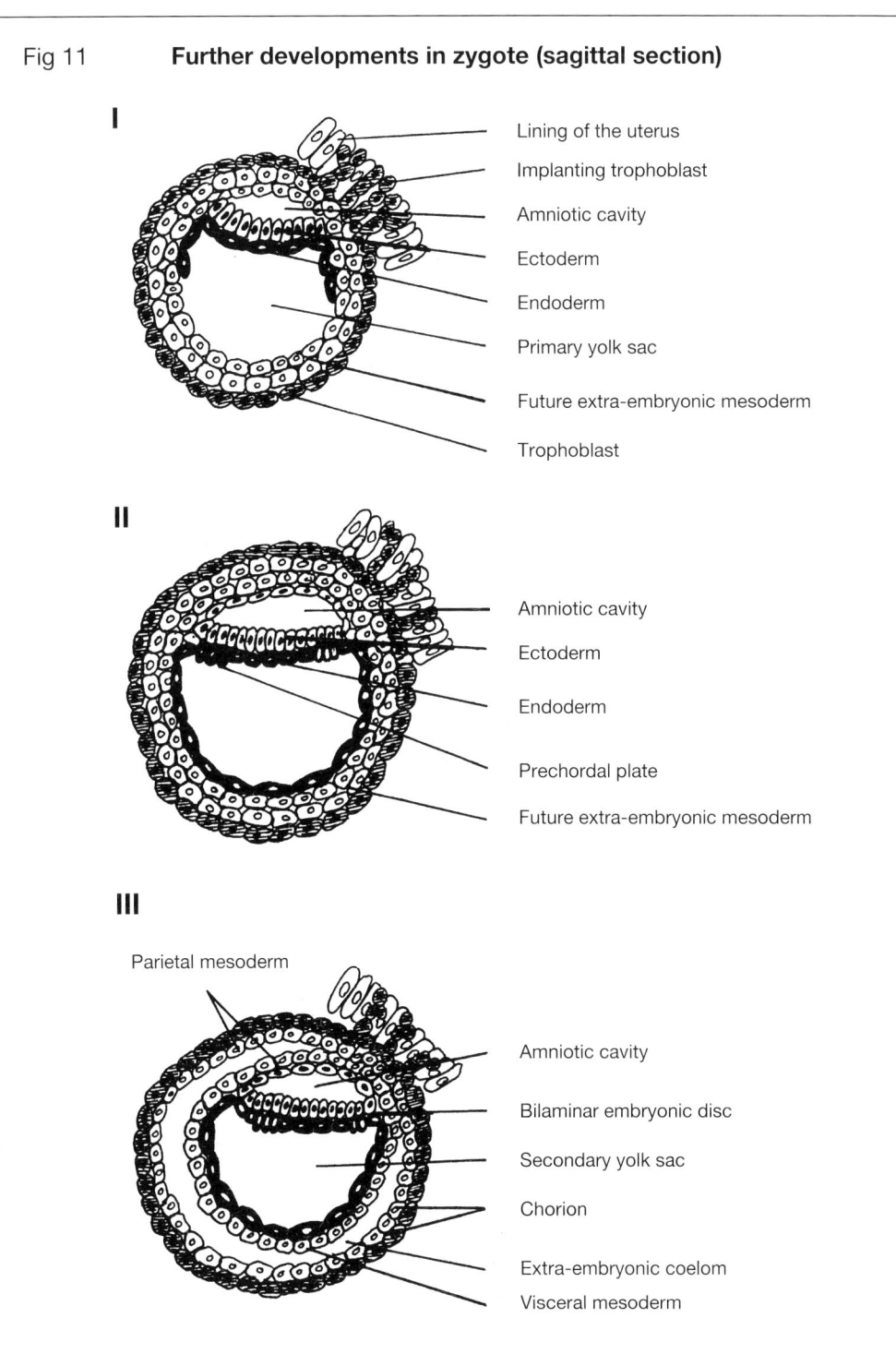

The endodermal cells in a localized area close to the future cephalic end of the bilaminar disc and in contact with the primary ectoderm next become columnar to form the prechordal plate. Its presence bestows on the disc an antero-posterior axis and a bilateral symmetry (Fig 11, II).

The primary yolk sac becomes separated from the trophoblast by the extra-embryonic mesoderm. Cavities develop in this mesoderm, become confluent, and lead to the formation of the extra-embryonic coelom, except in the caudal part of the embryonic disc where the amnion remains adjacent to the trophoblast. In this area the placenta and umbilical cord will later develop (Fig 11, III).

Extra-embryonic mesoderm has thus become divided by the extra-embryonic coelom into a layer covering the yolk sac (now termed secondary) and the amniotic cavity on the one hand, and a layer lining the inside of the trophoblast. The layer covering the amniotic cavity and lining the trophoblast is known as the parietal mesoderm, while that covering the yolk sac is termed visceral mesoderm. Parietal mesoderm, together with the trophoblast, forms the chorion.

Fifteen days after fertilization the human bilaminar embryonic disc develops a midline caudal thickening of ectoderm. This is the beginning of the development of the primitive streak from which, by a process of migration of cells which come to lie between the ectoderm and the underlying endoderm, the third germ layer of the embryo, namely the intra-embryonic mesoderm, is derived (Fig 12, I). The intra-embryonic mesoderm eventually becomes continuous with the extra-embryonic mesoderm on the periphery of the now trilaminar embryonic disc (Fig 12, II). Part of the intra-embryonic mesoderm extends forward to surround the prechordal plate and to form the septum transversum on its cranial aspect. No mesoderm is found between the ectoderm and endoderm of the prechordal plate. With the later development of the head fold the prechordal plate becomes the buccopharyngeal membrane (Fig 13). At the caudal end of the embryo behind the primitive streak, as in the case of the prechordal plate, contact between ectoderm and endoderm occurs without intervening mesoderm. This is the cloacal membrane. From the region anterior to the edge of the primitive streak new developments now give rise to a forward rod-like midline proliferation of cells between the bilaterally arranged intra-embryonic mesoderm. This is covered by ectoderm above and endoderm below. This notochord, or primitive body axis, comes into contact with the caudal edge of the prechordal plate. It becomes relatively insignificant in later human development. Cavities form in the intra-embryonic mesoderm on the periphery of the embryonic disc. These cavities later become confluent to form a horseshoe shaped, tube-like, intra-embryonic coelom, lined by mesothelium, which extends around the anterior edge of the prechordal plate (buccopharyngeal membrane), where it constitutes the primordium of the pericardium. It extends caudally on both sides to open into the extra-embryonic coelom. The pericardial portion of the intra-embryonic coelom, together with the surrounding mesoderm, forms the cardiogenic area. The remaining intra-embryonic mesoderm of the septum transversum anterior to the cardiogenic area is the site of origin of the liver.

Fig 12   **Surface view of ectoderm with amnion removed**

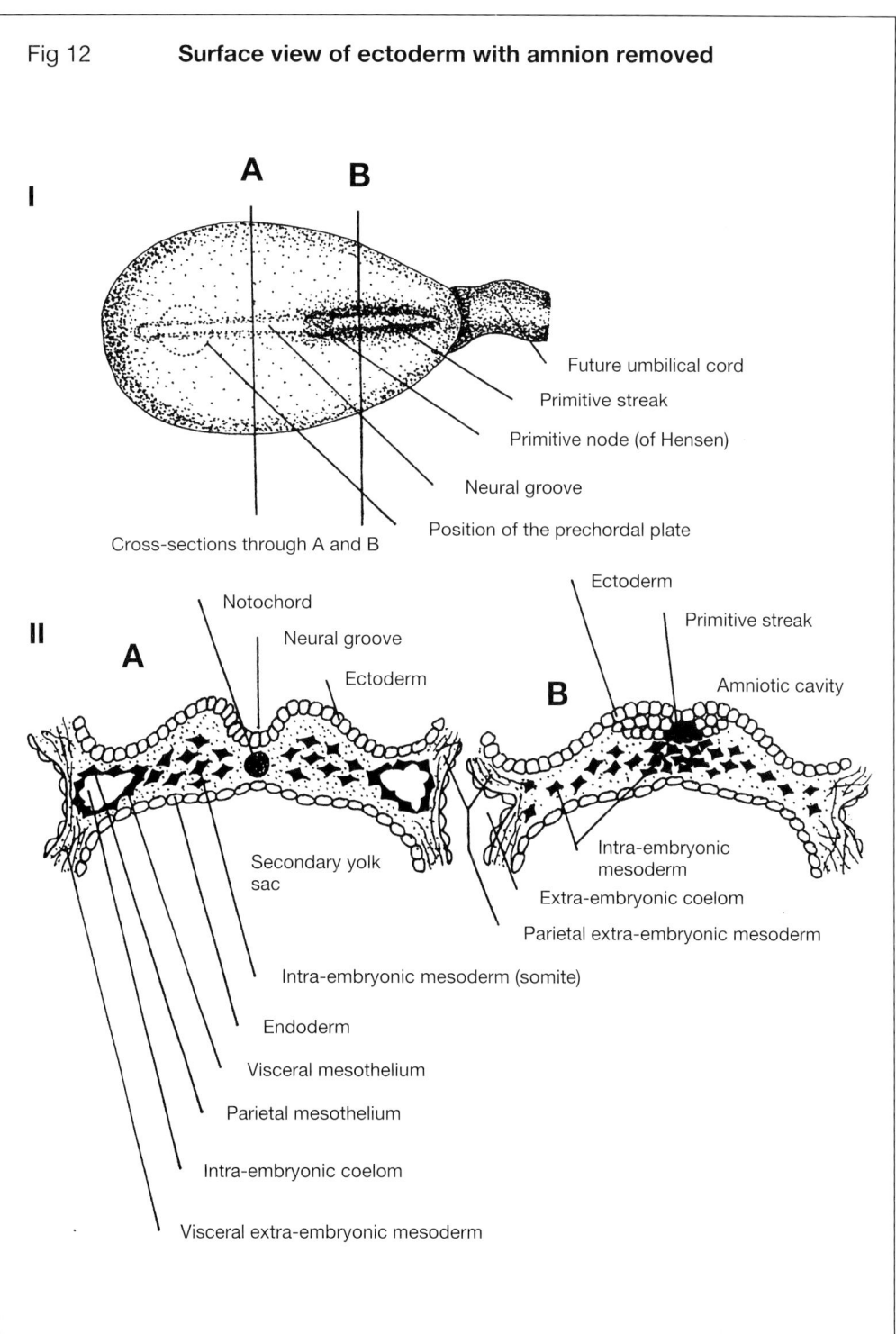

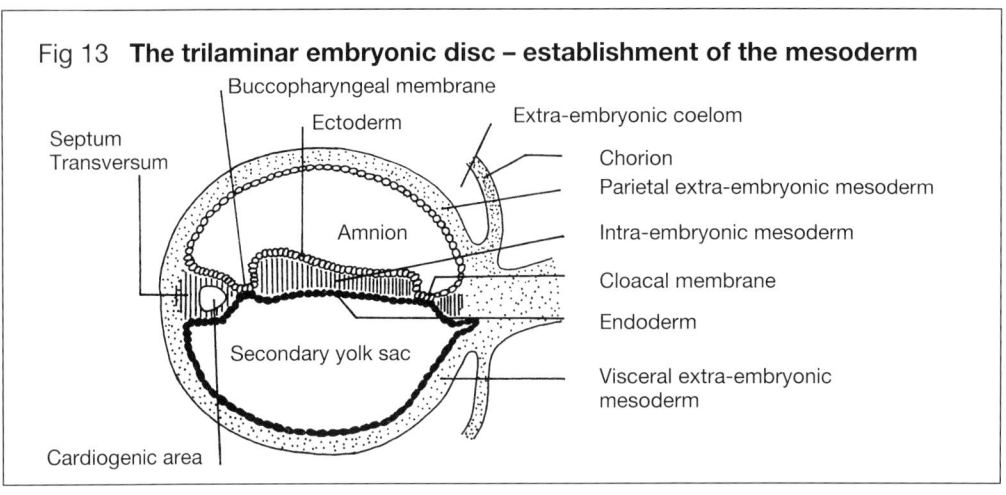

Fig 13 **The trilaminar embryonic disc – establishment of the mesoderm**

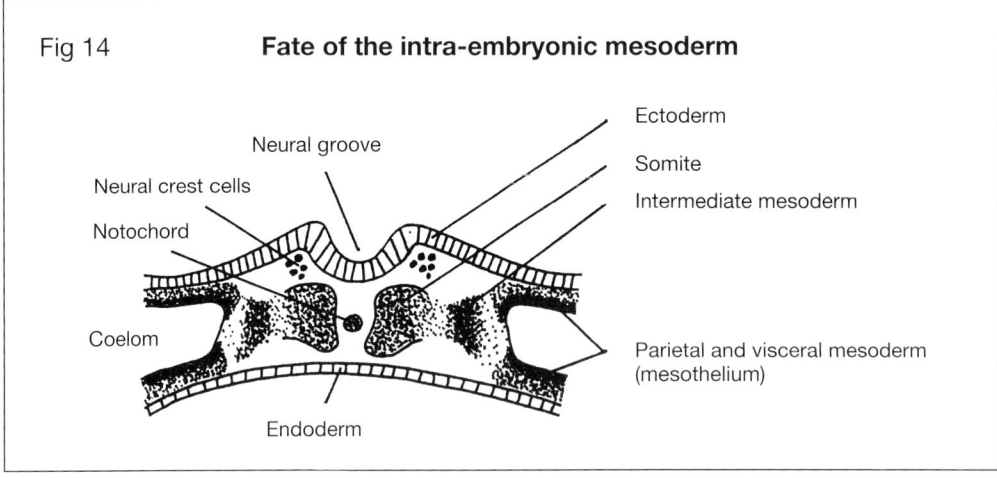

Fig 14 **Fate of the intra-embryonic mesoderm**

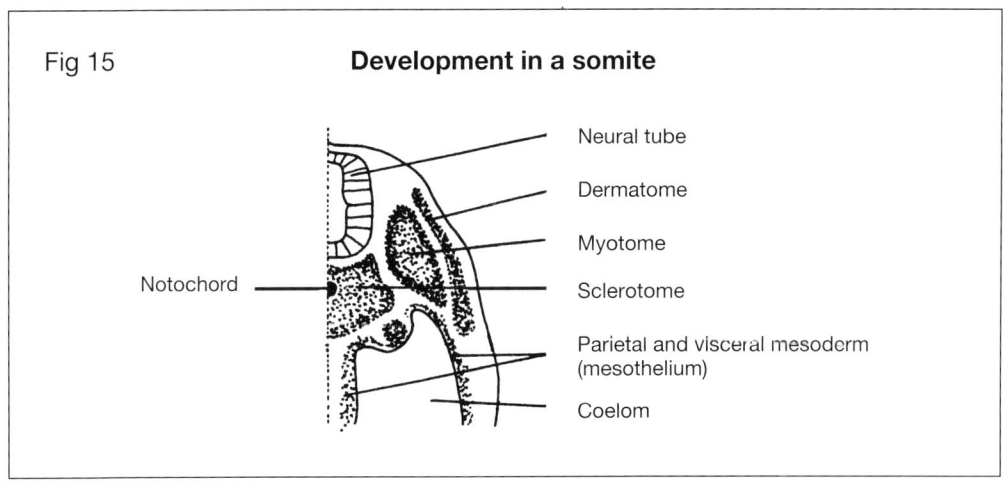

Fig 15 **Development in a somite**

Fate of the primary embryonic tissues

The ectoderm gives rise to the following tissues and structures:
1. The skin of the entire body together with the skin derivatives, namely hair, nails, and the epithelial components of sweat glands, sebaceous glands, and the mammary glands.
2. The epithelium of the oral mucosa including the gingiva, floor of mouth, tongue, palate, and associated glands.
3. Olfactory epithelium of the nose and the epithelial lining of the paranasal sinuses.
4. The epithelium of the terminal parts of the urinary and genital tracts and of the lower part of the anal canal.
5. The dental laminae which give rise to the enamel organs and thus the enamel of the teeth.
6. The outer layer of the tympanic membrane, the anterior epithelial layer of the cornea, and the lens of the eye.
7. The central and peripheral nervous systems.

The endoderm gives rise to the following:
1. The epithelium of the digestive tract not derived from ectoderm.
2. The epithelial components (parenchyma) of the thyroid and parathyroid glands, the thymus, pancreas, and liver.
3. The lining epithelium of the middle ear cavity (including the inner layer of the tympanic membrane) and of the pharyngotympanic (Eustachian) tube.
4. The epithelial lining of the alveoli of the lungs, bronchi, trachea, and larynx.
5. The lining epithelium of most of the female urethra, and of the lower part of the vagina, part of the male urethra together with associated glands, for example the prostate, and the urinary bladder.

With the exception of certain minor tissues and structures not mentioned above, all the remaining parts of the body arise from mesoderm. Early in development the mesoderm can be divided into three components, as follows (Fig 14):
1. The initially laterally placed mesodermal epithelia (the lateral plate mesoderm) which line the intra-embryonic coelom and form the parietal and visceral mesothelial linings of the pericardial, pleural, and peritoneal cavities.
2. The intermediate mesoderm which plays a major role in the development of the gonads, kidneys, and adrenal glands. Peritoneal mesothelium also plays a part in some of these very complex developments.
3. The paired somites on either side of the notochord. The somites, in addition to other derivatives, give rise to a loose tissue which fills the spaces between the germ layers. This tissue (mesenchyme) also derives contributions from the parietal and visceral intra-embryonic mesoderm.

In the cervical, thoracic, lumbar, and sacral areas, one spinal nerve supplies one somite. The number of somites therefore corresponds to the number of spinal nerves (31). In the coccygeal area the number of somites initially outnumbers the spinal nerves but most degenerate. The final number of spinal nerves is 31, divided into 8 cervical nerves, 12 thoracic nerves, 5 lumbar nerves, 5 sacral nerves, and 1 coccygeal nerve.

The first cervical somites are, however, not the most cranial somites to develop. The occipital somite (supplied by the hypoglossal nerve) and the pre-occipital (pre-otic) somites (supplied by the third, fourth, and sixth cranial nerves) develop in an even more cranial position.

The somites (initially 42-44 pairs) initially show epithelial characteristics but as development proceeds part of the somites form a more loosely arranged tissue called mesenchyme (Fig 15). This mesenchyme partly migrates medially (the sclerotome) to surround the notochord and is concerned with the development of the axial skeleton (vertebrae and intervertebral discs). From the remaining part of a somite a dorsolateral dermatome develops (future dermis and subcutaneous tissue), as well as a myotome which gives rise to skeletal muscle. Some authors group the latter two derivatives of a somite together as a dermomyotome.

No somites form in the mesoderm anterior to the notochord but paraxial mesoderm (the mesoderm on either side of the notochord) extends forward to the cephalic end of the embryo in the form of a loose mass of mesenchyme. This mesenchyme, which is largely derived from the somites, receives a contribution from neural crest cells (neuro-ectoderm) during development of the central nervous system and is known as ectomesenchyme.

Mesenchyme is a multipotential primitive "packing tissue" between the germ layers and gives origin to:

1. Different varieties of blood cells.
2. Connective tissues proper, as well as bone, cartilage, dentine, cementum, periodontal ligament, and alveolar bone.
3. The visceral musculature generally, as well as the myocardium and the muscular coat of the blood vessels.
4. The endothelium of the blood vessels and the endocardium.
5. The spleen, lymph glands, and lymph vessels.

Selected bibliography

1. Arey, L. B. (1974) *Developmental Anatomy*, revised 7th edition. Philadelphia: W. B. Saunders Company.
2. Bhaskar, S. N. (editor) (1991) *Orban's Oral Histology and Embryology*, 11th edition. St Louis: Mosby-Year Book, Inc.
3. Hamilton, W. J., Boyd, J. D. and Mossman, H. W. (1972) *Human Embryology*, 4th edition, revised by Hamilton, W. J. and Mossman, H. W. Cambridge: W. Heffer and Sons Ltd.
4. Mjör, I. A. and Fejerskov, O. (editors) (1986) *Human Oral Embryology and Histology.* Copenhagen: Munksgaard.
5. Osborn, J. W. (editor) (1981) *Dental Anatomy and Embryology.* Oxford: Blackwell Scientific Publications.
6. Sperber, G. H. (1989) *Craniofacial Embryology*, 4th edition. London: Wright.
7. Ten Cate, A. R. (1989) *Oral Histology*, 3rd edition. St Louis: C. V. Mosby Company.
8. Williams, P. L. and Wendell-Smith, L. P. (1978) *Basic Human Embryology*, 2nd edition. Tunbridge Wells, Kent, England: Pitman Medical Publishing Co. Ltd.

Review questions

1. Give a brief description of a spermatozoon and an ovum.
2. Discuss the formation of a zygote from the male and female gametes.
3. Discuss the changes which take place when a zygote is transformed into a trilaminar embryonic disc.
4. Discuss the fate of the primary embryonic tissues ectoderm, mesoderm, endoderm and mesenchyme.

4. Epithelium

General remarks

An epithelium may be structured in two ways and may have two main functions. It may be arranged in the form of a sheet of cells covering a surface or lining a cavity (protective functions), or in the form of solid cords, tubules or follicles which developed from an epithelial surface and have a secretory, excretory, or absorptive function. The two functions of epithelia are not always clearly demarcated, since a lining or covering epithelium often contains cells with a secretory function.

An epithelium is composed of cells which are arranged in one or more layers. The spaces between the cells contain a small amount of glycosaminoglycan which is rich in cations, especially calcium which is important for intercellular adhesion.

By means of a light microscope it often appears as if adjacent cells are connected by means of intercellular "bridges". Probably the best example of this appearance is to be found in the stratum spinosum (prickle cell layer) of the skin or the oral mucous membrane. Electronmicroscopically, however, these cell "bridges" are seen to be specializations of the cell membrane (desmosomes) rather than a direct cytoplasmic intercellular contact.

When viewed through a light microscope epithelial cells are seen to be separated from the underlying connective tissue by a thin basement membrane. Electronmicroscopically the basement membrane consists of three layers, namely a lamina lucida (light zone) immediately subjacent to the cell membranes of the basal epithelial cells and containing glycoprotein (laminin) with a thickness of approximately 45 nm, and secondly a lamina densa (dark zone) consisting largely of type IV collagen associated with proteoglycan molecules on both sides and approximately 55 nm wide. Fibronectin is found mainly on the connective tissue side of the lamina densa. The third component of the basement membrane is the reticular lamina deep to the lamina densa and containing fine anchoring filaments (possibly type V collagen). This layer binds the underlying connective tissue to the epithelium, especially in the region of the hemidesmosomes. The lamina lucida and the lamina densa are collectively known as the basal lamina, and are of epithelial origin, while the reticular lamina is the product of connective tissue fibroblasts and consists largely of fine argyrophilic reticulin fibers. When treated with solutions containing silver salts, it appears black.

Epithelial cells which cover moist surfaces (like the oral cavity, or line ducts, usually have smooth surfaces while other epithelial cells may have cytoplasmic projections of a relatively complex nature, the cilia. Motile cilia are found in, for example, the respiratory passages and in the uterine tube. Flagella have the same structure as cilia but are present as only one or two per cell and are longer than cilia. The spermatozoon with its long tail is an example of a cell with a single movable flagellum.

A second type of surface specialization of an epithelial cell is a striated (brush) border. This appearance is due to the presence of numerous thin, closely-packed cytoplasmic extensions known as microvilli. This type of epithelium is found where absorption is a primary function, as in the small intestine.

Types of epithelium

Epithelia are classified into two main types on the basis of the number of cell layers present. Epithelia composed of only one cell layer are termed simple, and all cells have contact with the basal lamina, whereas epithelia consisting of two or more cell layers are termed stratified, and only the basal cell layer has contact with the basal lamina. Further subdivision of epithelia depends on the shape of the cells, as follows (Fig 16):

Simple epithelia	Stratified epithelia
Simple squamous epithelium	Stratified squamous epithelium
Simple cuboidal epithelium	Stratified cuboidal epithelium
Simple columnar epithelium	Stratified columnar epithelium
Pseudostratified (columnar) epithelium	Transitional epithelium

Pseudostratified epithelium is a modification of simple epithelium, since not all the cells reach the surface.

Simple epithelia

Simple squamous epithelium

The cells of a simple squamous epithelium are flattened. In a surface view they are usually large cells with a clear cytoplasm and with oval or round nuclei, which may be centrally or eccentrically situated. The cell borders are usually indistinct and irregular. In cross-section the cytoplasm is unobtrusive, except in the vicinity of the nucleus. Simple epithelia are not found where protection, absorption, or secretion

Fig 16 Epithelium

Simple epithelium

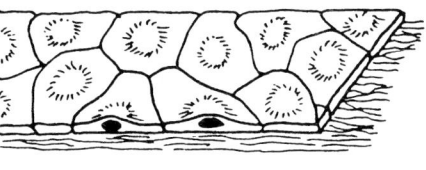

Squamous epithelium

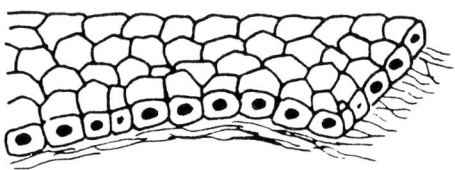

Cuboidal epithelium

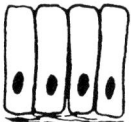

Unmodified columnar epithelium

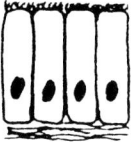

Columnar epithelium with cilia

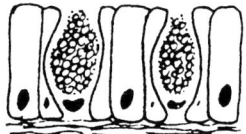

Columnar epithelium with goblet cells

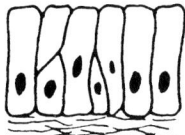

Unmodified

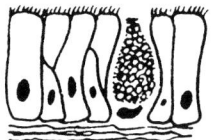

With cilia and goblet cell

Pseudostratified columnar epithelium

Stratified epithelium

Squamous epithelium

Orthokeratinized epithelium is shown on the left, parakeratinized epithelium on the right.

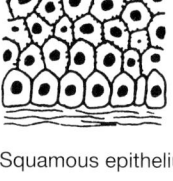

Cuboidal epithelium

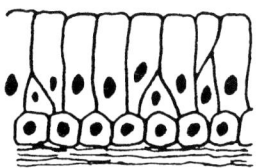

Columnar epithelium

Transitional epithelium

is a function of the epithelium but are typical of surfaces where diffusion or filtration occurs. Typical sites where this type of epithelium is found are the kidney (Bowman's capsule), the alveoli of the lungs, and as an endothelial lining of blood capillaries.

Simple cuboidal epithelium

In a surface view simple cuboidal epithelial cells are smaller than squamous cells and of a more uniform shape. The cells are roughly hexagonal in outline. In a side view the cells have a square shape with a round centrally situated nucleus. When, however, the cells are arranged around a lumen of a duct or a gland, they are pyramidal in shape. The cytoplasm of cuboidal cells varies from clear to granular. Simple cuboidal cells are, for example, found in some tubules of the kidney and in the thyroid gland.

Simple columnar epithelium

Simple columnar epithelial cells are similar to cuboidal cells in a surface view. In a side view, however, the cells are elongated with a round or oval nucleus, usually situated towards the basal end of the cell. The apical ends of the cells may be smooth or specialized in the form of cilia or microvilli. The cytoplasm may be clear or contain granules, or secretory vesicles, depending on the area where the cells are situated.

Some mucous cells are specialized columnar cells which are interspersed between other columnar cells in the mucosa of the intestines or in the respiratory passages. Secretory vesicles (granules, droplets) containing mucigen (the precursor of mucin and therefore mucus), are present in large numbers in the distal ends of these cells and are responsible for the name commonly given to these cells, goblet cells. The contents of the cell are discharged in due course and the process of synthesizing the secretion is repeated. Columnar cells are generally found in organs where the epithelium has the triple functions of protection, absorption, and secretion.

Pseudostratified (columnar) epithelium

This type of epithelium consists of columnar cells which are closely pressed together. Due to this arrangement there is a loss of the typical rectangular shape and not all the cells reach the surface. All the cells do, however, make contact with the basal lamina. Some of these cells have a broad basal end and are narrow apically. Others have a broad apical end and a narrow base. Still others may have a spindle shape. Nuclei are found in the broadest part of the cell and are consequently found at different levels. This type of epithelium is, for example, found in the trachea where, due to the presence of surface cilia, the name of pseudostratified ciliated columnar epithelium is used. Mucous producing goblet cells are also present in this epithelium.

Stratified epithelia

In all the stratified epithelia a layer of cuboidal or low columnar cells forms a basal layer in contact with the basal lamina. One or more layers of cells are found above the basal cells and the classification of the epithelium depends on the nature of the topmost layer of cells.

Stratified squamous epithelium

The number of cell layers above the basal cells is not the same in all regions where this type of epithelium is found but the consecutive cell layers follow a fairly uniform arrangement. The basal cell layer (the germinal layer or stratum germinativum) is covered by several layers of angular cells. Close to the basal cells, these angular cells are relatively small but they increase in size towards the surface. These cells have intercellular contact by means of prominent desmosomes which bestow a prickly appearance on these cells (prickle cell layer or stratum spinosum).

Some authors use the term stratum germinativum to embrace both the prickle cell layer and the basal cell layer, since mitotic cell division, although primarily a feature of the basal cells, also occurs in the prickle cell layer.

Closer to the surface of the epithelium the cells flatten, desmosomes become less conspicuous and a shrinkage of cells may occur. The nuclei may become flattened and appear pyknotic. The term parakeratin is used to describe such a surface layer (stratum corneum), which is fairly widespread in the mouth.

In those areas of the mouth subjected to normal masticatory friction, for example the gingiva and hard palate, the epithelium consists of a similar basal cell layer and a prickle cell layer. The cells of the next more superficial layer, however, become slightly flattened and contain many keratohyalin granules. Consequently, this layer of cells is termed the granular cell layer (stratum granulosum). Keratohyalin is intimately concerned with the synthesis of keratin (a fibrous protein) found in the upper layer (stratum corneum) of the epithelium. The cells containing keratin (orthokeratin) have a scaly appearance and have lost their nuclei. It should be emphasized that the uppermost cell layer of both parakeratinized and orthokeratinized stratified squamous epithelia is known as the stratum corneum, and these two types of epithelia often intermingle.

Stratified squamous epithelium is discussed in more detail in the chapter dealing with the oral mucosa.

Stratified cuboidal epithelium

This type of epithelium is fairly limited in the body and is found in, for example, the ducts of salivary and sweat glands.

Stratified columnar epithelium

In a stratified columnar epithelium, which is comparatively rare, surface columnar cells rest on a basal cell layer with one or more layers of associated cubical cells. The columnar cells do not have contact with the basal lamina. This type of epithelium is found, for example, in the olfactory mucosa and in areas where a stratified squamous epithelium adjoins a pseudostratified columnar type. An example of such an area is the junction between the oropharynx and the nasopharynx.

Transitional epithelium

Transitional epithelium cannot clearly be classified in any of the other categories. The basal cells are similar to those found in stratified squamous epithelium. These basal cells are separated from the cells of the uppermost layer by angular cells which become pear-shaped towards the surface. The cells of the surface layer are often described as dome-shaped and one of these cells may cover two or three of the underlying pear-shaped cells. The outstanding characteristic of this epithelium is its ability to stretch, and consequently it is found in an organ like the urinary bladder.

It must always be borne in mind that epithelia are avascular, and the cells are therefore dependent for their nutrition on the underlying vascular connective tissue.

Selected bibliography

1. Alberts, B., Bray, D., Lewis, J., Raff, M., Roberts, K. and Watson, J. D. (1989) *Molecular Biology of the Cell*, 2nd edition. New York: Garland Publishing, Inc.
2. Cole, A. S. and Eastoe, J. E. (1988) *Biochemistry and Oral Biology*, 2nd edition. London: Wright.
3. Kelly, D. E., Wood, R. L. and Enders, A. C. (1984) *Bailey's Textbook of Microscopic Anatomy*, 18th edition. Baltimore: Williams and Wilkins.

4. Krause, W. J. and Cutts, J. H. (1986) *Concise Text of Histology*, 2nd edition. Baltimore: Williams and Wilkins.
5. Lavelle, C. L. B. (1988) *Applied Oral Physiology*, 2nd edition. London: Wright.
6. Leeson, T. S., Leeson, C. R. and Paparo, A. A. (1988) *Text/Atlas of Histology*. Philadelphia: W. B. Saunders Company.
7. Osborn, J. W. and Ten Cate, A. R. (1983) *Advanced Dental Histology*, 4th edition. Bristol: Wright PSG.

Review question

Classify epithelia and briefly describe each type.

5. Connective tissue, including bone and cartilage

General remarks

The arrangement of cells and intercellular substance in connective tissue is quite different from that of epithelia. The cells are not arranged in the form of cords or layers but are commonly widely separated and do not have the same relatively intimate intercellular contact. In some instances contact does exist but only at the ends of long cell processes. The intercellular substance is much more prominent than in the epithelia and forms the major component of most connective tissues.

The stem cell of all connective tissues is an irregular cell with branching processes, a vesicular nucleus, and a finely granular cytoplasm (Fig 17). These cells belong to the tissue known as mesenchyme, an embryonic tissue which gives origin to ordinary connective tissue, bone, and cartilage. These cells, often referred to as undifferentiated mesenchymal cells, undergo differentiation to form a variety of cell types, such as fibroblasts, osteoblasts, chondroblasts etc.

In this chapter the general characteristics of ordinary connective tissue will be discussed, followed by bone and cartilage. It must be borne in mind that many of the elements discussed under ordinary connective tissue are also found in bone and cartilage, and the common origin of these tissues must be remembered.

Connective tissue

The intercellular component (matrix) of connective tissue consists of fibers, ground substance, and tissue fluid.

Fibers

Connective tissue fibers are of three types, namely collagenous (white), reticular, and elastic.

Collagenous (white) fibers

Collagenous fibers are widespread. They possess almost no elasticity, are dissolved by weak acids, and produce the soluble derived protein, gelatin (animal glue), when heated with water. The name collagen is derived from the Greek and means glue producer. Collagen fibers usually run an irregular wavy course with much branching. They are made up of bundles of very fine fibrils (unit fibrils) which lie parallel to one another and do not branch (Fig 18). The unit fibrils show transverse bands spaced at 64 nm along their length. They are composed of tropocollagen macromolecules, approximately 260-300 nm long and 1.5 nm in width, which in turn consist of three polypeptide chains arranged in a triple helix and linked by hydrogen bonds. The polypeptides contain hydroxyproline and hydroxylysine and are rich in proline and glycine. The transverse banding is due to the fact that the macromolecules lie parallel to one another and overlap in stepwise fashion with a 64 nm periodicity.

Reticular fibers

Reticular fibers are similar to collagen fibers but are finer, are not arranged in bundles, and tend to form delicate networks. They are not seen in routinely prepared sections but can be demonstrated with silver-containing stains. These fibers are made up of the same unit fibrils as collagen and possess the same 64 nm periodicity.

Elastic fibers

Elastic fibers are as a rule thinner than collagen fibers and branch and anastomose freely, forming networks. They cannot be demonstrated by routine procedures but color specifically when special dyes such as orcein and resorcin-fuchsin are used. Elastic fibers, which do not show any crossbanding, are composed of bundles of microfibrils embedded mainly in the periphery of an amorphous component, elastin. The amino acid component of elastic fibers differs in a few important aspects from that found in collagen (or reticular) fibers.

Ground substance

The ground substance, which stains metachromatically with toluidine blue and in which the cells and fibers of connective tissue are embedded, is an amorphous transparent gel-like material which consists of differing proportions of glycoproteins and proteoglycans in different tissues and varies in viscosity. It contains a high proportion of water bound to proteoglycans and long chain carbohydrates. This bound water acts as a medium for the exchange of metabolites, nutrients, and gases between tissue cells and blood. The principal proteoglycan of loose connec-

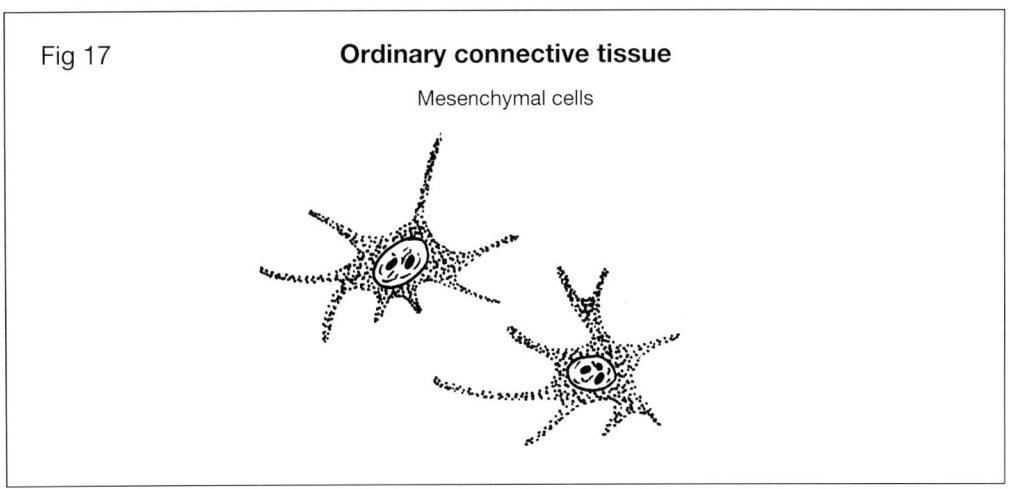

Fig 17 — **Ordinary connective tissue**
Mesenchymal cells

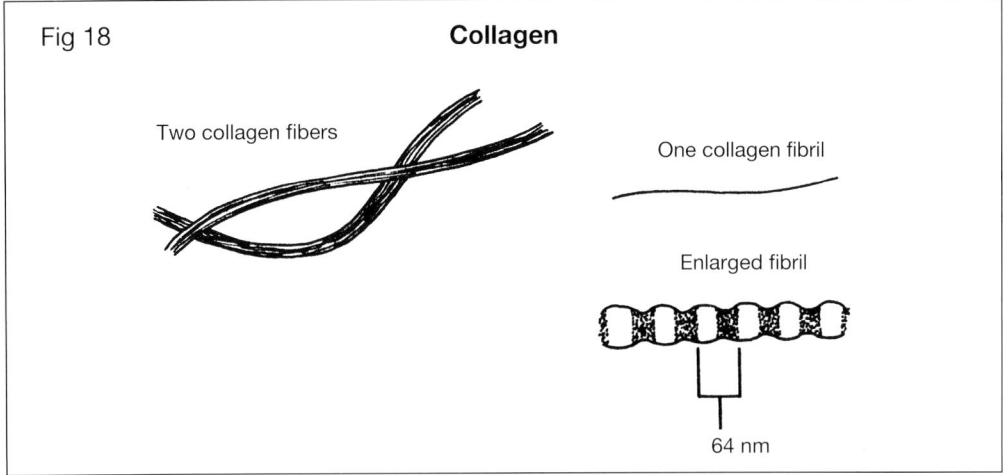

Fig 18 — **Collagen**

Two collagen fibers
One collagen fibril
Enlarged fibril
64 nm

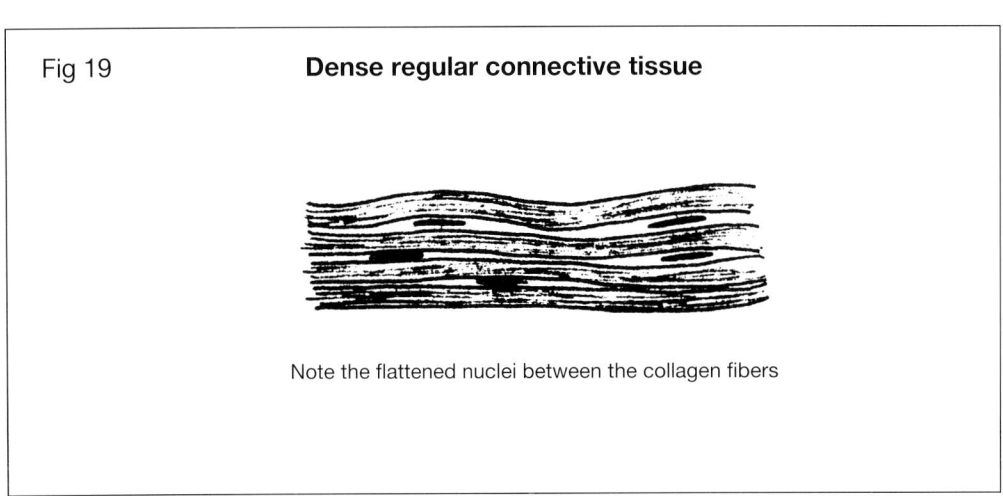

Fig 19 — **Dense regular connective tissue**

Note the flattened nuclei between the collagen fibers

tive tissue is hyaluronic acid. This is the reason why the enzyme hyaluronidase diminishes the viscosity and permeability of connective tissue.

Fibroblasts

Fibroblasts are present in all fibrous connective tissues and are responsible for synthesizing the precursors of collagenous, reticular, and elastic fibers and for producing the amorphous ground substance. In loose connective tissue they are comparatively large, flattened cells with branching processess and pale oval nuclei. In denser connective tissue these cells often lie in intimate contact with the fibers but they are never directly continuous with the fibers. Macrophages (histiocytes) are also very abundant in connective tissue, while plasma cells and mast cells may also be present.

Specific types of connective tissue are discussed next.

Reticular tissue

Reticular connective tissue, as seen, for example, forming the framework of lymph nodes, consists of reticular cells which are stellate in shape (similar to mesenchymal cells) and with cytoplasmic processes often extending along reticular fibers. The reticular cell is the equivalent of the fibroblast of other tissues and produces the same reticular fibers of loose connective tissue.

Dense, irregular fibrous tissue

Collagen forms the basis of this tissue. The collagen bundles form a branching network, while spindle-shaped fibroblasts are found between the collagen bundles. Elastic fibers form an extensive network among the collagen fibers. The perichondrium, periosteum, dermis, and capsules of various organs are examples of such a tissue.

Dense, regular fibrous tissue

This type of connective tissue is found in tendons, ligaments, and aponeuroses and consists of regularly arranged, densely packed collagen bundles (Fig 19). The fibroblasts, which are flattened between the bundles, lie with their long axes parallel to the fiber bundles. Their nuclei are similarly flattened.

Areolar tissue

In areolar tissue, collagen fibers and a few elastic fibers are found embedded in an almost fluid-like ground substance (Fig 20). It is very widely distributed in the body, occurring in the mesenteria, the omenta, subcutaneously, the lamina propria

Fig 20　**Areolar tissue**

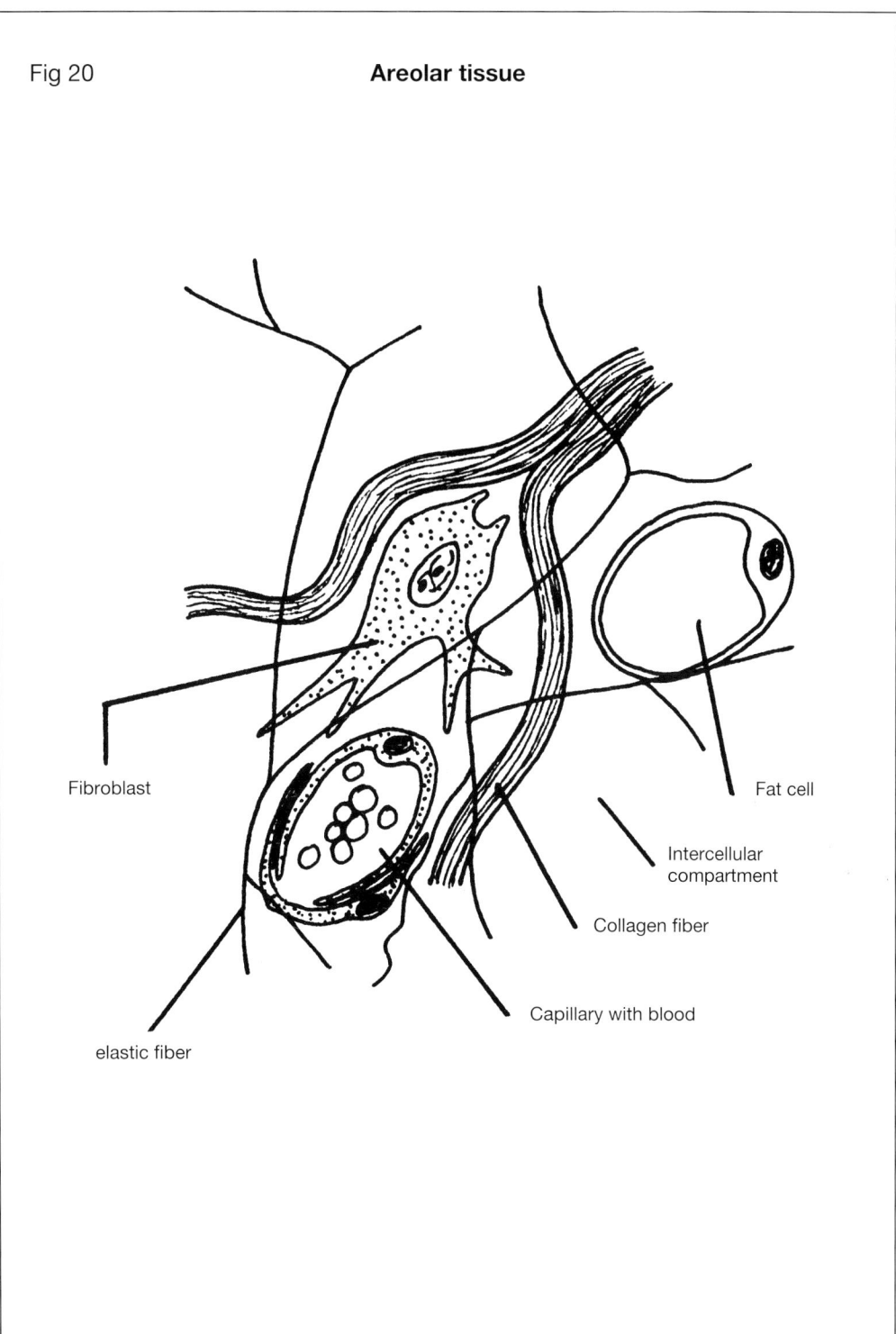

and submucosa of the oral epithelium, superficial and deep fascia, and as part of the framework of most organs and surrounding blood vessels and nerves. The collagen fibers branch extensively and run in all directions. Fibroblasts are widely distributed in areolar tissue. Fat cells may be present in varying numbers and, when abundant, the tissue is named adipose tissue.

Bone

Ordinary connective tissue, as described above, forms a part of almost every organ of the body. Bones form the skeleton which acts as a support for the body, protects the soft parts and, by attachment of individual bones or groups of bones to muscles, act as levers to make locomotion and other body movements possible. Bone consists of cells and fibrous organic matrix. Increasing mineralization of bone is accompanied by a reduction in the relative quantity of organic matrix to such an extent that the mineral component is eventually able to maintain the architecture and hardness of bone in the living body or in a dried state.

The development of bone (osteogenesis)

Bone forms in two ways. In the embryo some bones are formed in mesenchyme (intramembranous bone formation) while others, especially the long bones, are preceded by a temporary cartilagenous model which is transformed into bone (endochondral bone formation). Both the chemical and morphologic aspects of osteogenesis are discussed in detail in the chapter on calcium metabolism and bone mineralization.

The morphology of bone

All the bones have a similar chemical composition but show differences in form and internal structure to adapt to specific functions (Fig 21).

The outer layer of any bone is known as cortical or compact (dense) bone. This cortical bone is denser than the inner cancellous (alternatively, spongy or trabecular) bone. Cortical bone is composed of many vascular canals (Haversian canals) which anastomose with one another by oblique and transverse communications. Bone is laid down in concentric layers (lamellae) around the Haversian canals. Osteocytes ("resting" former osteoblasts) are found in small oval spaces, or lacunae, between or in the lamellae. Thin cytoplasmic processes of the osteocytes are found in radiating canaliculi which serve to interconnect lacunae across the lemellae. The concentric lamellae, the cells, and the central canal constitute an osteon or Haversian system. In a long bone the Haversian canals usually

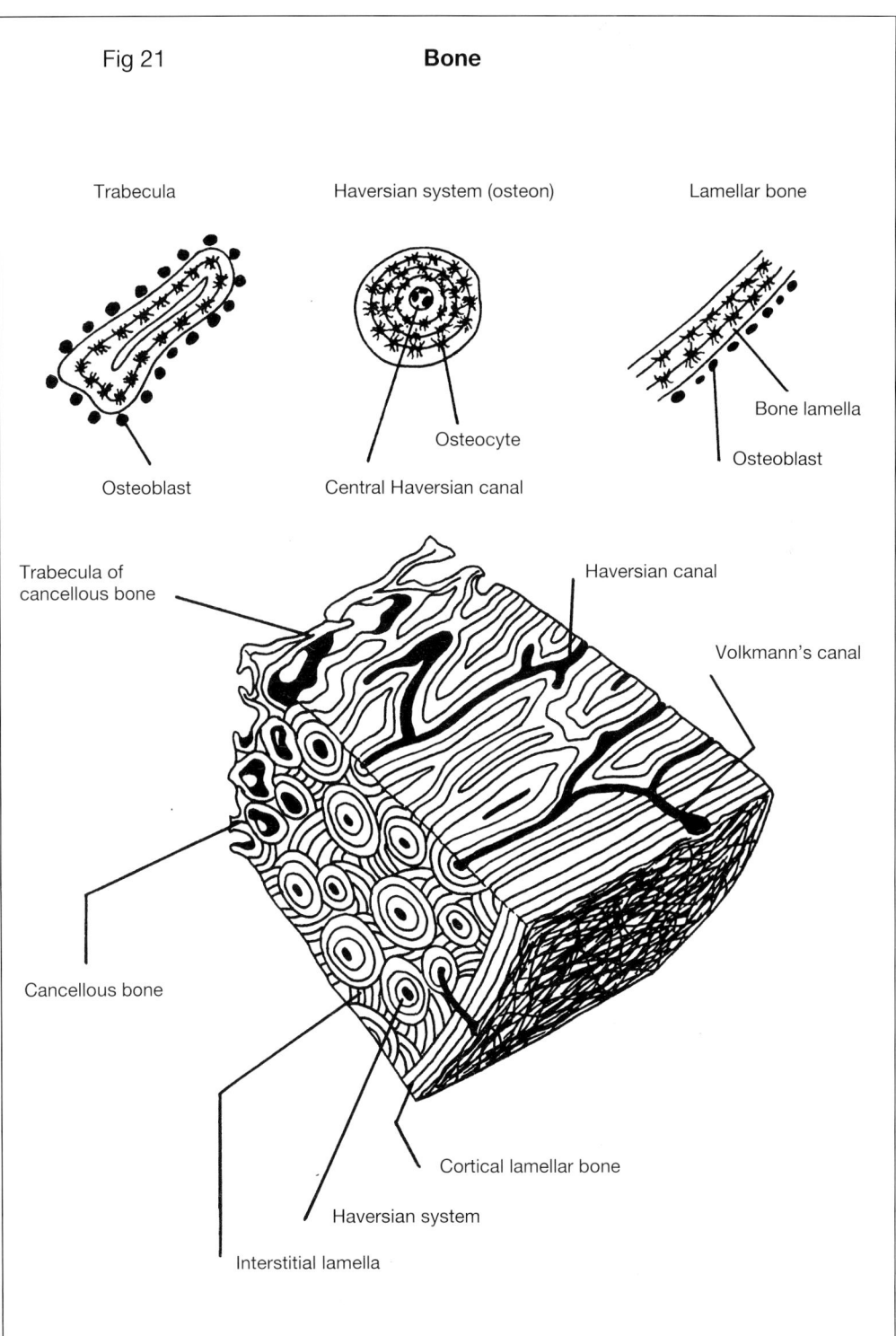

Fig 21 **Bone**

run parallel with the long axis of the bone and communicate with one another, with the marrow (medullary) cavity, and with the outer (periosteal) surface of the bone by means of nutrient (Volkmann's) canals.

The connective tissue matrix of an osteon has a unique arrangement. The collagen fibers in each consecutive concentric lamella follow a spiral course around the central canal but the spiral of each lamella is in the opposite direction to that of the adjacent lamellae. The spirals of adjacent lamellae therefore cross one another.

Although cortical bone consists mainly of Haversian systems, growth and remodelling processes result in the presence of interstitial lamellae, in the form of more irregular lamellae of bone, between osteons. These interstitial lamellae are the result of Haversian systems which have been partially resorbed, followed by the formation of new osteons, in the remodelling process.

In the outer layers of cortical bone the lamellae course parallel with the surface and form relatively solid layers, or sheets, of bone. In a long bone these outer lamellae are known as circumferential lamellae. Although present to a varying extent, the circumferential lamellae are not as clear on the marrow cavity aspect of the cortical bone.

Cancellous bone consists of irregular bars (trabeculae) of bone of varying size which unite and branch to form a network of spicules between which bone marrow is present. The trabeculae are of a more simple construction than the osteons and consist only of varying numbers of irregular lamellae with associated osteocytes and canaliculi. In a long bone the inner cancellous bone merges with the cortical bone on the periphery and forms almost the entire outer margin of the marrow (medullary) cavity.

The bone marrow

Bone marrow, although strictly speaking not part of bone as a tissue, is functionally and morphologically such an integral part of bone that a description is appropriate. Bone marrow exists in two forms, red marrow and yellow (fatty) marrow, the names being derived from the colour in a fresh state. Both have a framework of reticular connective tissue. Red marrow is very vascular and is concerned with hemopoiesis, deriving its name from the presence of erythrocytes and their pigmented precursors, but a variety of other elements are also produced, such as platelets, granular leukocytes, and many of the lymphocytes. In yellow marrow the hemopoietic tissue has been replaced by fatty tissue with a concomitant decrease of reticular tissue. During fetal life and for at least the first year of life, red marrow is present in all bones but steadily decreases in quantity until it is present only in the epiphyses of the long bones, in the ribs, the vertebrae, the bones of the skull, and in the sternum of the adult human.

The periosteum

The periosteum consists of a layer of fibrous connective tissue investing the bones, except on their articular surfaces. It consists of two layers, an outer coarse fibrous layer which is vascular and contains many nerves, and an inner cellular layer which contains a loose network of elastic fibers and is osteogenic by means of osteoblasts, especially during the period of appositional growth. Collagenous fibers of the periosteum become trapped within the outer lamellae of developing bone and these Sharpey's fibers then have an anchoring function. The endosteum, which is similar in function to periosteum but thinner, lines the marrow cavity.

Cartilage

Ordinary connective tissue consists of cells and fibers embedded in a semi-fluid ground substance. In cartilage the ground substance is of firm consistency and contains a polysaccharide–protein compound (more specifically, the glycosaminoglycans chondroitin 4-sulphate, chondroitin 6-sulphate and some hyaluronic acid) as major components. Some authors use the term chondromucoid for the above complex. This substance is PAS-positive, is basophilic and colors metachromatically with toluidine blue. In the matrix of cartilage, collagenous fibers form a network similar to that seen in areolar tissue.

Cartilage forms a major part of the skeleton of the embryo but is limited in the adult to isolated areas such as the knee joints (semilunar cartilages), the ribs (costal cartilages), the thyroid, the trachea and bronchii, the epiglottis, the nose, and the auricle. Cartilage lacks a nerve and blood supply and derives its nutrients from surrounding tissues.

During chondrogenesis the chondroblasts, which are of mesenchymal origin, are responsible for laying down the fibers of the matrix and thereafter the ground substance. With an increase in the amount of intercellular material, each cell becomes enclosed in an individual compartment (a lacuna) and becomes a chondrocyte. As the intercellular material continues to increase, the chondrocytes move further apart. The cartilage cells can initially undergo mitosis but lose this ability with increasing maturation and hardness of the tissue. Further peripheral growth is then only possible through appositional chondrogenesis in the cellular layer of the perichondrium.

Three types of cartilage are described, namely fibrous cartilage, hyaline cartilage, and elastic cartilage.

Fibrous cartilage

Fibrous cartilage (or fibrocartilage) may be considered to represent a transition between dense connective tissue and hyaline cartilage (Fig 22). It consists of a network of dense collagenous fiber bundles between which chondrocytes are found in lacunae surrounded by a small quantity of ground substance. The chondrocytes lie singly, in pairs or short rows between the fiber bundles and it is often difficult to distinguish between fibrous cartilage and dense connective tissue. The presence of chondrocytes in lacunae is, however, diagnostic. Fibrous cartilage lacks a perichondrium and merges into other tissues such as dense fibrous tissue, bone or hyaline cartilage. This type of cartilage is found in the intervertebral discs.

Hyaline cartilage

Collagenous fibers in hyaline cartilage do not form bundles but are relatively evenly distributed as a fine network of slender fibrils in a firm ground substance (Fig 22). The matrix of hyaline cartilage appears homogeneous, since the fibers and the ground substance have the same staining characteristics. The chondrocytes are found in lacunae in the almost glass-like matrix, hence the term hyaline cartilage. Hyaline cartilage is found covering the articular surfaces of bones in the joints, in the nose, trachea, bronchii, the larynx, and as costal cartilages.

A sheath of fibrous tissue, the perichondrium, encloses hyaline cartilage except on articular surfaces of bones. The outer layers of the perichondrium are similar to the surrounding areolar tissue and is vascular. It contains fibroblasts and collagenous and elastic fibers. On the cartilage side the perichondrium becomes more cellular and this is the chondrogenic zone in which chondroblasts are found and which blends imperceptibly with the cartilage itself.

Elastic cartilage

Elastic cartilage is similar to hyaline cartilage but contains a higher proportion of branching elastic fibers. It is more flexible than hyaline cartilage and is found in the epiglottis, pharyngotympanic tube, and in the external ear.

Fig 22 Cartilage

Fibrous cartilage

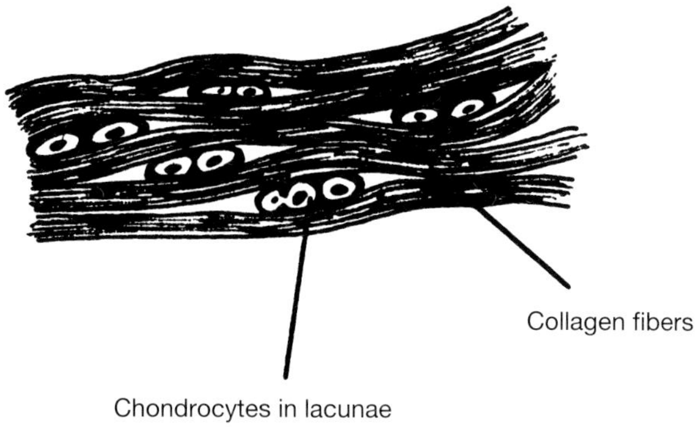

- Collagen fibers
- Chondrocytes in lacunae

Hyaline cartilage

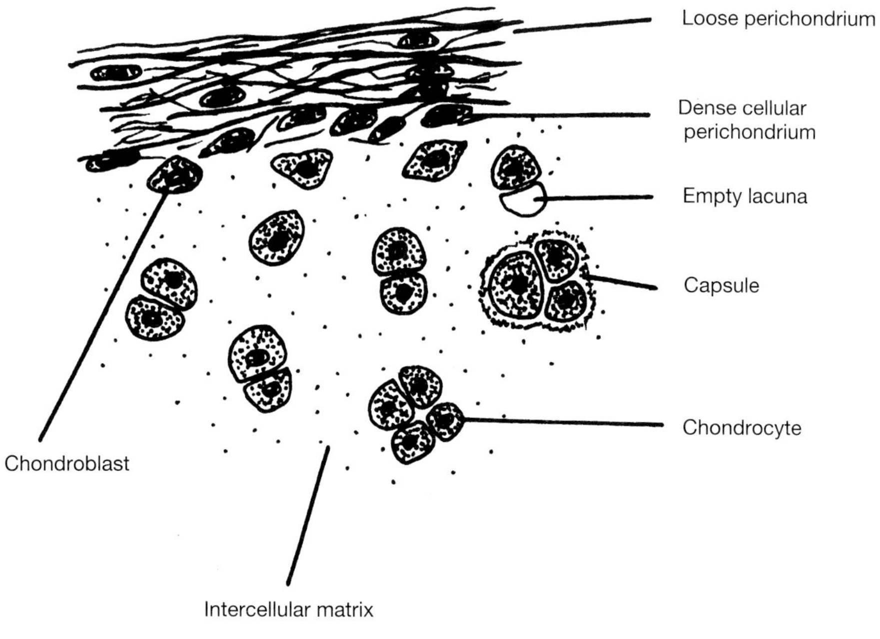

- Loose perichondrium
- Dense cellular perichondrium
- Empty lacuna
- Capsule
- Chondrocyte
- Chondroblast
- Intercellular matrix

Selected bibliography

1. Cole, A. S. and Eastoe, J. E. (1988) *Biochemistry and Oral Biology,* 2nd edition. London: Wright.
2. Kelly, D. E., Wood, R. L. and Enders, A. C. (1984) *Bailey's Textbook of Microscopic Anatomy,* 18th edition. Baltimore: Williams and Wilkins.
3. Krause, W. J. and Cutts, J. H. (1986) *Concise Text of Histology,* 2nd edition. Baltimore: Williams and Wilkins.
4. Leeson, T. S., Leeson, C. R. and Paparo, A. A. (1988) *Text/Atlas of Histology.* Philadelphia: W. B. Saunders Company.
5. Ten Cate, A. R. (1989) *Oral Histology,* 3rd edition. St Louis: C. V. Mosby Company.

Review questions

1. Give a brief description of the fibers found in ordinary connective tissue.
2. Classify ordinary connective tissue and mention sites in the body where each is found.
3. Briefly describe cortical bone.
4. Briefly describe cancellous bone.
5. Briefly describe bone marrow.
6. Briefly describe the periosteum.
7. Briefly describe chondrogenesis.
8. Classify cartilage and mention sites in the body where each is found.

6. Blood

General remarks

The embryonic multipotential connective tissue, mesenchyme, contains stellate cells which are in many instances interconnected by means of cytoplasmic cell processes. These cells differentiate and give rise to the formation of blood, lymph, associated vascular canals, ordinary connective tissue, bone, and cartilage. The intercellular substance in blood and lymph is fluid, in connective tissues it varies in degree of fluidity and contains fibers, in bone it is fibrous and mineralized, and in cartilage it is firm but pliable.

Blood and lymph constitute the fluid tissues of the body. Their functions include the transport and distribution of oxygen, nutrients, and the products of the endocrine glands in the body, and the removal of waste products and toxins which are dealt with in various ways. Both contain a fluid matrix, the blood plasma, and various types of cells. Circulating blood cells are of two main types, namely non-nucleated erythrocytes (red blood corpuscles), and nucleated white cells, the leukocytes. Leukocytes are of different types and are classified as granular (polymorphonuclear) leukocytes which include neutrophil leukocytes, eosinophil (acidophil) leukocytes, and basophil leukocytes, and the non-granular leukocytes which include lymphocytes and monocytes.

Red blood corpuscles (erythrocytes)

Red blood corpuscles are biconcave discs with a diameter of approximately 7-8 µm (Fig 23). The bilateral hollow confers on the corpuscles a central light zone which may at first glance be mistaken for a nucleus when viewed through a light microscope. Nuclei are, however, not present in mature red blood corpuscles in mammals.

The cytoplasm of red blood corpuscles contains hemoglobin which combines with oxygen in the lungs to form oxyhemoglobin. In the tissues of the body, oxyhemoglobin is reduced and oxygen becomes available for metabolic activity. Hemoglobin is also important for the transport of carbon dioxide from the tissues back to the lungs for disposal in exhaled air. When stained by means of the Wright stain, or by hematoxylin and eosin, red blood corpuscles have an orange color.

White blood cells (leukocytes)

Leukocytes are true cells containing nuclei and cytoplasmic organelles. Some show ameboid movement. In blood smears stained with hemotoxylin and eosin, the leukocytes are readily distinguishable from red blood corpuscles but special staining techniques are employed for a critical examination.

Granular (polymorphonuclear) leukocytes

The granular leukocytes are, as the name implies, characterized by the presence of cytoplasmic granulation (Fig 23). Each member of this group of leukocytes contains specific granules which can be readily distinguished. A second characteristic of these cells is the presence of a nucleus with more than one lobe, hence the term polymorphonuclear. The following descriptions of granular leukocytes is based on the appearance with Wright's staining method.

Neutrophil leukocytes

The nucleus of a neutrophil contains from two to five irregular lobes connected by thin chromatin threads (Fig 23). The cytoplasm contains numerous granules which are of two types. The majority of granules are fine and may be light pink in color. Scattered amongst the fine granules are larger granules which stain a reddish purple color. The granules consist largely of antibacterial enzymes such as lysozyme and are responsible for phagocytosis. The neutrophil is more active than any other blood cell, migrates by means of an ameboid movement and forms the body's first line of defense against bacterial invasion. The cells are approximately 8 µm in diameter in a fresh state and up to 12 µm in a dried smear preparation.

Eosinophil leukocytes

The nucleus of an eosinophil is usually bilobed with a connecting filament (Fig 23). The cytoplasm contains many round membrane-bound granules which vary in color from orange-red to brilliant red. The granules are lysosomes and contain a high content of peroxidase. Eosinophils are very active in phagocytosis and increase markedly in number in allergic states such as asthma and hay fever. The cells are about the same size as the neutrophils.

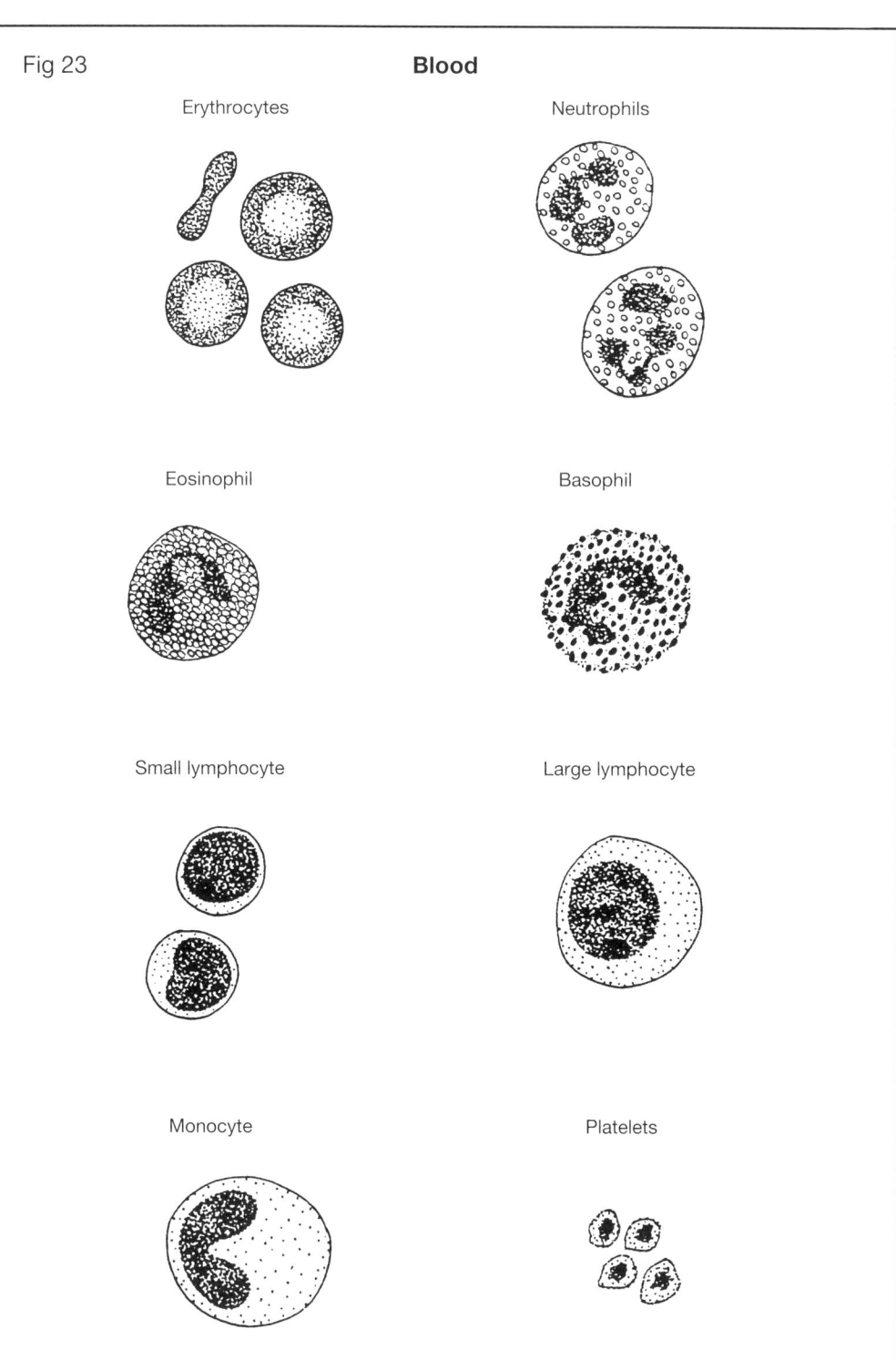

Fig 23 Blood

Basophil leukocytes

The basophils are the rarest of the leukocytes and are about the same size as the neutrophils. The nucleus is narrowed in one or two places but does not show the pinched-off appearance of the other granular leukocytes (Fig 23). The basophils contain granules which frequently appear irregular in size and shape in stained preparations, often creating the illusion that they lie outside the cell membrane. They assume a deep blue to violet color. The granules are not as specifically antibacterial as those of the other granular leukocytes and contain a platelet activating factor, prostaglandins and chemotactic factors.

Non-granular leukocytes

Lymphocytes

The majority of lymphocytes are relatively small with a diameter of approximately 6-8 μm although larger forms are found (Fig 23). A notable feature of the lymphocytes is a large densely-staining nucleus which may have a slight indentation on one side and has a purplish blue color in stained preparations. The rim of cytoplasm around the nucleus is lightly basophilic and may contain fine granules.

Monocytes

The monocytes are large cells and vary in diameter from 9-12 μm in a fresh state to 20 μm in some preparations (Fig 23). Although the cells are somewhat similar to lymphocytes, the light blue cytoplasm is more abundant and the nucleus is often eccentrically placed. The nucleus is oval or kidney-shaped and stains less intensely.

Functions of the lymphocytes

Lymphocytes are primarily concerned in the immune response which consists of two types, a humoral or antibody-mediated immune response, and a cell-mediated immune response. In the humoral response, B lymphocytes (bursa-dependent lymphocytes), react with a foreign antigen by proliferating and become plasma cells which synthesize a specific antibody (immunoglobulin) against that antigen. B lymphocytes and plasma cells are present in great numbers in the germinal centres of lymph nodes. This stimulatory reaction is relatively slow and is known as the primary reaction.

Other B lymphocytes which have been exposed to the antigen do not differentiate into plasma cells but remain as "memory cells". When these cells are once again exposed to the same antigen, rapid proliferation occurs and results in the formation of large numbers of plasma cells and a consequent rapid production of large amounts of the original specific antibody. The individual is immune to that antigen. This is termed the secondary reaction.

Antibodies consist of separate groups of immunoglobulins such as IgA, IgD, IgE, IgG, and IgM and possess the ability to bind with antigens, usually bacteria, followed by microbial membrane damage, cell lysis, and phagocytosis by macrophages. Activation of the complement system further leads to chemotaxis, whereby polymorphonuclear leukocytes are attracted to the area, and kinin production which leads to increased vascular permeability and therefore signs of inflammation.

The cell-mediated immune response is dependant on T lymphocytes (thymus-dependent lymphocytes) which are found both in peripheral lymphoid tissue and in the circulating blood. In the lymph nodes they are present between the follicles as well as in the deeper parts of the cortex. When T lymphocytes are activated by an antigen they do not secrete free antibody but proliferate and produce differentiated cells, which are active over a short range against foreign cells such as transplanted tissues and bacteria. These cells produce lymphokines and other substances which attract polymorphonuclear leukocytes, other lymphocytes, and macrophages to the area (chemotaxis), inhibit macrophage migration from the area and stimulate macrophage activity.

In the humoral response (stimulation of B lymphocytes, formation of plasma cells and antibody production) the T lymphocytes and macrophages play an important role. The mode of communication between T and B lymphocytes is, however, poorly understood but the macrophages, after phagocytosis, process and present the antigen to the lymphocytes in an immunogenic form.

A distinct population of lymphocytes involved in the cell-mediated immune response are the natural killer (NK) cells. They lyse sensitive cells on first contact without prior antigen sensitization. The NK cells do not arise from lymphoid stem cells which produce B and T lymphocytes and their role in immunology is not clear.

Further aspects of the immune response are discussed in Part I, Chapter 26 and in Part II, Chapter 23.

Monocytes are found in the circulating blood but leave the bloodstream and differentiate into tissue macrophages (histiocytes) which play an important role in phagocytosis. These cells further play an important part in the immune response (as described above).

Blood platelets

Circulating blood contains very large numbers of non-nucleated bodies, the platelets (Fig 23). They respresent cell fragments derived from a precursor cell in the bone marrow but are bounded by a typical cell membrane and are approximately 2 µm in diameter. The cytoplasm of a platelet stains blue with Wright's stain and consists of a central darker zone containing numerous granules (the granulomere) and a paler peripheral area which contains very fine granules (the hyalomere). Platelets are concerned with the production of the enzyme thromboplastin (the intrinsic pathway). Thromboplastin (tissue factor) is also produced by the membranes of damaged cells (the extrinsic pathway) and both these pathways lead to the transformation of prothrombin into thrombin which transforms fibrinogen into fibrin, the essential steps in normal blood clotting. Thromboplastin is also present in the plasma. Platelets also have the ability to aggregate and "plug" a small leak in a blood vessel.

Blood plasma

Blood plasma is a homogeneous weakly alkaline fluid which contains virtually all the mineral elements and many of the organic substances present in the body. It is composed of water, solids and dissolved gases. The constituents can be summarized as follows:

1. *Water:* 90-92 %.
2. *Inorganic substances:* sodium, potassium, calcium, phosphates, chloride, and bicarbonate.
3. *Organic substances:* lipids, including fatty acids, triglycerides, phospholipids and steroids; the proteins albumin and globulin, as well as amino acids, hormones, urea, uric acid, and creatinine.
4. *Dissolved gases:* O_2, CO_2, N_2.

Plasma constitutes approximately 55 % of the total blood volume, while the formed elements (red blood corpuscles, cells etc.) account for 45 % of the volume.

Lymph

Lymph contains the same constituents as are found in blood plasma, though in different concentrations, and consists of that part of the plasma which leaves the arterial end of the blood capillaries to enter the intercellular compartment but is not reabsorbed at the venous end of the capillaries. It is collected in the tissue spaces by fine capillary vessels called lymphatics which return the lymph to the blood circulation via the lymph nodes. Lymph contains numerous lymphocytes and some granular leukocytes.

Blood values

Blood volume

An adult male with a body mass of 70 kg has approximately 5 l of circulating blood, composed of 2.3 l of formed elements (red blood corpuscles, cells etc.) and 2.7 l blood plasma. The ratio of blood to body mass is therefore approximately 71 ml/kg. In females the ratio is approximately 65 ml/kg. Blood constitutes 7.5-8.0 % of total body mass.

Red blood corpuscles

(a) Count in adult males: 5.4×10^6 per µl (mm^3) (ranges $4.7 - 6.1 \times 10^6$).
(b) Count in adult females: 4.8×10^6 per µl (mm^3) (ranges $4.2 - 5.4 \times 10^6$).
(c) No differences in blood count are noticeable in the sexes before puberty.
(d) Count in newborn babies: frequently higher than 6.2×10^6 per µl.
(e) Count increases with severe muscular exercise due to mobilization of formed elements from blood depots.
(f) The count is higher in persons living at high altitudes. Persons living at, for example, 4500 m above sea level may have counts as high as 7.0×10^6 per µl. This is due to stimulation of erythropoiesis by an oxygen shortage in the inhaled air.
(g) The life span of a red blood corpuscle is 110-120 days.

Leukocytes

Count

The total white cell count in adults may vary from 4.0 to 10.0×10^6 per µl. In newborn babies the white cell count can be as high as 20.0×10^6 per µl but gradually descreases until adult values are attained at the age of approximately 10 years.

Differential white cell count (ranges in parentheses)

Cell type	Mean concentration per µl (mm^3)	approximate % of circulating white cells
Neutrophils	4500 (2600-7000)	40-70
Eosinophils	150	0.5-6.0
Basophils	40	0-1.0
Lymphocytes	2500 (1500-4000)	20-55
Monocytes	300	2-8

Blood platelets

The average platelet count is approximately 2.5×10^6 per µl (ranges from 1.5 to 4.0×10^6).

The average life span of a platelet is approximately 10 days.

Selected bibliography

1. Krause, W. J. and Cutts, J. H. (1986) *Concise Text of Histology*, 2nd edition. Baltimore: Williams and Wilkins.
2. Lavelle, C. L. B. (1988) *Applied Oral Physiology*. 2nd edition. London: Wright.
3. Leeson, T. S., Leeson, C. R. and Paparo, A. A. (1988) *Text/Atlas of Histology*. Philadelphia: W. B. Saunders Company.
4. West, J. B. (ed) (1991) *Best and Taylor's Physiological Basis of Medical Practice*, 12th edition. Baltimore: Williams and Wilkins.

Review questions

1. Discuss red blood corpuscles with reference to the following:
   (a) Morphology;
   (b) Functions; and
   (c) Count per µl and factors which influence the count.
2. Classify the leukocytes and briefly describe the morphologic features of each type.
3. Discuss the functions of the granular leukocytes.
4. Discuss the functions of lymphocytes.
5. Give a brief description of the blood plasma.
6. Tabulate the normal differential white blood cell count.

7. Muscle

General remarks

Muscle fibers differ from connective tissue fibers both in structure and in function. The latter are situated in the intercellular compartment, they do not possess any contractile ability, and their function is to bind structures together or as a filling material. Muscle fibers, on the other hand, are elongated cells with a highly developed contractile ability. The function of muscle fibers (muscle cells) is to move parts of the body. A minimum amount of intercellular substance is present in muscle tissue.

All types of muscle share the following characteristics:

(a) the fibers are elongated with obvious nuclei,
(b) the cytoplasm (sarcoplasm) stains red by means of eosin and contains myofibrils which course parallel to the long axis of the fibers,
(c) the fibers are enclosed by a sarcolemma (the cell membrane of the muscle cell).

Three types of muscle can be distinguished on morphologic grounds, namely smooth muscle (involuntary), striated skeletal muscle (voluntary), and striated cardiac muscle (involuntary). Smooth muscle and cardiac muscle are under autonomic control. Skeletal muscle is under the direct control of higher centers via the central nervous system (motor cortex of the brain).

Smooth muscle

When smooth muscle fibers are formed, the mesenchymal stem cells elongate and assume a spindle shape. The resulting fibers vary in length from 20 µm in the walls of small blood vessels to 500 or 600 µm in the pregnant uterus. On either side of the expanded central area containing the nucleus, the fiber thins to end in a rounded point (Fig 24).

The myofibrils in a smooth muscle fiber are very fine and not easily observed. The nucleus has an elongated oval shape and lies in a small area of granular cytoplasm. The sarcolemma is very delicate and the fibers very seldom branch.

The above characteristics are not all clearly seen in longitudinal sections of smooth muscle (Fig 25). The cells are usually arranged in densely packed bundles with indistinct cell borders. The elongated nuclei are obvious but the individual myofibrils are only seen on very careful examination. In a cross-section the rounded outlines of each cell are clearly seen. Depending upon the level at which each cell is sectioned, one sees either the centrally situated nucleus surrounded by a small rim of cytoplasm or only the dotted myofibrils on either side.

Smooth muscle is widely distributed in the body and is found in glands, the walls of blood vessels, and hollow organs such as the digestive, urinary, reproductive, and respiratory tracts, and in the skin and other areas. It is not usually covered by distinct connective tissue layers but mingles with surrounding areolar or reticular tissue.

Striated skeletal muscle

In contrast to smooth muscle, striated skeletal muscle develops from a mesenchymal stem cell precursor through an intermediate myoblast stage. According to some authors, a myoblast elongates, accompanied by rapid mitotic division of the nucleus, resulting in a multinucleated muscle fiber. Others believe that the multinucleated fiber results from fusion of separate myoblasts. Whatever the case, it is obvious that the final striated muscle consists of multinucleated fibers which are thicker than smooth muscle fibers, maintain a constant diameter and have blunt ends. Nuclei are found on the periphery of the fiber and the sarcolemma is clearly seen.

The most striking feature of striated skeletal muscle is the banding or cross-striations of the myofibrils seen in a longitudinal section (Figs 26, 27). The following bands or stripes are present. The A bands are dark and relatively wide. The adjacent I bands are a little narrower and appear light. A dark narrow Z line runs through the center of the I band. The lighter H band in the middle of the A band is transected by a thin dark M line. The segment between two successive Z lines is called a sarcomere and represents the contratile unit of the myofibril. The corresponding crossbands of adjacent myofibrils are responsible for the striated appearance of the entire fiber.

Myofibrils are composed of smaller units, the myofilaments, which are of two types. The thick myosin filaments are the main component of the A band, while the thinner actin filaments extend from either side of the Z line as far as the H band.

Skeletal muscles usually extend from one bony structure (the origin) to another which is movable (the insertion) and are attached to the bones by means of tendons.

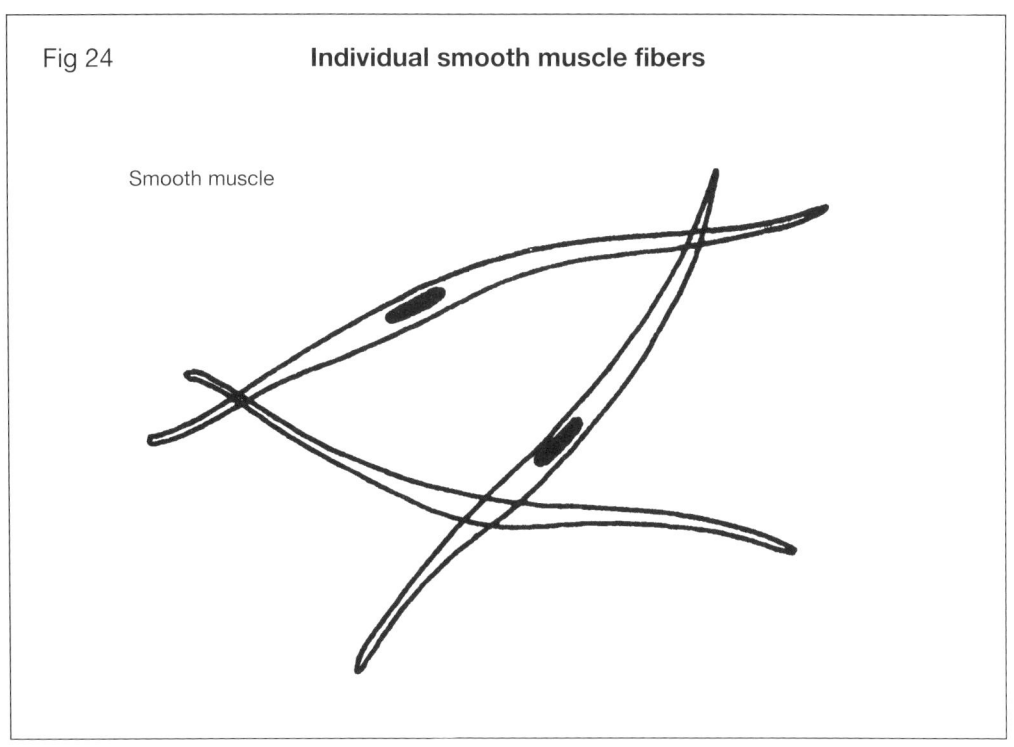

Fig 24 — Individual smooth muscle fibers

Smooth muscle

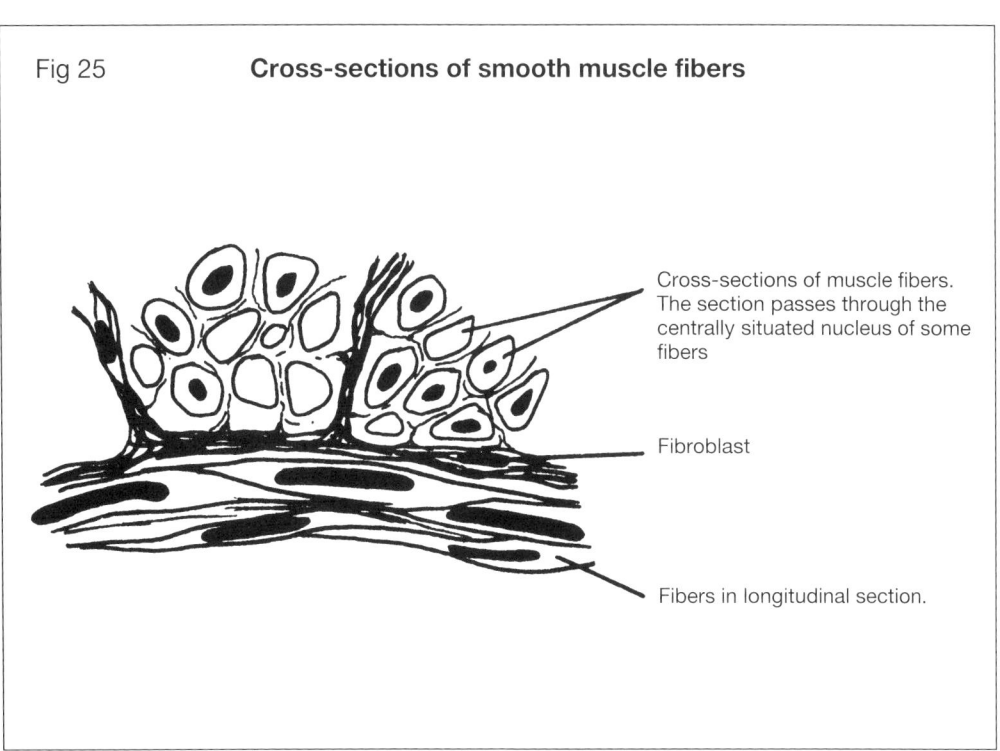

Fig 25 — Cross-sections of smooth muscle fibers

Cross-sections of muscle fibers. The section passes through the centrally situated nucleus of some fibers

Fibroblast

Fibers in longitudinal section.

Fig 26

Striated skeletal muscle

A group of fibers

One fiber consisting of many myofibrils. One myofibril is pulled out of the bundle in the sketch

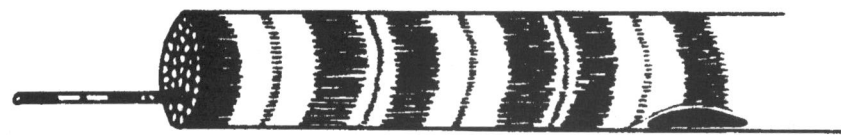

A myofibril highly magnified

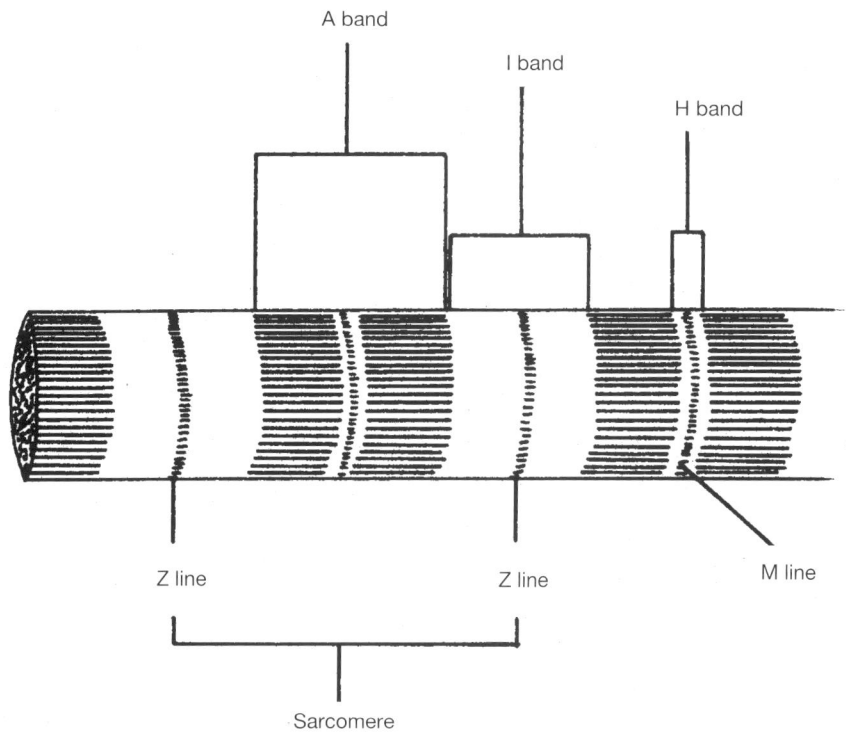

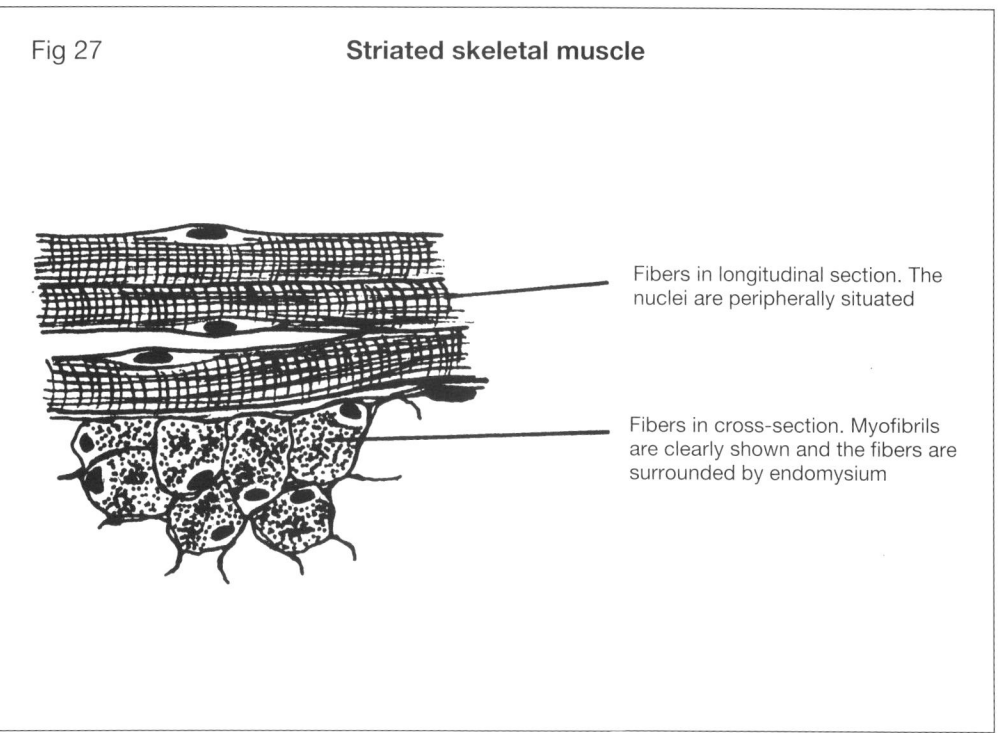

Fig 27 **Striated skeletal muscle**

Fibers in longitudinal section. The nuclei are peripherally situated

Fibers in cross-section. Myofibrils are clearly shown and the fibers are surrounded by endomysium

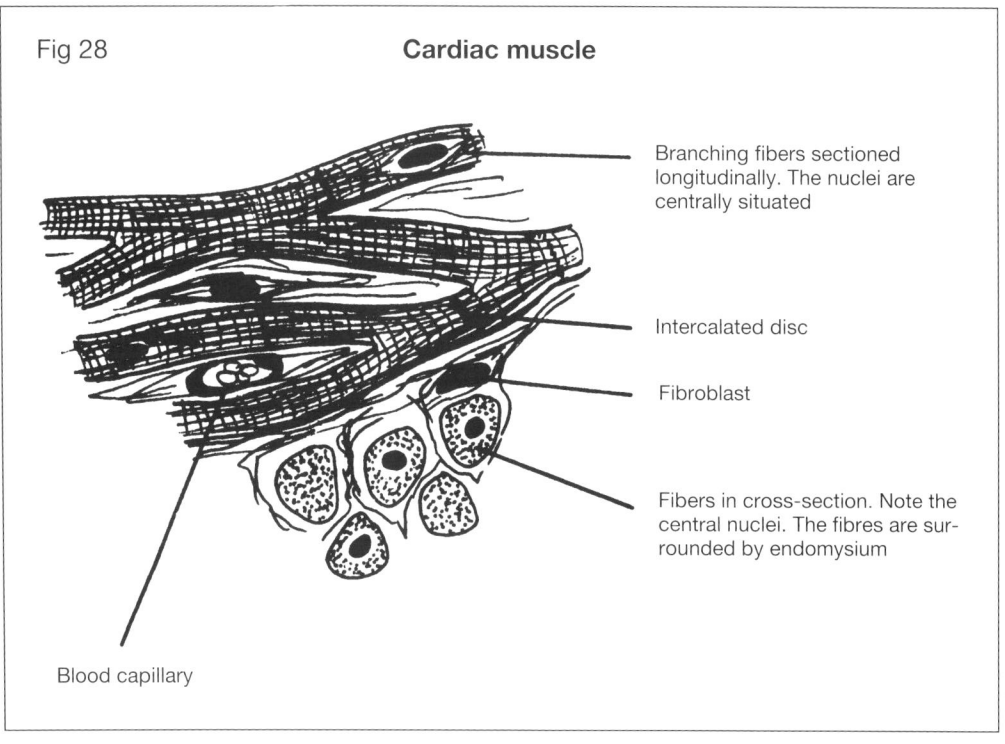

Fig 28 **Cardiac muscle**

Branching fibers sectioned longitudinally. The nuclei are centrally situated

Intercalated disc

Fibroblast

Fibers in cross-section. Note the central nuclei. The fibres are surrounded by endomysium

Blood capillary

An individual muscle fiber (the cell with its contained myofibrils) is enclosed by a delicate connective tissue, the endomysium. Groups of fibers are arranged in a bundle which is surrounded by a perimysium. The connective tissue enclosing a group of muscle fiber bundles is termed the epimysium.

Cardiac muscle

The fibers (cells) of cardiac muscle show extensive branching and interconnect freely (Fig 28). The nuclei of the fibers are centrally situated (as in smooth muscle) and show a banding pattern which is similar to that seen in striated skeletal muscle. The banding pattern is, however, less distinct. Where the muscle fibers meet, special membrane junctions, the intercalated discs, are apparent at the level of the Z lines.

Selected bibliography

1. Kelly, D. E., Wood, R. L. and Enders, A. C. (1984) *Bailey's Textbook of Microscopic Anatomy,* 18th edition. Baltimore: Williams and Wilkins.
2. Krause, W. J. and Cutts, J. H. (1986) *Concise Text of Histology,* 2nd edition: Baltimore: Williams and Wilkins.
3. Leeson, T. S., Leeson, C. R. and Paparo, A. A. (1988) *Text/Atlas of Histology.* Philadelphia: W. B. Saunders Company.

Review questions

1. Describe the characteristics which all muscle tissue have in common.
2. Classify muscle tissue and give a short description of each type mentioned.

8. Transition from embryo to fetus

The development of the embryo and the fetus during the normal gestation period of 280 days can be divided into three main phases. The initial period of the ovum lasts from fertilization to the 7th or 8th day (implantation), followed by the embryonic period which lasts from the beginning of the 2nd week to the end of the 8th week. The fetal period follows and ends at the termination of normal pregnancy.

The end of the 2nd month therefore marks the end of the embryonic period and the commencement of the fetal period of intra-uterine life. The transition from the embryo to the fetus is characterized by the development in the embryo of all the body elements which can be recognized in the adult.

The 2nd month is a period of major development. The embryo lengthens from approximately 5 mm crown-rump (C-R) length at the beginning of the 2nd month to about 30 mm at the end of the 2nd month. The head and trunk are distinct at the beginning of the 2nd month. Somites are still present but are not prominent surface structures. The limb buds become visible by the end of the 4th week. The upper limb buds appear first at the level of the pericardial swelling, while the buds of the lower limbs appear slightly later in a position slightly caudal to the attachment of the umbilical cord. The head shows a sharp ventral flexure. The neck has not yet developed although the neck segments (somites 5-12) have been present for approximately 1 week. The optic and otic vesicles are visible. At the commencement of the 2nd month the lens of the future eye can be distinguished as an area of thickened ectoderm over the optic vesicle. These optic (lens) placodes give origin to the lens and quickly lose any connection with ectoderm. Pigment appears in the optic vesicles of the embryo of 10 mm C-R length and rapidly become a distinctive feature which is readily visible through the transparent ectoderm.

In the embryo with a C-R length of 10 mm (35 days) the pharyngeal area shows considerable changes. The maxillary processes which formed from the mandibular arches grow medially under the optic vesicles and unite with the lateral nasal folds. The first pharyngeal groove, which is destined to form the external acoustic meatus, deepens and later becomes surrounded dorsally by six small swellings which eventually form the auricle. Three of these swellings develop in relation to the mandibular arch and three in relation to the second (hyoid) arch. Opinions differ on the relative future contribution of each of the abovementioned arches to the auricle but it does seem that the hyoid arch is responsible for most of the auricle, while the mandibular arch swellings eventually form only the tragus and its immediate surroundings. The external acoustic meatus and the auricle migrate cranially during the 2nd month from their initial cervical location through differential growth. The upper limb bud is now divided into an arm, a forearm, and a hand, while that of the lower limb still has a paddle shape.

After 40 days (13 mm C-R length) the head of the embryo has significantly increased in size, the ventral flexure of the head has decreased and the forebrain vesicles are well developed. The first pharyngeal groove has deepened and the auricular swellings are more clearly apparent. The hand shows division into fingers while the bud of the lower limb becomes divided into thigh, lower leg, and foot. The developing heart can be clearly seen through the transparent pericardial swelling.

At the age of 45 days, when the embryo has attained a C-R length of 17 mm, the head has enlarged further and development of the neck has resulted in a further reduction in the ventral flexure of the head. The eyelids appear as folds of ectoderm and mosoderm above and below the lens and the development of the auricle is complete. Development of the facial complex is rapid. The fingers have clearly separated and the toes become separated before the end of the 2nd month.

At the end of the second month ($\pm$ 30 mm C-R length) the embryo is definitely human in appearance.

Selected bibliography

1. Hamilton, W. J., Boyd, J. D. and Mossman, H. W. (1972) *Human Embryology*, 4th edition, revised by Hamilton, W. J. and Mossman, H. W. Cambridge: W. Heffer and Sons Ltd.
2. Sperber, G. H. (1989) *Craniofacial Embryology*, 4th edition. London: Wright.

Review question

Give a review of human development during the 2nd month of intra-uterine life.

9. Endocrinology

The endocrine system consists of several glands which secrete hormones which are absorbed directly into the bloodstream and thereby exert a major influence on a diversity of body functions. The following are the main endocrine glands:

   A. Thyroid gland
   B. Parathyroid glands
   C. Pituitary gland
   D. Thymus gland
   E. Adrenal (suprarenal) glands
   F. Pancreas
   G. Gonads

The thyroid gland

The thyroid is a vascular U-shaped gland located on the anterior and lateral aspects of the thyroid cartilage in association with the larynx and the upper part of the trachea. The two lateral lobes are united by means of an isthmus (narrowing) which lies on the upper rings of the trachea.

The gland consists of many follicles filled with thyroglobulin, which is the precursor of thyroxin (the active thyroid hormone). This hormone is of major importance in the regulation of metabolic processes in the body.

Hypothyroidism and hyperthyroidism result from abnormal functioning of the thyroid gland.

Hypothyroidism results from a decreased activity of the thyroid gland with a concomitant low blood level of thyroxin. If the condition arises at a very early age, cretinism will result. This condition is characterized by a dry skin, a dwarf-like stature and mental retardation. In an adult, hypothyroidism results in myxedema, a condition in which a subcutaneous edema occurs, accompanied by a dryness of the skin, loss of hair, and an increase in body weight. Hypothyroidism responds well to administration of thyroxin.

Hyperthyroidism is marked by an accelerated metabolic rate. Persons with this condition are abnormally energetic, need little sleep, have an increased heart rate, and a moist skin. In more serious cases exophthalmos results. Goitre (an increased size of the gland) can be present in both hypothyroidism and hyperthyroidism.

Calcitonin, a hormone secreted by the parafollicular cells (C cells) of the thyroid gland, is the hypocalcemic hormone of the body and is secreted when the plasma calcium ion concentration increases. It inhibits osteoclastic activity and intestinal calcium and phosphate absorption.

The parathyroid glands

There are four parathyroid glands which are found on the posterior aspect of the thyroid. Their secretions are essential for life. Parathyroid hormone (parathormone) regulates the levels of calcium and phosphate in the blood by influencing the reabsortion of phosphates in the kidneys. The calcium level in the blood is similar to the phosphate level. When the phosphate level in the blood is decreased by the action of parathormone, a concomitant increase in calcium level occurs (the product of the phosphate and calcium levels is very nearly constant). Under these conditions, calcium is not reabsorbed and is excreted in the urine, resulting in decalcification of the skeleton. This may also occur in pathologic states of the parathyroid gland in which excessive parathormone is secreted (hyperparathyroidism). The activity of the parathyroid glands is regulated by the calcium level in the blood. When the blood level is low as a result of an increased demand for calcium, for example during pregnancy, parathyroid hormone secretion is increased resulting in withdrawal of calcium from the skeleton in favor of the fetus. This process can be prevented by vitamin D, which facilitates the absorption of calcium from the alimentary tract.

Damage to the parathyroid glands may occur during surgical therapy to the thyroid and results in hypoparathyroidism in which tetany occurs. This condition is marked by pain in the hands and feet and a hyperexcitability of both the peripheral and central nervous system, manifesting as a generalized tremor or convulsions.

The pituitary gland

The pituitary gland is often referred to as the "master gland". It is situated in the cranial cavity and exerts control over a large number of other glands and body functions. It controls the functioning of the adrenal glands, as well as the thyroid gland and the gonads. The hormones of the pituitary gland are of major importance for virtually all the physiologic processes in the human body, including reproduction, pregnancy, and lactation. Growth hormone (somatotrophin) and diabetogenic hormone (an antagonist to insulin) which causes a rise in blood sugar level, are further examples of pituitary gland activity. All the abovementioned functions are the results of activity of the anterior lobe of the gland.

The posterior lobe is responsible for the production of the antidiuretic hormone which controls urine production. Decreased activity of this hormone results in excessive production of urine and a general dehydration of the body.

The thymus gland

The thymus gland attains its maximum size by the age of 2 years and has virtually disappeared at puberty. It is located superiorly in the thoracic cavity and is approximately 50 mm long and 30 mm broad before it starts to decrease in size. The function of the thymus gland is not quite certain but it is known to influence sexual development and has an important function in the immune system (T lymphocytes).

The adrenal glands

The adrenal glands are moderately sized yellowish structures which lie in close relation to the upper pole of each kidney. They consist of a medulla and a cortex which exert different functions through their respective hormones.
One of the most important hormones of the medulla is epinephrine (adrenaline). Epinephrine duplicates the effect of sympathetic stimulation of an organ to a very large extent and is necessary for a swift physiologic response to crisis situations, for example shock or the so-called fight-or-flight reaction. It causes an increase in the blood glucose level and, in addition, a vasodilatation in the skeletal muscles, while at the same time causing a vasoconstriction in the skin, the mucous membranes, and in internal (splanchnic) organs. It increases both the heart rate and the force of contraction, resulting in an increased minute volume and blood pressure. It causes relaxation of the muscles of the intestines and of the bronchioles, while at the same time increasing the depth of respiration.

The steroid hormones of the adrenal cortex also perform vital functions. Hydrocortisone, a hormone with very many effects and which is often used excessively and possibly erroneously in therapeutic procedures, is not completely understood, but has a marked anti-inflammatory effect by, for example, decreasing capillary permeability (thereby inhibiting inflammatory exudate and leukocyte migration from blood vessels), and reducing the peripheral leukocyte count. Aldosterone, another steroid, has to do with electrolyte balance in the body, while the androgens (male sex hormones) are secreted in both sexes and influence growth.

The pancreas

The pancreas is an elongated organ with a fish-like appearance. The head of the pancreas is situated in the loop of the duodenum, while the body is hidden by the stomach. It has both exocrine and endocrine functions. It produces pancreatic juices for digestion (exocrine function),and the hormone insulin (endocrine function) which acts on the liver to promote glucose uptake and inhibit glucose production and release. It therefore leads to a lowering of blood glucose level (hypoglycemia). An insulin deficiency causes diabetes mellitus.

Glucagon, also a pancreatic hormone, has the opposite effect to insulin by increasing the blood glucose level (hyperglycemia) by hepatic glycogenolysis. Epinephrine has the same effect and it is therefore apparent that a delicate balance is maintained.

The gonads

In the male the gonads are represented by the testes and in the female by the ovaries. In addition to producing the spermatozoa and the ova respectively, the gonads secrete hormones. The testes lie in the scrotum outside the body because normal body temperature is too high for normal sperm production. Testosterone is the main hormone of the testes and is responsible for the onset of secondary sex characters at puberty, such as growth of facial hair, deepening of the voice, and muscular development.

The ovaries produce different hormones at different stages of the reproductive cycle. Estrogen is responsible for the changes associated with puberty, while a second hormone, progesterone, stimulates and maintains the uterus during pregnancy.

Selected bibliography

1. Krause, W. J. and Cutts, J. H. (1986) *Concise Text of Histology*. 2nd edition. Baltimore: Williams and Wilkins.
2. Lavelle, C. L. B. (1988) *Applied Oral Physiology,* 2nd edition. London: Wright.
3. West, J. B. (ed) (1991). *Best and Taylor's Physiological Basis of Medical Practice,* 12th edition. Baltimore: Williams and Wilkins.

Review questions

1. Describe the location of the endocrine glands.
2. Briefly discuss the main functions of the following hormones:

    (a) Parathormone
    (b) Calcitonin
    (c) Epinephrine
    (d) Hydrocortisone
    (e) Insulin
    (f) Testosterone
    (g) Estrogen
    (h) Progesterone

10. Growth and development

The adolescent growth spurt is a constant feature of development in all children but may vary in intensity, duration, and in time of occurrence from one child to another.

In boys the peak increase in length is approximately 100 mm, and 80 mm in girls, occurring between the ages of 11 and 15 in boys, and between 10 and 13 years in girls.

Virtually all skeletal and muscular dimensions are affected by the growth spurt but not in equal measure. Increase in length during this period of maximum growth is mainly due to growth of the trunk and not of the lower limbs. The spurt in muscular growth commences approximately 3 months after the increase in length, while weight increase reaches a peak after a further 3 months.

Mass at birth averages 3.4 kg (3000 million times the mass of the ovum) and is not greatly affected by the mother's diet, except in cases of severe malnutrition. The maximum rate of mass increase is attained shortly after birth, after which it gradually lessens. One year after birth, body mass has trebled and after 4 years the baby weighs four times its birth mass. Thereafter there is an average mass increase of 2 - 3 kg/year until puberty. During the growth spurt boys may become heavier by as much as 20 kg and girls by 16 kg.

General differences between the sexes

Many of the sex differences seen in human adults are the result of differential growth patterns at adolescence.

The greater relative width of the hips in females, and of the shoulders in the male, are largely due to differential stimulation of cartilage cell growth (signs of sexual dimorphism in the pelvis are distinct as early as the 4th month of intra-uterine life).

The greater length of the male lower limbs, relative to the trunk, is the result of the longer male prepubertal period, since the legs grow faster than the trunk during this period.

Other sex differences become apparent at an earlier age. The male forearm is relatively longer than the upper arm, this difference being established at birth. The relative lengths of the second and fourth fingers is a further sex difference. The second finger of females is frequently of the same length, or longer, than the fourth finger, while in males it is usually shorter than the fourth finger.

Female children are more advanced than their male counterparts in skeletal development and in development and eruption of the dentition, but the dental differences appear to be largely confined to the secondary dentition.

Skeletal differences already become apparent during fetal life. The relative retardation in development in the male is apparently linked to genetic factors related to the Y chromosome. A person with Klinefelter syndrome (sex chromosome status XXY) has virtually the same skeletal development as a normal male child (XY), while development in a person with Turner syndrome (XO) is similar to that of a normal female, at least till puberty.

At birth, male skeletal development is approximately 4 weeks in arrears, relative to development in females, and males maintain a state of relative retardation until the end of the adolescent growth spurt. This accounts for the fact that females pass through adolescence and reach adult size earlier than males.

Changes in physiologic function

Major changes in physiology take place during the adolescent growth spurt. The changes are possibly more drastic in boys and result in greater muscular development, stamina, and other changes. Before adolescence boys are slightly stronger than girls, possibly due to the fact that a mesomorph body type is more common in boys, but this difference becomes more pronounced at adolescence. After adolescence males have larger hearts and lungs, relative to their body size, compared to females, as well as a larger oxygen-carrying capacity of their blood (a red cell count of 5.4×10^6/ml compared to 4.8×10^6/ml in females). The red cell count at birth is $6 - 7 \times 10^6$/ml but it decreases to a count which is only slightly higher than in adults within the first 2 weeks.

Although the above changes during adolescence take place in an orderly fashion in an individual, differences exist between individuals regarding the age at which the changes occur. Children of the same chronologic age may exhibit different stages of development and may reach puberty and adolescence at different ages. Chronologic age is therefore not a very reliable indicator of developmental stage, and means of assessing physiologic maturity are necessary. The two most commonly used measures of physiologic maturity are skeletal age and dental age.

Skeletal age

Skeletal age is the indicator which is perhaps more often used for exact assessments of maturity and is determind by assessing skeletal development by means of radiographs. The long bones pass through several distinct developmental stages, for example, the appearance of centers of ossification which increase in size, and the appearance of epiphyseal centers at various ages. The final shape and size of a long bone are attained when the epiphyses unite with the main mass of the bone (diaphysis) and cartilage cell formation and endochondral osteogenesis ceases. All the above changes may be observed by means of radiographs.

Determination of the skeletal age of a child is based on a comparison of development, as seen in that child, with a standard atlas in which the average stages of development of large numbers of children of known specific chronological ages are shown. The hand and wrist are most commonly used in the above regard, since many individual bones are present (much information) and it is an anatomically safe area to irradiate.

Dental age

Assessment of dental age as an indicator of the stage of physiologic maturity is of practical value to a dentist and is commonly used. The first method of assessing dental age is based on a comparison of the erupted teeth in the mouth with standard tables of eruption dates.

Between the ages of 6 months and $2\frac{1}{2}$ years the primary dentition erupts. Very little information is obtained regarding eruption status between the appearance of the last primary tooth and the appearance of the first secondary tooth at approximately $6\frac{1}{2}$ years. The secondary teeth erupt according to a fairly fixed pattern till approximately 13 years of age, when the second molar tooth appears. The third molar is inconsistent in its time of appearance.

A more refined method for assessing dental age is based on radiographs of unerupted teeth in which the stage of mineralization is compared to existing diagrams in which average stages of mineralization at different chronologic ages are shown.

Factors which may influence growth

Hormonal factors

The pre-adolescent period

The rate at which length increases accelerates until the age of 4 months of intra-uterine life (a maximum of 1.5 mm per day). Thereafter the rate decreases until approximately 4 years after birth (length at birth averages 500 mm), followed by a brief increase at approximately 6 years. The existence of this juvenile growth spurt is doubted by many authors. After the age of 6 years the rate of increase diminishes until 13 years, after which there is a rapid acceleration in the rate of increase in length to approximately 17 years. After this age it rapidly diminishes to a nil figure. Daily variation in length is largely dependent on the width of the intervertebral discs. As the day progresses, the mass of the upper parts of the body tends to compress the discs and it is not uncommon for the length of an adult person to decrease by as much as 20 mm during a day.

The main hormone which controls growth from adolescence to puberty, is the somatotrophic (growth) hormone of the pituitary gland. This hormone determines growth up to the period of rapid adolescent growth, which in turn is induced by steroid hormones. Under experimental conditions, administration of growth hormone stimulates growth of muscle and other fat-free tissues, while animals not receiving the hormone tend towards obesity and generally accumulate less protein while on the same diet.

The thyroid hormone (thyroxin) also plays an important role in the growth period but thyroid activity gradually lessens between birth and adolescence, when a brief increase in activity may occur for a year or two. Hypothyroidism leads to a retardation of growth as reflected in skeletal and dental maturation, as well as in growth of the brain, and obesity. Small individual variations in growth rate may be partly attributable to activity of the thyroid gland.

Adolescence

The period of rapid growth during adolescence is controlled by the steroid hormones of the adrenal glands and the gonads. Two of the three main groups of adrenal hormones, namely hydrocortisone and aldosterone (important for electrolyte balance in the body), maintain fairly constant blood levels from the time of birth. The third group, the androgens (neutral 17-ketosteroid, a male sex hormone), only appear at adolescence in active quantities and is presumably responsible for the entire female growth spurt, and for that part of the male growth spurt which is not due to testosterone (also an androgen) produced by the testes. There is little doubt that testosterone is responsible for the increase in size and strength of the muscle cells of the male, as well as for the higher red cell count.

Gonadotrophic hormones (pituitary gonadotropins) are only present in demonstrable levels shortly before adolescence and cause growth of the ovaries and testes. In due course the secretion of estrogen (female hormone) and testosterone (male hormone) is stimulated, leading to development of most of the secondary sex characters. It is, however, thought that secretion of limited quantities of testosterone already takes place from the 2nd month of intra-uterine life, stimulating development of the penis and scrotum.

Most of the facial dimensions change during the adolescent growth spurt. It appears, however, that mandibular growth rate more closely approximates the rate of increase in length during this period than any other part of the face.

Body type and genetic factors

Dimensions of people, especially in the same age group, may be compared. Anthropometry (measurement of bodily dimensions) enables one to classify humans into different body types, for example an ectomorph type (tall and thin), an endomorph type (average length and muscular). The French terms suggested for these body types in 1822 are, in the same order, cèrèbrale, digestif, and musculaire. Late in the 19th century attempts were made to link intelligence to a specific body type. The results were disappointing, but a slight yet consistent low positive relation was found between length and intellectual ability, while a slight but consistent low negative relation existed between laterality and intellect.

Ectomorphs are taller than mesomorphs from the 4th year, while mesomorphs are heavier than ectomorphs from the 2nd year of life. Endomorphs do not exhibit a constant growth pattern.

It must always be borne in mind that the length, mass, or body type of an individual reflects an interaction between environmental and genetic factors. Possession of a genotype for a large body does not imply that such a body type is inevitable, although very likely. When, for example, an immature limb bone from a fetal or newly born mouse is transplanted subcutaneously in an inbred mouse of the same strain (thereby largely preventing rejection), it will develop to almost an adult size in spite of the fact that it has never functioned. An inherent growth pattern which is genetically determined is responsible for the above phenomenon. Changes which occur as a result of function can be seen as merely the finishing touches in this instance.

Phenotypic expression is, however, largely dependant on genetic interaction and the environment. A study of Japanese immigrants to Hawaii is often quoted as an example of the effects of the environment on the growth pattern. The Japanese males emigrated to Hawaii and subsequently married Japanese women. It was found that the children resulting from these marriages were generally significantly bigger than their parents. The sons were, for example, on the average, 41 mm taller than their fathers. This is an example of how a favourable environment can positively influence phenotypic expression. A negative influence is just as likely. Many examples of genetic effects on growth and development are known, for example, it is known that menarche (the first menstrual period) may occur within 2 months of one another in monozygotic sisters living apart, while a difference of as much as 10 months may be found in non-identical sisters, according to chronologic age. The same effect of genetic influence is found in other forms of growth and maturation, for example, in eruption of teeth and the sequence in which they mineralize.

Race, climate, and seasons

Racial differences in the growth pattern, exist. While many of these differences are regarded by many to be genetically influenced, nutritional factors undoubtedly play a major role.

In contrast to what is generally believed, extremes of climate do not appear to have a major influence on growth rate. The average age at menarche of relatively well-nourished Nigerian schoolgirls is 14.3 years (hot climate), while it is 14.4 years in Eskimo girls. Burmese girls, subjected to an average summer day temperature of 44°C, reach menarche at an average age of 13.2 years, a figure which is practically the same as the average in Europe.

Genetic differences are undoubtedly responsible for the fact that people of Negroid origin living in West and East Africa and in the USA enjoy an advanced growth status compared to White races. This is reflected in a comparatively advanced skeletal maturation at birth, an advantage maintained for at least the first 2 years of life. These children show an advanced motor development and reach milestones, such as sitting upright and crawling, before their White counterparts. This advantage in motor development appears to be lost by approximately the 3rd year, possibly as a result of detrimental nutritional factors, although it is stated that the secondary teeth erupt approximately 1 year earlier in people of Negroid descent.

A seasonal effect on human growth rate is described. Growth in height is reported to be faster in spring, while an increase in mass is more pronounced in autumn. In many children, however, these seasonal differences are insignificant.

Nutrition and general socio-economic conditions

Malnutrition retards growth. This has been repeatedly shown in conditions of armed conflict accompanied by famine. Children, however, possess great recuperative powers, provided the adverse conditions are not too severe or do not last too long. Growth is retarded during short periods of malnutrition but takes place at an unusually fast rate once conditions normalize until the genetically determined rate for the particular chronologic age is reached. Girls appear to be more resistant to adverse conditions of malnutrition and illness and appear to recover at a faster rate than boys.

Children from different socio-economic backgrounds show differences in average body size at all ages, those from upper levels appearing to be advantaged regarding stage of maturity. It was reported in 1964 that differences in height between children of professional classes and of unskilled labourers was approximately 25 mm at the age of 3 years, and 37 - 75 mm at adolescence. Mass did not show the same differences. Furthermore, girls from the more favored groups reached menarche 2 - 3 months before the others while earlier eruption of the secondary dentition is also reported in children belonging to the former group.

Several factors may be responsible for the above differences but nutrition and related factors such as regular meals, sleep, regular exercise, and general domestic organization, which are generally indicative of higher socio-economic conditions, are perhaps the most important.

Minor illnesses, such as measles, influenza, middle ear infection, and even pneumonia do not influence the growth rate in well-nourished children but are reported to adversely affect under-nourished children.

Differences in Western socio-economic conditions are becoming less pronounced and an important factor in this regard is family size. Generally speaking, children from very large families are smaller and lighter than children from small families.

During the last hundred years (or more) a consistent tendency has appeared for children of all ages to be bigger, and to attain adult size earlier, than previously. Moreover, the length of adults was reported in 1964 to have increased over the previous 100 years at a rate of 10 mm per decade or 25 mm per generation in Europe. The reasons for this tendency are not clear but improved living conditions are usually regarded as an important factor.

Selected bibliography

1. Harrison, G.A., Weiner, J.S., Tanner, J.M. and Barnicott, N.A. (1977) *Human Biology*, 2nd edition. London: Oxford University Press.
2. Osborn, J.W. (ed) (1981) *Dental Anatomy and Embryology*. Oxford: Blackwell Scientific Publications.
3. Shaw, J.H., Sweeney, E.A., Capucino, C.C. and Meller, S.M. (1978) *Textbook of Oral Biology*. Philadelphia: W.B. Saunders Company.
4. Sinclair, D. (1975) *Human Growth after Birth*, 2nd edition. London: Oxford University Press.
5. Williams, P.L. and Wendell-Smith, C.P. (1978) *Basic Human Embryology*, 2nd edition. Tunbridge Wells, Kent, England: Pitman Medical Publishing Company.

Review questions

1. Describe the general sex differences between males and females.
2. What is meant by the terms skeletal age and dental age?
3. Discuss the influence of hormones on growth and development during the adolescent period.
4. Discuss the effects of nutrition and general socio-economic conditions on growth and development.

11. Development of the nervous system

Development of the head fold (Refer to Chapter 3)

The enlarging embryonic disc bulges into the amniotic cavity (Fig 29). This results in the development of a cranial fold (head fold), a caudal fold (tail fold), and lateral body folds, as a result of which the former lateral borders of the disc approach one another ventrally.

These changes have a noticeable influence on the secondary yolk sac. That part of the yolk sac which lies adjacent to the endoderm is gradually incorporated within the embryo and develops into the future alimentary tract. The development of the head and tail folds divides the developing alimentary tract in three parts, namely a foregut in the head fold, a hindgut in the tail fold, and a midgut which is connected to the remaining part of the yolk sac by means of a canal which gradually becomes smaller.

With the development of the body folds, the embryonic end of the umbilical connection is gradually displaced to a position immediately anterior to the cloacal membrane.

The amniotic cavity enlarges and gradually envelopes the entire embryo, together with the primitive alimentary tract. The developments in the head region of the embryo will now be discussed.

The embryonic ectoderm and endoderm are in contact (the prechordal plate) immediately anterior to the notochord, and this plate is the future buccopharyngeal membrane. Development of the head fold causes this area, together with the cardiogenic area and the septum transversum, to be located under the anterior end of the notochord and to be separated from the notochord by the anterior part of the foregut. At this stage the buccopharyngeal membrane lies in a small hollow in the antero-inferior end of the embryonic head region. This hollow, the primitive mouth or stomodeum, is lined by ectoderm and is separated from the endodermally lined foregut by the buccopharyngeal membrane. A further consequence of the formation of the head fold is that the mesoderm, which was previously located in front of the cardiogenic area, now lies posterior to it and ventral to the foregut. This mesoderm, the septum transversum, is caudally penetrated by the liver bud (future parenchyma and ducts of the liver) from the foregut, serves as an early partial diaphragm and gives rise to the early great veins in this area.

At this stage the developing brain formes a bulge anterior to the stomodeum and the rest of the primitive nervous system lies dorsal to the notochord.

Fig 29 **Embryo after development of the head fold**
(midsagittal section)

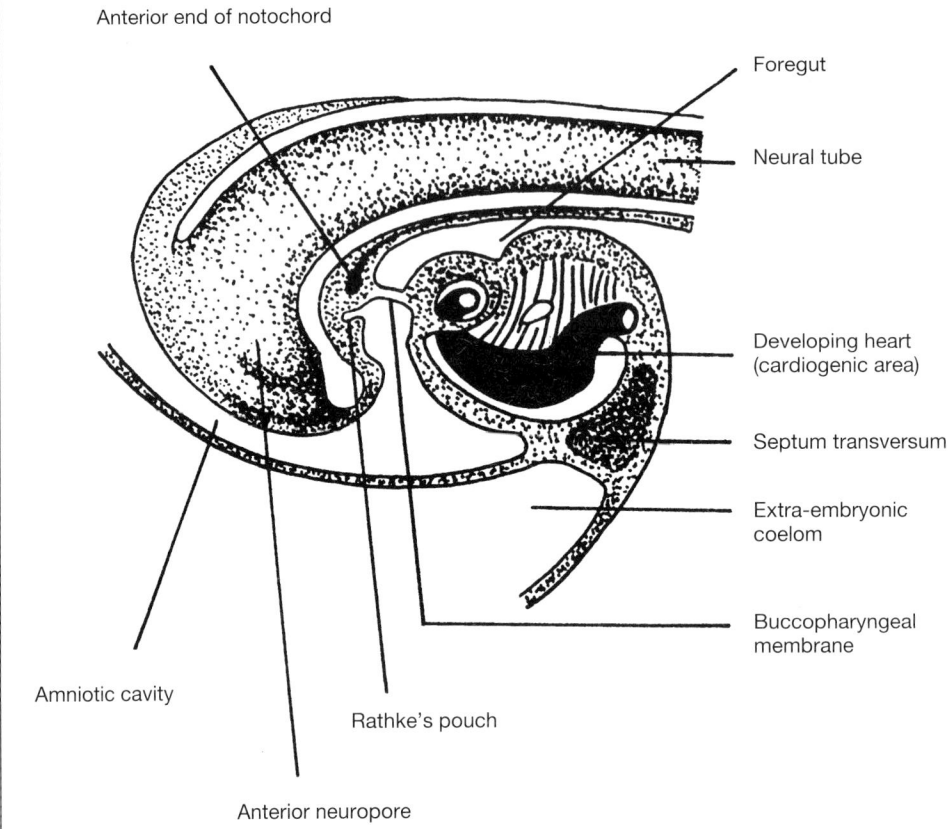

Anterior end of notochord

Foregut

Neural tube

Developing heart (cardiogenic area)

Septum transversum

Extra-embryonic coelom

Buccopharyngeal membrane

Amniotic cavity

Rathke's pouch

Anterior neuropore

Development of the central nervous system

The central nervous system develops from the elongated axially situated neural plate (neurectoderm), an area of ectoderm on the dorsal surface of the embryo. A groove, the neural groove (Fig 30), develops in the neural plate and deepens to such an extent that the lateral edges, or crests, of the groove meet in the midline and fuse to form a neural tube (neural canal). The edges of the neural groove initially fuse in its central part, leaving cranial and caudal openings through which the tube communicates with the amniotic cavity. The cranial opening, the anterior neuropore, closes at the 20-somite stage (26 days), while the caudal opening, the posterior neuropore, closes at the 25-somite stage (28 days) when the embryo is approximately 3.4 mm long.

The central nervous system (the brain and spinal cord) takes origin from the ectodermal lining (neurectoderm) of the neural tube. Neural crest cells, initially arranged in the form of a specialized elongated band of ectoderm on the periphery of the neural plate and later on the crests of the neural groove, are not incorporated in the neural tube but migrate in segmental fashion to lie dorsolateral to the neural tube, deep to the surface ectoderm.

Sensory cells and fibers of the peripheral (somatic) nervous system (with a few exceptions), as well as most of the peripheral cells of the autonomic nervous system, develop from neural crest tissue. In older embryos, neural crest tissue can be divided into the following primordia: trigeminal, facial, and auditory (vestibulocochlear), glossopharyngeal and vagal complex, occipital, and spinal. These primordia, except the occipital primordium which evidently disappears, give origin to the sensory cells of the cranial nerve ganglia and, through segmentation of the spinal primordium, to the dorsal spinal ganglia. In addition, the cells of the neural crest differentiate into Schwann cells, pigment cells, meninges, and even into odontoblasts, according to some authors. Neural crest cells migrate extensively and further take part in the formation of ectomesenchyme, a pluripotential tissue which will be discussed in a later chapter.

At an early stage the neural tube shows a division into an expanded cephalic part, the future brain, and an elongated caudal part, the future spinal cord (Fig 31).

The cephalic part develops three additional dilatations which are separated by constrictions. These represent the primary brain vesicles. Their central cavities become the ventricles and aqueduct of the brain, while their walls represent the three primitive divisions of the brain. The most anterior dilatation is the forebrain vesicle, the prosencephalon; the middle dilatation is the mesencephalon, the midbrain; and the caudal dilatation is the rhombencephalon, the hindbrain.

The prosencephalon develops two lateral telencephalic vesicles which in later development will become the cerebral hemispheres (Fig 32). The remaining original part of the prosencephalon is termed the diencephalon. The optic nerve, the neural component of the retina of the eye, the thalamus, the hypothalamus and the posterior lobe of the pituitary gland (pars nervosa) develop from the dience-

Fig 30   **Development of the neural tube and neural crest cells**

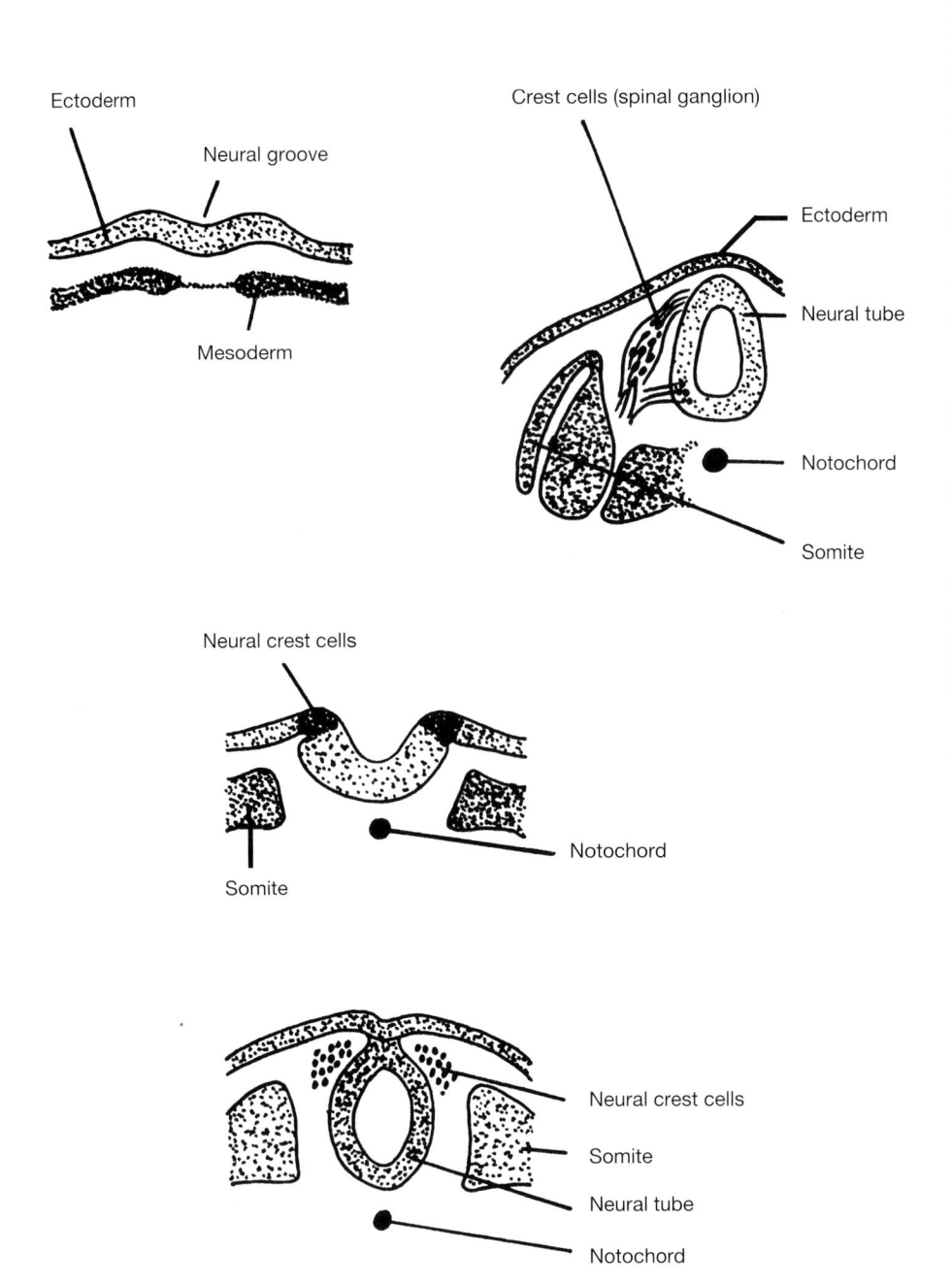

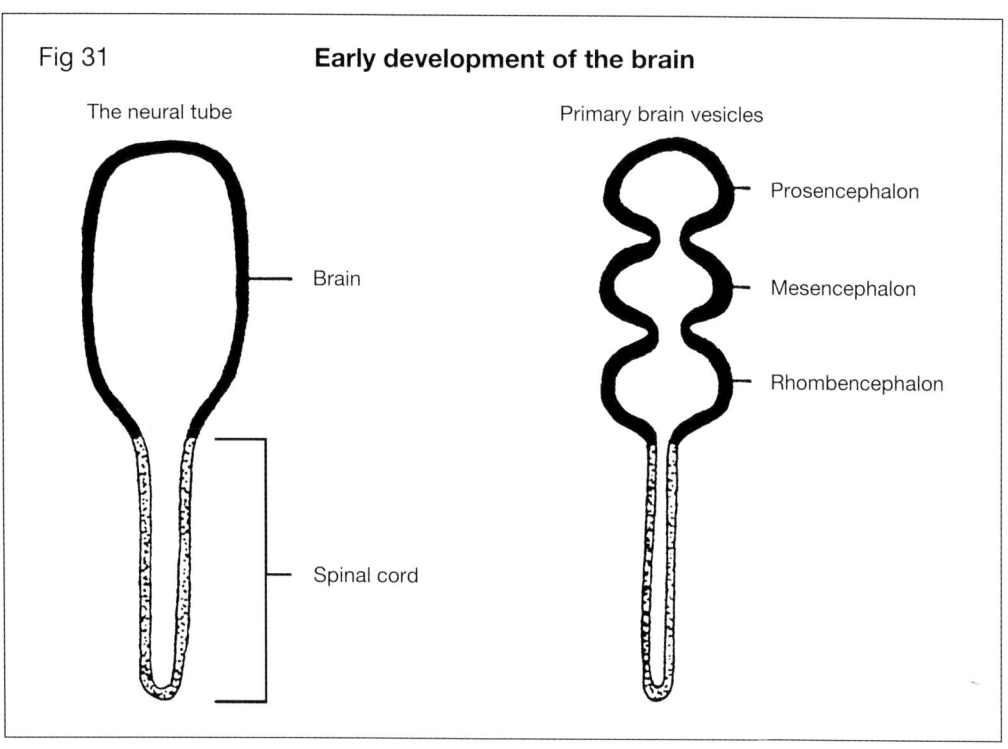

Fig 31 Early development of the brain

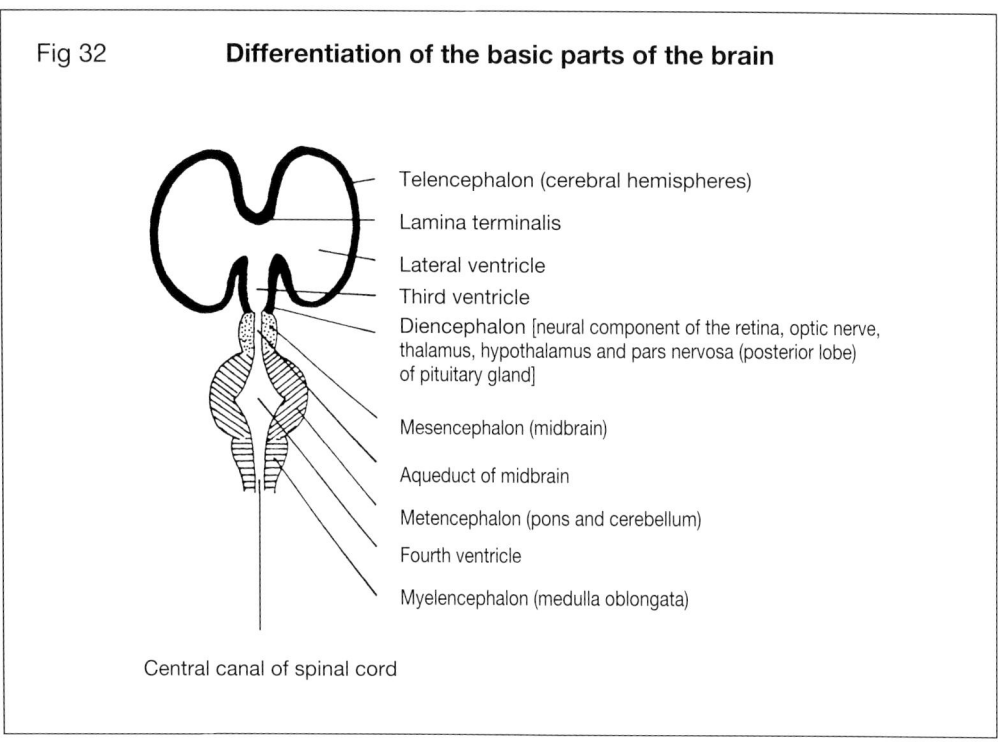

Fig 32 Differentiation of the basic parts of the brain

phalon. In further development, the mesencephalon (the future colliculi, tegmentum, and crura cerebri) shows relatively little structural change. The rhombencephalon in later development can be subdivided into a cephalic metencephalon (the future pons and cerebellum), which is connected to the mesencephalon by means of a narrow canal, the cerebral aqueduct, and a caudal myelencephalon (the future medulla oblongata), which is continued into the spinal cord.

During early development the future brain bends at three points, the primary flexures (Fig 33). The first bend (cephalic flexure) takes place in the mesencephalon, the second (cervical) flexure forms in the myelencephalon in the same ventral direction as the cephalic flexure, while the third (pontine) flexure forms in the metencephalon. The pontine flexure is in the opposite direction and is actually a ventral bend in the floor of the pons. The parts of the adult brain develop on the model thus created.

After completion of the neural tube it extends along the dorsal length of the embryo. The lateral walls thicken and become divided on each side into a dorsal and a ventral part by a horizontal groove, the sulcus limitans (the central canal of the early spinal cord is initially a dorsoventral slit).

Three layers can be distinguished in the wall of the differentiating neural tube, namely an inner ependymal layer which is the germinal layer for the cells of the nervous system (neurons, as well as the supporting cells, the astrocytes and the oligodendrocytes), an intermediate mantle layer into which neuroblasts migrate from the ependymal layer and differentiate into neurons, and an outer marginal layer which countains the protoplasmic processes of the developing neurons. In the adult nervous system the ependymal layer persists as the ependyma (lining epithelium), the mantle layer becomes the grey matter and the marginal layer becomes the white matter. In parts of the nervous system, especially in the cerebrum and the cerebellum, cells of the mantle zone migrate to the surface of the marginal zone to form a superficial layer of grey matter.

The cell bodies aggregate in groups in certain areas of the nervous system to form the nuclei of the cranial and spinal nerves (Fig 34). The group in the dorsal part (alar lamina) of the neural tube are involved with the sensory component of the somatic nervous system, while those in the ventral part (basal lamina) are involved with the motor component. Enlargement of these groups is responsible for the formation of the sulcus limitans. The cell bodies of the alar lamina are the second afferent cells of the sensory tract (the first arise in the dorsal spinal ganglia outside the spinal cord).

Neurons of the basal lamina send their processes through the marginal zone and these fibers are grouped on the surface of the brain and spinal cord to form the anterior, or motor, components of the cranial or spinal nerves. At the same time, neurons which developed in dorsal ganglia originally formed by neural crest cells, send their processes in two directions, peripherally and centrally. The centrally directed fibers penetrate the developing central nervous system to synapse with cells of the alar lamina and form the sensory, the posterior or afferent, root of a cranial or spinal nerve. The peripherally directed fibers are responsible for conducting impulses to the central nervous system from the periphery.

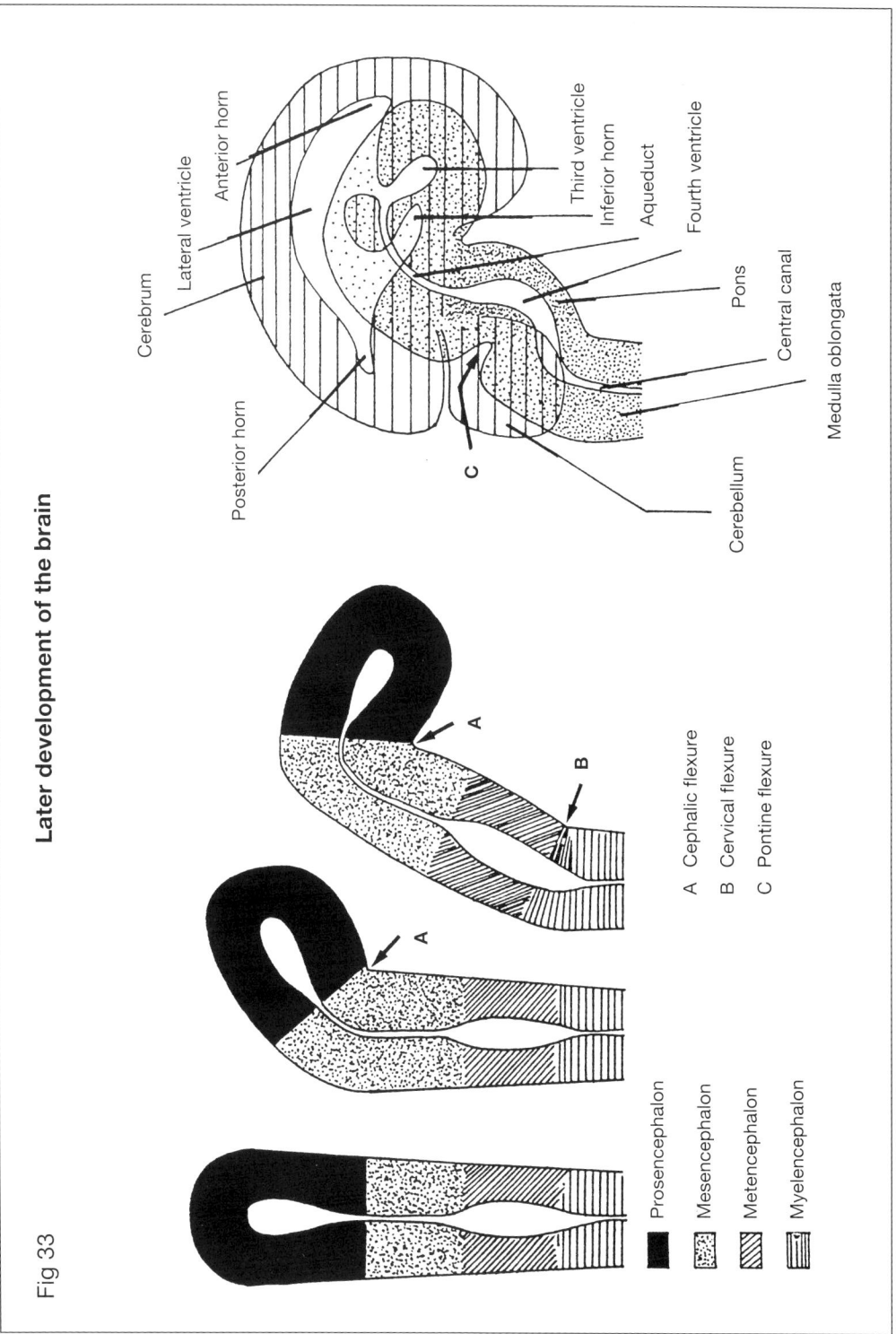

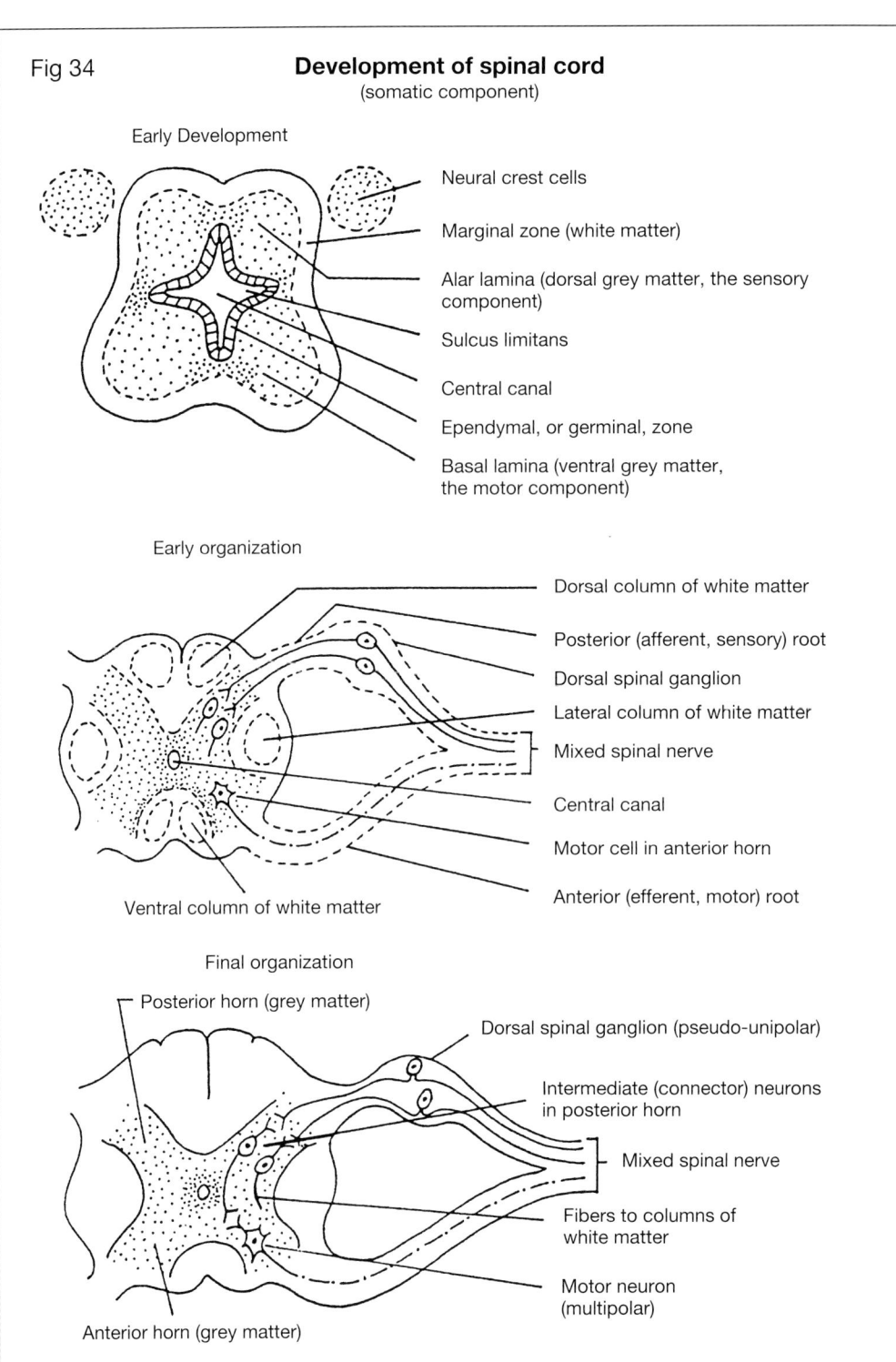

The spinal cord and peripheral nerves

A cross-section of the developed spinal cord shows two prominent areas, the peripheral white matter and the more centrally situated grey matter. The grey matter is of a darker color, since it contains fewer nerve fibers with myelin sheaths, and more nerve cells, than the white matter.

The grey matter is arranged in the form of an H and surrounds the central canal. The anterior limbs of the H are the anterior horns of the grey matter, and contain many multipolar cells from which fibers leave the spinal cord by way of the anterior roots of the spinal nerves. The posterior horns of the grey matter are thinner and receive the centrally directed processes of the sensory cells of the dorsal spinal ganglia via the posterior roots.

Intermediate, or connector, neurons in the posterior horns relay some incoming impulses to anterior horn cells, while others are conducted to higher centers via tracts such as the anterior and lateral columns of the spinal white matter. In the thoracic part of the spinal cord a lateral horn of grey matter is found. These lateral horns contain the cell bodies of neurons from which processes leave the spinal cord via the anterior horns to form the sympathetic tracts of the autonomic nervous system. The white matter is divided into anterior, lateral, and posterior columns which contain nerve fibers and form the ascending and descending tracts of the spinal cord.

There are 31 pairs of spinal nerves, consisting on each side of 8 cervical, 12 thoracic, 5 lumbar, 5 sacral, and 1 coccygeal nerve. Note that the spinal cord in the adult only extends to approximately the level of the inferior border of the first lumbar vertebra. There has therefore been a certain degree of retraction of the cord as development proceeded.

The nerves are bilaterally attached to the spinal cord by means of anterior (ventral) and posterior (dorsal) roots. The anterior roots contain motor (efferent) fibers, while the posterior roots contain sensory (afferent) fibers. Close to, or inside, the intervertebral foramina, the two roots unite to form a mixed spinal nerve. The dorsal spinal ganglion is situated in the posterior root and contains the cell bodies of the sensory fibers.

Spinal nerve trunks divide into anterior and posterior primary divisions, or rami, shortly after being formed. Nerves to the skin and voluntary muscles are conducted via these rami. Spinal nerves also contain sympathetic nerves to blood vessels and glands, while some sacral nerves also contain a parasympathetic component.

The autonomic nervous system contains the same components as the somatic nervous system. The difference lies therein that the axon of the intermediate (connector) neuron leaves the central nervous system to form a synapse with an efferent neuron in a peripheral ganglion, and not with an anterior horn cell as in the somatic system. Axons of the intermediate neurons of the autonomic system are therefore termed preganglionic fibers. They possess a myelin sheath and are known as white rami communicantes. Axons of postganglionic fibers do not possess any myelin and are known as the grey rami communicantes.

Selected bibliography

1. Hamilton, W.J., Boyd, J.D. and Mossman, H.W. (1972) *Human Embryology*, 4th edition, revised by Hamilton, W.J. and Mossman, H.W. Cambridge: W. Heffer and Sons Ltd.
2. Moore, K.L. (1988) *Essentials of Human Embryology*. Toronto: B.C. Decker, Inc.
3. Osborn, J.W. (ed) (1981) *Dental Anatomy and Embryology*. Oxford: Blackwell Scientific Publications.
4. Sadler, T.W. (1985) *Langman's Medical Embryology*, 5th edition. Baltimore: Williams and Wilkins.
5. Sperber, G.H. (1989) *Craniofacial Embryology*, 4th edition. London: Wright.

Review questions

1. Discuss the development of the brain.
2. Discuss the development of a spinal nerve.
3. Discuss the origin and fate of neural crest tissue.

12. General functional and histologic aspects of the nervous system

General remarks

The nervous system consists of the brain and spinal cord (central nervous system) and the nerve fibers which conduct impulses from the central nervous system to the periphery (efferent fibers), and from the periphery to the central nervous system (afferent fibers). These fibers, as well as nerve cells (neurons) in autonomic ganglia outside the central nervous system, constitute the peripheral nervous system.

The brain is a large and complex organ which fills the cranial cavity. It is divided into five distinct parts, namely the medulla oblongata which is continuous with the spinal cord, the pons, the midbrain, the cerebellum, and the cerebral hemispheres. The first three parts are collectively termed the brain stem. The cerebellum and the cerebral hemispheres each consist of two lobes marked by many surface folds, separated by fissures. The brain stem is not clearly divided into two halves.

The brain, like the spinal cord, consist of grey and white matter. In the cerebellum and cerebrum the grey matter is found peripherally and constitutes the cortex, while the white matter is located centrally. The grey matter consists of neurons, while the white matter is made up of nerve fibers enclosed by myelin, a lipoprotein material. The distinction between grey and white matter in the brain stem is not as clear. Nerves which are anatomically linked to the brain are termed cranial nerves.

The spinal cord extends from the caudal end of the medulla at the level of the foramen magnum to the level of the first lumbar vertebra, and has an average length of 450 mm in the adult. In the spinal cord the white matter is arranged peripherally, while the grey matter is centrally located and is H-shaped on section. The grey matter surrounds the central canal of the spinal cord which is continuous with the ventricular system of the brain. Nerves arising anatomically from the spinal cord are termed spinal nerves.

Functions of the nervous system

The nervous system can be divided into two functional components:

1. The somatic nervous system is concerned with reception of stimuli of an exteroceptive nature, for example pain impulses from the periphery, as well as impulses resulting from touch and temperature changes. Impulses concerned with smell, taste, sight, and hearing (the special senses) may also be included in this group.

   The somatic nervous system is further concerned with proprioceptive impulses arising from ligaments, tendons, muscles, and joints. These are necessary for the co-ordination of posture and movement. The regulation of masticatory forces through reflex activity is a result of proprioceptive feedback from the muscles of mastication, the temporomandibular joints, and the periodontal ligaments of the teeth.

   Generation of motor impulses to the skeletal muscles is a further very important activity of the somatic nervous system.

2. The autonomic nervous system is concerned with interoceptive impulses arising from viscera and blood vessels, and the regulation of involuntary activities of, for example, cardiac muscle, glands, and internal organs. It consists of central and peripheral parts. The central parts are located in the cerebral cortex, hypothalamus, cerebellum, brain stem, and spinal cord, while the peripheral parts consist of ganglia and plexuses with both peripheral and central connections (Fig 35).

   The efferent component contains sympathetic and parasympathetic fibers.

The sympathetic part of the autonomic nervous system consists of nerve fibers from the thoracic and lumbar regions of the spinal cord (thoracolumbar outflow), while parasympathetic fibers originate in the brain and sacral parts of the spinal cord (craniosacral outflow). Any organ of the body with an autonomic nerve supply receives motor stimuli from both parts (sympathetic and parasympathetic) and a fine balance is necessary for homeostasis since the two parts have an antagonistic action in many respects.

Each neural pathway to organs and tissues under autonomic control involves two neurons. One arises in the brain or spinal cord and ends in a peripheral ganglion. In the ganglion a synapse is formed with a neuron which ends in the organ which is supplied. The former neuron is termed a preganglionic fiber while the latter which ends in the effector organ is termed the postganglionic fiber.

The basic arrangement of nerve cells and their processes in the somatic nervous system and the autonomic nervous system are very similar, but a difference exists in the location of the effector cells. The cell body of a typical somatic motor cell is found in the anterior horn of the grey matter (in the case of the spinal cord), while the corresponding cell of the autonomic system, which is, in effect, the same as the

Fig 35   **General arrangement of the autonomic nervous system**

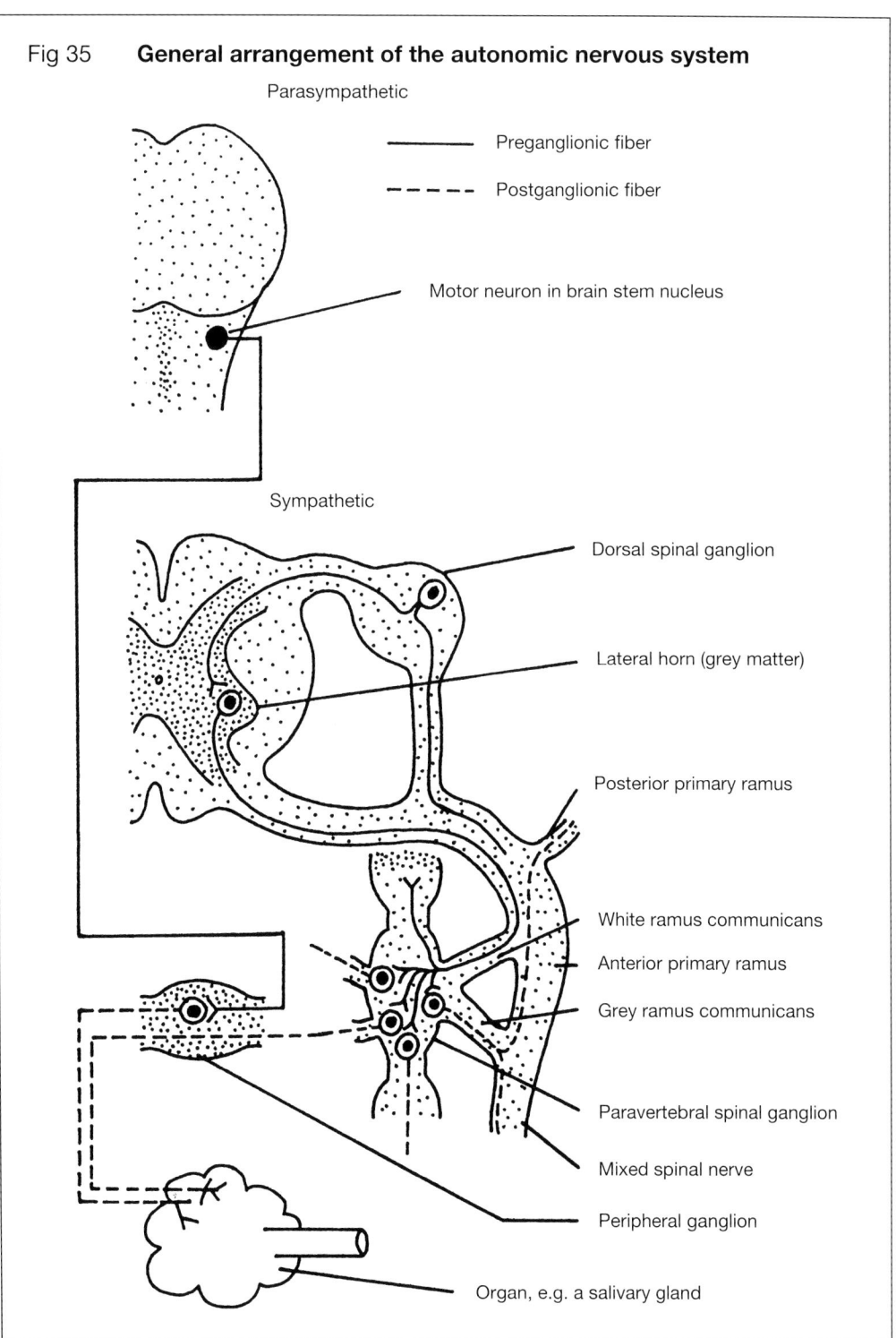

intermediate, or connector, neuron of the somatic nervous system, is situated in a ganglion outside the central nervous system. The ganglia of the sympathetic nervous system are arranged in a row next to the vertebral column (the paravertebral sympathetic chain), while those of the parasympathetic system lie close to the organ supplied.

The autonomic supply of the lacrimal gland, submandibular gland, sublingual gland, parotid gland, and the mucous membrane of the mouth are of major importance. The sympathetic supply of these structures consists of preganglionic fibers which arise from T1 and T2 and course to the superior cervical ganglion. Postganglionic fibers arise in the superior cervical ganglion and pass through the sphenopalatine ganglion to end in the lacrimal gland and the glands of the hard and soft palates, and through the otic ganglion to end in the parotid gland and probably the glands of the cheeks and lips, without forming any synapses in these parasympathetic ganglia. In a similar way, sympathetic fibers pass through the submandibular ganglion (parasympathetic) to end in the submandibular and sublingual glands, and the glands of the floor of the mouth, without forming any synapses. In the parasympathetic pathway, synapses are formed in all the parasympathetic ganglia mentioned above. The autonomic innervation of the salivary glands is discussed in detail in the chapter on saliva (Part II, Chapter 23).

All impulses which are conducted to the central nervous system via afferent fibers arise at peripheral nerve endings.

Arrangement of neurons in the spinal cord

The arrangement of neurons in the spinal cord can best be illustrated by describing the neural pathways followed in a simple reflex arc. A pain impulse arising from a nerve ending in a finger, for example from a pin prick, is conducted via an afferent neuron (a dendrite) to a dorsal spinal ganglion in which the cell body (sensory neuron) of that dendrite is located.
The conducting axon leaves this cell body and enters the posterior horn of grey matter in the cord via the posterior nerve root where a synapse is formed with an intermediate (connector) neuron. The intermediate neuron synapses with an anterior horn cell (efferent or motor neuron) and the motor impulse leaves the spinal cord via an axon in the anterior nerve root. The motor impulse can in this way reach the muscles of the arm to jerk the hand away.

In most cases, however, an elaborate series of intermediate neurons are involved. The reflex as described above is a short-circuiting pathway for fast reaction. Intermediate neurons are, however, also responsible for conducting impulses to other levels of the spinal cord and to the brain stem and cerebral cortex via, for example, the spinothalamic tracts, for further processing, conscious perception, and relay to other parts of the body for other reactions to stimuli.

Histology of nervous tissue

The nervous system consists of functional units, the neurons, which are highly differentiated specialized cells which have lost the ability to divide and have a very limited regenerative capacity.

A neuron consists of a cell body, the perikaryon, one or more branching afferent processes (dendrites) which conduct an impulse to the perikaryon, and an efferent process (axon or axis cylinder) which conducts the impulse away from the perikaryon (Fig 36). The dendrites and the axons are known as the nerve fibers. Collateral branches may be found along the length of a fiber, while terminal branches occur at the ends of the fibers. A synapse is an area where the terminal branches of one axon come into intimate contact with the dendrites or cell bodies of other neurons without protoplasmic continuity being established. It is the site where a nerve impulse is transmitted from one neuron to another.

The cell body (perikaryon)

The cell body of a neuron is a cytoplasmic expansion containing a nucleus which is usually large and lightly colored with a prominent nucleolus and fine chromatin granules. The cytoplasm contains many angular, darkly colored (basophilic when stained by means of hematoxylin and eosin) granules. These granules constitute the Nissl substance, which consists of aggregates of rough endoplasmic reticulum, while ribosomes are also present in large numbers. Nissl substance is present in the proximal part of dendrites but not in the axon hillock or the axon itself. The cell body of a neuron further contains a Golgi apparatus and mitochondria while fine neurofibrils are found throughout the neuron.

Neurons are classified according to the number of cell processes present. In this way, unipolar neurons have only an axon and lack dendrites. These neurons are very rare. Bipolar neurons have a single axon and dendrite, usually at opposite poles of the cell body, and are present in olfactory epithelia, and in the retina. In pseudo-unipolar neurons, found in the cranial and spinal ganglia, the dendrites and axons have united to form a single process. The combined process divides into two processes of which one serves as an axon, which enters the grey matter of the central nervous system where synapses are formed with other neurons, while the other acts as a dendrite and receives stimuli from the periphery. The latter resembles an axon. Multipolar neurons are common and have multiple dendrites and one axon and are found, for instance, in the anterior horn of grey matter.

The axon cylinder is the functional part of a nerve fiber. It is a thin cytoplasmic extension of the cell body and contains many neurofibrils which are continuous with fibrils in the cell body of the neuron. The axon cylinder is surrounded by a series of cells which form a membrane-like covering, the neurilemma. Within the central nervous system the neurilemma is formed by oligodendrocytes, supporting cells. Outside the central nervous system the neurilemmal sheath is formed by

Fig 36 **A neuron**

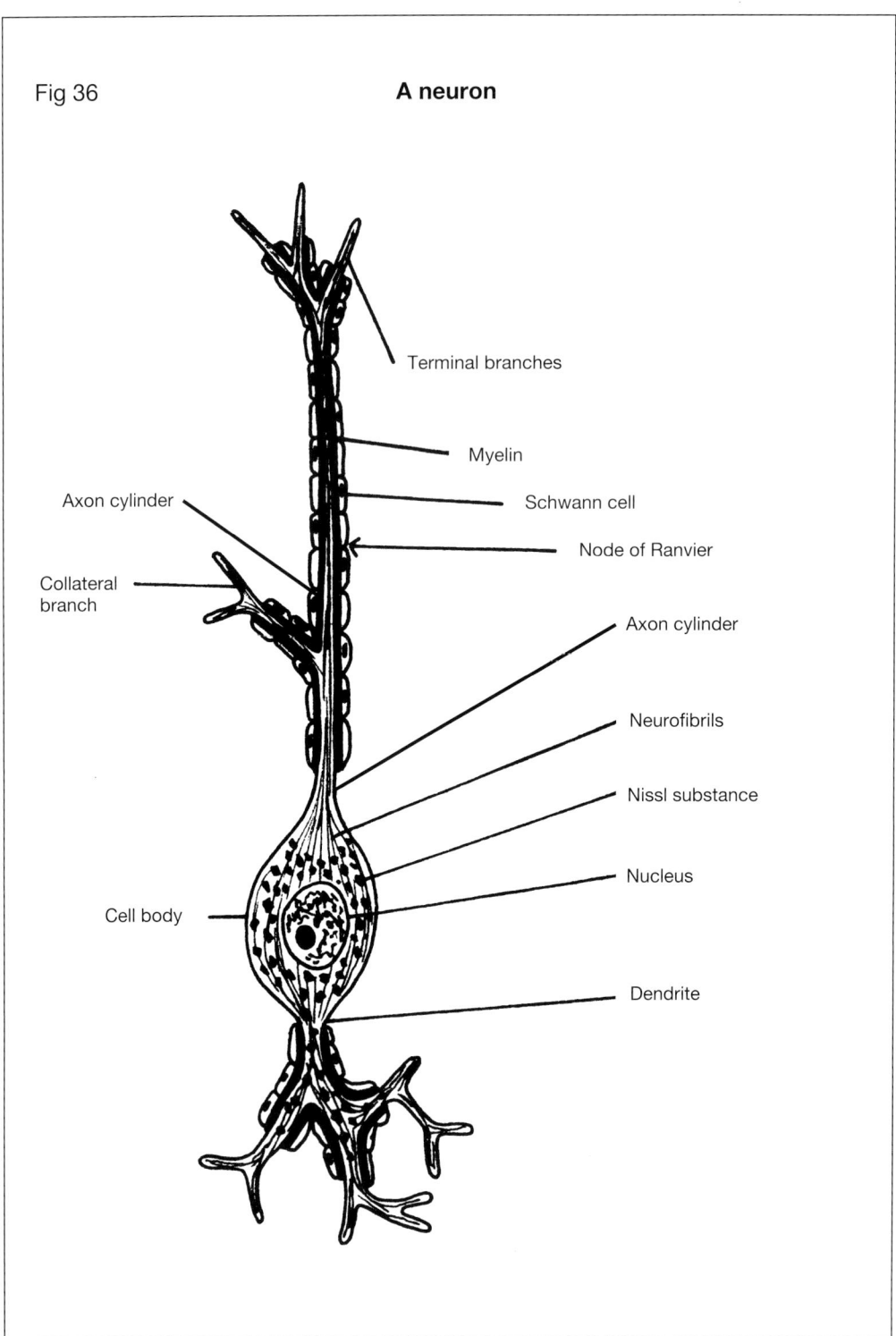

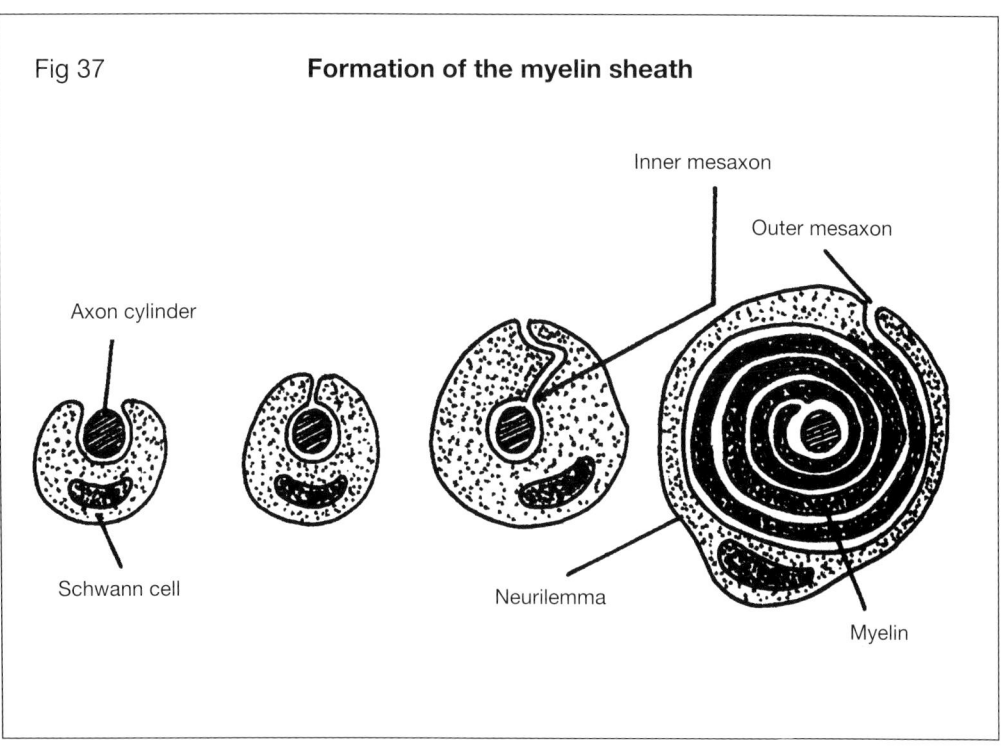

Fig 37 — Formation of the myelin sheath

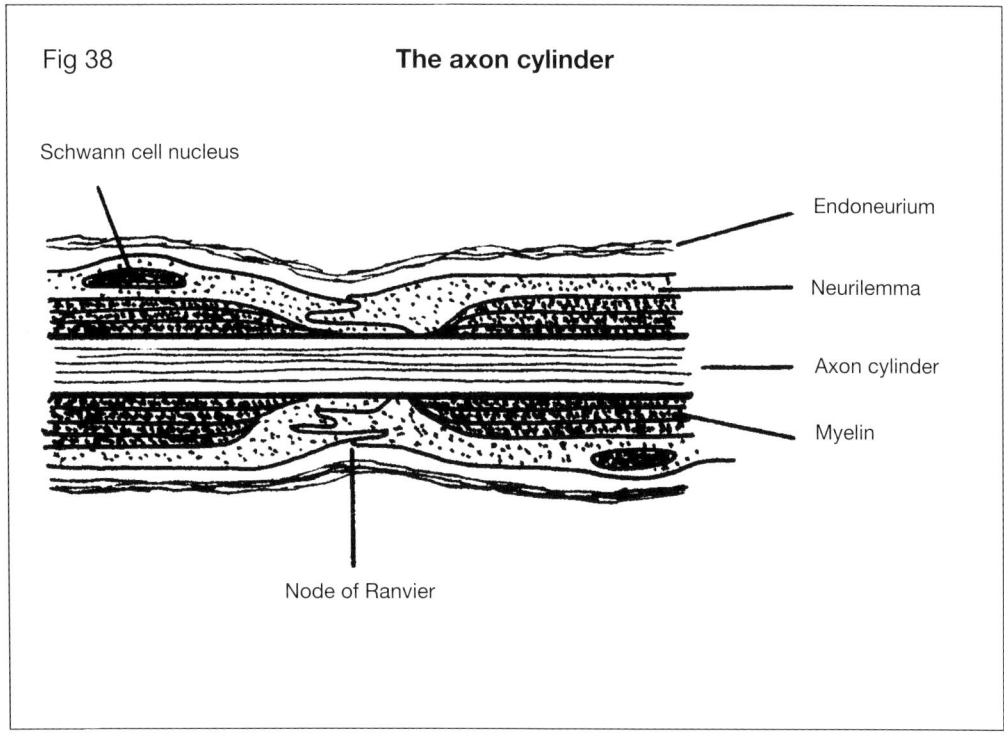

Fig 38 — The axon cylinder

Schwann cells which are embryologically derived from neural crest tissue (Fig 37). Both oligodendrocytes and Schwann cells form myelin, a lipoprotein, which has a concentrically layered appearance, since these cells are revolved around the axon cylinder while that part of the cell membrane in contact with the cylinder remains static. In this way an inner and an outer mesaxon forms.

The thickness of the myelin sheath may vary considerably and it is sometimes doubtful whether the terminal part of a peripheral axon cylinder in muscles or glands possesses any myelin. Both axons and dendrites may be enveloped by myelin and it is virtually impossible to distinguish between them for this reason. Many nerves do not, however, possess an obvious myelin sheath and, on this basis, a distinction is made between myelinated (medullated) and unmyelinated nerve fibers. The former are found in the tracts of the central nervous system and in peripheral nerves, while the latter are primarily found as postganglionic fibers of the autonomic nervous system.

Small gaps, the nodes of Ranvier, in the enveloping myelin sheaths are found spaced at regular intervals, approximately 0.5 mm apart, along the length of the nerve fiber (Fig 38). These nodes are points where individual Schwann cells meet and each segment between two nodes contains a Schwann cell nucleus.

An individual axon, with its neurilemma, is surrounded by a delicate connective tissue, the endoneurium, while small groups of axons are surrounded by further delicate connective tissue, the perineurium. The entire nerve trunk is enveloped by a thicker connective tissue sheath, the epineurium.

The trigeminal nerve (Cranial nerve V)

There is a marked similarity between a spinal nerve and the trigeminal nerve, both in embryological development and in morphology.

The trigeminal nerve leaves the anterior surface of the pons by means of a small motor root containing fibers which supply the muscles of mastication, and a separate larger sensory root which constitutes the main sensory system of the oral cavity and related structures (Fig 39).

The sensory fibers originate in neurons located in the trigeminal ganglion. Efferent fibers from this ganglion enter the pons via the sensory root, while afferent (peripheral) fibers form the three divisions of the nerve outside the ganglion, namely the ophthalmic, maxillary, and mandibular divisions.

The motor root joins the mandibular division immediately outside the foramen ovale and course peripherally together with the mandibular division. Motor fibers do not have any anatomical connection to the trigeminal ganglion.

The trigeminal nerve has deep connections in the brain stem (Figs 40, 41). The sensory nuclei of the nerve are divided into a mesencephalic nucleus in the mid-

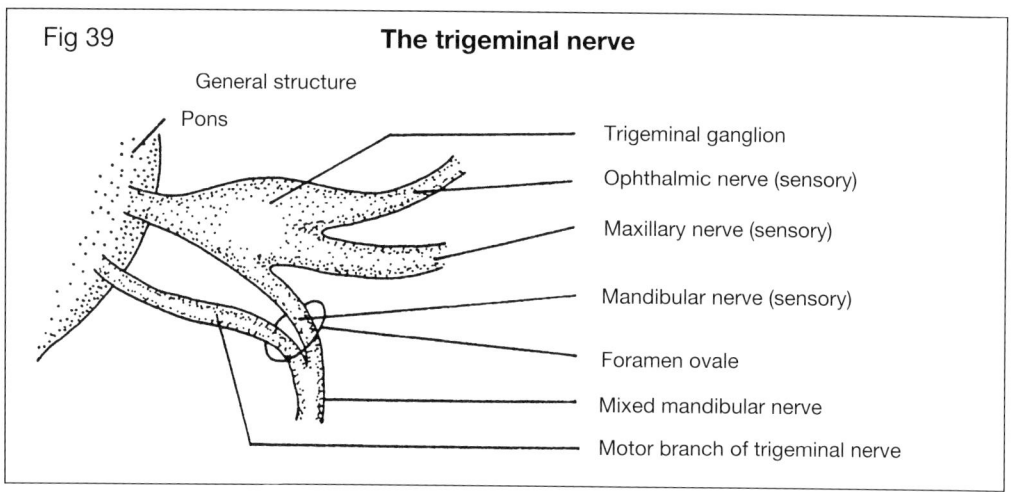

Fig 39 — The trigeminal nerve (General structure)

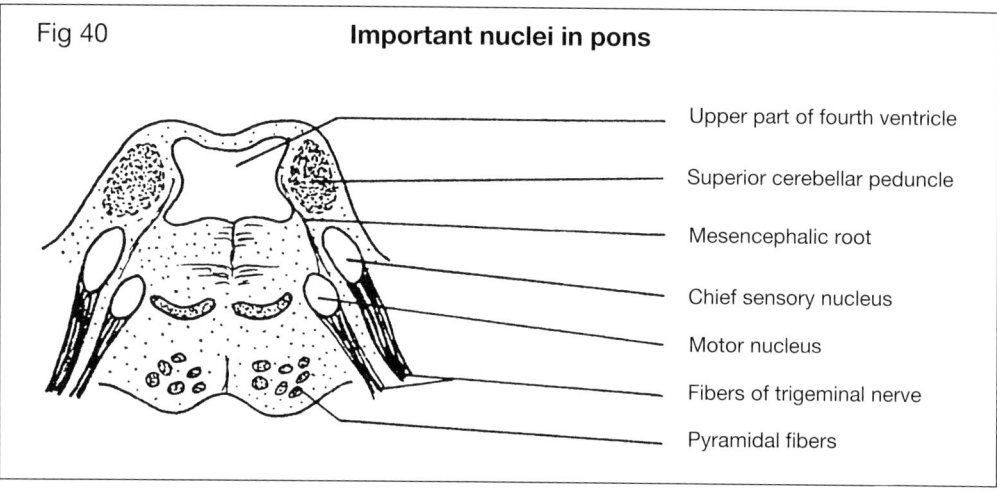

Fig 40 — Important nuclei in pons

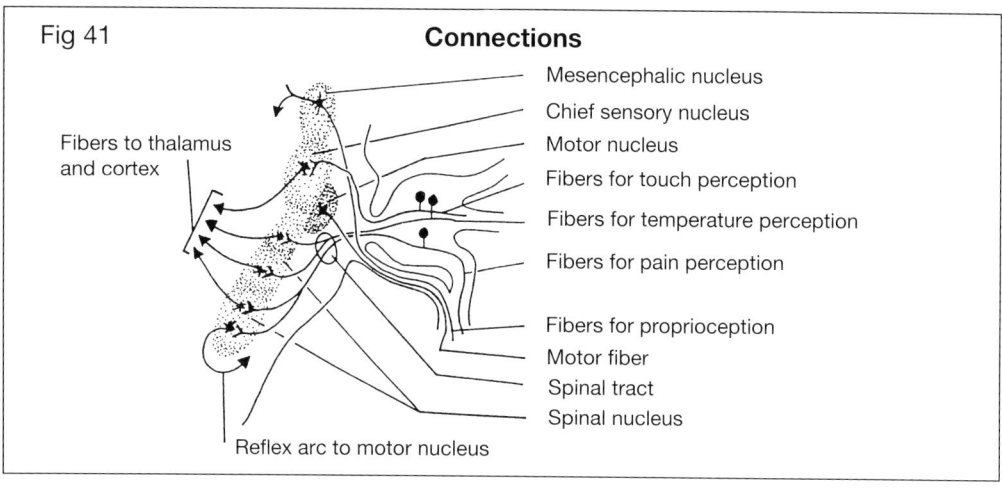

Fig 41 — Connections

brain, the chief sensory nucleus in the pons, and a spinal nucleus which is in fact a caudal extension of the chief sensory nucleus and ends at the level of the second cervical segment of the spinal cord (Fig 41). The axons of the spinal nucleus may be regarded as the equivalent of the intermediate neurons of the spinal cord. These neurons have connections with the post-central (sensory) gyrus of the cerebral cortex via the thalamus, or indirectly with the cortex of the frontal lobe, as well as with the motor nucleus which lies on the medial side of the chief sensory nucleus for reflex action. Fibers from the motor nucleus leave the pons in the motor root, as already described. The location of the motor nucleus, relative to the sensory nucleus, is very reminiscent of the arrangement of the motor and sensory horns of grey matter in the spinal cord and they also arise from the basal and alar laminae of early development.

The mesencephalic nucleus receives proprioceptive impulses from the temporomandibular joints, the muscles of mastication, the periodontal ligaments of the teeth, and from the mouth as a whole. It is believed that the cells of the mesencephalic nucleus, probably of neural crest origin, failed to migrate from the neural tube in early development, unlike the cells constituting the ganglion. The peripheral distribution of the proprioceptive fibers is via the motor root of the trigeminal nerve and they do not pass through the ganglion.

Efferent fibers from the trigeminal ganglion terminate in the chief sensory nucleus or in the spinal nucleus. The fibers which course caudally towards the spinal nucleus form the spinal tract of the trigeminal nerve which lies superficially within the medulla oblongata.

The spinal nucleus is cranio-caudally divided into three subnuclei, the subnucleus rostralis, the subnucleus interpolaris, and the subnucleus caudalis which is continuous with the dorsal horn of grey matter in the spinal cord.

Fibers ending in the chief sensory nucleus are concerned mainly with touch perception, while the spinal nucleus is concerned with temperature sensitivity (subnucleus rostralis and subnucleus interpolaris) and pain perception (subnucleus caudalis). Axons of neurons from all the sensory nuclei project, in the first instance, mainly to the thalamus.

Selected bibliography

1. Dixon, A.D. (1986) *Anatomy for Students of Dentistry,* 5th edition. Edinburgh: Churchill Livingstone.
2. Kelly, D., Wood, R.L. and Enders, A.C. (1984) *Bailey's Textbook of Microscopic Anatomy,* 18th edition. Baltimore: Williams and Wilkins.
3. Krause, W.J. and Cutts, J.H. (1986) *Concise Text of Histology,* 2nd edition. Baltimore: Williams and Wilkins.
4. Lavelle, C.L.B. (1988) *Applied Oral Physiology*, 2nd edition. London: Wright.

Review questions

1. Give a brief account of the functions of the nervous system.
2. Describe the differences between the autonomic nervous system and the somatic nervous system.
3. Describe the functioning of a reflex arc.
4. Describe a neuron.
5. Discuss the similarities between the trigeminal nerve and an ordinary spinal nerve.

13. Development of the face

The initial rapid development of the head, face, and oral cavity, compared to the more caudal parts of the embryo, is characteristic of the more advanced state of development which the cranial region maintains for a large part of prenatal development. Differences in the rate of development are responsible for the fact that the embryonic disc appears pear-shaped with an expanded future head region already at an early stage. A further indication of the more rapid cranial development is that the three primary embryonic germ layers in the head region already commence specific developments by the 3rd week while the caudal germ layers are still differentiating in the 4th week. The advanced development of the head region is also responsible for the fact that the head forms one half of the total length of the embryo after 2 months. Postcranial growth is, however, faster in later stages of development and the head forms one quarter of the length of the body at birth, while constituting only one eighth of body length in the adult.

It can be said that the face begins to develop during the 3rd week of embryonic life when the embryo is approximately 3 mm long.
At this stage the prechordal plate appears in the bilaminar embryonic disc. After formation of the head fold, the prechordal plate (future buccopharyngeal membrane) lies in a hollow, the stomodeum (or primitive mouth).

The stomodeum is an invagination of the ectoderm on the ventral surface of the future head of the embryo. In the deepest part of the stomodeum, the lining ectoderm is in contact with the endoderm of the foregut. There is no intervening mesoderm and the combined ectoderm and endoderm constitutes the buccopharyngeal membrane which lies in the approximate area of the future palatine tonsils.

At about the end of the 3rd week, a structure develops which is not usually associated with the oral cavity. A pouch develops in the roof of the stomodeum immediately anterior to the buccopharyngeal membrane. This is Rathke's pouch, which deepens towards the brain. Ectodermal cells in this pouch proliferate and migrate in the direction of the ventral aspect of the forebrain and differentiate to form the anterior lobe of the pituitary gland. It later loses all continuity with the oral ectoderm.

Several structures which will later give rise to the face and different parts of the mouth and nasal cavity appear around the stomodeum on the outer surface of the embryonic head (Fig 42, I and II). Above the stomodeum, the forebrain swelling forms a prominent bulge, while a series of bilateral ridges, the pharyngeal arches, appear in the neck region. These arches represent regions of proliferation and development in the underlying mesenchyme.

On the ventral aspect of the embryo, the pericardial swelling forms a prominent bulge which houses the developing heart. In a ventral view, the pericardial swelling partly conceals the ventral aspects of the pharyngeal arches. Only the first (mandibular) arch can be observed to cross the midline above the level of the pericardial swelling. The mandibular arch gives origin on either side to the maxillary processes which then form the lateral boundaries of the primitive mouth. The buccopharyngeal membrane ruptures during the 4th week, after which the mouth is in continuity with the foregut.

Development of the face during the 2nd month is dominated by changes which lead to the formation of the nose. During the 5th week, when the embryo is approximately 6 mm long, bilateral localized thickenings of epithelium form on the anterior surface of the head above the mouth (Fig 42, III). These olfactory (nasal) placodes are initially uniformly elevated above the surface but rapidly form central hollows, the olfactory pits.

Although some authors believe that the pits form as a result of peripheral epithelial proliferation in the placodes, thus leaving a central depression, it is generally accepted that the epithelium of the placodes proliferates dorsally above the stomodeum. There is uncertainty whether this proliferating epithelium retains a major contact with the epithelium of the primitive mouth, but indications are that a contact is maintained in the form of a sagittally elongated epithelial septum, the so-called nasal fin. This lies between the forming medial and lateral nasal folds and is continuous with the epithelial lining of the roof of the mouth. Later, probably as a result of lack of nutrients to the central cells of the proliferating epithelium, cellular degeneration results in pit formation.

Raised inverted U-shaped ridges appear on both sides of the nasal pits and are known as the medial and lateral nasal folds (Fig 43, I). The medial nasal folds together with the intervening inferior part of the forebrain swelling, form the frontonasal process. Each lateral nasal fold borders on a maxillary process and separates the nasal pits from the developing eyes.

All the elements which will eventually form the face are now present, namely the eyes (the lens placodes develop at the same time as the olfactory placodes but in a more dorsal position), the lateral nasal folds, the frontonasal process, and the maxillary and mandibular processes.

Although this chapter deals primarily with the development of the face, it is fitting to discuss the development of the primary palate, since the development of the upper lip region and the primary palate are inter-related.

In the head region, the ectomesenchyme migrates in two streams. The cranial stream forms the mesenchyme of the frontonasal process while the more caudal stream enters the lateral nasal and maxillary processes (Fig 44). The facial swellings therefore result from proliferation of underlying ectomesenchyme. By diffe-

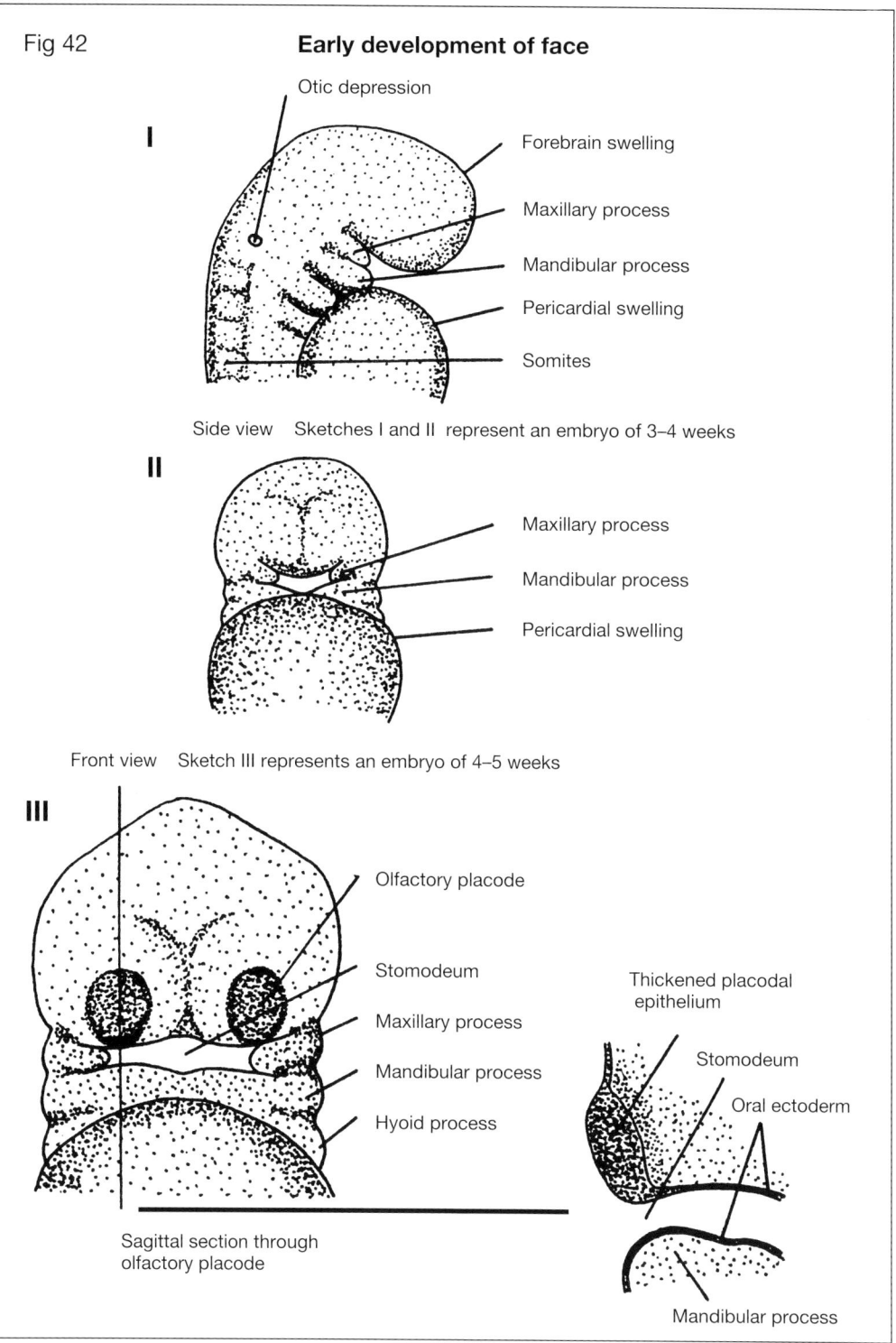

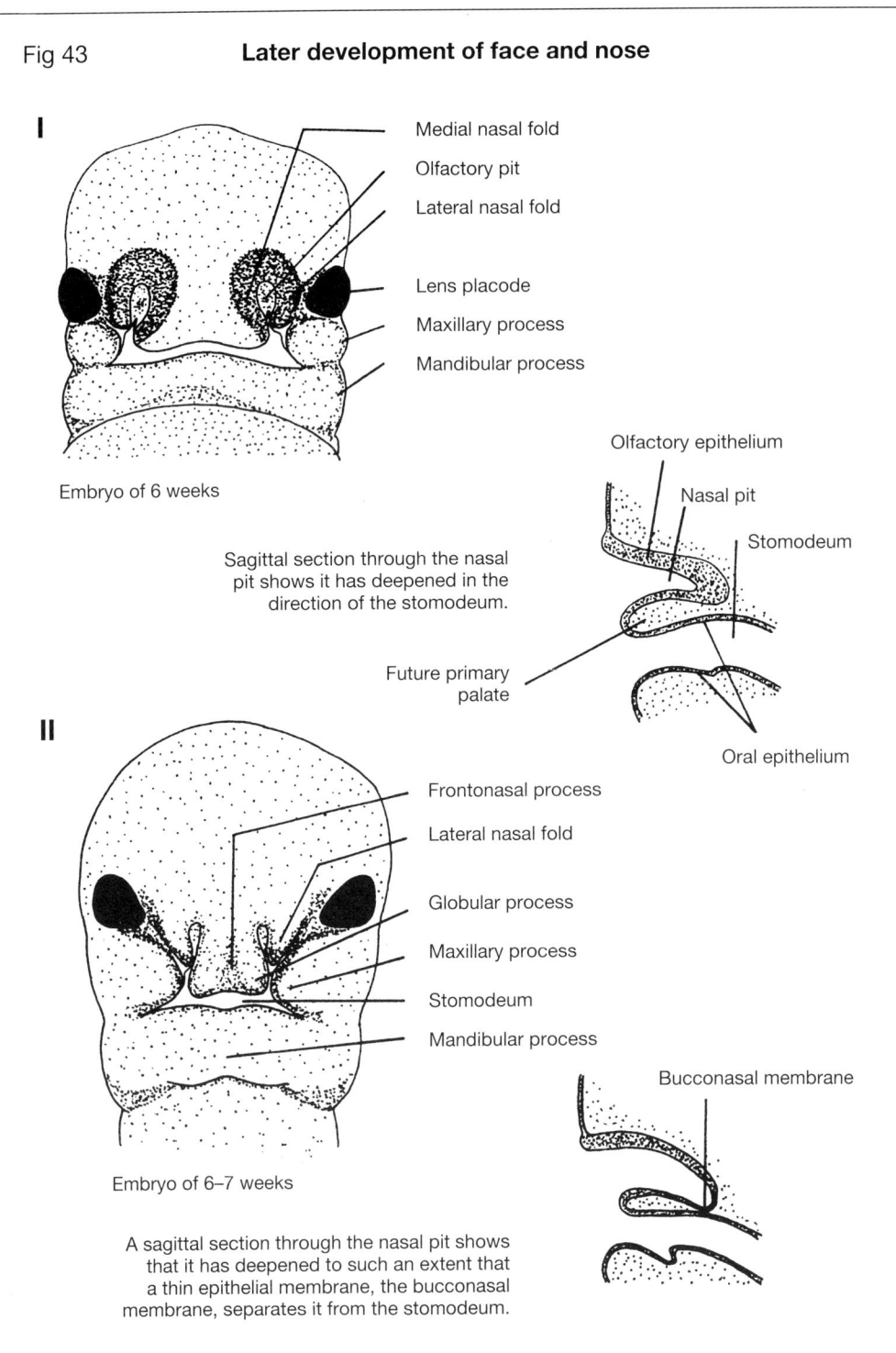

Fig 43 Later development of face and nose

rential growth of the mesenchyme, grooves between the facial processes are eliminated (fusion and/or merging), resulting in smooth facial contours. Clefts, for example congenital clefts of the lips, are the result of insufficient growth in the area of the developmental grooves and result from adverse genetic and/or environmental factors.

The nasal pits deepen and form nasal sacs, bordered by the medial and lateral nasal processes. The maxillary processes grow forwards and medially towards one another above the stomodeum and become continuous with the lower lateral edges of the lateral nasal folds. The frontonasal process grows downwards at a faster rate than the lateral nasal folds and comes to lie between, and on the same level as, the maxillary processes where their inferolateral parts are known as the globular processes (Fig 43, II).

Differences exist regarding development in the floor of the nasal sac. The older view was that facial processes are initially completely separated from one another. As they become apposed, epithelial fusion occurs between them, followed by degeneration of the fusion epithelium and the establishment of mesenchymal continuity between the processes.

On the other hand, some authors believe that epithelial continuity does not exist between the dorsally proliferating epithelium of the nasal sac and the epithelium of the roof of the primitive mouth. According to this view, the medial nasal folds (and later the globular processes of the frontonasal process), the lateral nasal folds and the maxillary processes are united by mesenchyme in the depths of the primitive face, being merely separated by superficial grooves. Invading mesenchyme then merely fills these processes from the dorsal aspect (mergence of processes as opposed to fusion), thus establishing mesenchymal, and superficial epithelial, continuity between the different processes.

A third view, perhaps more generally accepted at present, is that epithelial continuity exists from the beginning between the medial nasal folds, lateral nasal folds, and maxillary processes in this area. According to this view, the anterior part of the epithelial septum thus formed is penetrated from all sides by mesenchyme (see earlier remarks on the nasal fin) and is eliminated.

In the dorsal part of the nasal sac the placodal epithelium in the meantime continues to proliferate in the direction of the posterior part of the roof of the primitive mouth and in this way the nasal sac deepens. The sac ultimately becomes separated from the roof of the mouth by a thin membrane, the bucconasal membrane, which ruptures to form the posterior nares (nasal openings). The posterior part of the primitive nasal cavity is now in communication with the primitive oral cavity.

The area of mesenchymal invasion between the anterior and posterior nares is the primary palate which thus forms the first separation between the stomodeum and the primitive nasal cavities (Figs 44, 45). More superficial mesenchymal continuity across the midline, between the centrally placed globular processes and the maxillary processes below, and between the maxillary processes below and the la-

Fig 44   Migration of ectomesenchyme

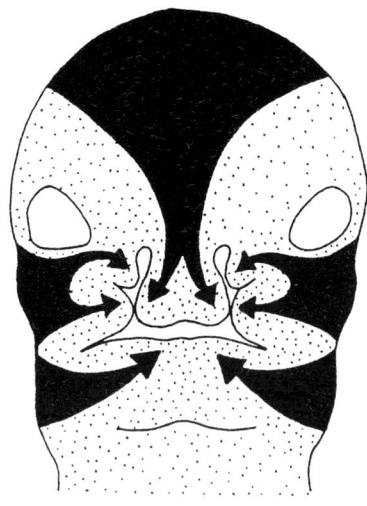

Streams of ectomesenchyme migrate into all the facial processes and the pharyngeal arches. The presence of mesenchyme is essential for fusion, or mergence of embryonic processes.

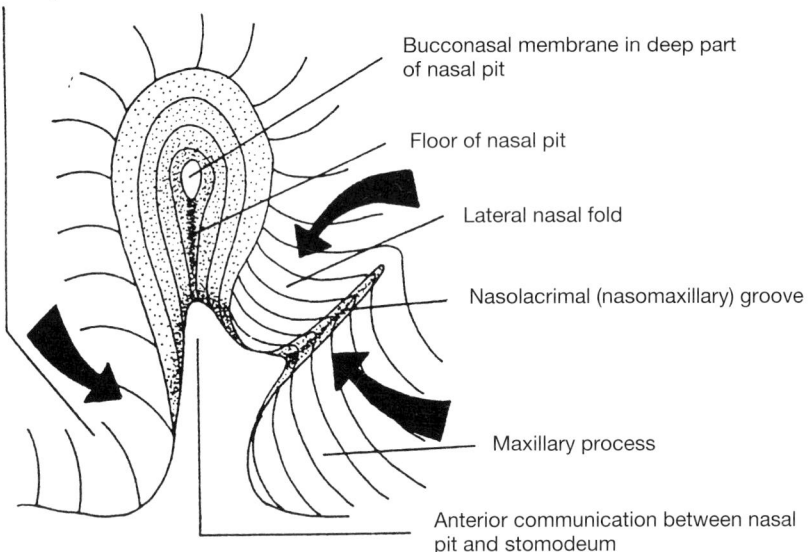

Anterior view of left nasal pit

The sketch represents the processes which fuse to form the upper lip.
Infiltration beneath the surface will eliminate the grooves between processes.

Fig 45   Development of the primary palate and upper lip
(anterior view)

I

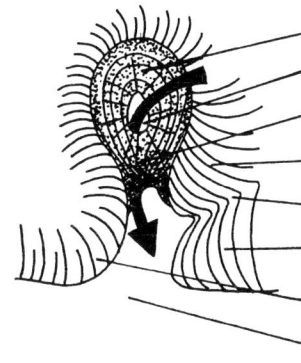

- Nasal pit
- Posterior naris
- Floor of nasal pit
- Lateral nasal fold
- Nasolacrimal groove
- Maxillary process
- Globular part of frontonasal process
- Stomodeum

II

Establishment of the primary palate

Sketch I represents the nasal pit after the bucconasal membrane has ruptured. The black arrow indicates the communication between the nasal pit and the stomodeum. Anteriorly the nasal pit still communicates with the primitive mouth between the three facial processes (frontonasal, lateral nasal fold, and maxillary process).

Sketch II indicates how the infiltration of ectomesenchyme (black arrows) promotes fusion of the facial processes and elimination of grooves. The primary palate is formed in the floor of the nasal pit.

III

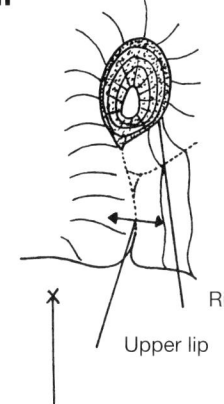

Sketch III represents mergence of ectomesenchyme beneath the surface to establish the upper lip. The outer nasal opening is formed.

- Rim of anterior naris (outer nasal opening)
- Upper lip
- Stomodeum

Fig 46 Final development of the face

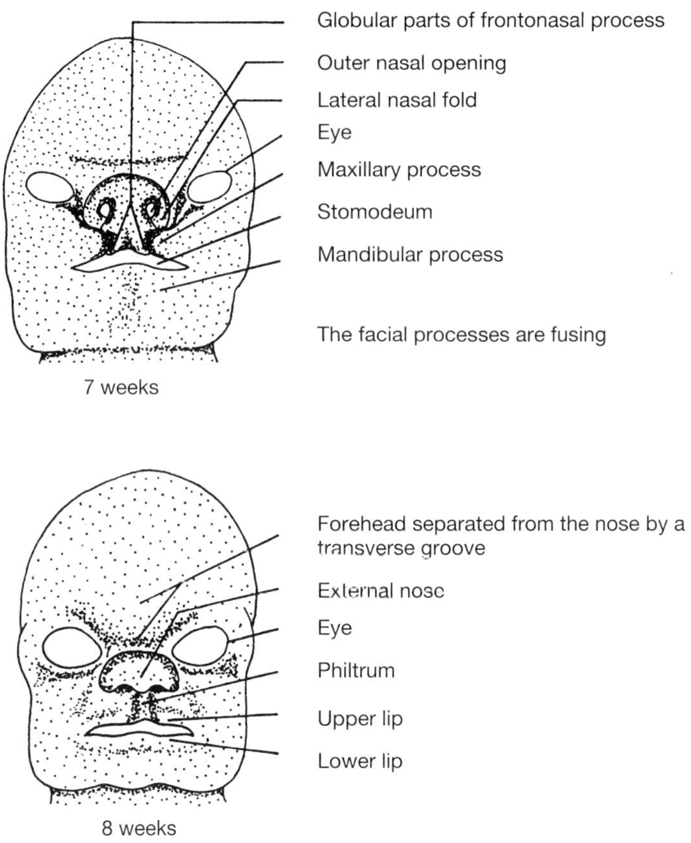

7 weeks
- Globular parts of frontonasal process
- Outer nasal opening
- Lateral nasal fold
- Eye
- Maxillary process
- Stomodeum
- Mandibular process

The facial processes are fusing

8 weeks
- Forehead separated from the nose by a transverse groove
- External nose
- Eye
- Philtrum
- Upper lip
- Lower lip

Fig 47 Development of facial profile

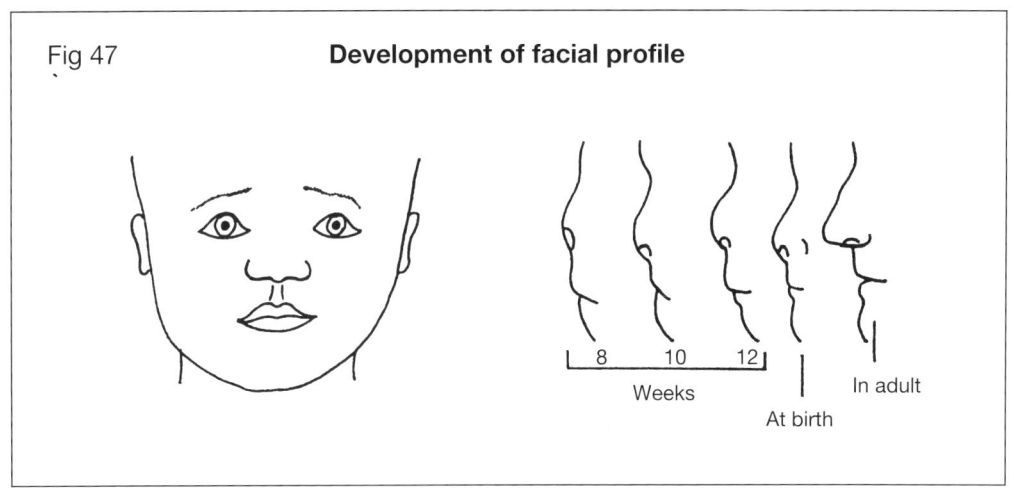

8 10 12
Weeks
At birth
In adult

teral nasal folds at a higher level, creates the upper lip. This develops between the 5th and the 7th weeks of intra-uterine life (Figs 44, 45).

At this stage the fate of the embryonic facial processes should be briefly described. The lateral nasal folds give origin to the alae and lateral part of the external nose; the medial nasal folds and the intervening frontonasal process form the central parts of the external nose, namely the elevated dorsal part, the tip (apex), and the anterior mobile part of the nasal septum. Controversy still surrounds the formation of the philtrum of the upper lip but it is widely accepted that it is formed by the globular parts of the frontonasal processes which, by merging with the two lateral maxillary processes, form the upper lip. The maxillary processes form the major part of the upper lip and the cheeks. The lower lip, chin, and surrounding soft tissues are formed by the mandibular process.

Mesoderm of the second pharyngeal arch (the hyoid arch) invades the subcutaneous tissues of the face, head, and front parts of the neck to form the facial muscles, buccinator, occipitofrontalis, and platysma, while mesoderm of the third, fourth, and sixth pharyngeal arches migrates to surround the cranial parts of the foregut to form the muscles of the soft palate, pharynx, and larynx.

The groove between the maxillary processes and the lateral nasal folds extends from the developing eye to the mouth. It is believed that this groove is associated with the development of the nasolacrimal duct at a deeper level, possibly as a result of mechanisms similar to those leading to the formation of the neural tube. This groove is therefore termed the nasolacrimal, or nasomaxillary, groove.

Differential growth in facial development (Figs 46, 47) causes the eyes to move closer to one another from their initial lateral position, while the ears move cranially from a more cervical position of origin. The two external nasal openings move closer together and the external nasal prominence and the forehead form. The face has a distinctly human appearance after 50 days when the embryo has a length of 22 mm.

Selected bibliography

1. Bhaskar, S.N. (ed) (1991) *Orban's Oral Histology and Embryology,* 11th edition. St. Louis: Mosby-Year Book, Inc.
2. Hamilton, W.J., Boyd, J.D. and Mossman, H.W. (1972) *Human Embryology,* 4th edition revised by Hamilton, W.J. and Mossman, H.W. Cambridge: W. Heffer and Sons Ltd.
3. Moore, K.L. (1988) *Essentials of Human Embryology.* Toronto: B.C. Decker, Inc.
4. Mjör, I.A. and Fejerskov, O. (eds) (1986) *Human Oral Embryology and Histology,* 1st edition. Copenhagen: Munksgaard.

5. Sadler, T.W. (1985) *Langman's Medical Embryology,* 5th edition. Baltimore: Williams and Wilkins.
6. Sperber, G.H. (1989) *Craniofacial Embryology,* 4th edition. London: Wright.
7. Stark, R.B. (1973) "Development of the face", *Surgery, Gynecology and Obstetrics,* 137, 403-408.
8. Streeter, G.L. (1948) Developmental horizons in human embryos. Description of age groups XV, XVI, XVII and XVIII, being the third issue of the Carnegie Collection. *Carnegie Institution of Washington Publications,* 575, 32, no. 211, 133-203.

Review questions

1. Give a brief account of the development of the embryonic facial processes.
2. Describe the role of ectomesenchyme in the mergence and fusion of facial processes.
3. Describe the formation of the primary palate.
4. Describe the development of the upper lip.

14. Development of the palate and the nasal cavity

After the buccopharyngeal membrane ruptures (at approximately 4 weeks), the roof of the primitive mouth is anteriorly formed by oral ectoderm and posteriorly by foregut endoderm. A thin layer of mesenchyme separates the ectodermally lined part of the primitive mouth from the forebrain but the notochord intervenes between the endoderm and the nervous system.

After formation of the primitive nasal cavities and rupture of the bucconasal membrane (at approximately 6 weeks), each primitive nasal cavity opens into the primitive mouth by means of a posterior naris behind the primary palate (Fig 48). The greater part of the primitive mouth is still, however, at this stage a combined oronasal cavity and extends backward beyond the posterior nares. The final oral and nasal cavities only become separate entities once the secondary palate is established.

The combined oronasal cavity expands superiorly and the primitive nasal cavities deepen with enlargement of the posterior nares. A median part of the frontonasal process extends backwards beyond the posterior nares from the posterior aspect of the primary palate and serves as a primary nasal septum.

When the embryo is 6 weeks old, the mesenchyme of the maxillary processes proliferates upwards in the lateral walls of the combined oronasal cavity to fuse with the mesenchyme of the median primitive nasal septum. This is probably followed by a downwards proliferation of the primary nasal septum to give rise to the secondary nasal septum in which cartilage and bone later develops. At this stage the lower loosely hanging edge of the septum is in contact with the dorsal surface of the developing tongue which starts developing rapidly at the end of the 1st month.

While the nasal septum is forming, the maxillary processes (Fig 49, B) proliferate to form processes on the same level as the primary palate. These palatal processes, or shelves, are initially vertically orientated on either side of the tongue (Fig 49, C and D).

Several factors may be responsible for a rapid movement (probably within hours) of the shelves from a vertical to a horizontal position very early in the 3rd month. The upward movement of the shelves, which is initially prevented by the intervening tongue, may be caused by differential mitotic activity in the shelves,

Fig 48 **Primary palate**

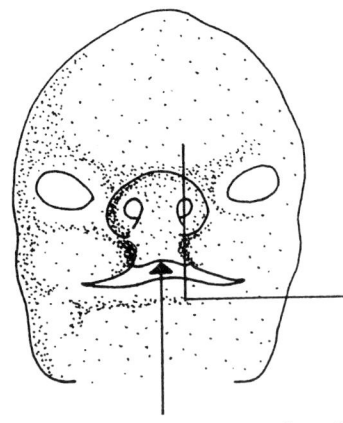

The sketch below represents a view of the medial half of the left nasomaxillary complex as seen after a sagittal section through the left nasal opening and primary palate.

The sketch below represents a view of the roof of the primitive mouth.

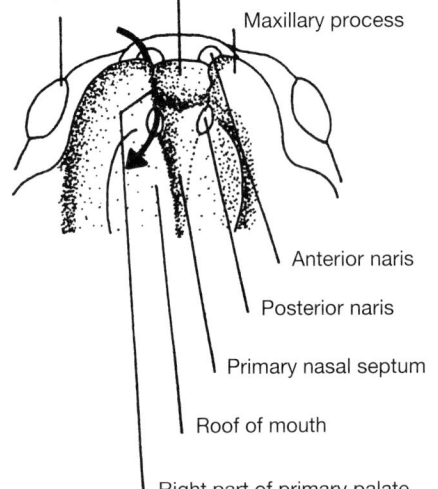

Eye
Frontonasal process
Maxillary process
Anterior naris
Posterior naris
Primary nasal septum
Roof of mouth
Right part of primary palate

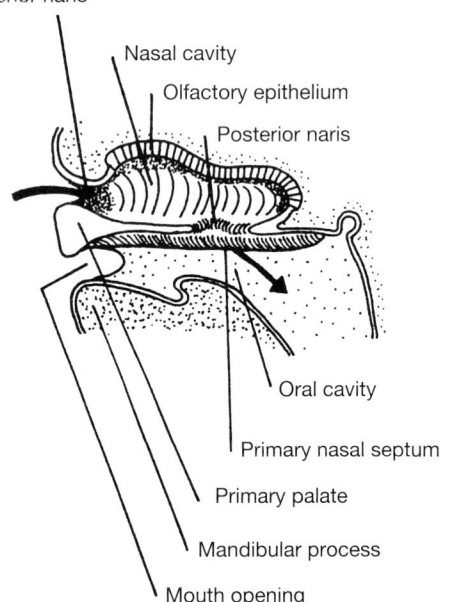

Anterior naris
Nasal cavity
Olfactory epithelium
Posterior naris
Oral cavity
Primary nasal septum
Primary palate
Mandibular process
Mouth opening

The arrow extends through the right nasal opening and nasal cavity into the oral cavity.

Fig49 **Initial development of the secondary palate**

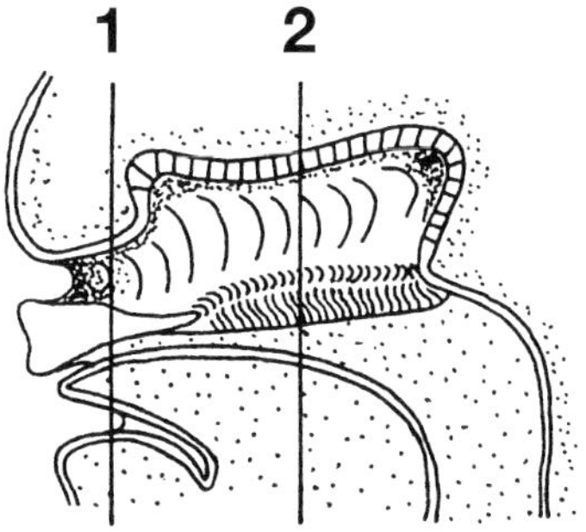

Sketch A represents the appearance of structures seen in coronal plane 1 through the primary palate.

Sketches B, C and D represent successive developmental stages in coronal plane 2 (behind the primary palate).

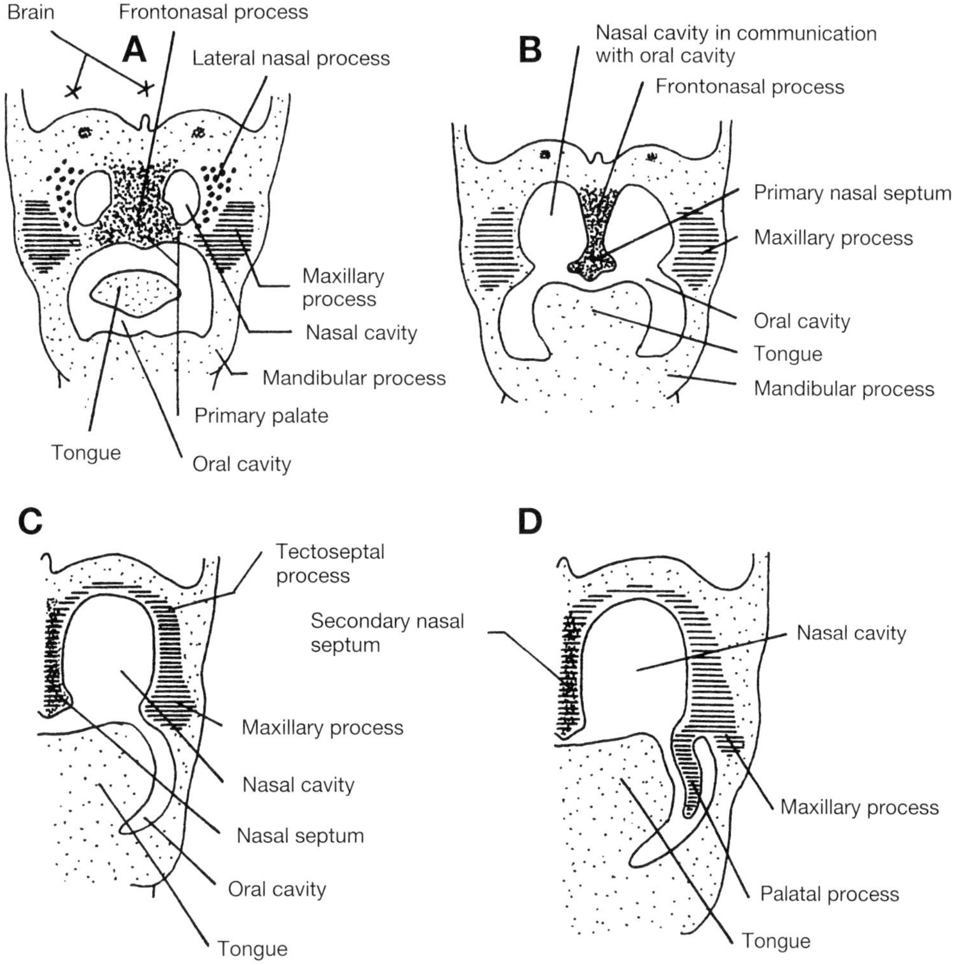

A
- Brain
- Frontonasal process
- Lateral nasal process
- Maxillary process
- Nasal cavity
- Mandibular process
- Primary palate
- Tongue
- Oral cavity

B
- Nasal cavity in communication with oral cavity
- Frontonasal process
- Primary nasal septum
- Maxillary process
- Oral cavity
- Tongue
- Mandibular process

C
- Tectoseptal process
- Secondary nasal septum
- Maxillary process
- Nasal cavity
- Nasal septum
- Oral cavity
- Tongue

D
- Nasal cavity
- Maxillary process
- Palatal process
- Tongue

mandibular movement, an upward flexing of the head away from the pericardial bulge, thereby lowering the tongue and increasing intra-oral space, an intrinsic shelf force, and other factors.

With further growth of the nasal septum and the now medially directed palatal processes, they come into close proximity in the median plane behind the primary palate. Fusion occurs, firstly between the palatal processes and the deep part of the primary palate anteriorly, and then, from before backwards, with each other in the midline and with the lower edge of the nasal septum above (Fig 50). When the epithelial surfaces of the palatal processes and of the septum meet, they adhere to each other. This is followed by cell fusion, epithelial cell degeneration, programmed cell death, and secondary mesenchymal invasion to consolidate the union of the processes, followed by merging of the processes in the soft palate region and uvula. In this way the secondary palate is formed as a final separation between the oral and nasal cavities.

Closure of the palate, which is completed after approximately 12 weeks, also results in the posterior nasal openings, which were initially situated immediately behind the primary palate, being transferred to a more posterior position behind the secondary palate on both sides of the nasal septum.

Ossification of the hard palate later occurs from centers in the premaxillary area (primary palate), the maxilla, and the palatine bone. This is fully dealt with in the following chapter.

A small opening, the nasopalatine canal, persists for a short while between the oral and nasal cavities on either side of the nasal septum between the palatal processes and the primary palate. Superficial epithelial fusion eventually obliterates the oral openings of these canals but they persist on a deeper level as the incisive canals.

While the above changes are occurring, protuberances develop on the lateral walls of the nasal cavities (Fig 50, II). These will later form the superior, middle, and inferior conchae (turbinates) of the nose.

Olfactory epithelium, which arises in the epithelium of the nasal placodes, and subsequently in the epithelial lining of the nasal sacs, migrates to the roof of the nasal cavities on either side of the septum. Neuroblasts in this olfactory epithelium develop into bipolar neurons which develop axons in the direction of the corresponding side of the brain where they synapse with neurons in the olfactory bulb after passing through the cribriform plate. The olfactory bulb neurons give origin to the secondary olfactory fibers which form the olfactory tracts in the brain.

The maxillary sinus develops in the 4th month in the middle meatus of the lateral nasal wall as an outpouching of the nasal mucous membrane. At birth these sinuses are only a few millimeters in diameter, although fully pneumatized, but they enlarge throughout life with eventual invasion of the maxillary alveolar process, especially after the loss of teeth.

The sphenoidal and ethmoidal sinuses also start developing before birth at approximately 4 months, but become fully pneumatized only after birth.

Fig 50 **Later development of the secondary palate**

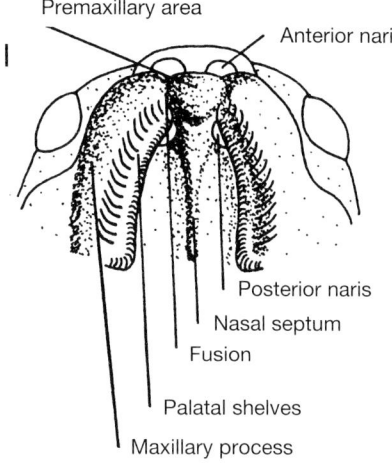

I — Premaxillary area, Anterior naris, Posterior naris, Nasal septum, Fusion, Palatal shelves, Maxillary process

Sketches I, II and III show successive stages in the fusion of the secondary palate.

Sketch I shows horizontally developing palatal shelves fusing with the premaxillary part of the frontonasal process. The primary posterior naris has been partly included in the final nasal cavity.

Sketches II and III show fusion of the secondary palate from before backwards in the midline.

Coronal sections show the establishment of the secondary palate and fusion with the nasal septum.

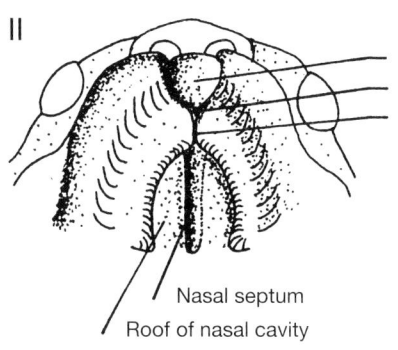

II — Frontonasal process (premaxilla), Nasopalatine canal, Partial fusion of secondary palate, Nasal septum, Roof of nasal cavity

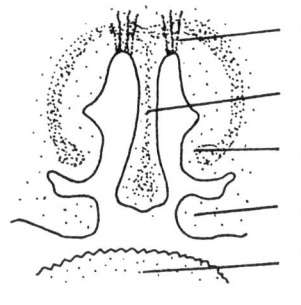

Olfactory nerve, Nasal septum, Inferior concha, Palatal shelf, Tongue

Coronal section

III

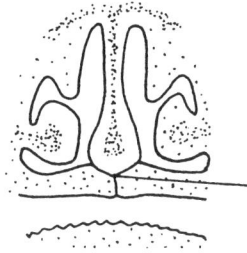

Fusion of nasal septum with palate

Coronal section

Midpalatal suture between fusing palatal processes

Selected bibliography

1. Bhaskar, S.N. (ed) (1991) *Orban's Oral Histology and Embryology,* 11th edition. St. Louis: Mosby-Year Book, Inc.
2. Hamilton, W.J., Boyd, J.D. and Mossman, H.W. (1972) *Human Embryology,* 4th edition revised by Hamilton, W.J. and Mossman, H.W. Cambridge: W. Heffer and Sons Ltd.
3. Mjör, I.A. and Fejerskov, O. (eds) (1986) *Human Embryology and Histology,* 1st edition. Copenhagen: Munksgaard.
4. Moore, K.L. (1988) *Essentials of Human Embryology*. Toronto: B.C. Decker, Inc.
5. Sperber, G.H. (1989) *Craniofacial Embryology,* 4th edition. London: Wright.

Review questions

1. Give an account of the development of the secondary palate after the establishment of the primary palate.
2. Write brief notes on the following:
   (a) Development of the olfactory nerves.
   (b) Development of the maxillary sinus.
   (c) What is meant by "fusion" of embryonic processes?

15. Development of the septomaxillary complex

In the previous two chapters it was seen how the blending of mesenchyme of the maxillary processes, the medial and lateral nasal folds, and of the frontonasal process, lead to the formation of the embryonic upper lip, primary palate, nasal septum, nasal cavities, and secondary palate. This chapter deals with further development in the septomaxillary complex with the emphasis on the origin and development of bony structures in the region.

The nasal capsule

Before centers of ossification appear, the skeleton of the face consists of two cartilages, namely the nasal capsule which surround the nasal cavities to a large extent, and the cartilages of Meckel in the second (mandibular) pharyngeal arch. These play a role in mandibular development (to be discussed in the following chapter).

The nasal capsule develops in the mesenchyme of the maxillary processes (lateral to the nasal cavities) and of the primitive nasal septum, which is of frontonasal origin. It is continuous with the chondrocranium, which forms the early cartilagenous base of the skull, and which arises mainly in occipital sclerotomal mesenchyme which extends into the floor of the developing brain (Fig 51). Endochondral ossification in the chondrocranium will later give rise to the cranial base. The nasal capsule is, therefore, virtually a part of the chrondrocranium and occupies an intermediate position between the neurocranial skeleton and the later facial skeleton (the maxilla and mandible). Further important components of the chondrocranium are the otic capsules. These cartilagenous structures surround the vestibulocochlear apparatus and later ossify to form the petrous and mastoid portions of the temporal bones.

The fully developed cartilagenous nasal capsule consists of a septal part within the nasal septum, and two lateral wings which embrace the nasal cavities (Fig 51). The lower lateral free edges of the capsule are curved medially. The end of the 2nd month marks the beginning of rapid development in and around the nasal capsule.

Fig 51   Developments in and around the nasal capsule

Early skeleton of the head

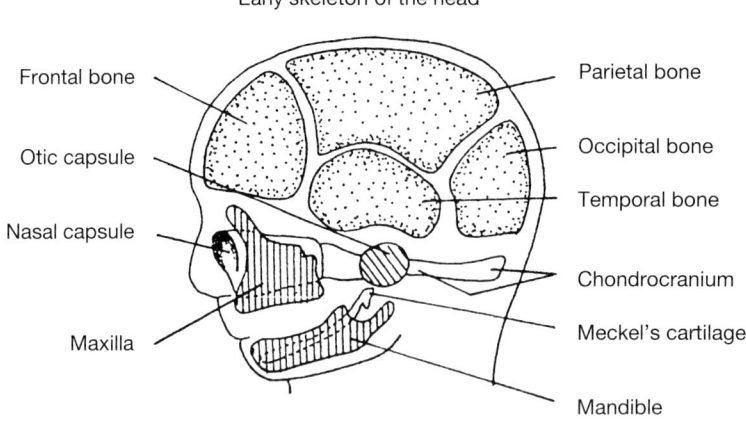

- Frontal bone
- Otic capsule
- Nasal capsule
- Maxilla
- Parietal bone
- Occipital bone
- Temporal bone
- Chondrocranium
- Meckel's cartilage
- Mandible

Coronal section through nasal capsule

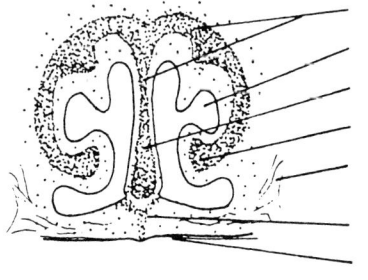

- Nasal capsule
- Nasal cavity
- Septal part of nasal capsule
- Inferolateral free edge of capsule
- Maxillary fibrocellular condensation
- Fusion of palatal shelves with septum
- Palate

Further development seen in a coronal section

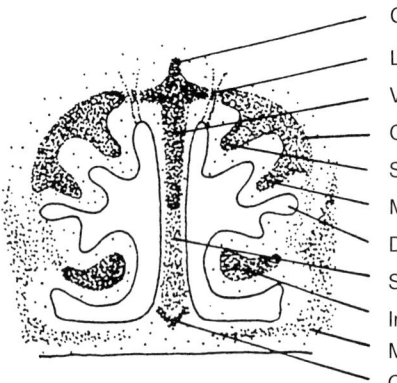

- Crista galli
- Lamina cribrosa
- Vertical plate of ethmoid bone
- Orbital plate of ethmoid bone
- Superior concha
- Middle concha
- Developing maxillary sinus
- Septal cartilage
- Inferior concha
- Maxillary ossification
- Ossification of vomer

For convenience of description, the developments in the above sketch are represented as taking place in one plane at the same time. This is, of course, not the case.

The orbital plate of the ethmoid bone develops in the upper part of the lateral wing of the capsule and will later be invaded by the ethmoidal sinus (Figs 51, 52). The superior and inferior turbinates develop as part of the orbital plate. The vertical, or perpendicular, plate of the ethmoid bone develops in the upper posterior septal portion of the capsule and later protrudes into the anterior cranial fossa as the crista galli. Bony union of these two parts of the ethmoid bone around the filaments of the olfactory nerves is completed at about the 6th year and gives rise to the cribriform plate. Cartilage persists in the anterior part of the septum and forms the cartilagenous nasal septum. The medially curved lower free edge of the nasal capsule ossifies to form the inferior turbinate (concha) on the lateral wall of the nasal cavity.

The inferior turbinate remains as a separate bone. That part of the capsule lying between the middle turbinate and the inferior turbinate atrophies and is invaded by the developing maxillary sinus. All the bony structures described above develop by endochondral ossification.

The maxilla, premaxilla, lacrimal, and palatine bones, and the vomer develop outside the nasal capsule by intramembranous ossification.

The maxilla

The maxilla develops in the maxillary process which is a derivative of the first (mandibular) pharyngeal arch, and is composed entirely of fibrocellular (mesenchymatous) tissue until approximately the 6th week of intra-uterine life. The maxillae ossify intramembranously (Fig 52).

Ossification of the maxilla commences in the 7th week, slightly later than that of the mandible, from a center which appears in a band of fibrocellular tissue on the outside of the nasal capsule, close to the point where the anterior superior alveolar branch is given off by the infraorbital nerve, and slightly above the site of origin of the future canine tooth enamel organ from the dental lamina.

Ossification spreads from this centre in five main directions, as follows (refer to the chapter on the jaws):

1. upwards to form the frontal process of the maxilla;
2. backwards to form the zygomatic process;
3. inwards to form the palatal process;
4. downwards to form the alveolar process; and
5. forward towards the midline to form the facial surface of the maxilla which is involved in premaxillary development.

In an early stage of development, the bony maxilla looks like a vertically arranged curved plate of bone which is convex medially. The developing maxilla forms a

Fig 52   **Development of the maxilla and the premaxillary region**

Coronal section through the palate and maxillary ossification

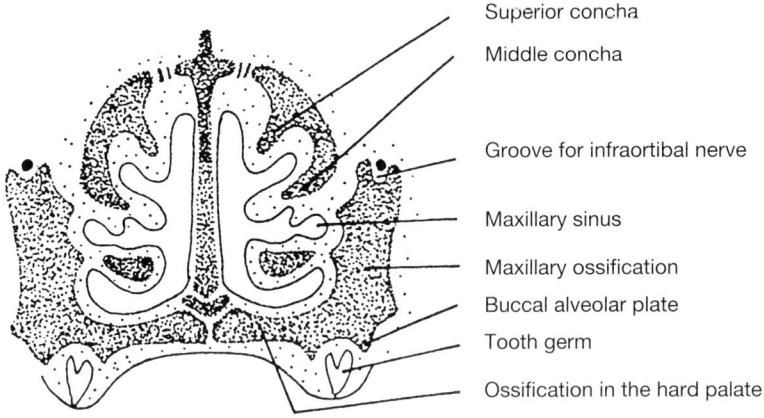

- Superior concha
- Middle concha
- Groove for infraorbital nerve
- Maxillary sinus
- Maxillary ossification
- Buccal alveolar plate
- Tooth germ
- Ossification in the hard palate

Anterior view of maxilla and premaxillary region before birth

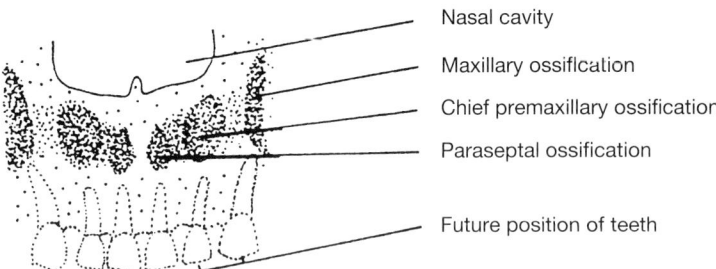

- Nasal cavity
- Maxillary ossification
- Chief premaxillary ossification
- Paraseptal ossification
- Future position of teeth

The sketch below shows fusion between premaxilla and maxilla before birth

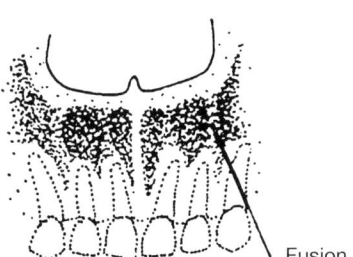

Fusion

The sketch below shows the palatal aspect of the palate after birth

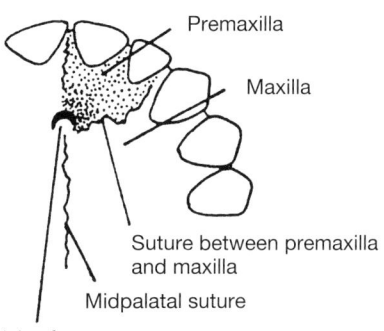

- Premaxilla
- Maxilla
- Suture between premaxilla and maxilla
- Midpalatal suture
- Incisive foramen

furrow for the infraorbital nerve and, by downward growth, an outer alveolar plate in relation to the future tooth germs. The maxilla grows upwards, backwards, inwards, and medially to form the bony parts mentioned previously. In the region where ossification of the horizontal shelves of the hard palate commences from the main bony mass, a downward extension of bone forms the medial alveolar plate. In this way a furrow is created for the developing tooth germs. In later development this furrow is partitioned into alvcoli for the individual teeth. Further growth of the maxilla to adult proportions is mainly due to remodeling of the bone.

The premaxilla

Embryologically, the premaxillary region develops mainly in the frontonasal process, but the formation and existence of a separate premaxillary bone in humans is a controversial subject.

The premaxillary area apparently begins to ossify after the maxilla proper, with the appearance of "ossification centers" on each side of the midline (Fig 52). The center which is responsible for the main mass of the bone on each side appears first above the area of future development of the primary lateral incisor tooth and is followed by the appearance of a second (paraseptal) center in a slightly more medial position. The two centers on one side subsequently fuse with one another and with the main maxillary ossification of that side. The latter fusion is completed on the facial surface before birth, while a bilateral suture can still be observed between the premaxilla and the maxilla on the palatal surface after birth. These sutures extend from the incisive foramen towards the alveolar region between the lateral incisor and canine teeth. A permanent midline suture exists between the two premaxillae. This suture is continuous with the midpalatal (midsagittal) suture between the two horizontal palatal processes of the bony maxillae.

According to other views, possibly more accepted at present, the premaxillary ossification on each side is merely an extension of maxillary ossification and the existence of a separate premaxillary bone is disputed.

In other primates and in lower animal forms, the suture between the premaxilla and the maxilla on the facial surface persists after birth, and they are therefore regarded as possessing a separate premaxillary bone.

The lacrimal bone

The lacrimal bone develops in mesenchyme anterosuperiorly to the nasal capsule.

The palatine bone

The palatine bone develops intramembranously on the medial side of the nasal capsule. Ossification commences in the vicinity of the maxillary tuberosity, upwards to form the vertical plate of the bone, and horizontally inwards to form the horizontal plate which meets the corresponding plate of the other palatine bone at the midpalatal suture. All the processes of the palatine bone are visible by the end of the 2nd month. The palatine bone forms the posterior boundary of the opening of the maxillary sinus.

The vomer

The vomer developes from two intramembranous ossification centers in the mesenchyme surrounding the lower more posterior parts of the cartilagenous nasal septum. The two halves of the ossifying vomer unite partially before birth and form a furrow which cups the inferior edge of the cartilagenous septum. The vomer increases in height posteriorly and, together with the vertical plate of the ethmoid and the septal cartilage, forms the definitive nasal septum.

The maxillary sinus

The maxillary sinus starts to develop at approximately 10 weeks as a small evagination of the middle meatus on the lateral nasal wall. It is initially separated from the maxilla by nasal capsule cartilage and only comes into direct relation to the bone when this cartilage atrophies.

The sinus gradually enlarges and hollows out the maxilla until it eventually occupies a fairly large space between the orbital part of the maxilla and its tooth-bearing area. The sinus only attains its adult size after the secondary teeth have erupted but may thereafter gradually expand into the alveolar process, especially after loss of teeth.

Selected bibliography

1. Bhaskar, S.N. (ed) (1991) *Orban's Oral Histology and Embryology*, 11th edition. St Louis: Mosby-Year Book, Inc.
2. Dixon, A.D. (1986) *Anatomy for Students of Dentistry*, 5th edition. Edinburgh: Churchill Livingstone.
3. Osborn, J.W. (ed) (1981) *Dental Anatomy and Embryology.* Oxford: Blackwell Scientific Publications.
4. Sperber, G.H. (1989) *Craniofacial Embryology*, 4th edition. London: Wright.

Review questions

1. Briefly describe the cartilagenous skeleton of the face and surrounding structures before ossification commences.
2. Write brief notes on the bony structures which develop in, and around, the cartilagenous nasal capsule.

16. Development of the mandible, the temporomandibular joint, and muscle

The mandible

Meckel's cartilage

The mandible develops in the first (mandibular) pharyngeal arch where it is preceded by Meckel's cartilage (Fig 53, I). In humans, Meckel's cartilage attains its full length after 6 weeks of embryonic life and extends from the region of the otic capsule to the midline of the mandibular arch in the form of an uninterrupted rod. Close to the midline, the ventral end of the cartilage is curved upwards in close proximity to the corresponding end of the cartilage of the other side, to which it is loosely connected by fibrocellular tissue. Fibrocellular tissue envelopes the rod in its entire length.

At this stage Meckel's cartilage is closely associated with the mandibular nerve and its branches (Fig 53, II). The main trunk of this nerve leaves the skull through the foramen ovale in a ventromedial position to the dorsal end of the cartilage and meets the cartilage at the junction of its dorsal one third with the remainder of the cartilage. At this point the nerve divides to form the lingual and inferior alveolar branches. The lingual nerve courses forwards on the medial aspect of the cartilage, while the inferior alveolar nerve crosses the cartilage and courses forwards on its lateral side before dividing into its mental and incisive branches.

The body of the mandible

Meckel's cartilage is closely associated with the development of the mandible. The mandible makes its first appearance as a band of dense mesenchymal fibrocellular tissue lateral to the incisive and inferior alveolar nerves (Fig 53, III).
During the 7th week, intramembranous ossification commences in the fibrocellular tissue in the angle formed by these two nerves (Fig 53, IV) and spreads backwards, providing a notch for the mental nerve, and along the lateral aspect of the inferior alveolar nerve. Subsequent spread of ossification gives rise to the mental foramen, and by forward extension, a trough is formed by outer and inner bony plates for the incisive nerve (Fig 54, I). This trough is later converted into fine canals for the filaments

Fig 53 — Development of the mandible

I

Sketch I shows the extent of Meckel's cartilage in an embryo after about 6 weeks.

Meckel's cartilage

II

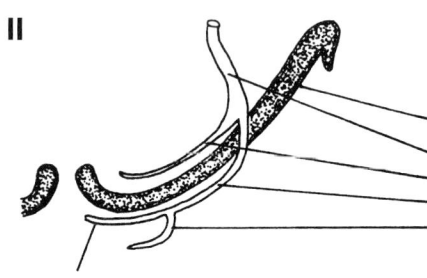

Sketch II shows the relation between the branches of the mandibular nerve and the left Meckel's cartilage.

Meckel's cartilage
Mandibular nerve
Lingual nerve
Inferior alveolar nerve
Mental nerve

Incisive nerve

III

Sketch III shows the mesenchymal fibrocellular condensation mapping the shape of the mandible before ossification.

Fibrocellular tissue

IV

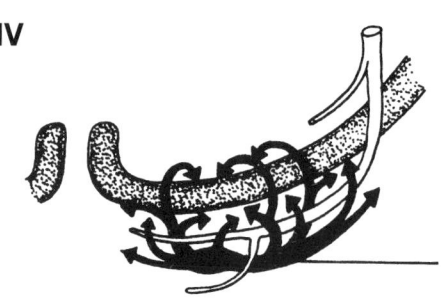

Sketch IV shows the directions (black arrows) of the spread of ossification in the body of the mandible from a point close to the angle between the incisive and mental branches of the mandibular nerve.

Directions taken by spreading ossification in the body of the mandible of the incisive nerve. Midline union between the two halves of the bony mandible takes place during the 1st year of life. By backward extension, outer and inner shallow plates of bone, which form a trough for the inferior alveolar nerve, are formed. Only later is this trough converted into the mandibular canal (Fig 54, II). All the changes described above take place in intimate association with the lateral aspect of Meckel's cartilage.

From the growth process described above, it is clear that spread of bone from a single ossification center is responsible for forming the body of the mandible, from the mandibular foramen posteriorly to the area of the symphysis anteriorly. The developing tooth germs are at this stage of development not enclosed by bone and are superficially located deep to the mucous membrane. Later, as a result of the formation of the buccal and lingual alveolar plates above the level of the roof of the alveolar canal, the developing teeth are found in a bony trough which is subsequently divided by transverse bony septae into alveoli for individual teeth.

The fate of the cartilage of Meckel

Meckel's cartilage is gradually enclosed by a lingual extension of ossification along the entire length of the body of the mandible (Fig 54, II and III). This lingual extension gradually surrounds the cartilage which is resorbed and replaced by bone. The ventral end of Meckel's cartilage may persist as small nodules of cartilage in the symphyseal fibrous tissue till the time of birth. The dorsal part of the cartilage not directly associated with the body of the mandible disappears, except for its fibrous perichondrium which differentiates to form the sphenomalleolar and sphenomandibular ligaments.

Persisting dorsal portions of the cartilage become isolated and ossify to form major portions of the malleus and the incus, while it is also suggested that the spine of the sphenoid bone and the lingula of the mandible may arise in Meckel's cartilage.

The development of the incus has a phylogenetic basis. In humans it is apparently formed in Meckel's cartilage, as described above, but in lower animal forms, for example in reptiles, it corresponds with part of the quadrate bone (upper jaw). The hinge joint between the malleus and the incus constitutes the primary (primitive) jaw joint and functions as such for a limited period. The stapes is largely formed in the cartilage of the second pharyngeal arch and is the first of the three ear ossicles to appear. The ossicles of the ear are the first bones in the body to reach adult proportions early in the 7th intra-uterine month.

The ramus of the mandible

As is the case in the body of the mandible, the bony ramus is preceded by fibrocellular tissue into which a direct spread of ossification from the body occurs (Fig 54, III). Development of the ramus is independent of Meckel's cartilage. The ramus, together

Fig 54

The lingual side of the developing mandible

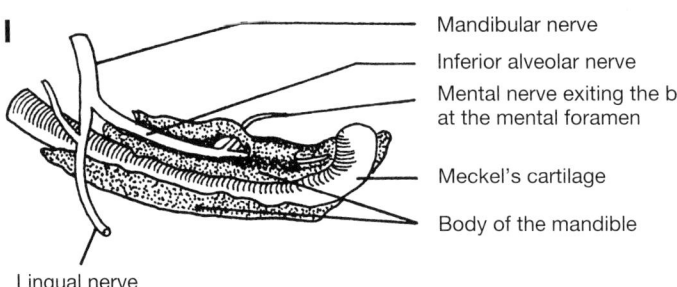

I
- Mandibular nerve
- Inferior alveolar nerve
- Mental nerve exiting the bone at the mental foramen
- Meckel's cartilage
- Body of the mandible
- Lingual nerve

Cross-section through developing mandible

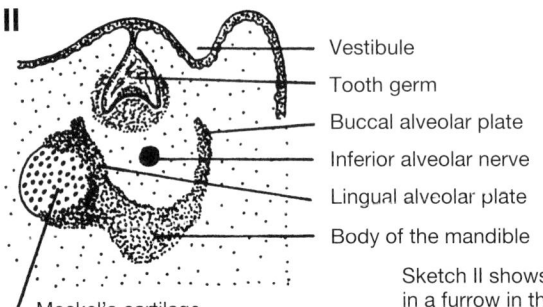

II
- Vestibule
- Tooth germ
- Buccal alveolar plate
- Inferior alveolar nerve
- Lingual alveolar plate
- Body of the mandible
- Meckel's cartilage

Sketch II shows how the developing tooth germs lie in a furrow in the bone, together with the nerve. Meckel's cartilage is resorbing.

Ossification of mandibular ramus

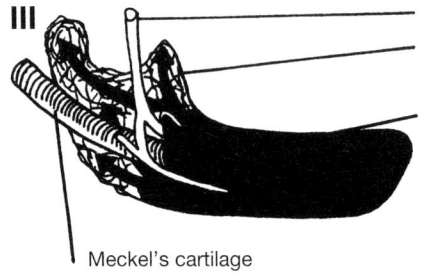

III
- Mandibular nerve
- Fibrocellular tissue
- Bony body of the mandible
- Meckel's cartilage

Sketch III indicates the mapping of the mandibular ramus by fibrocellular tissue prior to ossification. Black arrows indicate the spreading ossification from the body of the mandible. Meckel's cartilage lies outside the developing ramus.

Secondary cartilages

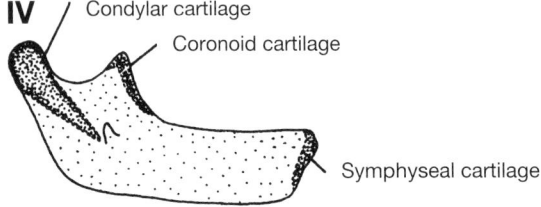

IV
- Condylar cartilage
- Coronoid cartilage
- Symphyseal cartilage

with its condylar and coronoid processes, is rapidly ossified and are distinct bony entities by the 10th week. The appearance of secondary cartilages influence subsequent development in these processes (Fig 54, IV).

The condylar secondary cartilage

Secondary, or accessory, cartilages, so-called because they form at a later stage than the primary (Meckel's) cartilage, form at three main sites in the developing mandible. The first, and largest to appear, is the condylar secondary cartilage. It makes its appearance as a rim of cartilage on the superolateral aspect of the bony condylar process. It rapidly develops into a cone-shaped mass with its apex close to the mandibular foramen. It occupies the entire condylar process. Endochondral ossification occurs in the cartilage which, except for a thin zone below the articular surface of the condyle, is largely replaced by bone by the 5th intra-uterine month. Cartilage is, however, present in a diminishing quantity below the articular surface for the first 20 years of life after which it is entirely replaced by bone. The secondary condylar cartilage was previously regarded as an important growth center for the elongating ramus.

The coronoid and symphyseal secondary cartilages

Cartilage forms for a short while on the crest and anterior edge of the coronoid process but all signs of this cartilage disappear before birth. Its function is unknown. The third of the secondary cartilages forms in the symphyseal region. The adjacent symphyseal cartilages are separated by symphyseal connective tissue and are eliminated early in the 1st year of life when the two halves of the bony mandible unite.

The temporomandibular joint

The temporomandibular joint develops between the mandibular condyle and the temporal bone of the skull (Fig 55). These developing bones are initially separated by a mesenchyme-filled space. The condylar secondary cartilage appears between the 10th and 12th weeks and grows towards the temporal bone. The intervening mesenchyme differentiates to form fibrous connective tissue in which parallel horizontal clefts appear adjacent to the bones, thereby giving rise to a central articular disc (meniscus) between the lower and upper joint cavities. Some authors are of the opinion that movement in the joint is necessary for optimum development of the joint cavities and that the concurrent development of lateral pterygoid muscle attachments to the disc is significant. It is stated that when developing joints are

Fig 55   **Development of the temporomandibular joint**

I
- Temporal fibrocellular condensation
- Mesenchyme-filled space
- Incus
- Primary jaw joint
- Malleus
- Meckel's cartilage
- Mandibular fibrocellular condensation

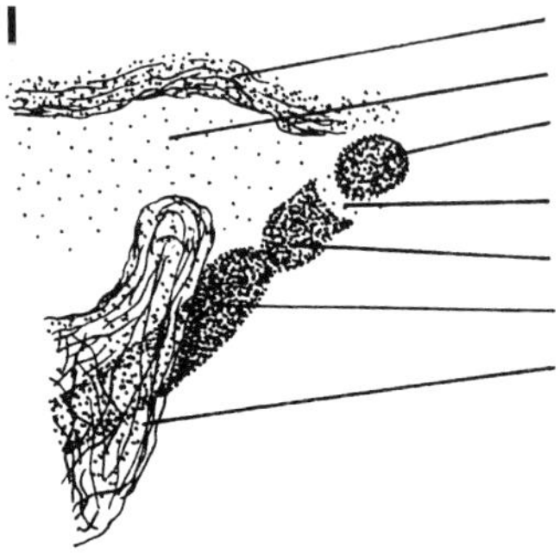

Sketch I shows the mandibular and temporal fibrocellular condensations approaching one another.
The malleus and incus will be taken up in the middle ear cavity.

II
- Upper joint cavity
- Meniscus (disc)
- Lower joint cavity
- Secondary condylar cartilage

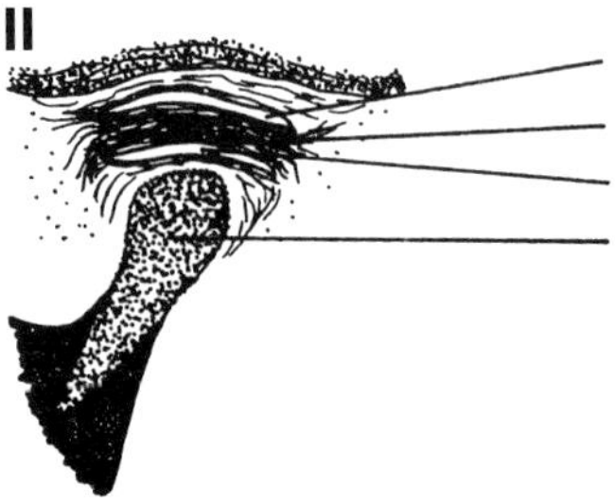

Sketch II shows compression and condensation of fibrocellular tissue between the approaching bones.
The joint cavities and the meniscus have formed.

III Temporal bone
- Upper joint cavity
- Meniscus
- Lower joint cavity
- Condylar cartilage
- Joint capsule

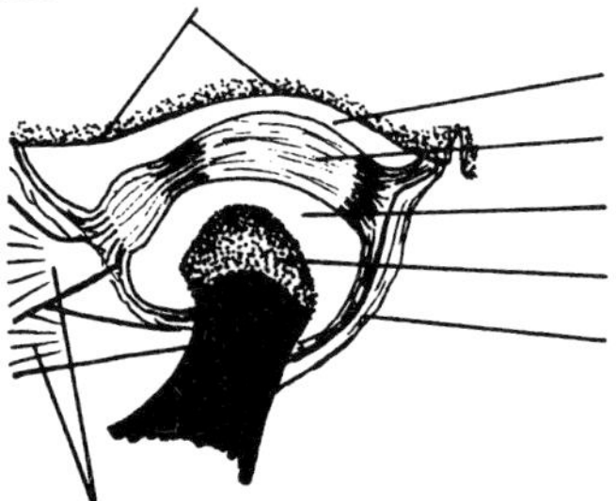

Sketch III shows the joint soon after birth.

Lateral pterygoid muscle

IV
Articular eminence
- Upper joint cavity
- Meniscus
- Squamotympanic fissure
- Lower joint cavity
- Joint capsule
- Condyle

Lateral pterygoid muscle

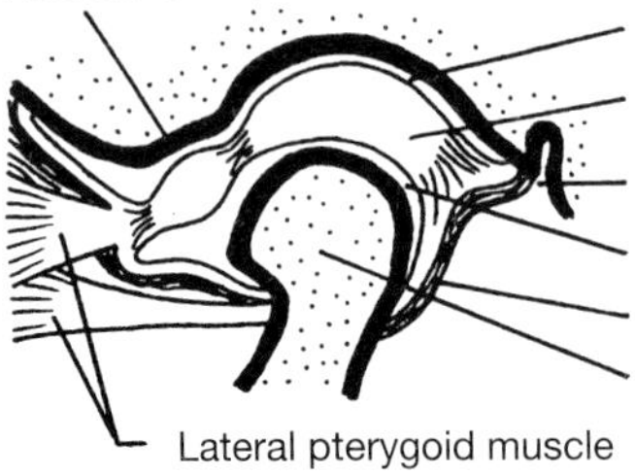

Sketch IV shows the joint in an adult.

immobilized, joint cavities do not form and ankylosis of the joint elements occurs. Condensation of the mesenchymic and subsequent formation of fibrous connective tissue around the joint is responsible for the formation of the capsule.

Immediately after birth the temporomandibular joint is a lax structure while the mandibular fossa and the articular eminence form a flat surface. The joint attains adult form by the 12th year of life.

The temporomandibular joint is a secondary development, both from an ontogenetic (embryological) and a phylogenetic (evolutionary) point of view. The primary joint between the malleus and the incus, which was described previously, is phylogenetically the primary joint and the homolog of the jaw joint of a typical reptile. With the evolutionary development of a middle ear, this joint lost its connection to the jaws and was replaced by the temporomandibular joint which represents a mammalian development.

Muscle

Skeletal muscle develops from mesenchymal condensations which differentiate to form elongated primitive muscle cells, or myoblasts. Myoblasts divide mitotically until approximately the middle of fetal life. It is generally believed that multinucleated fibers result from fusion of separate myoblasts.

The myofilaments, which are initially synthesized in the muscle cells, aggregate to form myofibrils. Muscle cells, together with their contained myofibrils and myofilaments, fuse to form a multinucleated muscle fiber. Myofilaments are of two types, depending on size and composition. The thick filaments (10 nm in diameter) are composed of myosin, while the thin filaments (5 nm in diameter) are composed of actin. Both myosin and actin are contractile proteins. Growth in size of muscle fibers in later fetal and postnatal life is through increase in size (hypertrophy) of individual fibers.

The muscles of the face, the masticatory muscles, and the laryngopharyngeal muscles arise in pharyngeal arch mesoderm. A postpharyngeal muscle mass gives origin to the trapezius and sternocleidomastoid muscles, while occipital somite myotomes migrate ventrally (the hypoglossal cord) to form the muscles of the tongue (except palatoglossus, which is formed from fourth arch mesenchyme). As can be expected, the mesoderm of the first arch gives origin mainly to the muscles of mastication. Second arch mesoderm is responsible for the muscles of facial expression, while mesoderm of the third, fourth, and lower arches form the musculature of the palate, pharynx, and larynx. The muscles of the face, mouth, and oropharynx are the first of the body muscles to develop.

Selected bibliography

1. Bhaskar, S.N. (ed) (1991) *Orban's Oral Histology and Embryology*, 11th edition. St Louis: Mosby-Year Book, Inc.
2. Mjör, I.A. and Fejerskov, O. (eds) (1986) *Human Oral Histology and Embryology*. Copenhagen: Munksgaard.
3. Osborn, J.W. (ed) (1981) *Dental Anatomy and Embryology*. Oxford: Blackwell Scientific Publications.
4. Scott, J.H. and Symons, N.B.B. (1982) *Introduction to Dental Anatomy*, 9th edition. Edinburgh: Churchill Livingstone.
5. Sperber, G.H. (1989) *Craniofacial Embryology*, 4th edition. London: Wright.

Review questions

1. Describe the morphology and location of Meckel's cartilage before mandibuar ossification commences.
2. Briefly describe the ossification of the mandible.
3. Write brief notes on the fate of Meckel's cartilage.
4. Discuss the temporomandibular joint with reference to the following:
    (a) embryological development,
    (b) morphology at birth, and
    (c) phylogenetic history.
5. Write brief notes on the embryology of the muscles of mastication, facial expression, and tongue.

17. Development of the pharyngeal arches and the tongue

The pharyngeal arches

A series of five or six bilateral swellings form around the ventral aspect of the embryonic neck during the 4th intra-uterine week. These segmentally arranged pharyngeal arches (Fig 56) are initially separated externally by pharyngeal grooves which rapidly disappear, except the first which lies between the first and second pharyngeal arches. This groove becomes modified to form the external auditory meatus and the middle ear cavity. On the pharyngeal aspect the grooves correspond to pharyngeal pouches. In fishes the pharyngeal arches support the gills and may be termed branchial arches in this regard. The first pharyngeal (mandibular) arch lies between the mouth and the first pharyngeal groove and is succeeded caudally by the second (hyoid) arch and the more caudal arches. The arches are covered externally by ectoderm while they are covered on the pharyngeal side by endoderm, except the first which is probably lined by ectoderm in its largest part.

Each arch is initially composed of a cartilage rod, derived from neural crest tissue, a mesodermal (muscular) component, a vascular component (the pharyngeal, or aortic arch, artery), and a nervous component which represents the cranial nerve supply to the muscular and mucosal derivatives of that arch. All the above structures are embedded in a mass of mesenchyme. Although derivatives of six arches are described, the most caudal two arches do not contribute a great deal to further development, except through their mesodermal (muscular) and aortic arch components. The latter aspect is fully dealt with in Part I, Chapter 25.

The pharyngeal arch cartilages and mesenchyme, except the fifth cartilages and mesenchyme which apparently do not contribute much to further development, and pouches differentiate to give rise to a variety of tissues and structures.

The pharyngeal arch cartilages (Fig 57)

The cartilage of the first arch (Meckel's cartilage)

The cartilage of Meckel forms the framework for the development of the mandible but does not contribute directly to this development. The persisting dorsal

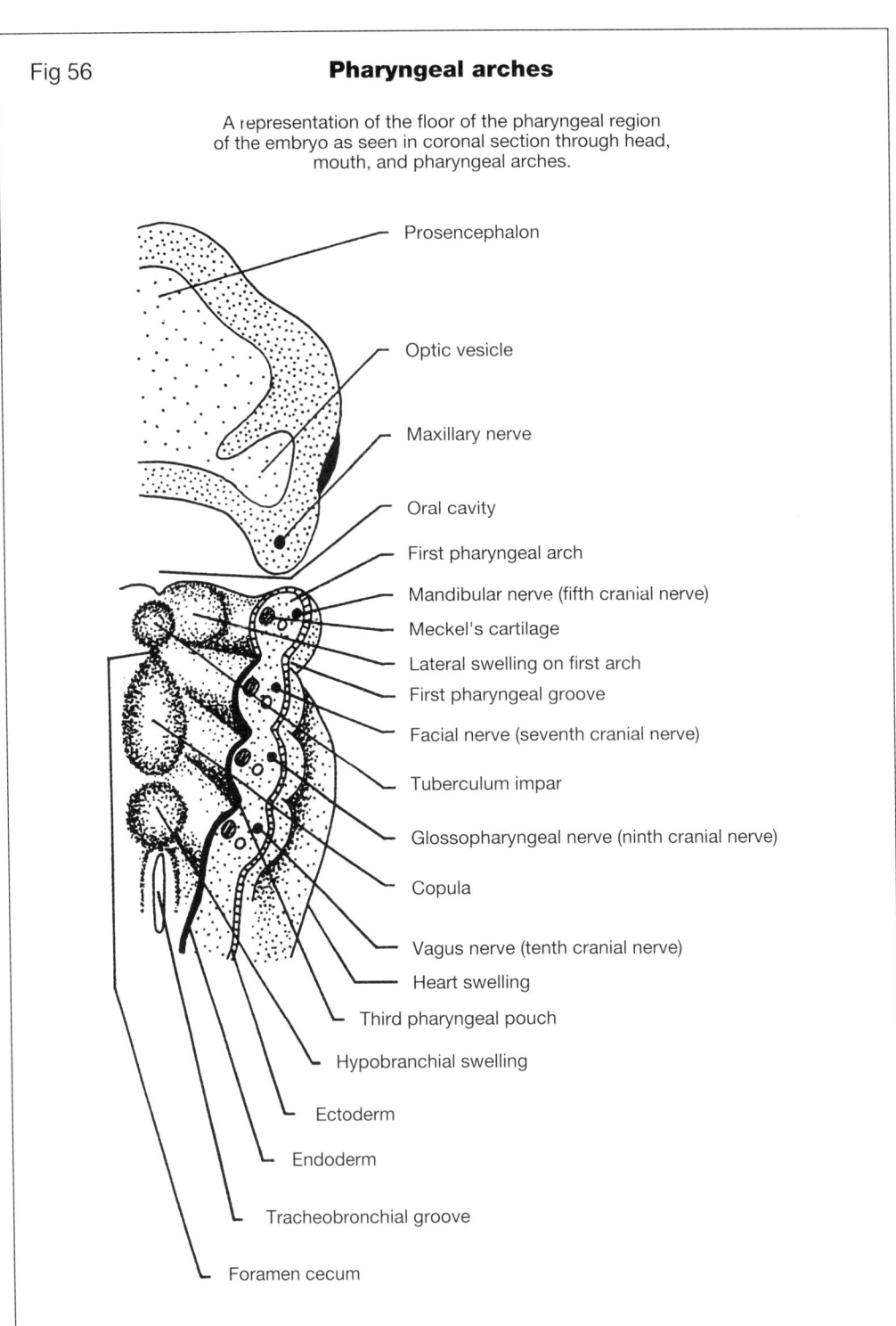

Fig 56 **Pharyngeal arches**

A representation of the floor of the pharyngeal region of the embryo as seen in coronal section through head, mouth, and pharyngeal arches.

portions of the two ear ossicles, the malleus, and the incus, while two ligaments arise from its fibrous perichondrium, namely the sphenomandibular ligament, which extends from the spine of the sphenoid bone to the lingula of the mandible, and the sphenomalleolar ligament, which becomes the anterior ligament of the malleus in the adult. It may also contribute to the spine of the sphenoid and the lingula.

The cartilage of the second arch (hyoid or Reichert's cartilage)

Dorsally this cartilage lies in close proximity to Meckel's cartilage and the dorsal end of the first pharyngeal groove and pouch. This fact is important in later development of the ear. From this cartilage and its perichondrium the greater part of the third ear ossicle, the stapes, develops, as well as the lesser horn, or cornu, and the upper part of the body of the hyoid bone, the styloid process and stylohyoid ligament.

The cartilage of the third arch

The cartilage of this arch gives origin to the major part of the body and greater horn, or cornu, of the hyoid bone.

The cartilage of the fourth arch

The thyroid cartilage probably arises in the cartilage of this arch.

The cartilage of the fifth arch

This is a transient structure which disappears rapidly.

The cartilage of the sixth arch

The laryngeal arytenoid and cricoid cartilages probably arise in this cartilage. Uncertainty exists regarding the origin of the tracheal cartilages.

Mesodermal (muscular) components of the pharyngeal arches

Mesoderm of the first arch

The chief muscles of mastication, namely the temporalis, masseter, medial, and lateral pterygoid muscles, as well as the mylohyoid, anterior belly of the digastric, tensor palati, and tensor tympani muscles arise in the first arch. The motor nerve to the muscles developing in the first arch is the mandibular division of the trigeminal (fifth cranial) nerve, while the ordinary sensory branches of the same nerve

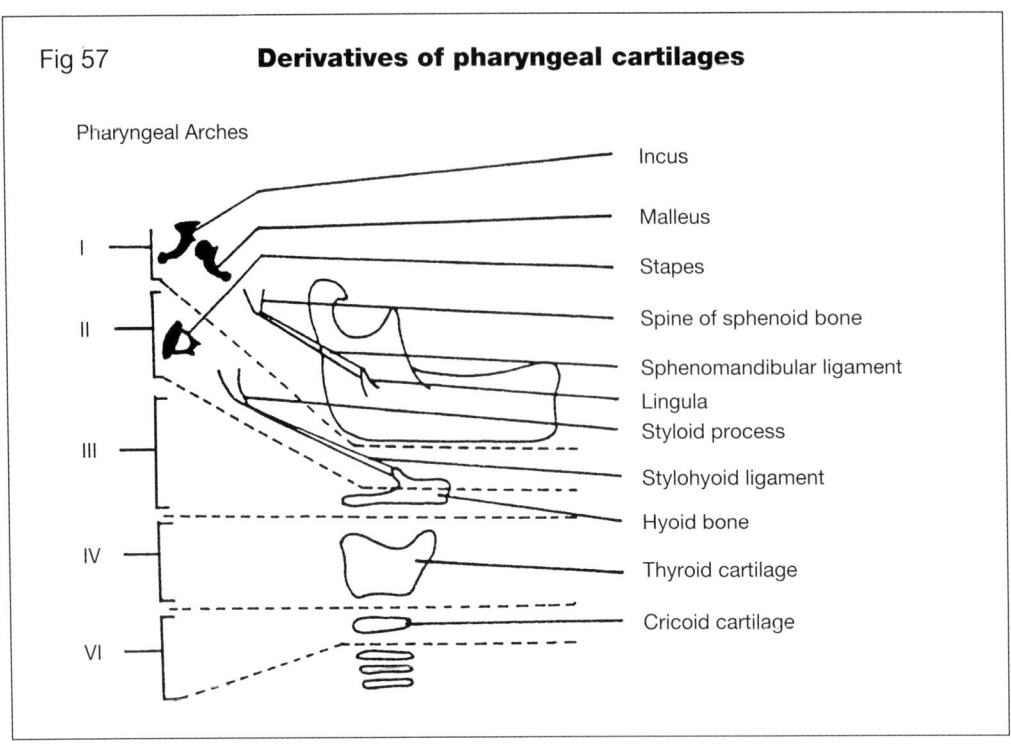
Fig 57 Derivatives of pharyngeal cartilages

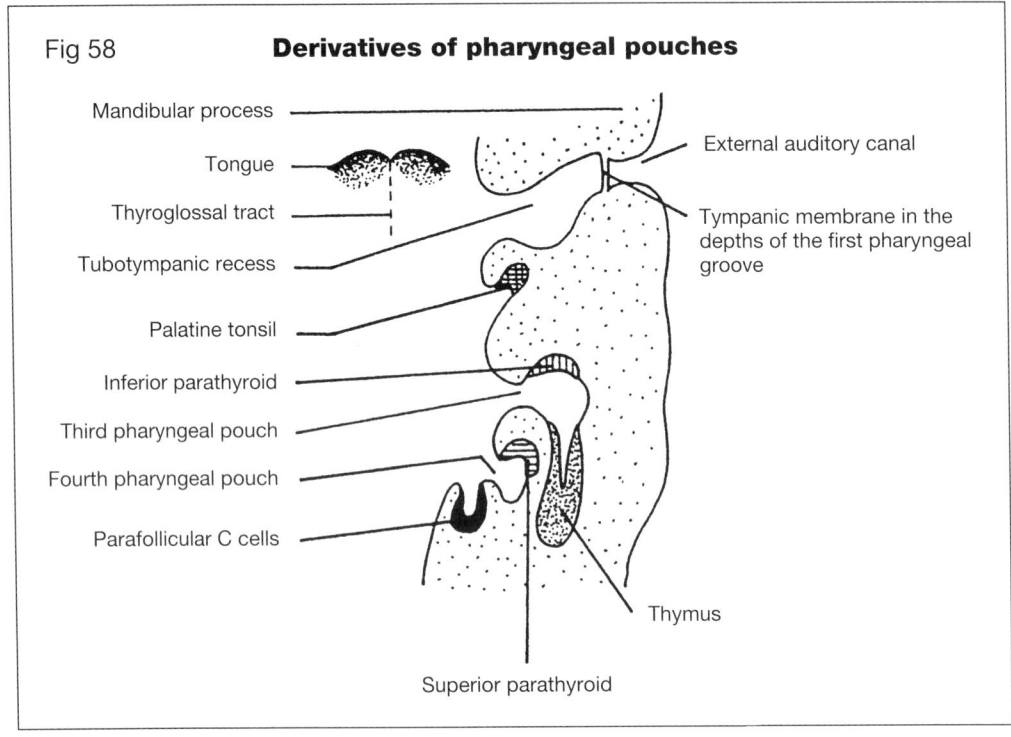
Fig 58 Derivatives of pharyngeal pouches

innervate the mandible and its associated mucous membrane, the gingiva, the mucous membrane of the anterior two-thirds, or oral part, of the tongue, the floor of the mouth, and the skin covering the lower part of the face. The special sensory (taste) innervation to the anterior two-thirds of the tongue is derived from a pretrematic branch of the facial nerve which invades this arch.

Mesoderm of the second arch

The mesoderm of this arch gives origin to the muscles of facial expression (the mimetic muscles), as well as the buccinator, extrinsic, and intrinsic auricular muscles, occipitofrontalis, platysma, posterior belly of digastric, stylohyoid, and stapedius muscles. The motor innervation of these muscles is derived from the facial (seventh cranial) nerve which serves the second arch.

Mesoderm of the third arch

The stylopharyngeus, and possibly parts of the upper pharyngeal muscles, arise in this arch and receive motor innervation through the glossopharyngeal (ninth cranial) nerve. This nerve also supplies ordinary and special sensory innervation to the posterior third, or pharyngeal part, of the tongue, including the taste buds of the circumvallate papillae.

Mesoderm of the fourth arch

The mesoderm of the fourth arch gives origin to the palatoglossus muscle of the tongue, the uvular and palatopharyngeus muscles, levator palati, cricothyroid, and pharyngeal constrictors. The motor supply to these muscles is through the superior laryngeal branch of the vagus (tenth cranial) nerve which is also responsible for the ordinary and special sensory innervation of the posterior third of the tongue.

Mesoderm of the sixth arch

The intrinsic muscles of the larynx form in this arch and receive their motor nerve supply through the recurrent laryngeal branch of the vagus.
    The ectodermally covered pharyngeal grooves flatten and disappear, except the first which deepens dorsally to form the external auditory meatus. This is surrounded by the auricular swellings which will later give rise to the auricle of the external ear. The tympanic membrane is formed in the depths of the external auditory meatus by the combined ectoderm of the meatus and the endoderm which lines the tubotympanic recess of the first pharyngeal pouch.
Major changes take place in the pharyngeal pouches and these are subsequently described.

The pharyngeal pouches (Fig 58)

The first pharyngeal pouch

The dorsal diverticulum of the first pouch (tubotympanic recess), which lies between the mandibular and hyoid arches, forms the middle ear, tympanic antrum, and the pharyngotympanic (Eustachian) tube which connects the middle ear with the pharynx. The internal ear partly arises in the otic placode. It was pointed out earlier that the ossicles of the ear develop in close association with the dorsal aspects of both the first pharyngeal groove and the corresponding pouch and therefore in close proximity to the ear. These ossicles migrate into the middle ear cavity to assume their adult positions.

The ventral part of the first pharyngeal groove is increasingly occupied by the developing tongue.

The second pharyngeal pouch

The palatine tonsil develops in the dorsal diverticulum of the second pouch.

The third pharyngeal pouch

In the dorsal diverticulum of the third pouch the inferior parathyroid gland arises, while the thymus gland develops in its ventral diverticulum.

The fourth pharyngeal pouch

The superior parathyroid gland develops in the dorsal diverticulum of this pouch. This gland is initially inferior to the inferior parathyroid gland and the thymus, but the latter glands pass the superior parathyroid gland in their downward migration into the neck to their final positions.

The fifth pharyngeal pouch

The fifth pharyngeal pouch develops as an outpouching of the fourth pouch and gives origin to the calcitonin-producing cells which are incorporated in the thyroid gland as parafollicular cells (C cells, light cells). Calcitonin and parathormone are antagonistic in their action.

The tongue

Several bulges can be observed on the ventral aspect of the pharyngeal region and these later give rise to the tongue (Fig 59).

The tongue is mainly formed by the first, second, and third pharyngeal arches. The anterior two-thirds of the tongue (the body and tip, or oral part) arises in the

4th week in three swellings associated with the first arch. These swellings consist of two lateral, or lingual, swellings and a midline swelling, the tuberculum impar. The posterior third of the tongue (the base, or pharyngeal part) arises in the ventromedial parts of the second, third, and part of the fourth arches which coalesce to form an enlarged midline swelling, the copula. A lower midline swelling, the hypobranchial eminence, which is regarded by many to be a subdivision of the copula, later gives rise to the epiglottis. The thyroid gland arises in the midline from the endoderm on the dorsal surface of the tongue between the tuberculum impar and the copula. The thyroid diverticulum, an endoderm-lined duct, migrates into the neck as the thyroglossal duct which differentiates to form the thyroid gland. It loses all connection with the surface of the tongue but its site of origin is marked throughout life by a shallow pit, the foramen cecum. This lies in the apex of the terminal sulcus which is a shallow V-shaped groove with its apex directed towards the base of the tongue. It is situated immediately behind the row of circumvallate papillae. In rare instances thyroid tissue persists on the dorsal surface of the tongue and is known as the lingual thyroid. The terminal sulcus separates the anterior two-thirds of the tongue from the posterior third.

The tuberculum impar is a temporary structure and contributes little to the anterior two-thirds of the tongue. It is rapidly overgrown by the lateral lingual swellings which proliferate and fuse to form the body of the tongue. The base of the tongue is formed largely by growth of the copula. The tracheobronchial groove, which develops caudal to the hypobranchial eminence, is the site of origin of a large part of the respiratory system.

Fungiform papillae develop on the dorsal surface of the anterior two-thirds of the tongue at about 11 weeks, while filiform papillae develop later in the same region and are not fully differentiated until after birth. The 8-12 large circumvallate papillae form anterior to the terminal sulcus between the 2nd and 5th intra-uterine months.

At the time of birth the mucous membrane on the posterior third of the tongue develops small pits which, after infiltration by lymphatic tissue, give rise to the lingual tonsillar tissue.

Inductive interaction between epithelial cells and invading nerve cells gives origin to the taste buds which are innervated by the vagus and glossopharyngeal nerves, and the chorda tympani (facial nerve). The taste buds are predominantly found in the mucosa of the dorsal surface of the tongue but are also found on the palate and in other regions of the mouth. Recognizable taste buds are only found at approximately 14 weeks of intra-uterine life.

The muscles of the tongue have a dual origin. The intrinsic muscles (transverse, longitudinal, and vertical) probably arise in situ in pharyngeal arch mesenchyme, while the extrinsic muscles (palatoglossus, styloglossus, genioglossus, and hyoglossus) arise in the occipital somite region in relation to the origin of the hypoglossal nerve. This muscle mass, or hypoglossal cord, migrates forwards under the mucosa around the sides of the pharynx to invade the developing tongue. The hy-

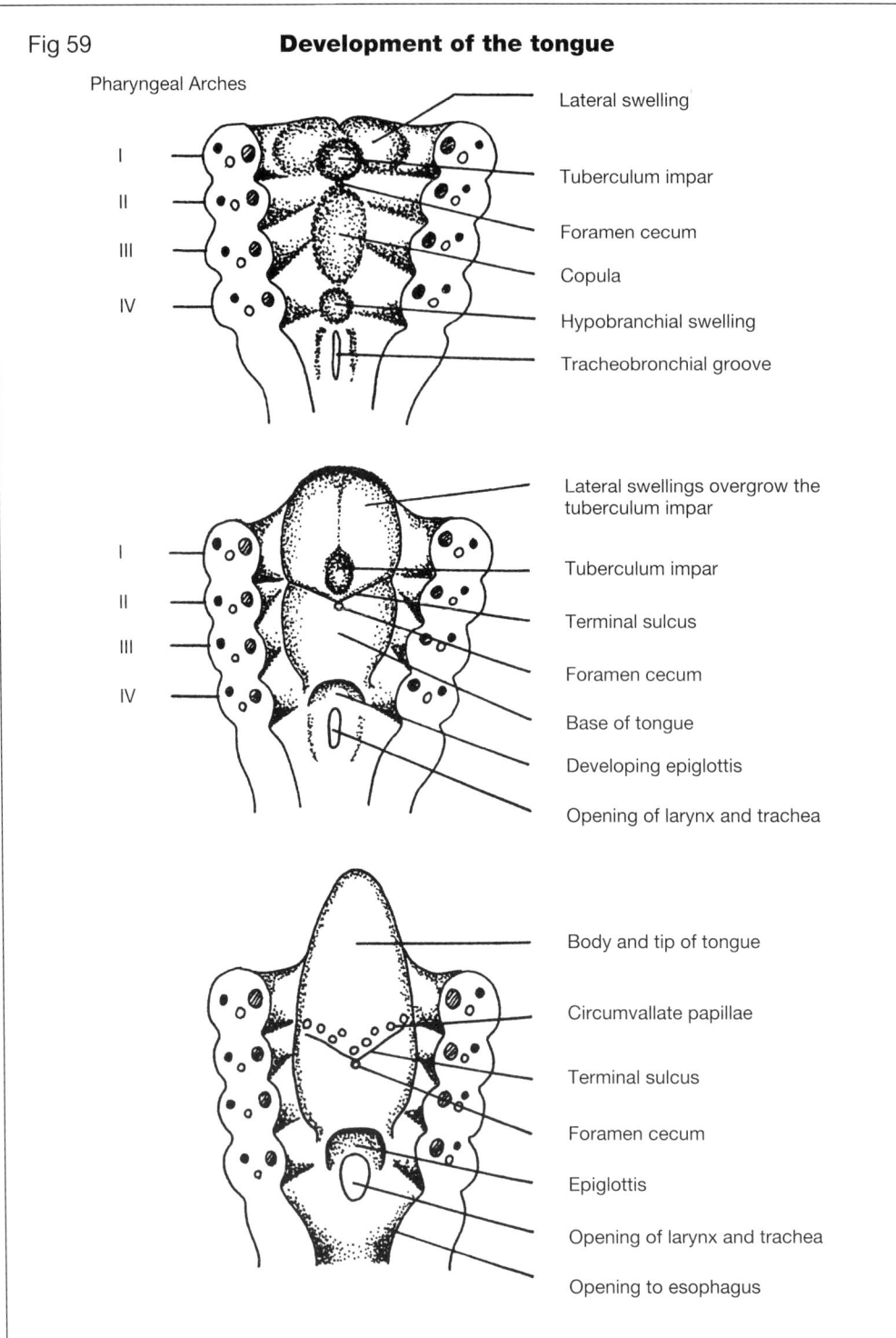

Fig 59 Development of the tongue

poglossal nerve accompanies this migration and provides the motor innervation to these muscles, with one exception. The palatoglossus muscle is supplied by the pharyngeal branch of the vagus nerve.

The tongue rapidly enlarges and doubles in all dimensions between birth and adulthood. The tongue grows more rapidly in size than the remainder of the oral cavity and initially occupies almost the entire mouth. The tongue has almost reached adult size by the 8th year and it is thought that the initial rapid enlargement is related to the suckling needs of the infant. The tongue may be abnormally small (microglossia) or exceptionally large (macroglossia).

Nerve supply of the epithelium of the tongue

The nerve supply of the tongue epithelium may be explained by a description of the nerve distribution seen in gilled animals. In vertebrates with functional gills, or branchiae, the sensory parts of each branchial (pharyngeal) nerve are distributed to both the cranial and caudal parts of each gill cleft, or trema. Fibers to the cranial parts are termed the pre-trematic branches (exclusively sensory), while post-trematic branches are distributed to the caudal parts. The motor branch of a branchial nerve is included with sensory fibers in the post-trematic branch. It is therefore obvious that two branchial nerves are represented in each branchial arch the post-trematic branch of the nerve of the same arch, and the pre-trematic branch of the nerve of the succeeding arch. Muscle tissue in each arch is supplied by fibers of post-trematic origin, while the associated endodermal derivatives derive their sensory supply from fibers in both the pre- and post-trematic branches, as described above.

The ordinary sensory nerve supply of the epithelium of the anterior two-thirds of the human tongue is from the trigeminal nerve through the lingual branch of its mandibular division (post-trematic branch of the trigeminal nerve, the cranial nerve of the first pharyngeal arch). The special sensory (taste) fibers to the anterior two-thirds are derived from the chorda tympani, the pre-trematic branch of the facial nerve in the second arch. The posterior part of the tongue has a more complicated nerve supply. It is thought that mesenchyme of the third arch, together with the associated epithelium and its sensory nerve supply, overgrows the tissue of the second arch and merges with the first arch during development. This creates the major part of the base of the tongue which is thus supplied by the glossopharyngeal nerve of the third arch for both ordinary and special sensation. A small part of the tongue adjacent to the epiglottis derives its sensory nerve supply from the vagus nerve of the fourth arch through the internal laryngeal branch of the superior laryngeal nerve. The mucous membrane on the ventral surface of the tongue and the floor of the mouth is supplied by the lingual nerve.

The reader is advised to consult the selected bibliography for other versions of the development of the tongue.

Selected bibliography

1. Bhaskar, S.N. (ed) (1991) *Orban's Oral Histology and Embryology,* 11th edition. St Louis: Mosby-Year Book, Inc.
2. Hamilton, W.J., Boyd, J.D. and Mossman, H.W. (1972) *Human Embryology,* 4th edition revised by Hamilton, W.J. and Mossman, H.W. Cambridge: W. Heffer and Sons, Ltd.
3. Mjör, I.A. and Fejerskov, O. (eds) (1986) *Human Oral Embryology and Histology.* Copenhagen: Munksgaard.
4. Moore, K.L. (1988) *Essentials of Human Embryology.* Toronto: B.C. Decker, Inc.
5. Scott, J.H. and Symons, N.B.B. (1982) *Introduction to Dental Anatomy,* 9th edition. Edinburgh: Churchill Livingstone.
6. Sperber, G.H. (1989) *Craniofacial Embryology,* 4th edition. London: Wright.
7. Ten Cate, A.R. (1989) *Oral Histology: Development, Structure, and Function,* 3rd edition. St. Louis: C.V. Mosby Company.

Review questions

1. Write brief notes on the following:
   (a) Structures which are derived from the pharyngeal arch cartilages.
   (b) Muscles which develop in the mesodermal component of the pharyngeal arches.
   (c) Developments which occur in the pharyngeal pouches.
2. Discuss the development of the external form of the tongue.
3. Explain the sensory innervation of the tongue by making use of knowledge of comparative anatomy.
4. Briefly describe the origin of the muscles of the tongue.

18. The mouth, pharynx, and nose

The mouth

The oral cavity (Fig 60) is bordered by the lips and cheeks anteriorly and laterally, the hard and soft palates superiorly, and the floor of the mouth inferiorly. It contains the tongue and the teeth and is continuous posteriorly with the oropharynx, by way of the fauces which consists of two vertical pillars, or mucosal folds. The palatine tonsils lie between the pillars on either side. The anterior pillar contains the palatoglossus muscle which links the soft palate with the tongue, while the posterior fold contains the palatopharyngeus muscle which links the palate to the pharyngeal musculature. The ducts of the parotid, submandibular, and sublingual salivary glands, as well as the very many minor salivary glands, open into the oral cavity.

The teeth and their supporting alveolar processes divide the mouth into a vestibular part between the teeth on the inside and the lips and cheeks on the outside, and a central greater oral cavity.

The vestibule of the mouth

The vestibule of the mouth is lined by mucous membrane (mucosa) and is related to the buccinator muscle in the cheeks and to the muscles of the lips. The parotid duct (Stensen's duct) opens into the vestibule through a foramen situated in the cheek opposite the maxillary second molar tooth in the adult. Folds of mucous membrane, or frenulae, may connect the cheeks and lips to the alveolar processes, and it is common to find distinct midline vertical frenulae connecting both the upper and lower lips to the alveolar process. The superior labial frenulum is sometimes excessively large and attaches to the palate, thus causing undesirable spacing between the central incisor teeth.

The tongue and the floor of the mouth

The parts of the tongue normally visible in the mouth are the dorsal surface, especially its anterior part, and the inferior, or ventral surface. The dorsal surface is marked by papillae of different types. A midline sublingual frenulum loosely connects the tongue with the floor of the mouth behind the central incisor teeth. Bilateral soft ridges are formed on either side of the sublingual frenum by the up-

Fig 60 **Mouth, pharynx and nose**

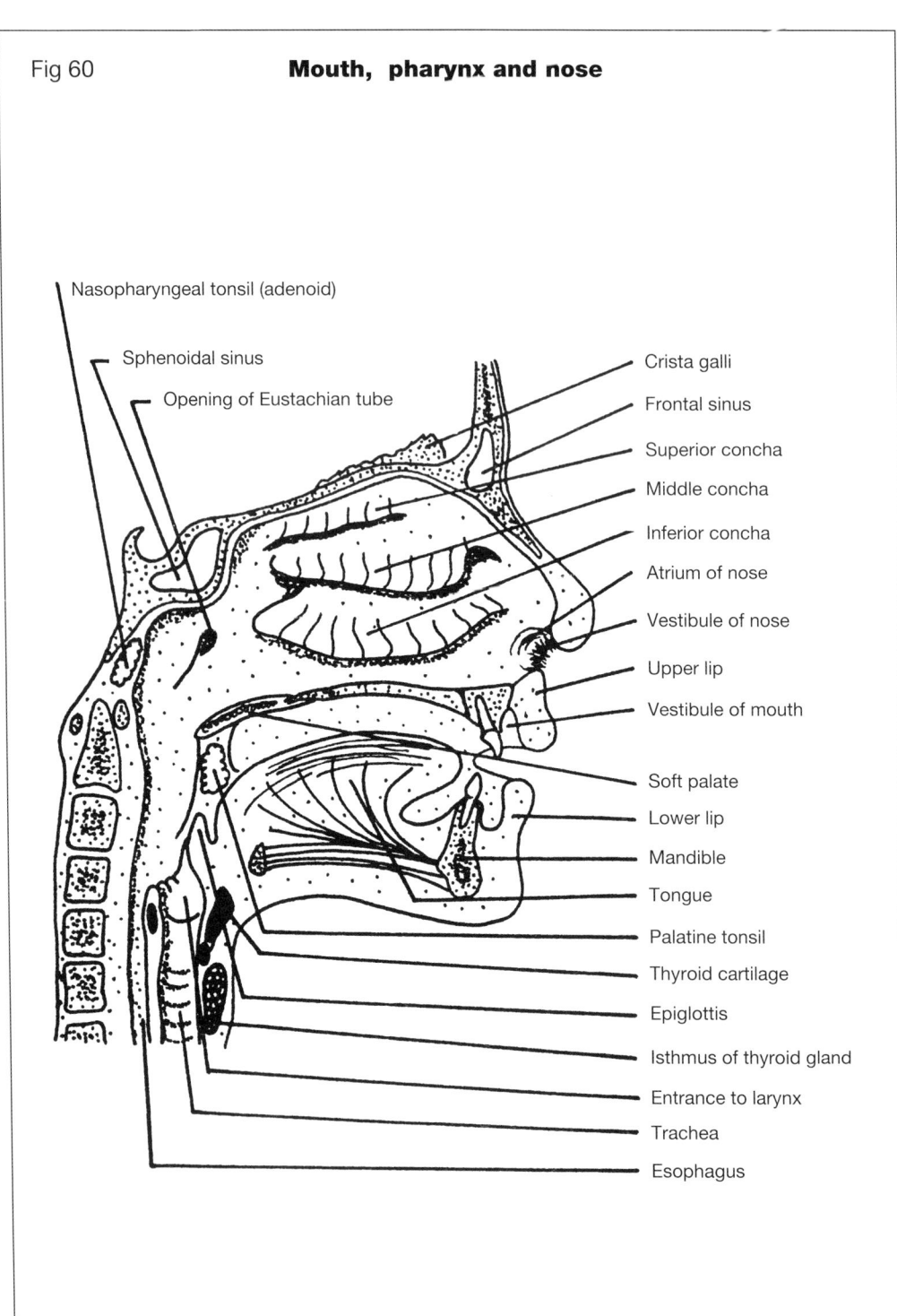

per margins of the sublingual glands which open into the mouth by means of 10-20 small ducts which pierce the overlying mucous membrane. The duct of each submandibular gland opens at the apex of a small papilla which lies below and behind the lower incisor teeth in the floor of the mouth.

The mucous membrane of the floor of the mouth is peripherally continuous with the gingiva on the internal aspect of the body of the mandible.

The retromolar area

The retromolar area extends from the distal aspect of the last mandibular molar tooth inferiorly to the distal aspect of the last maxillary molar tooth superiorly. In the mandibular part of the area, the mucous membrane and the underlying minor salivary glands form a retromolar cushion which covers the bony retromolar triangle. This cushion sometimes becomes inflamed during eruption of the third molar, especially in the case of impaction. Furthermore, this cushion and the palpable underlying bony triangle constitute an important landmark for the administration of a mandibular block injection.

The roof of the mouth

The hard and soft palates (Fig 61) constitute the roof of the mouth and are bordered anteriorly and laterally by the maxillary dental arch. Bilateral radiating ridges of mucous membrane, the rugae, are found in the more anterior part of the hard palate. These ridges are firmly attached to the underlying bone and can be used for identification purposes since an individual possesses a unique pattern of rugae, like fingerprints. The incisive papilla, which covers the opening of the incisive canal, lies in the midline of the palate immediately behind the central incisor teeth. The mucous membrane of the soft palate is attached to underlying muscles. A midline projection, the uvula, is found on the posterior margin of the soft palate. The uvula contains muscle fibers but its function, if any, is not clear. The soft palate is more vascular and sensitive than the hard palate.

The soft palate lies in contact with the dorsal surface of the tongue when at rest, and forms a seal between the mouth and the pharynx. During the swallowing process, the soft palate is elevated and makes contact with the posterior wall of the pharynx to isolate the nasal cavities from the pharynx.

The teeth and the gingiva

The upper and lower teeth, together with the supporting alveolar processes and gingiva, form the two U-shaped dental arches. There are no natural spaces between the teeth of humans.

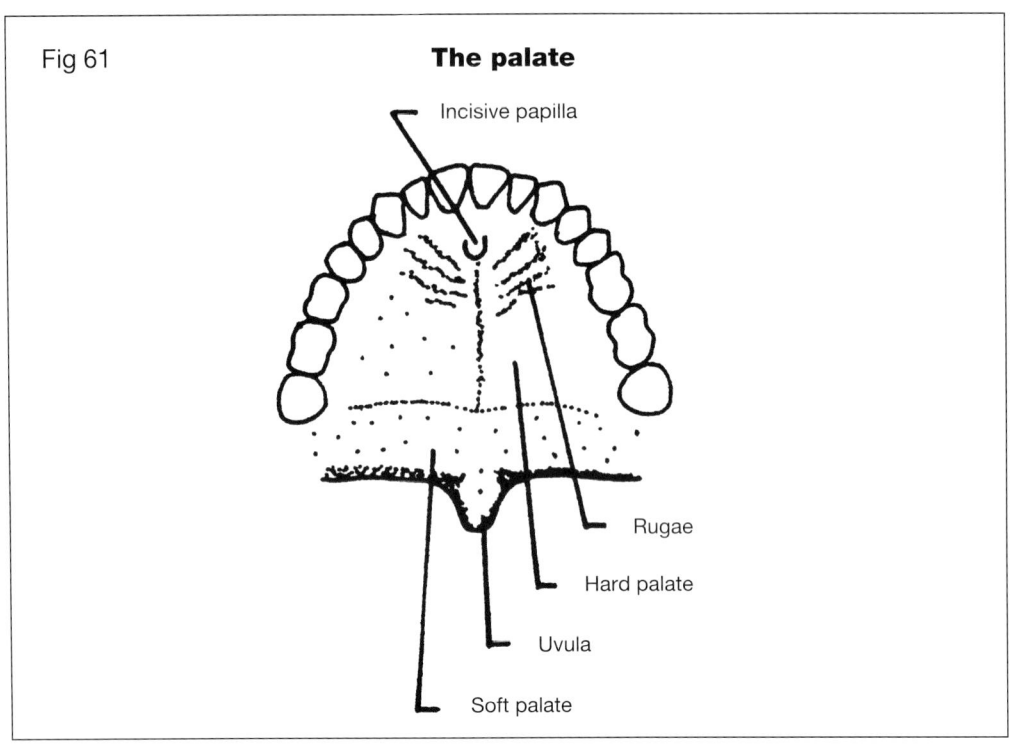

Fig 61 **The palate**

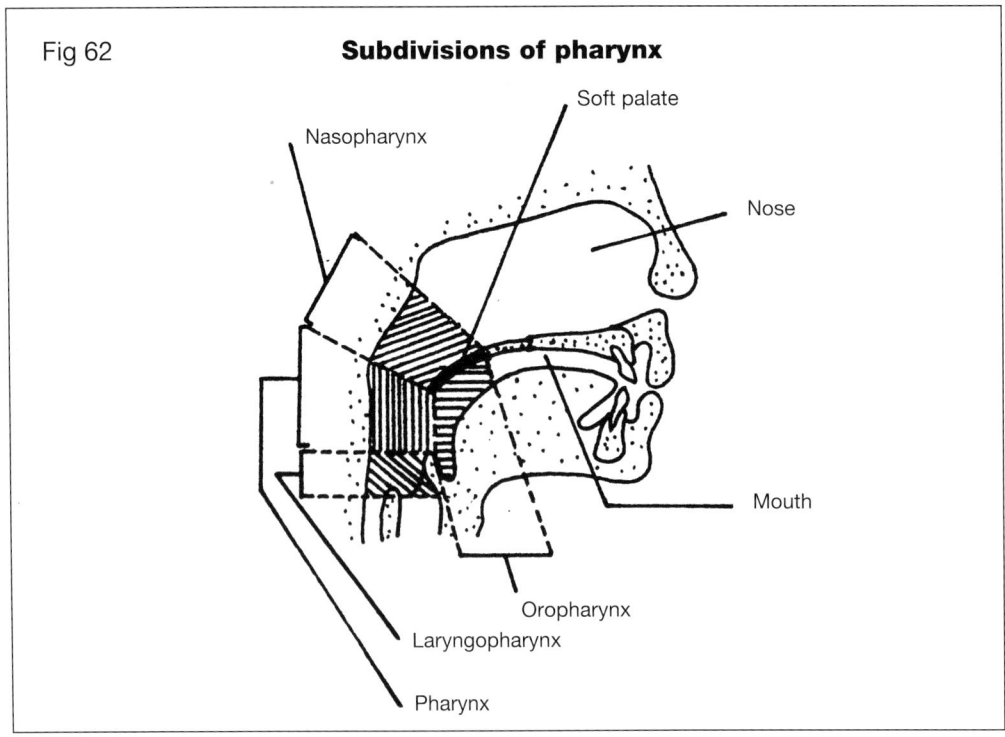

Fig 62 **Subdivisions of pharynx**

The pharynx

The pharynx is a tube-like structure which connects the mouth and nose with the esophagus and larynx below (Fig 62).

The nasopharynx is located behind the nasal cavities and is separated from the oropharynx by the soft palate during swallowing. The oropharynx lies behind the pharyngeal surface of the tongue and communicates with the oral cavity through the opening between the anterior (palatoglossal) pillars of the fauces. The laryngeal part of the pharynx lies behind the epiglottis and communicates with the openings of the larynx and esophagus.

The nose

The two nasal cavities are separated by the midline nasal septum which is composed of bone and cartilage. Various paranasal air sinuses open into the nasal cavity, for example, the frontal and maxillary sinuses. The soft anterior part of the external nose has a cartilagenous skeleton, while the bridge of the nose is formed by the nasal bones.

On its inner aspect the lateral wall of the nose can be divided into three areas from before backwards. The smooth anterior part of the cavity consists of the vestibule, which possesses hair, and the hairless atrium. The middle part of the lateral wall is divided by three horizontal turbinate processes, or conchae, into four slit-like grooves. These grooves are the inferior meatus below the inferior turbinate, the middle meatus between the middle and inferior turbinates, and the superior meatus between the middle and superior turbinates. The groove above the superior turbinate is the olfactory sulcus. The third and hindmost part of the lateral wall lies behind the turbinates and is again smooth and leads into the nasopharynx.

Selected bibliography

1. Dixon, A.D. (1986) *Anatomy for Students of Dentistry*, 5th edition. Edinburgh: Churchill Livingstone.
2. McMinn, R.M.H. (ed) (1990) *Last's Anatomy*, 8th edition. Edinburgh: Churchill Livingstone.
3. Scott, J.H. and Symons, N.B.B. (1982) *Introduction to Dental Anatomy*, 9th edition. Edinburgh: Churchill Livingstone.

Review questions

1. Describe what you would see during a thorough clinical examination of the mouth.
2. Describe the superficial anatomy of the lateral walls and roof of the nasal cavity.
3. What is the pharynx?

19. The digestive system

The digestive system (Fig 63) commences in the mouth and pharynx, and is continued as the alimentary canal, or tract, which consists of the esophagus, stomach, duodenum, jejunum, and the ileum (the latter three being known collectively as the small intestine), followed by the cecum and the colon, or large intestine, and the rectum and anal canal.

The esophagus

The esophagus courses through the neck and the thoracic cavity where it lies between the vertebral column posteriorly and the trachea in its upper part. It traverses the diaphragm and enters the abdominal cavity where it becomes continuous with the stomach. The total length of the esophagus in the adult is approximately 250 mm.

The stomach

The stomach is the most expanded part of the alimentary tract and is roughly J-shaped. It receives food through the esophagus, after which digestion commences by means of gastric juices, which contain enzymes, mucin, and hydrochloric acid. Digestion is facilitated by a churning motion caused by the stomach muscles. Nutrients are, however, not absorbed into the blood to an appreciable extent in the stomach.

The small intestine

The duodenum is a C-shaped part of the alimentary tract and is approximately 250 mm long in the adult. It receives the excretory ducts of the liver and the pancreas through which various enzymes, for example, lipase which aids in the digestion of fat, trypsin which is active in the digestion of proteins, and amylase which digests starches, enter the digestive tract. Absorption of nutrients into the blood occurs mainly in the duodenum. The jejunum and the ileum, which together measure 3 - 7 m in length and are extensively coiled in the central part of the abdominal cavity, will not be described further. The ileum is continuous with the large intestine at the cecum.

The large intestine

The cecum is the first part of the large intestine and receives the terminal part of the ileum on its left side. The vermiform appendix, a blind diverticulum of the cecum, is found approximately 2.5 cm below the ileal opening. The cecum, which lies on the right inferior side of the abdominal cavity, is continued upwards as the ascending colon, passes towards the left of the abdominal cavity as the transverse colon and then downwards as the descending colon. At the pelvic rim it becomes the pelvic colon. It then enters the rectum and finally the anus. The large intestine is 1.5 - 2 m long.

The absorption of water from the intestinal contents is a major function of the colon. The contents of the small intestine are in a highly fluid state when they reach the colon and consist of water, waste products, and secretions of the small intestine, pancreas, liver, and stomach. Should the absorption of fluid be inhibited, severe dehydration of the body may result.

Functions of the alimentary tract

The only function of the alimentary tract is the absorption of nutrients which results form the digestion of food by the action of enzymes. The stomach and duodenum are the main sites where enzymes are secreted, while nutrients are absorbed mainly in the small intestine. Fluids are absorbed from the intestinal contents in the colon. Blood vessels, lymphatics, and nerves reach the alimentary canal by means of the mesenteries. The abdominal cavity is lined by a thin mesothelium, the parietal peritoneum, while the abdominal contents are covered externally by the mesothelial visceral peritoneum. The peritoneal cavity is a potential space between these two layers.

Fig 63   **Digestive tract**

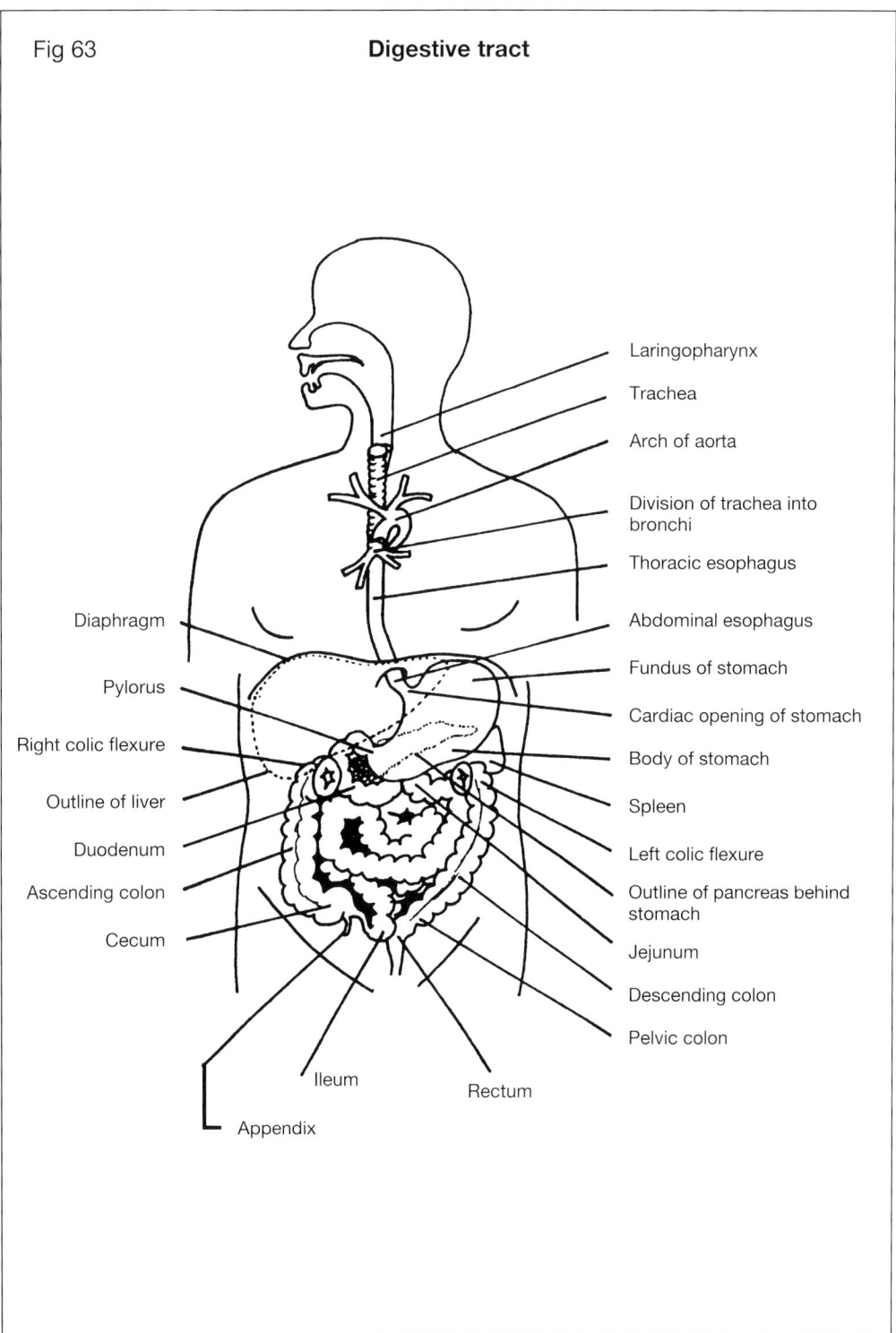

Selected bibliography

1. Dixon, A.D. (1986) *Anatomy for Students of Dentistry*, 5th edition. Edinburgh: Churchill Livingstone.
2. McMinn, R.M.H. (ed) (1990) *Last's Anatomy*, 8th edition. Edinburgh: Churchill Livingstone.

Review question

Describe the general arrangement of the digestive system.

20. The respiratory system

The functions of the respiratory system (Fig 64) are to transport oxygen-carrying air to the functional parts, or alveoli, of the lungs, the absorption of oxygen from the air, and the clearance of carbon dioxide from the blood. The respiratory system commence in the nose from where inhaled air passes through the pharynx and larynx to the trachea. The latter divides to form a left and a right main bronchus which enter the lung substance.

The trachea and the extrapulmonary bronchi

The trachea, which is approximately 12 cm long in the adult, arises in the neck and descends into the thoracic cavity on the anterior aspect, and slightly to the right, of the esophagus, and divides into the left and right main bronchi at the level of the fourth or fifth thoracic vertebra.

The right main bronchus is shorter, wider, and more vertically inclined than the left main bronchus and it is for this reason that foreign objects, such as fragments of tooth crowns or loose pieces of dental amalgam, quite readily enter the right bronchus when inhaled. This bronchus gives off a large segmental branch to the upper lobe of the lung before entering the hilum.

The left main bronchus reaches the hilum of the left lung after crossing the esophagus and does not branch as close to its point of origin from the trachea as the right bronchus.

The lungs

Each lung is attached to the mediastinum, which contains the heart, large blood vessels, esophagus and trachea, by means of a hilum through which the bronchi and pulmonary blood vessels reach the lungs. Inside the lungs the bronchi divide repeatedly to form a complex system of bronchioles which transport the air to the

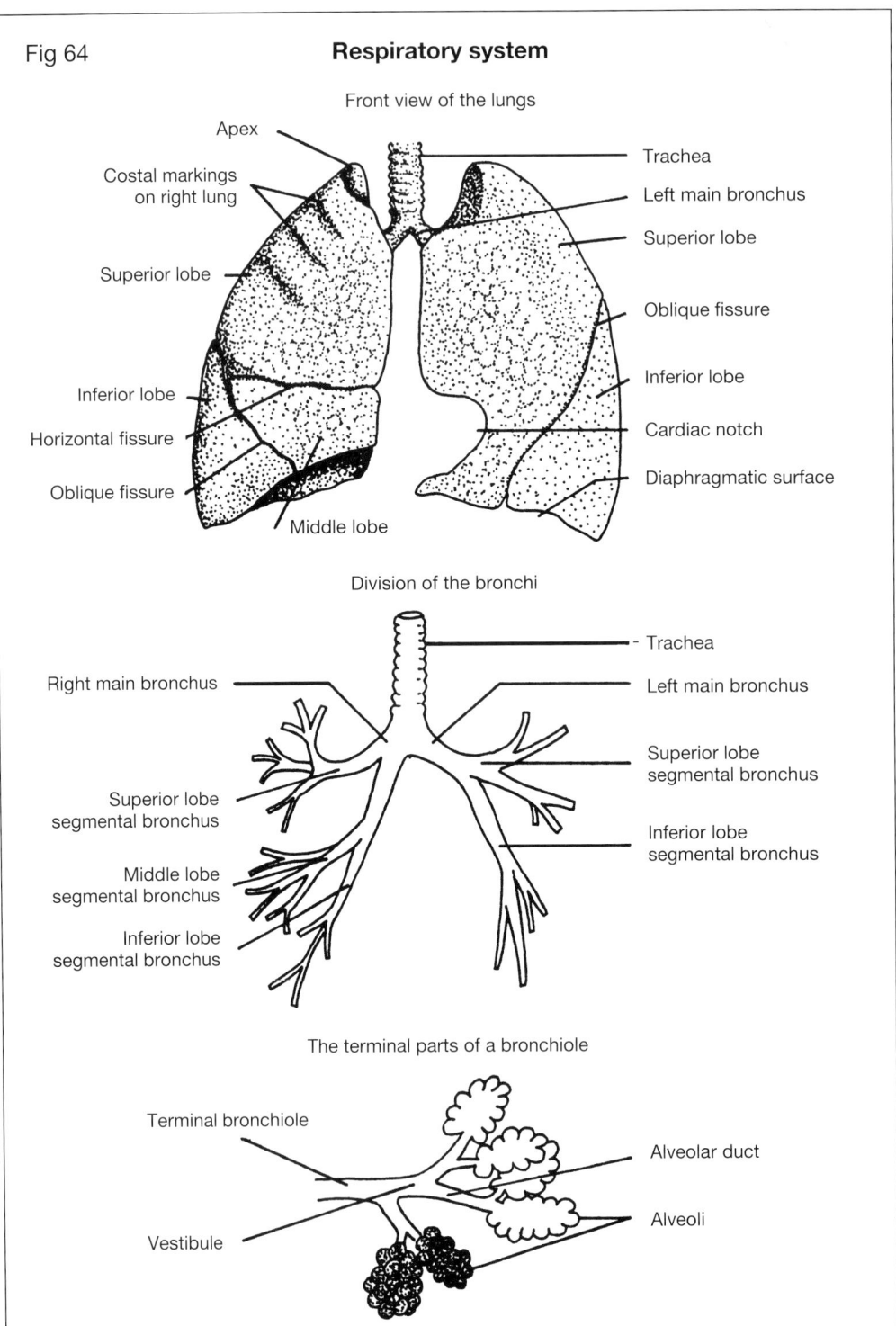

terminal lung alveoli. The larger bronchioles have cartilage rings in their walls, and in this respect resemble the bronchi and trachea, but this cartilage is lost in the smaller bronchioles. They do, however, possess more smooth muscle in their walls and this muscle has a regulatory function with regard to the flow of air to the alveoli.

The right lung is usually divided by fissures into three lobes, superior, middle, and inferior, while the left lung is divided into a superior and an inferior lobe.

The lungs are in contact with the pleura on virtually all surfaces. The parietal pleura lines the inner aspect of the thoracic cage, while the visceral pleura is closely applied to the surface of the lungs. In conditions of health there is never a separation between these two layers of the pleura and the so-called pleural cavity is only a potential space.

Air is normally inhaled into the lungs in two ways. Firstly, an upward and outward movement of the ribs increases the volume of the thoracic cage, thereby sucking air into the lungs and, secondly, a downward movement of the diaphragm has the same effect. Relaxation of all the muscles concerned in the above movements allows the lungs, which possess considerable elasticity, to return to the resting position and exhalation occurs.

When the body is at rest, blood is pumped through the lungs at a rate of 4 - 6 l/min. The alveolar capillaries are thin-walled and separated from the inhaled air by the very thin alveolar epithelium. Blood which returns to the lungs from the rest of the body has a low concentration of oxygen and a high concentration of carbon dioxide. Oxygen diffuses through the walls of the alveoli into the blood, while carbon dioxide leaves the blood in the same way to reach the air to be exhaled.

Selected bibliography

1. Dixon, A.D. (1986) *Anatomy for Students of Dentistry*, 5th edition. Edinburgh: Churchill Livingstone.
2. McMinn, R.M.H. (ed) (1990) *Last's Anatomy*, 8th edition. Edinburgh: Churchill Livingstone.

Review questions

1. Give a brief account of the anatomy of the respiratory system.
2. Briefly describe the processes involved in normal respiration.

21. Introduction to genetics

Chromosomes and genes

The physical link between successive generations consists of two cells, the male and female gametes, the spermatozoon, and the ovum, respectively. At fertilization the gametes form a single cell, the zygote. This is a first cell of a new individual. Although the spermatozoon is a smaller cell with much less cytoplasm than the ovum, it has long been recognized that the genetic influence of the two parents on an offspring is of equal magnitude and that the hereditary factors are confined to the nuclei. These factors were termed genes. Genes are arranged in orderly fashion on the chromosomes which are situated in the nuclei of all cells. A gene is a specific nucleotide sequence in a long series of neucleotides of the DNA of the chromosome.

The human karyotype (chromosome set)

The nuclei of the spermatozoon and ovum contribute the same number of chromosomes to a zygote. The number of chromosomes found in the gametes of specific animals is termed the haploid number, while those in a zygote constitute the diploid number. The diploid, or modal, number in a human zygote is 46, which also represents the number of chromosomes generally found in all human tissue cells, or somatic cells.

A study of human diploid chromosomes reveals that all the chromosomes, except two (in the male), are present in morphologically matching pairs. Chromosome pairs differ from other pairs in respect of size and shape. Chromosomes belonging to a pair are termed homologous chromosomes.

It is customary to designate numbers to chromosomes and this numerical classification is based on the relative lengths of arms, total size and often more subtle differences.

In a human female somatic cell all the chromosomes belong to distinct pairs, while two chromosomes in the male are different from one another. The smallest of the latter two chromosomes is the Y (male) chromosome, while the other is the

Fig 65 **The human chromosome set**

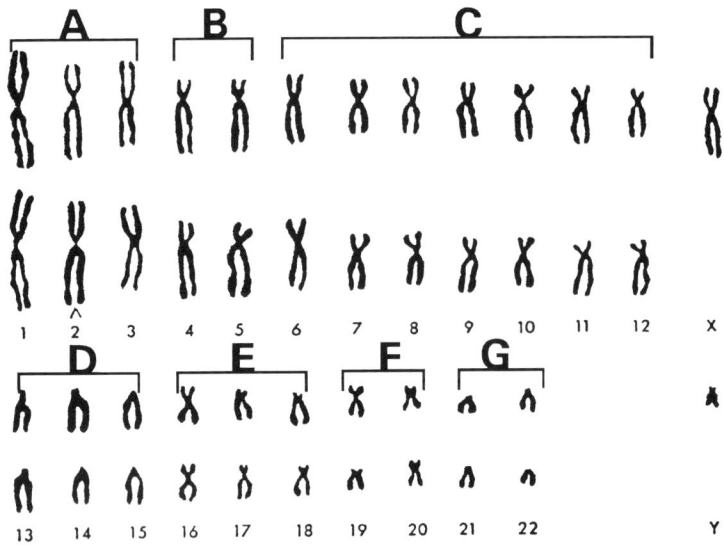

Arrangement of human chromosomes according to the Denver classification

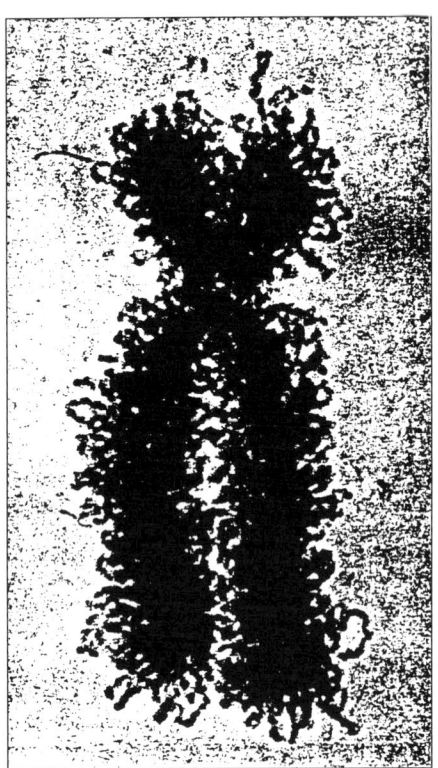

Electronmicrograph of chromosome number 12

X (female) chromosome. These are the sex chromosomes, which are represented in a female cell by a matching pair of X chromosomes and in the male by an XY configuration. The ordinary chromosomes are termed the autosomes. There are therefore 22 pairs of autosomes, plus a 23rd pair of sex chromosomes, in a human somatic cell. The way in which the gametes transfer their sex chromosomes to a zygote determines the sex of a new individual.

When observed in the metaphase of mitosis, when all the chromosomes are duplicated, a chromosome consists of a centromere, or primary constriction, and chromatid arms on both sides of the centromere. If these arms are of equal length, the chromosome is said to be metacentric. On the other hand, should the centromere not be in the middle of the chromosome, the arms would be of different lengths and the chromosome is termed submetacentric. If the constriction is located at the end of the chromosome, such a chromosome is acrocentric. According to the Denver system, chromosomes are classified according to the lengths of the arms and the position of the centromere. By this method, chromosomes are classified into seven groups (a - g) consisting of 22 pairs of homologous chromosomes, plus two sex chromosomes. The 22nd pair are the smallest (Fig 65).

Gene loci and alleles

A person possesses a pair of each type of chromosome, except the sex chromosomes in the male, and each chromosome has a number of loci (sites) which are occupied by different genes. Each chromosome has the same arrangement of genes as its homolog, and therefore every person has two of each kind of gene. These genes occupy homologous loci on homologous chromosomes. A locus is a specific position which a gene's unique nucleotide sequence occupies on a chromosome.

Genes which occupy homologous loci on homologous chromosomes are termed alleles, or allelomorphs, and may be identical or different. The term *allelomorph* is derived from the Greek words *allelon*, meaning "of one another", and *morphe*, meaning "form". An allelomorph is one of several, but related, genes which are inherited from the parents, or one of a group of related genes which can replace one another at the same locus. Alleles influence the same characteristic in an identical or different way. Alleles which have an identical effect are also termed iso-alleles.

A character or trait

A character or *trait* (from the French word meaning "a distinguishing feature") may be defined as any observable feature or characteristic of a developed individual and may include a biochemical characteristic, a cellular process, an anatomical feature, an organ function, or mental characteristic.

A person's characteristics, as defined above, are not determined exclusively by that person's genes, but are influenced to a greater or lesser extent by the environment in which development occurs. The gene content of a person's cells is termed the genotype of that person, while phenotype refers to the outcome of the interaction between that specific genotype and the environment. The latter term could refer to appearance or character. The genotype of a person is fixed at fertilization but the phenotype is potentially variable and depends on a life-long process of interaction between the genotype and the non-genetic environment. In the case of some characteristics, for example, blood group, eye color, and others, the limits between which variations can occur are extremely limited and no known environmental factors influence development. The outcome is determined solely by the genotype.

Most characteristics are the result of a complex interaction between many different genes, but a single gene may influence the development of more than one characteristic. Such a gene which influences more than one characteristic is termed pleiotropic, and the phenomenon is known as pleiotropism.

Phenocopy is a term used when the phenotype of an individual has been influenced by the non-genetic environment to such an extent that it resembles the phenotype of another individual with a different genotype but in whom environmental factors have not played any roll. For example, a child born to a mother who contracted rubella infection during early pregnancy may suffer from deafness and other congenital abnormalities which may mimic genetically determined varieties of the same conditions.

Homozygotes and heterozygotes

An individual is a homozygote for a particular gene when he possesses an identical allelomorph on both members of a chromosome pair. A person is a heterozygote for a specific gene when the allelomorphs are not identical. A person whose blood group is indicated by the letter GAGA (A group) is therefore a homozygote for these blood group genes, while a person whose blood group is indicated as being GAGB (AB group) is heterozygous.

Dominance, recessivity, and co-dominance

These three terms may be explained by making use of blood group inheritance. When a blood group A is the result of the two alleles GAG, the person is a heterozygote for the A gene, while the G, for practical purposes, represents the gene for an O group. In this instance the GA is dominant and determines the phenotypic expression in the same way as if the genotype were GAGA. The G appears inactive and is termed recessive. The G gene only expresses itself phenotypically when it is present in the homozygous form GG (O group). Similarly, GB is dominant when the other allele is G and the heterozygote GBG results in the blood group B.

When the heterozygote GAGB is considered, a new relationship results. Both alleles manifest dominantly and the result is a blood group AB. This phenomenon is termed co-dominance.

Situations may also be present where dominance is incomplete or where the expression of one allele is completely unaffected by the other.

Inheritance which is bestowed by genes linked to autosomes is termed autosomal inheritance.

Human sex chromosomes and sex determination

The nuclei of human somatic cells, for example, cells of the buccal mucosa, show sex differentiation in interphase. This is shown by the so-called Barr body which is found in 30 - 50% of cells of females. It is, however, unobtrusively present in all cells. The Barr body is a small particle (approximately 1μm in diameter) present on the inner aspect of the nuclear membrane, and cells in which it can be seen are termed sex chromatin positive. Various factors influence the visibility and size of this body, as well as the number of Barr bodies present. The percentage of cells showing the Barr body is reduced in pregnant females for a short period before and after birth of a child; it is absent in newborn females, and corticosteroid therapy reduces their incidence and size.

Each Barr body is thought to represent one X chromosome. When two X chromosomes are present (the normal female configuration), one X chromosome remains partially inactivated and condensed (the Barr body) while the other is essential for metabolic activities other than for sex determination and remains unravelled.

The sex of an individual is determined at the time of fertilization. A male produces spermatozoa which possess either an X or a Y sex chromosome, while the female ovum can possess only an X sex chromosome. The male is therefore a heterogamete, while a female is a homogamete. If a spermatozoon with an X chromosome fertilizes an ovum, the result is a female embryo, while a male embryo will result from the combination of the Y and X chromosomes during fertilization.

Sex-linked inheritance

Hemophilia, a sex-linked recessive trait, is a well-known example of a sex-linked inheritance. It is a defect of the blood-clotting process and is caused by a deficiency of antihemophilic globulin (Factor VIII) in the blood plasma and consequently an inability to form intrinsic thromboplastin. It is a serious condition in which a relatively minor injury, for example, a tooth extraction, may result in a major hemorrhage.

A male has a single X chromosome. The gene which predisposes to hemophilia is linked to this chromosome. If he possesses this gene he will be a "bleeder" since he possesses no "normal" dominating allele. If the gene for hemophilia is absent he will be completely free of the condition. A female who has one affected X chromosome is a carrier for the condition and does not clinically manifest as a hemophiliac.

The inheritance of hemophilia is illustrated by the following examples:
(The affected X chromosomes are indicated by small print)

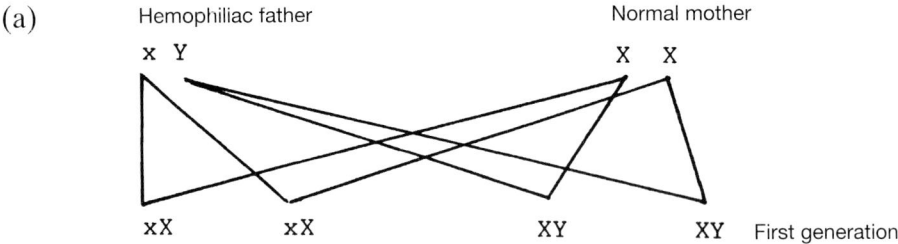

The first generation produces two carrier daughters in whom the recessive gene for hemophilia is dominated by the non-hemophiliac allele, and two sons who are completely free of the condition.

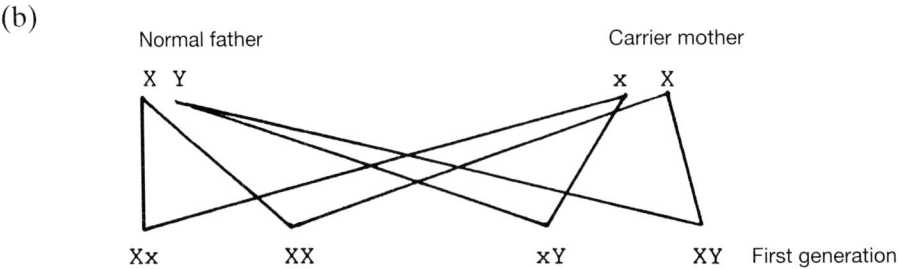

In this example, the first generation produces one carrier daughter, one normal daughter, one hemophiliac son, and one normal son.

Because of the rarity of the condition it is unlikely that two families in which hemophilia occurs will inter-marry. If this should happen, the following may occur:

(c)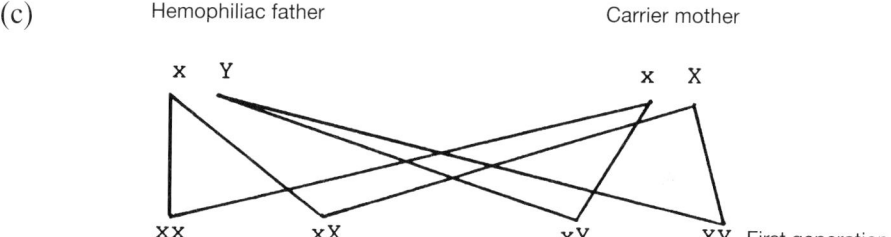

The first generation may produce one hemophiliac daughter, one carrier daughter, one hemophiliac son, and one normal son. It is therefore possible for females to be born with this defect.

Selected references

1. Cole, A.S. and Eastoe, J.E. (1988) *Biochemistry and Oral Biology,* 2nd edition. London: Wright.
2. Harrison, G.A., Weiner, J.S., Tanner, J.M. and Barnicott, N.A. (1977) *Human Biology,* 2nd edition. London: Oxford University Press.
3. Kelly, D., Wood, R.L. and Enders, A.C. (1984) *Bailey's Textbook of Microscopic Anatomy,* 18th edition. Baltimore: Williams and Wilkins.
4. Krause, W.J. and Cutts, J.H. (1986) *Concise Text of Histology,* 2nd edition. Baltimore: Williams and Wilkins.
5. Osborn, J.W. (ed) (1981) *Dental Anatomy and Embryology.* Oxford: Blackwell Scientific Publications.
6. Singer, S. (ed) (1978) *Human Genetics.* San Francisco: W.H. Freeman and Wilkins.

Review questions

1. Give a brief description of human chromosomes with reference to the following:
   (a) the haploid and diploid number;
   (b) the appearance of a chromosome in metaphase; and
   (c) homologous loci and homologous chromosomes.
2. Explain the terms genotype, phenotype, and phenocopy.
3. Give examples of genetic dominance, co-dominance and recessivity.
4. Write brief notes on the Barr body.
5. How is the sex of an embryo determined?
6. Give an example of sex-linked inheritance.

22. Blood groups

General remarks

When red blood corpuscles from one person are mixed with the serum of another individual, the red cells become clumped, or agglutinated, in some combinations. Landsteiner is reported to have described experiments in 1900 in which the above reactions were noted, and he was able to distinguish different groupings which we now identify as A, B, and O blood groups. A few years later the more rare AB group was recognized. These findings made safe blood transfusions possible and initiated the science of serology which plays an important role in clinical medicine, anthropology, genetics, and forensic identification.

Ten years after Landsteiner's findings it was determined that blood groups are inherited, but only 14 years later was the exact mechanism of inheritance of the ABO groups established. Many other blood group systems have been identified since the abovementioned discoveries, each of which is hereditary and independent of the ABO system. Blood group inheritance strictly obeys Mendelian laws and is of great value in cases of disputed paternity, in forensic science, and in the field of population genetics.

For many years it was believed that whole blood is the best transfusion since it restores blood volume as well as replenishing the red blood corpuscle count. Blood may, however, not be transfused indiscriminately from one person to another. The donor must be carefully selected otherwise the results could be fatal for the recipient, even though the individuals involved may belong to the same sex or family. A serious reaction may follow should the plasma of the recipient contain an agglutinin, or antibody, while the red blood corpuscles of the donor possess a reactive agglutinogen, or antigen. When unsuitable blood is used in a transfusion, the red blood corpuscles of the donor can agglutinate with serious, or even fatal, consequences. Blood groups which react in this way are termed incompatible.

The following table shows the reaction between sera and red blood corpuscles of different blood groups:

	Serum			
	O	A	B	AB
O	−	−	−	−
A	+	−	+	−
B	+	+	−	−
AB	+	+	+	−

Red blood corpuscles

+ = agglutination
− = no agglutination

The serum of group AB does not contain any antibodies (agglutinins) against the antigens (agglutinogens) of the red blood corpuscles of any other group. The red blood corpuscles of group O do not possess any agglutinogens and will not be agglutinated by the serum of any other group.

In summary, it may be stated that Group AB possesses both agglutinogens (A and B) on the red blood corpuscles, but no agglutinins (anti-A and anti-B) in the serum; group O possesses no agglutinogens, but both agglutinins; group A possesses agglutinogen-A and agglutinin anti-B, and group B possesses agglutinogen-B and agglutinin anti-A.

Although group O has in the past erroneously been regarded as a universal donor group, and group AB as a universal recipient group, such transfusions are never without risk. Group O blood possesses no agglutinogens on the red blood corpuscles and cannot, when donated, react with agglutinins of A or B, or AB blood groups because of their absence. In a relatively small transfusion of the above kind, the agglutinins of group O are diluted to such an extent in the blood of the recipients that no reaction will ensue. A large transfusion of whole blood of the O group can, however, result in a severe reaction. Group O is, therefore, strictly speaking not a universal donor group.

Group AB blood possesses both agglutinogens, but no agglutinins. Should a person with this blood group receive a small amount of foreign blood (A with agglutinin anti-B in the serum, or B with agglutinin anti-A), no reaction is likely to follow since the transfused agglutinins are diluted. A large transfusion of either of these two groups can, however, severely harm the red blood corpuscles of the recipient.

Tests for blood grouping can be simply illustrated as follows:

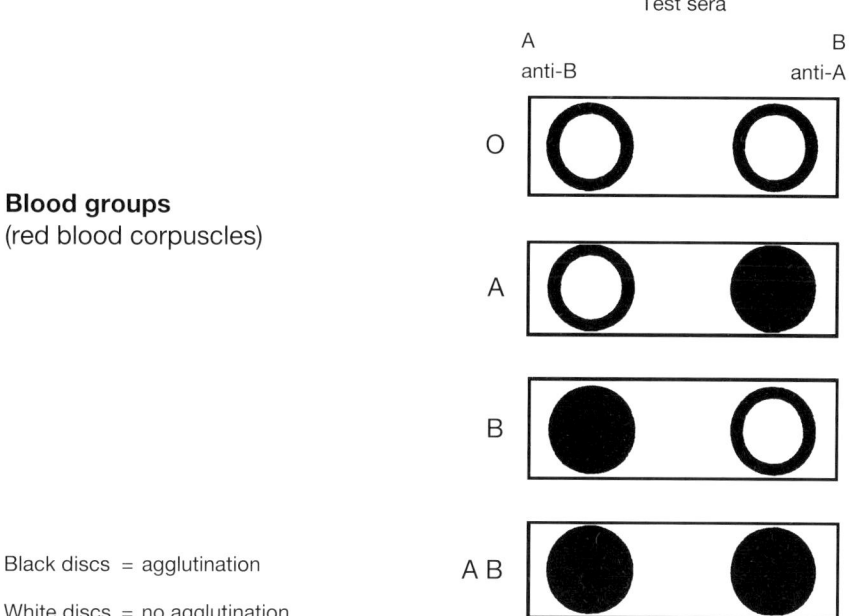

Blood groups
(red blood corpuscles)

Black discs = agglutination
White discs = no agglutination

Should not one of the test sera (of known agglutinin status) lead to agglutination, then the red blood corpuscles tested belong to group O (no agglutinogens). Should A serum agglutinate the red blood corpuscles, then they belong to group B; should B serum cause agglutination, then the corpuscles belong to group A, and if agglutination occurs in both the sera the red blood corpuscles belong to group AB.

The agglutinogens A and B do not, however, only occur on the cell membranes of red blood corpuscles but also in organs and in associated body fluids such as the salivary glands (saliva), the testes (semen), the kidneys (urine), and in the pancreas and liver.

The Rh (Rhesus) factor

Although cognisance was taken of blood groups, fatalities did occur after blood transfusions in the early days of this practice. Upon investigation, it was found that fatalities or serious complication occurred mainly in persons who had received a previous transfusion, or in women who had given birth to a child.

Research into the above phenomenon was based on findings by Landsteiner and

Wiener in 1940 that the injection of red blood corpuscles of the Rhesus monkey into rabbits resulted in the production of an agglutinin which was able to agglutinate the red blood corpuscles of the monkey as well as of a significant percentage of humans. The red blood corpuscles of a Rhesus monkey possess an agglutinogen which is able to stimulate the production of an agglutinin in the blood of a recipient who does not possess the agglutinogen.

Landsteiner and Wiener are also reported to have deduced that approximately 85% of white Americans possess this agglutinogen, or Rh factor (named after the Rhesus monkey). The red blood corpuscles of the remaining 15% do not possess the antigen. Blood which contains the Rh factor is termed Rh-positive, while other blood is Rh-negative.

When whole blood from an Rh-positive person is transfused into an Rh-negative person, the agglutinin, or anti-Rh factor (antibody) is produced in the serum of the recipient. This first contact merely causes the production of the agglutinin (immunization) and it is unusual for any reaction to occur. When, however, a second similar transfusion is performed, the consequences may be serious due to the action of the recipient's agglutinin on the agglutinogen of the transfused red blood corpuscles.

An Rh-negative mother who is bearing a first Rh-positive child is similarly stimulated to produce the agglutinin (anti Rh-factor) in her serum by red blood corpuscles or cell fragments crossing the placenta from the fetus. No reaction usually occurs. With a second Rh-positive child, however, the mother's anti-Rh factor may damage the red blood corpuscles of the child by a hemolytic process which may lead to prenatal death due to a severe hemolytic anemia. In less severe cases, the newborn child may suffer from anemia and severe jaundice, the so-called congenital hemolytic anemia, or hemolytic disease of the newborn. Treatment involves a total replacement of the child's blood by transfusion at birth. Fatalities are high in untreated cases.

In children who survive, blood pigment is deposited in the enamel and dentine of the developing primary teeth, causing a blue, brown, or green hue.

In most cases where an Rh-negative mother becomes sensitized by an Rh-positive fetus, the D-antigen is involved. This is the most reactive antigen of the Rh-positive group (at least five others are known).

The Rh factor is inherited as a dominant character, therefore an Rh-negative person can only have a dd genotype (homozygous Rh-negative). On the other hand, an Rh-positive person may have a DD or Dd genetic configuration, i.e. homozygous Rh-positive or heterozygous Rh-positive.

When both parents are homozygous Rh-positive there is, of course, no possibility of any reaction. When the mother is homozygous Rh-positive (DD) and the father homozygous Rh-negative (dd) there is again no reaction because the embryo must be Rh-positive (Dd). When the mother is Rh-negative (dd) and the father is Rh-positive (either Dd or DD), then the following combinations are possible:

(a)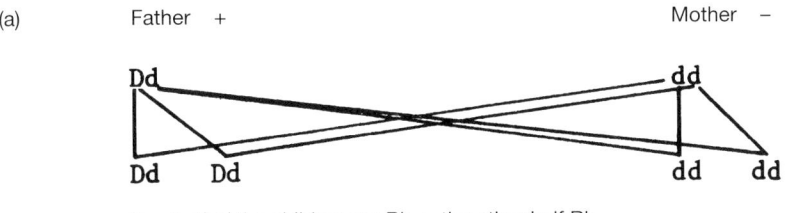

One half of the children are Rh +, the other half Rh -

(b)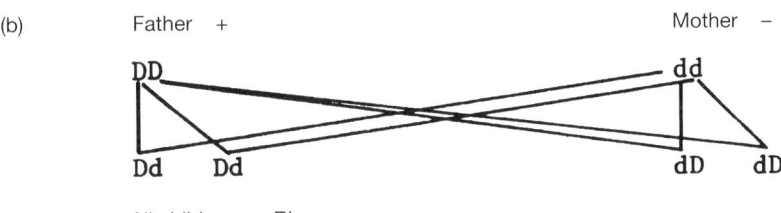

All children are Rh +

It is possible for an Rh-negative fetus to develop antibodies against an Rh-positive mother but the effects are slight.

Other minor blood group systems which have been described (more than 15 probably exist), include the MN system, the Henshaw and Hunter antigens, the Diego, Duffy, Kell, P, Lewis, and Lutheran systems, and the Gm-groups.

Some aspects of the genetics of blood

The ABO system

In order to distinguish between phenotypes and genotypes, three genotypes are indicated, namely GA, GB, and G. Combination of these genotypes in pairs result in six phenotypes, GAGA and GAG (phenotype A), GBGB and GBG (phenotype B), GAGB (phenotype AB) and GG (phenotype O). One cannot distinguish between phenotypes A and B, since a serum with anti-G agglutinin is not available.

The serology of the ABO system may be explained by the following table:

Phenotype	Genotype	Cell antigens	Serum antibodies
A	GAGA, GAG	A	anti-B
B	GBGB, GBG	B	anti-A
O	GG	–	anti-A, anti-B
AB	GAGB	A, B	none

Distribution of the ABO groups

In most populations the G-gene has a frequency of 50% or more. World-wide frequencies are estimated to be more or less as follows: GA (20%), GB (15%), G (62%) and GAGB (3%). American Indians have the highest frequency of G, while GA and GB are apparently completely absent in some tribes.

The ABO system does not play a significant role in natural selection, although it has been stated, for example, that group O persons are more susceptible to duodenal ulcers, while group A individuals are more resistant to this condition.

Selected bibliography

1. Harrison, G.A., Weiner, J.S., Tanner, J.M. and Barnicott, N.A. (1977) *Human Biology*, 2nd edition. London: Oxford University Press.
2. Osborn, J.W. (ed) (1982) *Anatomy, Biochemistry and Physiology.* Oxford: Blackwell Scientific Publications.
3. Singer, S. (1978) *Human Genetics.* San Francisco: W.H. Freeman and Company.
4. Stern, C. (1973) *Principles of Human Genetics.* San Francisco: W.H. Freeman and Company.
5. West, J.B. (ed) (1991) *Best and Taylor's Physiological Basis of Medical Practice*, 12th edition. Baltimore: Williams and Wilkins.

Review questions

1. Describe the reactions between the sera and the red blood corpuscles of different blood groups.
2. Write a short essay on the Rh factor.

23. Hemostasis

General remarks

Hemostasis is the arrest of bleeding. It may be brought about by the following factors:

1. External mechanical pressure.
2. Vascular spasm. This is a reflex vasoconstriction caused by the contraction of smooth muscle fibers in the walls of the vessel on either side of the injury. Pain impulses initiate this reaction via the sympathetic nervous system (the vascular response).
3. Formation of a platelet plug. Damage to the endothelium of a blood vessel causes the platelets to aggregate and to adhere to exposed foreign surfaces, notably to exposed collagen fibers, forming a plug.
4. Release of serotonin (5-hydroxytryptamine). Within 1 - 3 seconds of an injury, the clumped platelets produce various active substances, including serotonin, which causes further aggregation of platelets as well as constriction of vessels.
5. Clotting of the blood (Fig 66). This is discussed subsequently.

Blood clotting

When a blood vessel is damaged to such an extent that bleeding occurs, thromboplastin is formed via intrinsic and extrinsic systems, or pathways. It is perhaps more correct to talk about "thromboplastic activity", since the process whereby this activity develops is very complex and involves a large number of factors present in blood plasma in small amounts. The term "thromboplastin" does not refer to a single substance alone but will be used for the sake of convenience. Intrinsic thromboplastin is formed by factors present in the plasma upon contact with foreign surfaces, such as damaged tissue, while extrinsic thromboplastin formation is initiated by the release of thromboplastic precursors from damaged tissues.

The following factors are involved in the production of intrinsic thromboplastin: Factor XII (*Hageman* or *Contact Factor*) is activated and this leads to progressive activation of Factors XI (*plasma thromboplastin antecedent,* or *PTA*) and IX (*plasma thromboplastin component* (PTC), or *Christmas Factor*). In the presence of Factor VIII (*antihemophilic Factor*) and platelets, Factor X (*Stuart - Prower Factor*) is activated. In the presence of Ca++ ions (Factor IV) and plasma Factor V (*Labile Factor, Pro-accelerin or Accelerator globulin*), full thromboplastic activity results within 12 minutes.

Extrinsic thromboplastin production is initiated by the release of precursors by damaged tissues. These precursors react with Factor VII (*Proconvertin, serum prothrombin conversion accelerator, stable factor*) and result in activation of Factor X. In the presence of plasma Factor V and Ca++ ions, thromboplastic activity results within 20 minutes. The normal blood calcium level is more than adequate to enable normal blood clotting to take place.

In the presence of Ca++ ions, thromboplastin activates prothrombin, an inactive precursor, to form the enzyme thrombin. Prothrombin is an inactive circulating globulin which is formed in the liver and is present in the blood in a concentration of 0.4 g/l. The plasma concentration of prothrombin (Factor II) decreases after hepatectomy. Its formation is dependent upon an adequate supply of Vitamin K. Thrombin is an albumin which is not present in the circulating blood.

Thrombin catalyses the conversion of fibrinogen (Factor I), a soluble protein formed in the liver, into fibrin which is insoluble. In clotting blood, fibrin is found in the form of a loose network in which cellular elements of the blood can be caught up. The fibrin strands are very sticky and adhere to one another, to the tissues, and to other material (like a bandage), and form an excellent hemostatic agent. It takes 4 - 12 seconds from the activation of prothrombin to the appearance of fibrin.

A fresh blood clot is soft and jelly-like. It gradually hardens and contracts to slightly less than half its original size and serum is expressed in the process. The main difference between plasma and serum is the presence of fibrinogen in plasma, which can therefore coagulate. This is absent in serum.

Bleeding time

The bleeding time is determined by pricking the finger-tip. When a droplet of blood appears, it is dabbed lightly with filter paper until the filter paper is no longer discolored by the blood. The bleeding time is the period from the appearance of the droplet of blood to the moment the paper is no longer discolored. This is normally 2 - 6 minutes (range 1 - 9 minutes).

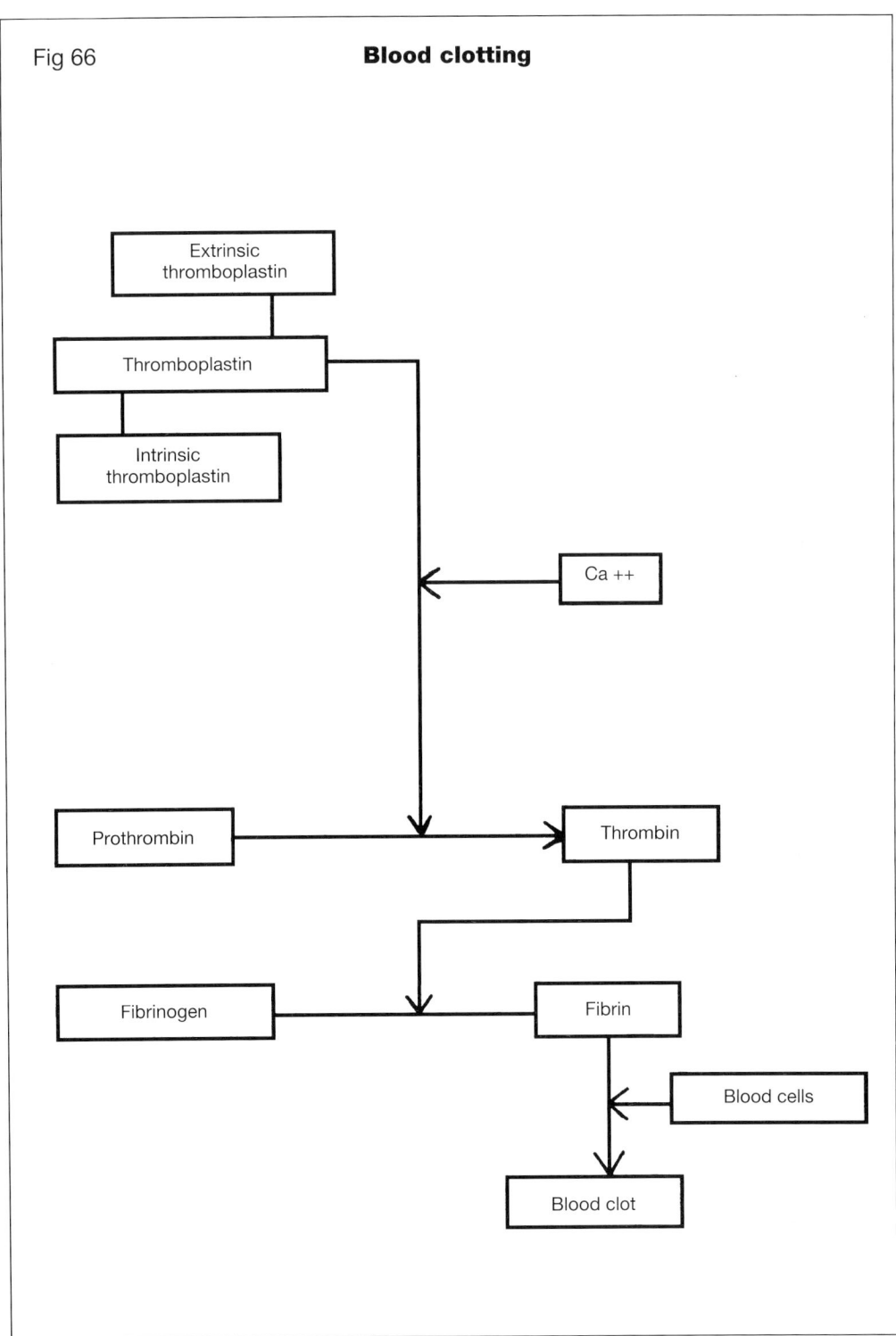

Clotting time

The clotting time is the length of time taken for blood to show the first signs of co-agulating from the moment it is collected. For the purpose of determining the clotting time, a small test tube which has been lightly lubricated on the inside with liquid paraffin, may be used. The blood is collected in this tube and gently tilted from side to side every 10 seconds. It is assumed that clotting has commenced when the contents are no longer liquid, a period of 5 - 10 minutes.

Blood clotting factors

I	Fibrinogen – forms fibrin
II	Prothrombin – forms thrombin
III	Thromboplastin (extrinsic) – catalyses the activation of Factor X
IV	Calcium ions – co-factor in the formation of thrombin and fibrin
V	Labile factor (Pro-accelerin, Accelerator globulin) – co-factor for Factor X
VII	Proconvertin (Serum prothrombin conversion accelerator, Stable factor) – co-factor for Factor X
VIII	Antihemophilic Factor – co-factor for Factor X
IX	Plasma thromboplastin component (Christmas Factor) – co-factor for Factor X
X	Stuart-Prower Factor (Intrinsic Thromboplastin – Factor) in conversion of prothrombin to thrombin
XI	Plasma thromboplastin Antecedent – factor in formation of Factor IX
XII	Hageman Factor (Contact Factor) – factor in formation of Factor XI
XIII	Fibrin-stabilizing Factor – catalyses the formation of fibrin

Selected bibliography

1. Cole, A.S. and Eastoe, J.E. (1988) *Biochemistry and Oral Biology,* 2nd edition. London: Wright.
2. Ganong, W.F. (1989) *Review of Medical Physiology,* 8th edition. Norwalk, Connecticut: Appleton and Lange.
3. West, J.B. (ed) (1991) *Best and Taylor's Physiological Basis of Medical Practice,* 12th edition. Baltimore: Williams and Wilkins.

Review question

Discuss the coagulation of blood.

24. The cardiovascular system

The heart and major blood vessels

The circulatory system consists of the heart, vessels (the arteries) which transport blood away from the heart, vessels (the veins) which return the blood to the heart from the tissues and organs, and a capillary network where exchange of nutrients, oxygen, carbon dioxide, waste products, hormones, and cellular elements takes place between the blood and the tissues. Movement of the blood is caused and maintained by contractions of the cardiac muscles.

The heart

The heart (Fig 67) is slightly assymetrically placed in the middle of the thoracic cavity where it is located more to the left side. It is enclosed by a pericardial sac which is partially covered on both sides by the pleura, and is connected to the great vessels which leave and enter the heart. It is attached to the diaphragm below.

The heart has four chambers: the right and left atria which receive veins, and the right and left ventricles which give origin to the aorta and the pulmonary arteries. The atria are relatively thin-walled, since their main function is to pump blood into the respective ventricles and they do not work against a large resistance. The walls of the ventricles are substantially thicker and more muscular. The right ventricle pumps blood to the lungs. The walls of the left ventricle are the thickest, since it is responsible for pumping blood to the rest of the body. Heart valves guard the atrio-ventricular openings as well as the openings of the pulmonary artery and the aorta.

Blood vessels of the heart

The blood vessels of the heart itself are the coronary arteries. There are two coronary arteries, a right and a left coronary artery, which arise from the aorta immediately outside the aortic valve. The right coronary artery supplies the whole of the right ventricle, except a small part of its anterior wall, as well as the back part of the interventricular septum and a part of the posterior wall of the left ventricle. The left coronary artery supplies the major part of the left ventricle, the anterior part of the septum, and a part of the anterior wall of the right ventricle. Blood is drained from the heart by the coronary veins which open into the right atrium.

The blood circulation after birth and the functions of the heart

Venous blood from the upper limbs and the head reaches the heart via the superior vena cava, while blood from the lower parts of the body returns to the heart via the inferior vena cava. Both these great veins enter the right atrium.

The pulmonary artery arises in the right ventricle and divides into a left and a right branch. These are responsible for transporting deoxygenated blood to the lungs where oxygen is taken up. Oxygen-rich blood is subsequently returned to the left atrium via the pulmonary veins.

The left ventricle pumps oxygenated blood to the organs and tissues through the aorta, which initially courses upwards before forming the aortic arch to the left, and then descends in the mediastinum. Important branches of the aorta are the following:

1. Two coronary arteries
2. The brachiocephalic artery. This artery, which courses to the right side of the body, commences in the arch of the aorta and subsequently divides into a right subclavian artery and a right common carotid artery. These branches respectively supply the right upper limb and the right side of the head.
3. The left common carotid artery which supplies the left side of the head.
4. The left subclavian artery to the left upper limb.

The aorta descends in the mediastinum as the thoracic aorta, passes through the diaphragm, and is then termed the abdominal aorta which eventually forms two main branches, the common iliac arteries, which supply the lower limbs. Various other branches arise in the thoracic and abdominal cavities to supply organs and tissues in these cavities.

Returning venous blood is received by the right atrium which contracts and pumps the blood through the right atrioventricular valve into the right ventricle. When the right ventricle contracts, this tricuspid valve closes and the blood in the ventricle is compressed for a moment before the pulmonary valve opens to allow the blood to enter the pulmonary artery. The left atrium, which contracts simultaneously with the right atrium, pumps oxygen-rich blood received from the lungs through the mitral valve, a bicuspid valve, into the left ventricle. The left ventricle contracts at the same time as the right ventricle and the aortic valve opens to allow the blood to escape into the aorta under considerable pressure. Both the pulmonary valve and the aortic valve each have three cusps. The aorta stretches when the blood enters it and the wave of expansion is transmitted along the arteries to the periphery as the pulse.

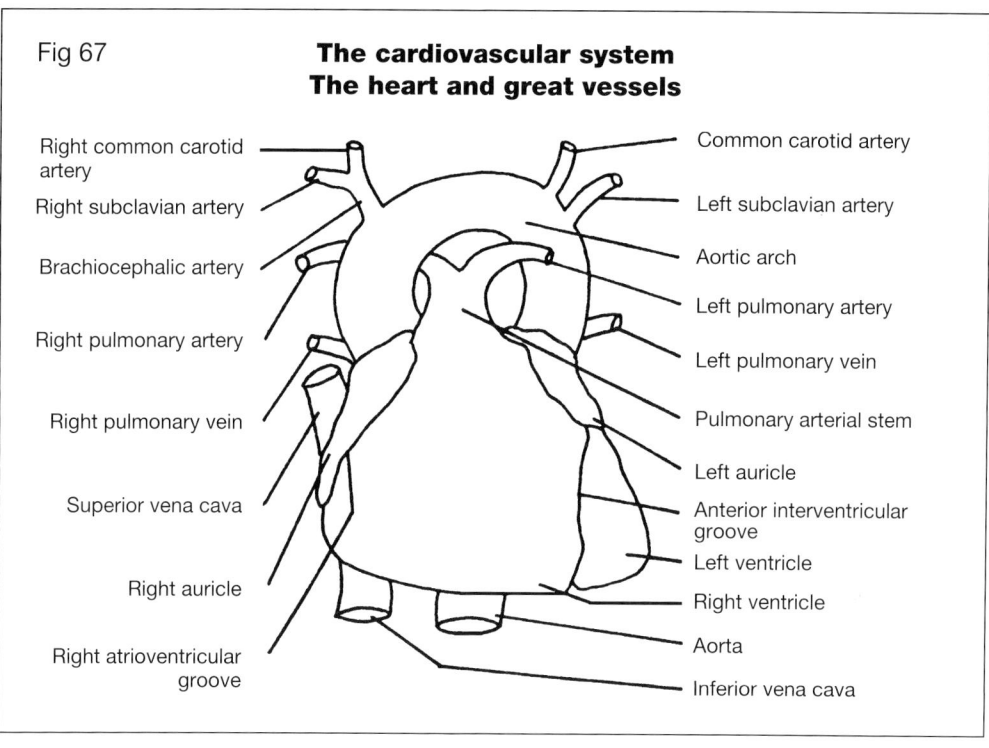

Fig 67 — The cardiovascular system: The heart and great vessels

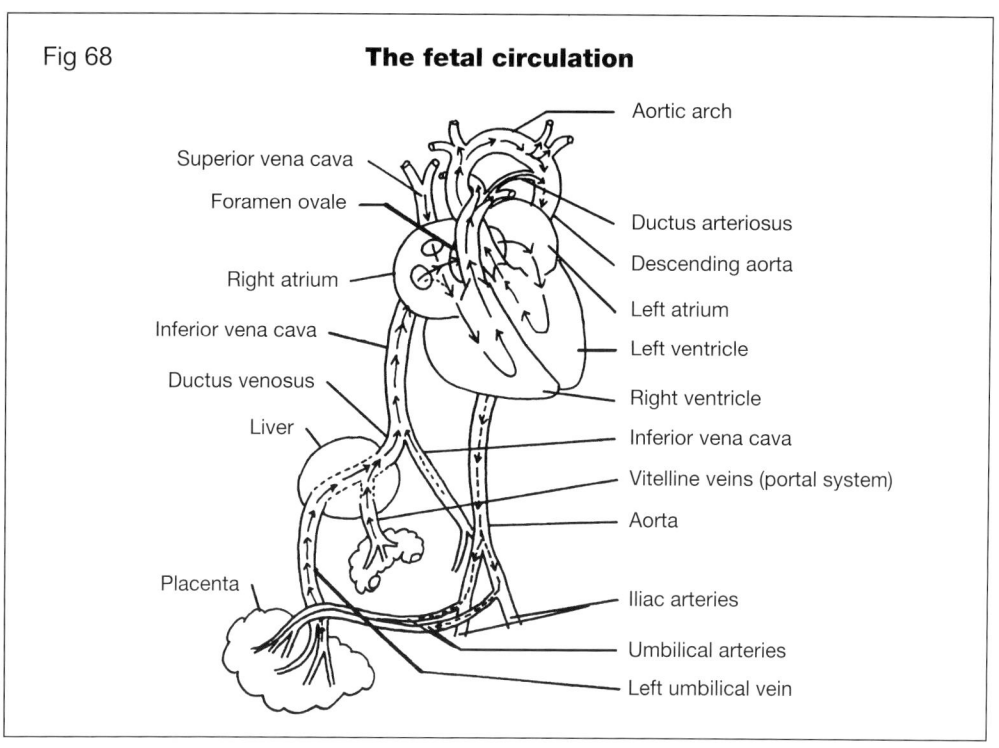

Fig 68 — The fetal circulation

The fetal blood circulation

Oxygen-rich blood flows from the placenta to the fetus via the umbilical veins (Fig 68). In later fetal life only the left umbilical vein persists. In the liver this blood is mixed with blood from the developing portal system (the vitelline veins from the digestive tract) and flows in the direction of the inferior vena cava via the hepatic veins and the ductus venosus. In the inferior vena cava the blood is mixed with blood from the lower parts of the body and enters the right atrium, where it is joined by the blood from the head and upper limbs which enters the atrium via the superior vena cava. Most of the blood which enters the right atrium from the inferior vena cava enters the left atrium directly via the foramen ovale, together with a small amount of blood from the lungs. A small amount of blood from the inferior vena cava, together with most of the blood from the superior vena cava, passes into the right ventricle via the right atrioventricular valve.

Oxygen-rich blood in the left atrium is pumped to the left ventricle and passes to the aorta, from where it is distributed mainly to the coronary arteries, the head, and the upper limbs. Relatively little blood reaches the lower limbs via the descending aorta. Blood from the right ventricle passes through the inactive lungs and reaches the descending aorta via the ductus arteriosus. The blood in the descending aorta reaches the digestive tract and the lower limbs and is returned to the placenta by two umbilical arteries.

Two major changes in the circulatory system take place at birth:

1. the foramen ovale closes due to the increased pressure in the left atrium as the lungs start functioning; and
2. the ductus arteriosus becomes obliterated, forming the ligamentum arteriosum, and all the blood received through the pulmonary arteries is distributed to the lungs from where it is returned to the left ventricle via the left atrium.

A fainting attack

The common fainting attack is termed a vasovagal episode and results from a transient cerebral anoxia. It may be brought on by emotional stress as caused by fear with, or without, accompanying pain, or may result from serious hemorrhage or long periods in a stationary standing position.

The vagus nerve has a continuous braking effect on heart rate which is increased on stimulation of the nerve. This vagal stimulation also results in a diminished force of contraction with a consequent drop in blood pressure.

Excitement or emotional stress may initially cause an acceleration in the heart rate (tachycardia) due to an increase in circulating epinephrine, but may have a stimulating effect on the vagus. This causes a slowing of the heart rate (brachycardia) with the consequences described above. These changes lead to cerebral anoxia and collapse. When the victim is in the horizontal position, the venous return to the heart is increased and is usually accompanied by a tachycardia. The blood pressure in the brain is restored and the patient regains consciousness. The symptoms and signs of a fainting attack are restlessness, dizziness, pallor, and sweating.

It is not inappropriate to briefly outline the management of a person who faints. It can be summarized as follows:

1. Place the patient in a supine position.
2. Ensure an open airway.
3. Feel the pulse. The initial brachycardia should be replaced by a tachycardia with recovery of the patient. The victim should regain consciousness within 1 minute.
4. Slowly bring the recovering patient to an upright position.

Selected bibliography

1. Dixon, A.D. (1986) *Anatomy for Students of Dentistry*, 5th edition. Edinburgh: Churchill Livingstone.
2. Lavelle, C.L.B. (1988) *Applied Oral Physiology*, 2nd edition. London: Wright.
3. Sadler, T.W. (1985) *Langman's Medical Embryology*, 5th edition. Baltimore: Williams and Wilkins.

Review questions

1. Describe the anatomy of the human heart.
2. Describe the functioning of the heart.
3. Describe the fetal blood circulation.
4. Describe the causation, signs, symptoms, and management of a fainting attack.

25. Development of the arterial circulation of the head and neck

As the embryo grows and lengthens, the heart is gradually displaced to a more caudal position. The development of the circulation of the head and neck commences with the formation of the first pharyngeal arches and their arteries, the first aortic arch arteries. These first aortic arch arteries, and all the aortic arch arteries which subsequently form, have a common origin in the aortic sac, which is an elongation of the bulbis cordis, before the heart undergoes internal division into four chambers (Fig 69).

The aortic arch arteries arise in situ in pharyngeal arch mesenchyme, course dorsally in each arch around the sides of the primitive pharynx, and join the ipsilateral dorsal aorta which also forms in situ in the mesenchyme alongside the notochord. The two dorsal aortae course caudally and join to form a single dorsal aorta between the primitive gut and the notochord. This dorsal aorta persists into later life as the descending aorta. The point where the dorsal aortae join is caudal to the last pharyngeal arches. As the heart becomes increasingly displaced caudally, and with the addition of further pharyngeal arches, five further pairs of aortic arch arteries form. The fifth aortic arch arteries are never well developed and are present for a very short time.

Further developments can be described as follows:

1. Both dorsal aortae grow cranially beyond the points where they are joined by the first aortic arch arteries. These extensions are the internal carotid arteries, which branch in the forebrain area to form the ophthalmic artery, and shortly afterwards, an anterior and middle cerebral artery, and an artery which initially represents the posterior cerebral artery. The latter artery later joins the definitive posterior cerebral branch of the basilar artery and itself becomes the posterior communicating artery. The above developments take place bilaterally and form the arterial circle of Willis (Fig 70).
2. The first and second aortic arch arteries later degenerate ventrally but remain dorsally for a time as a primitive maxillary artery and a hyoid artery, respectively (Fig 71 B). Both these arteries receive blood from the internal carotid artery. Close to its point of origin, the hyoid artery forms the stapedial arterial system, consisting of a stapedial artery, a mandibular artery, a temporary supraorbital artery, and an infraorbital artery. In due course both the first two aortic arch arteries lose their connection to the internal carotid arteries (Fig 71 C and 72).
3. The external carotid arteries arise as new branches (Fig 73) which grow cranially from the ventral parts of the third aortic arch arteries. The common stem of origin of the mandibular, infraorbital, and supraorbital arteries forms an anastomosis with the newly developed maxillary, or retromandibular, artery which, together with the superficial temporal artery, are the terminal branches of the external carotid artery. All connections to the internal carotid artery are subsequently lost when the outer parts of the primitive maxillary artery and the stapedial system become obliterated. The remnants of the first two pairs of aortic arch arteries probably contribute to the branches of the external carotid artery, especially the maxillary artery.

The external carotid artery (Figs 73, 75) gives origin on both sides to the superior thyroid, lingual and facial arteries. The initial supraorbital artery, which develops as a branch of the stapedial system, becomes the middle meningeal branch of the maxillary artery, while the final supraorbital artery develops as a new branch of the ophthalmic artery. The middle meningeal artery, together with the mandibular (inferior alveolar) artery, is given off by the first part of the maxillary artery (in relation to the neck of the mandible). Branches to the muscles of mastication and the buccal artery arise from the second part of the maxillary artery, in relation to the lateral pterygoid muscle, while the third part of the maxillary artery, related to the pterygomaxillary fissure and the pterygopalatine fossa, gives origin to the posterior superior alveolar artery. The maxillary artery is continued as the infraorbital branch.

The definitive supraorbital artery is a branch of the ophthalmic artery, which also gives origin to the central artery of the retina, the supratrochlear, dorsal nasal,

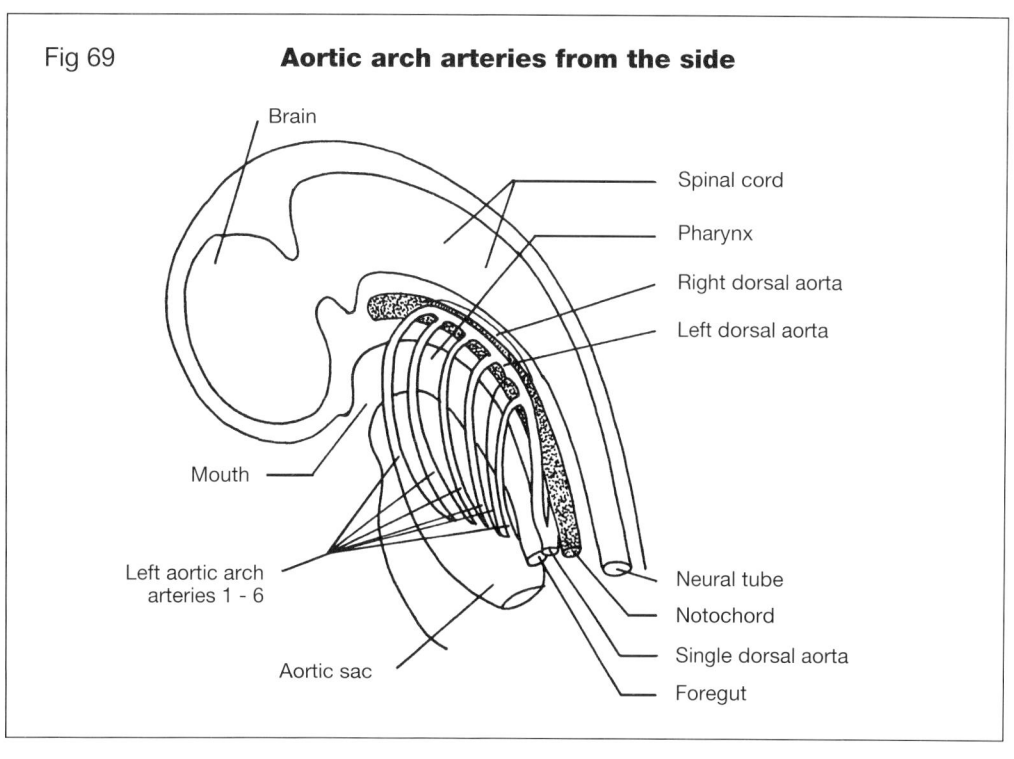

Fig 69 Aortic arch arteries from the side

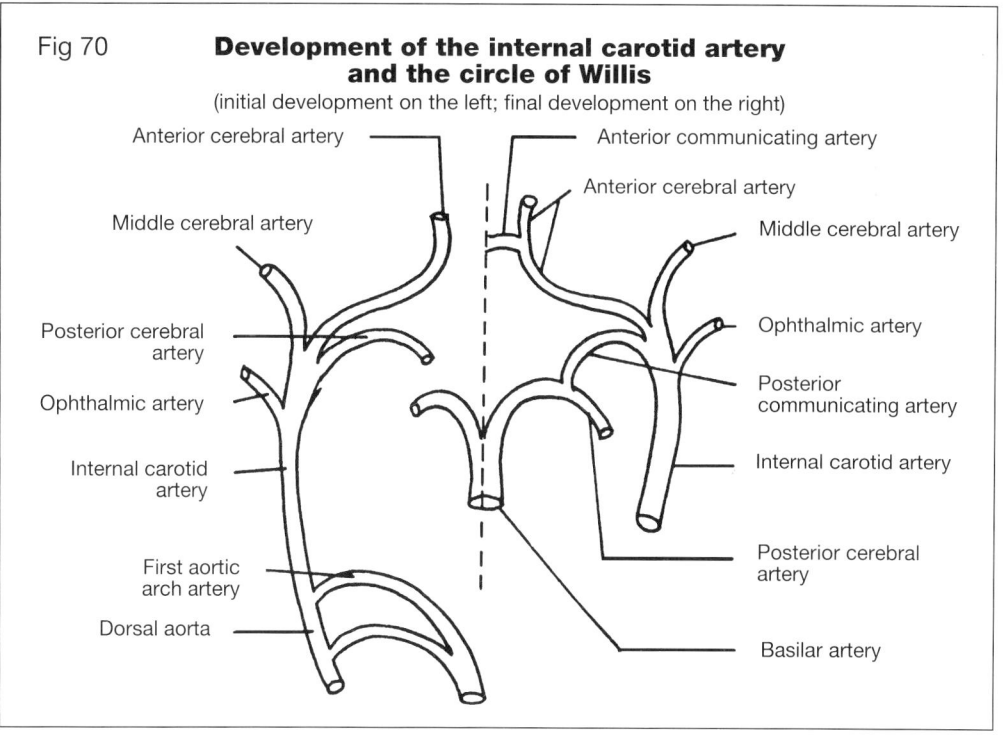

Fig 70 Development of the internal carotid artery and the circle of Willis
(initial development on the left; final development on the right)

201

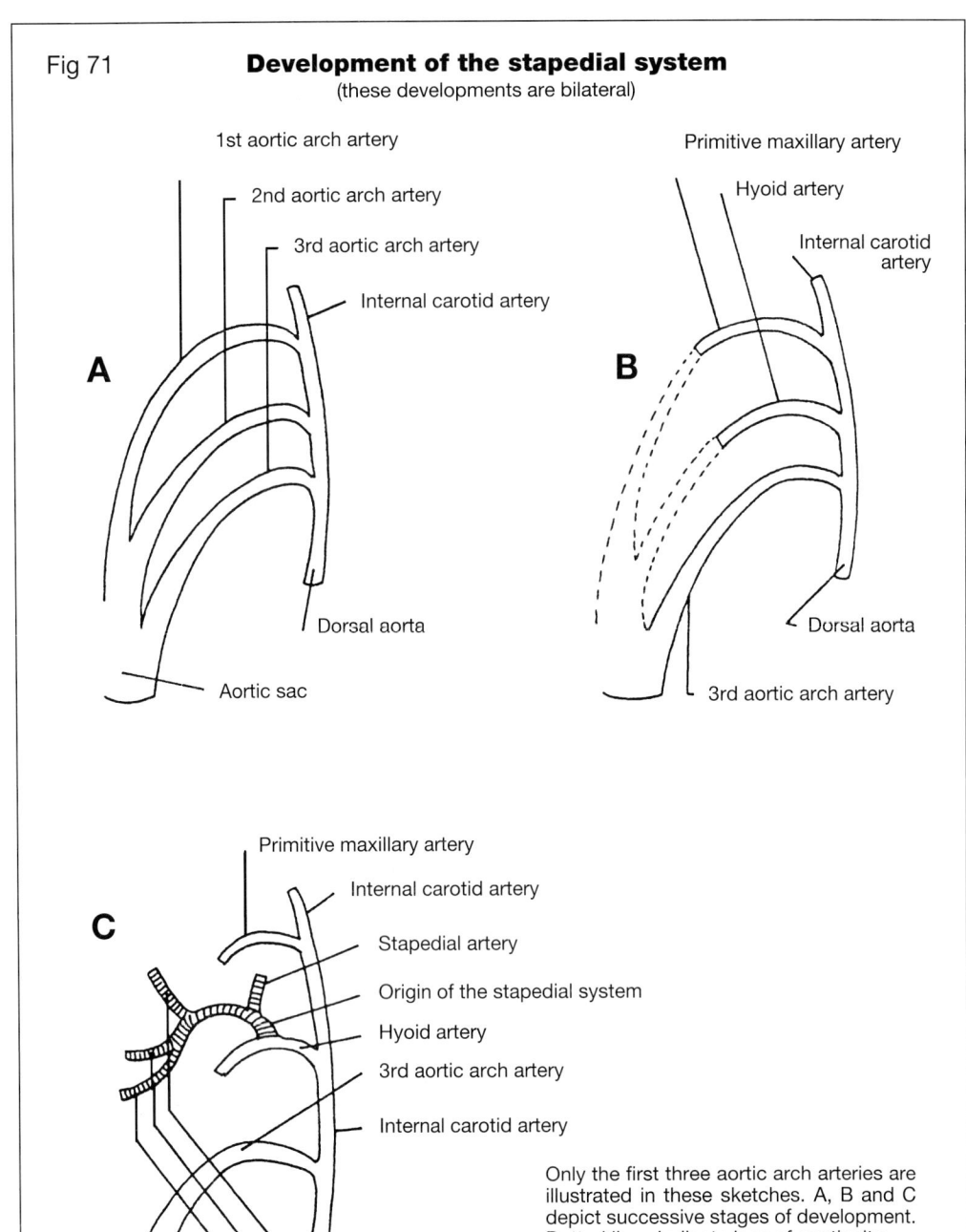

Fig 71 **Development of the stapedial system**
(these developments are bilateral)

Only the first three aortic arch arteries are illustrated in these sketches. A, B and C depict successive stages of development. Dotted lines indicate loss of continuity.

Fig 72 **Initial development of the internal carotid artery and the stapedial system (seen from the left side)**

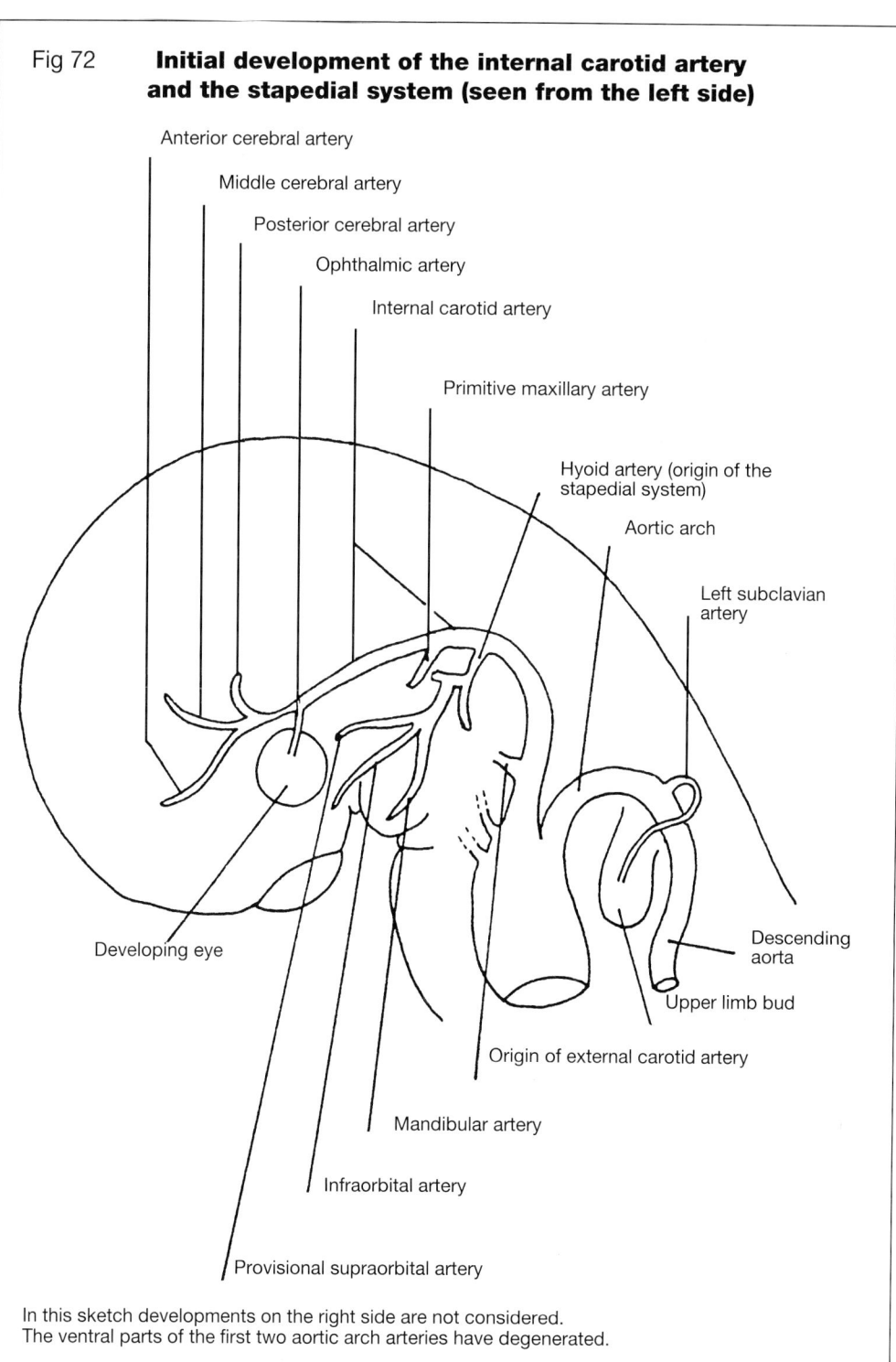

In this sketch developments on the right side are not considered.
The ventral parts of the first two aortic arch arteries have degenerated.

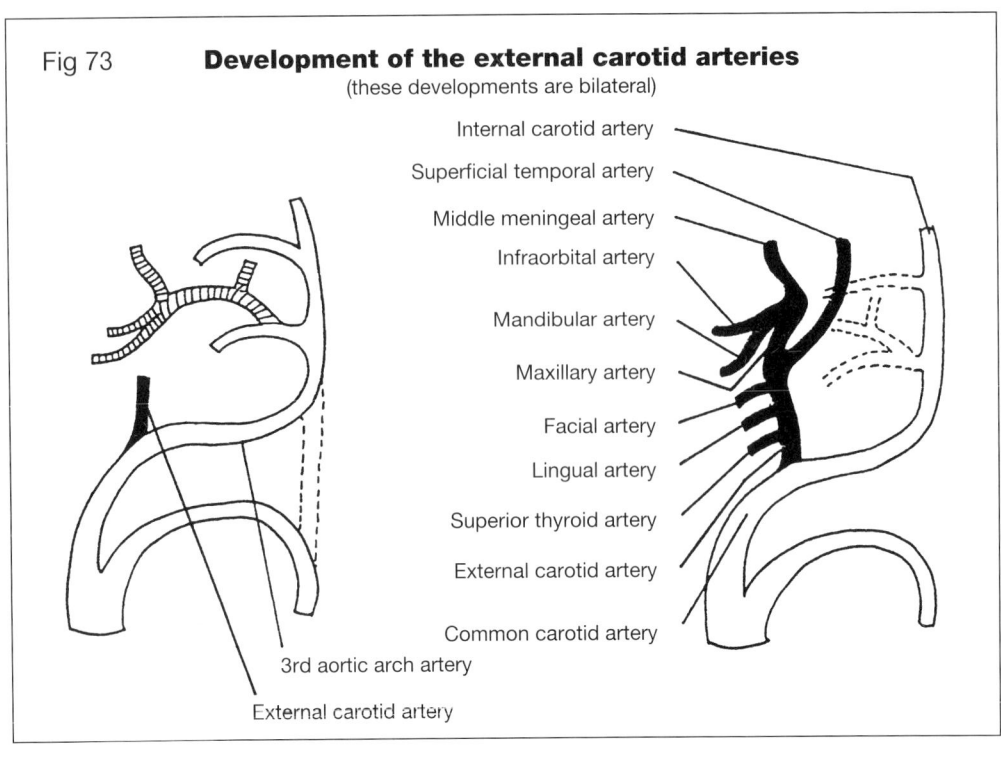

Fig 73 **Development of the external carotid arteries**
(these developments are bilateral)

- Internal carotid artery
- Superficial temporal artery
- Middle meningeal artery
- Infraorbital artery
- Mandibular artery
- Maxillary artery
- Facial artery
- Lingual artery
- Superior thyroid artery
- External carotid artery
- Common carotid artery
- 3rd aortic arch artery
- External carotid artery

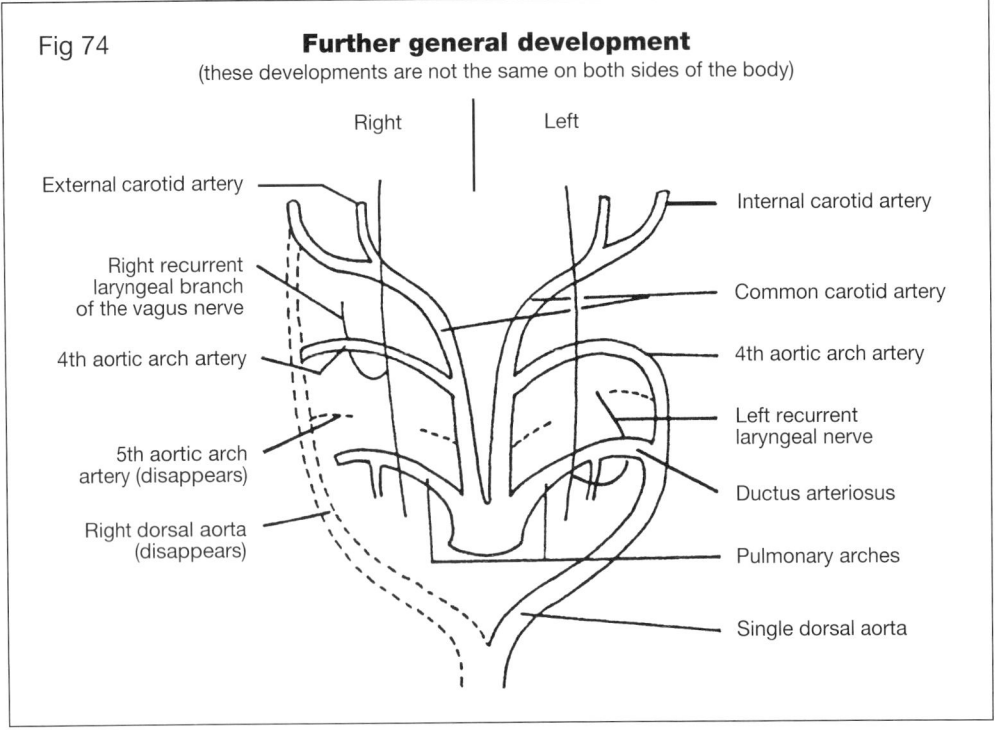

Fig 74 **Further general development**
(these developments are not the same on both sides of the body)

Right | Left

- External carotid artery
- Right recurrent laryngeal branch of the vagus nerve
- 4th aortic arch artery
- 5th aortic arch artery (disappears)
- Right dorsal aorta (disappears)
- Internal carotid artery
- Common carotid artery
- 4th aortic arch artery
- Left recurrent laryngeal nerve
- Ductus arteriosus
- Pulmonary arches
- Single dorsal aorta

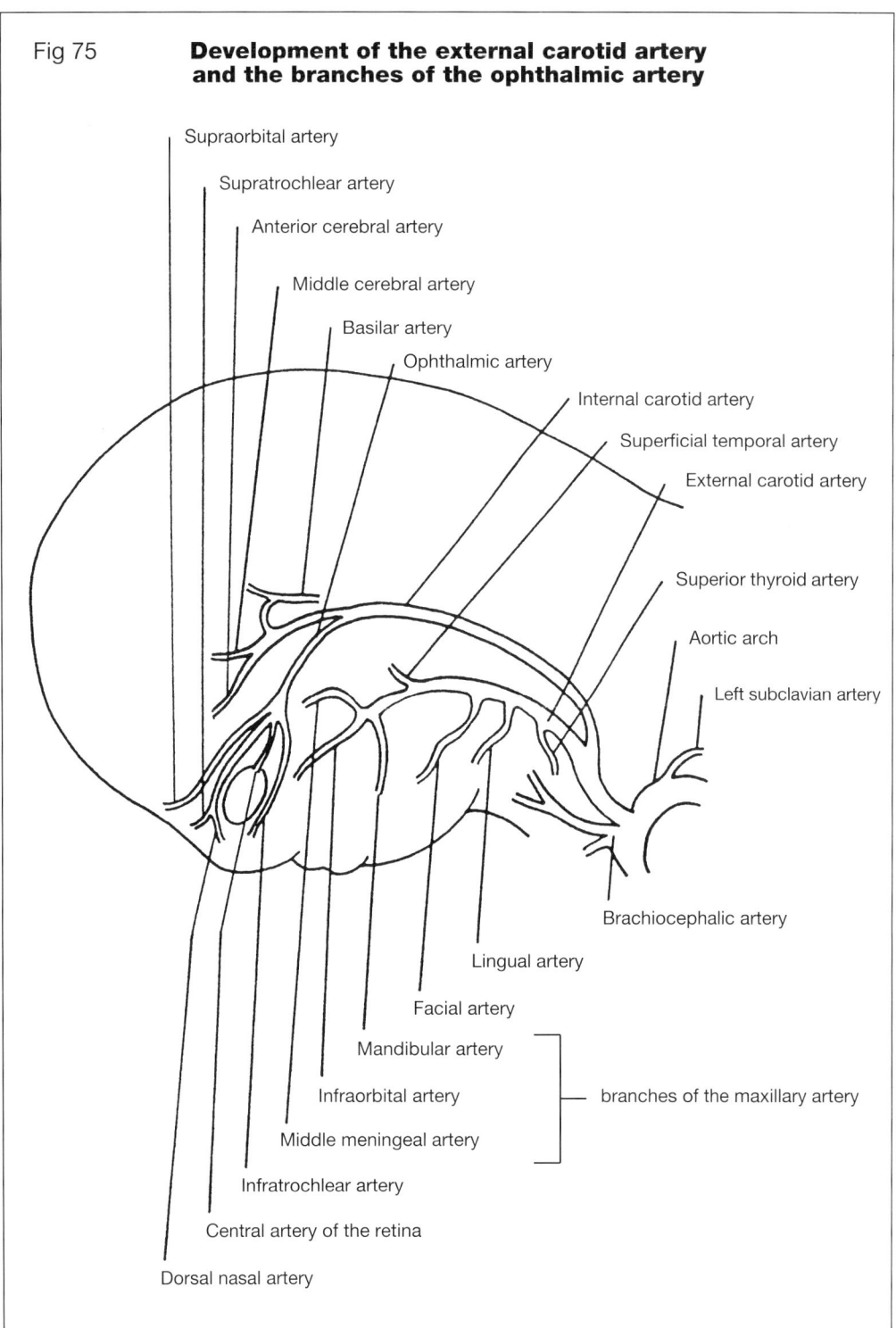

Fig 75 **Development of the external carotid artery and the branches of the ophthalmic artery**

lacrimal, infratrochlear, ciliary, palpebral, and ethmoidal arteries (Fig 75). After formation of the external carotid arteries, the proximal portions of the third aortic arch arteries become the common carotid arteries (Fig 76).

The part of the dorsal aortae between the points where the third and fourth aortic arch arteries join them disappears on both sides. Subsequently the entire dorsal aorta on the right side, between the points where it is joined by the fourth aortic arch artery and its point of confluence with the left dorsal aorta, disappears. This leaves a single dorsal aorta on the left side.

The sixth aortic arch arteries give off branches to the lungs on both sides and are termed the pulmonary arches. The lateral part of the right pulmonary arch disappears, and consequently blood can only enter the dorsal aorta via the left fourth and sixth arch arteries. The terminal part of the left pulmonary arch persists till birth as the ductus arteriosus, after which it remains as the ligamentum arteriosum which connects the pulmonary artery to the arch of the aorta.

The right horn of the aortic sac elongates and becomes the brachiocephalic (previously called the innominate) artery (Fig 77). This gives origin to the third and fourth aortic arch arteries which subsequently form the right common carotid and the right subclavian artery, respectively. The left subclavian artery arises as a new trunk from the arch of the aorta close to the point where the fourth aortic arch artery initially joined the dorsal aorta.

The fourth aortic arch artery on the left side becomes the definitive arch of the aorta which, after internal septation of the heart and attachment of the pulmonary artery to the right ventricle, gives origin to the brachiocephalic artery on the right side. The right ventricle therefore supplies blood to the lungs, while the left ventricle supplies the remainder of the body.

The recurrent laryngeal branches of the vagus nerves (Fig 74) are initially found in relation to the sixth aortic arch arteries on both sides. With the disappearance of the distal part of the right sixth arch artery, this nerve is displaced cranially till further migration is prevented by the intact fourth aortic arch artery. No similar developments take place on the left side. With caudal displacement of the heart and the great vessels, the right nerve assumes a recurrent course around the subclavian artery, while the left nerve follows a recurrent course around the ductus arteriosus (the ligamentum arteriosum after birth).

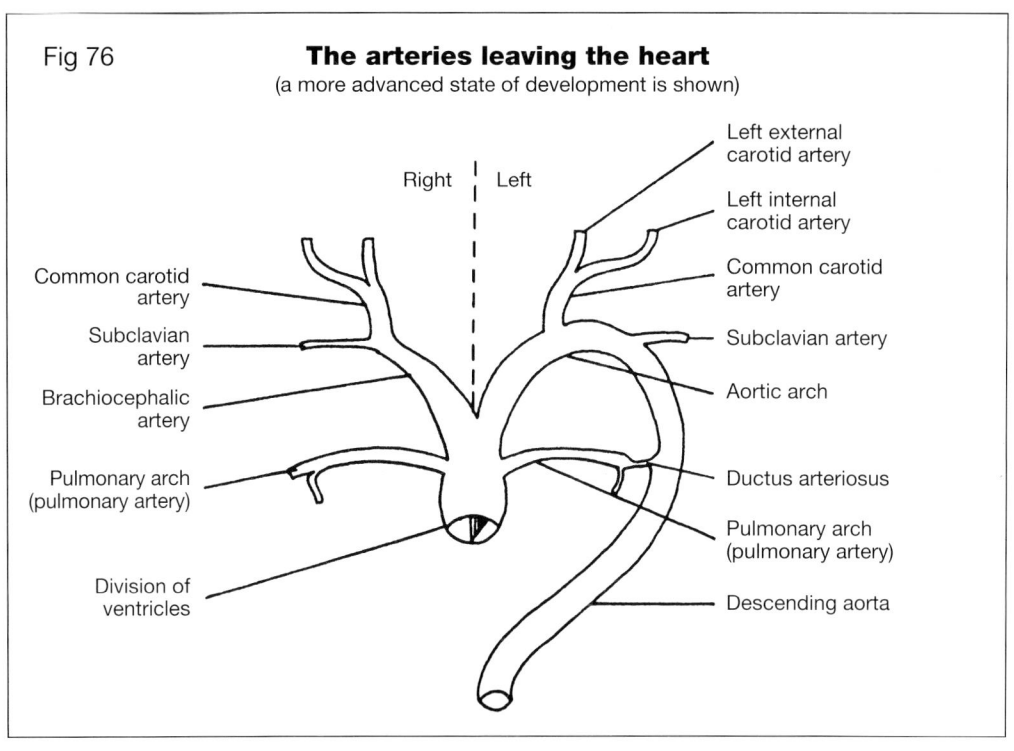
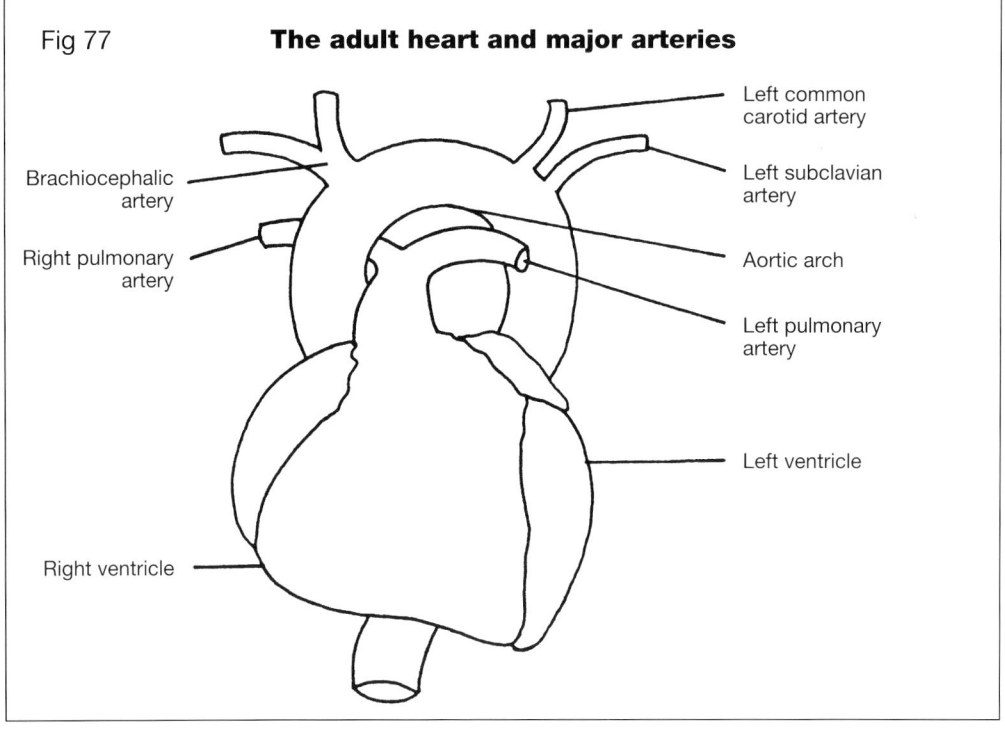

Selected bibliography

1. Dixon, A.D. (1986) *Anatomy for Students of Dentistry*, 5th edition. Edinburgh: Churchill Livingstone.
2. Hamilton, W.J., Boyd, J.D. and Mossman, H.W. (1972) *Human Embryology*, 4th edition, revised by Hamilton, W.J. and Mossman, H.W. Cambridge: W. Heffer and Sons Ltd.
3. McKenzie, J. (1958) The first arch syndrome. *Archives of Diseases in Childhood*, 33, 477 - 486.
4. Osborn, J.W. (ed) (1981) *Dental Anatomy and Embryology*, Oxford: Blackwell Scientific Publications.
5. Patten, B.M. and Carlson, B.M. (1974) *Foundations of Embryology*, 3rd edition. New York: McGraw-Hill Book Company.
6. Sperber, G.H. (1989) *Craniofacial Embryology*, 4th edition. London: Wright.

Review questions

1. Describe the origin and the course followed by the aortic arch arteries.
2. Describe the development of the internal carotid arterial system.
3. Describe the development of the external carotid artery and its main branches.
4. Describe the fate of the dorsal aortae.
5. Describe the development of the definitive arch of the aorta and the brachiocephalic artery.

26. Lymphatic drainage of the scalp, face, oral cavity, and associated structures

The lymphatic system consists of lymph vessels and lymphoid tissue which may be diffusely arranged or gathered into localized masses, the lymph nodes, through which lymph percolates before being returned to the blood circulation. Lymph is collected from virtually all the tissues of the body, a major exception being the central nervous system.

Lymph, a colorless substance, is essentially interstitial fluid, containing a small amount of protein, which has passed from blood capillaries into tissue spaces and has failed to return to the blood vascular system. Lymph is collected in blind-ended capillaries with a thin endothelial lining, surrounded by a small amount of connective tissue, and are more permeable than blood capillaries. The capillaries branch freely and form a network in the tissues and organs. The capillaries drain into larger collecting lymph vessels, or ducts. These vessels are larger and thicker than the capillaries, being composed of an endothelial lining and a smooth muscle coat with surrounding connective tissue. Valves are numerous along the course of the bigger lymph vessels and allow lymph to flow in a specific direction, for example from the head to the neck. Lymph from the larger lymph vessels is finally gathered into two main lymphatic trunks which return the lymph to the venous circulation.

The two major lymphatic trunks are the thoracic duct on the left side, and the right lymphatic duct. Both join the venous system close to the point of confluence of the internal jugular and subclavian veins on either side of the body. The thoracic duct drains lymph from the entire body, except the right half of the head, the right upper limb, and part of the right half of the mediastinum. These parts are drained by the right lymphatic duct.

The lymph nodes are collections of lymphoid tissue enclosed by a capsule of fibrous connective tissue (Fig 78). The nodes are usually kidney-shaped with a hilum on one side through which blood vessels enter and leave the node. Fibrous cortical trabeculae extend into the node from the capsule, while medullary trabeculae extend into the node from the hilum. The branching trabeculae provide a framework for the node, while the spaces between the trabeculae contain a network of reticular

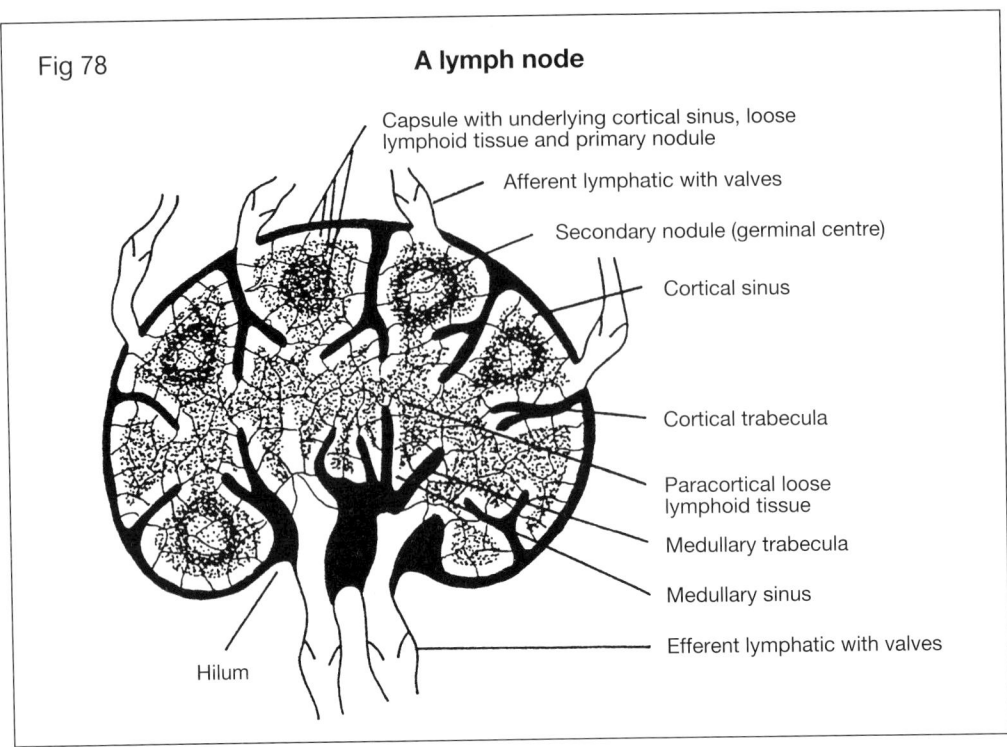

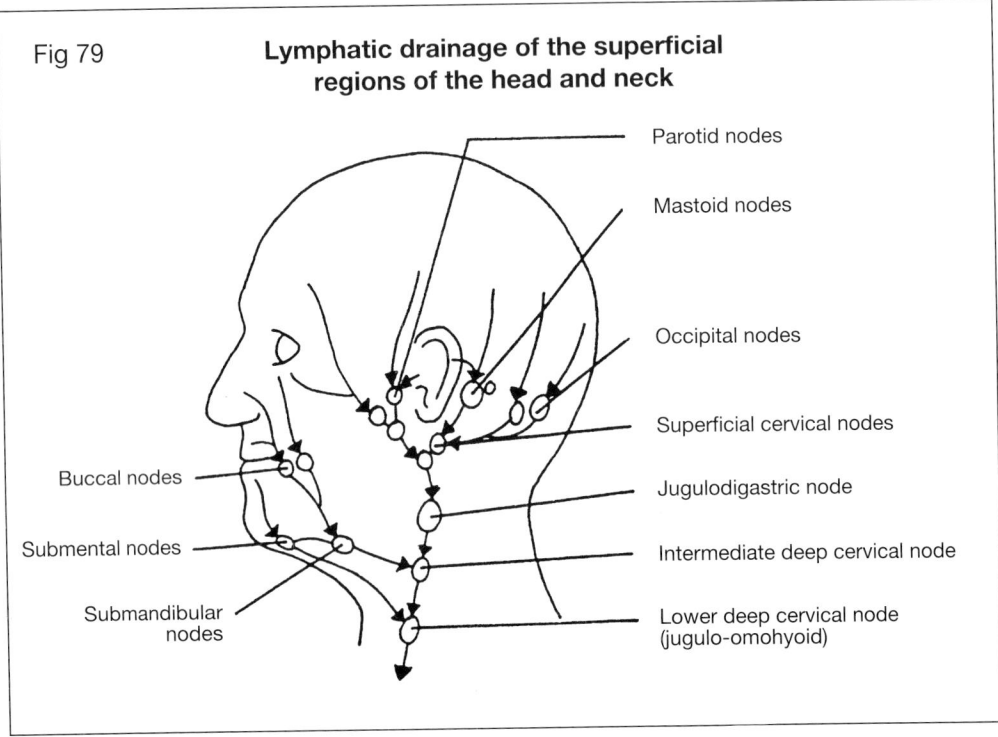

fibers. The lymph nodes are often arranged in groups along the course of the lymph vessels.

The cortical area of a lymph node is divided into a superficial area, the outer cortex, and a deep cortex, or paracortical area. The superficial cortical area contains nodular lymphatic tissue while the deep area contains diffuse lymphatic tissue. The medulla surrounds the hilum and consists of loose lymphatic tissue arranged in cords. The cords contain large numbers of lymphocytes, plasma cells, and macrophages. Between the capsule and the cortical area is the subcapsular sinus, which is continuous with cortical sinuses alongside the cortical trabeculae. These sinuses are, in turn, continuous with the medullary sinuses which are found alongside the medullary trabeculae.

A lymphatic nodule may be of uniform appearance consisting only of relatively small lymphocytes and, if present, represents inactive lymphoid tissue. These primary nodules are rare and only present at the time of birth before antigenic stimulation has taken place. When lymphocytes are actively forming, a paler central area, the germinal center, is seen. Such a nodule is often referred to as a secondary nodule. The germinal centers consist of antigen-stimulated lymphocytes (B lymphocytes) and therefore represent an area of antigen production. In general, B lymphocytes and plasma cells are found in the nodular tissue while T lymphocytes are more common in the paracortical and medullary areas. In the medullary lymphoid tissue, which is arranged in cords, large numbers of lymphocytes, plasma cells, and macrophages are found. It is therefore obvious that the lymph nodes, and lymphoid tissue in general, play a vital role in the response of the body to antigenic substances (the immune reaction). In addition, the lymphatic system is responsible for maintaining the normal filtration-absorption balance in the tissues and controls the fluid pressure in the interstitial spaces. If the lymphatic system is overloaded by the presence of excessive tissue fluid, or if the vessels are obstructed, edema will result. The lymph nodes are usually rapidly involved during infective processes (lymphadenitis). Lymph is, furthermore, a vehicle for the spread (metastasis) of malignant cells.

Afferent lymph vessels enter the subcapsular sinus and thence through the cortical and the medullary sinuses to the hilum, from where lymph leaves the node via slightly larger efferent vessels.

The lymphatic drainage of the superficial parts of the head and face is subsequently described (Fig 79).

Lymph vessels from the lower lip course through the soft tissues of the chin and initially end in the submental nodes, which lie posterior to the symphysis of the mandible on both sides of the midline.

Lymph from the angles of the mouth, upper lip, side of the nose, anterior part of the cheek, and part of the lower eyelid initially drains primarily into the submandibular nodes. These lymph vessels may pass through buccal nodes which lie on the buccinator muscle close to the angle of the mouth, and through mandibular nodes which are found on the outer surface of the mandible close to the facial vein and artery. The submandibular nodes lie in intimate relation to the submandibular salivary gland under cover of the posterior part of the body of the mandible.

From the lower and upper eyelids, the frontal and temporal regions of the scalp, the skin of the lower part of the auricle and of the zygomatic region, lymph drains to the parotid nodes which are embedded superficially in the fascia of the gland.

The scalp of the parietal and occipital regions and the skin of the rest of the auricle, are drained by mastoid and occipital nodes. The former are found at the site of insertion of the sternocleidomastoid muscle, and the latter at the site of insertion of the trapezius muscle.

Lymph from the submental nodes courses to the submandibular nodes, or directly to lower deep cervical nodes (the jugulo-omohyoid group). These lie close to the point where the internal jugular vein is crossed by the omohyoid muscle. Lymph from the submandibular nodes enters the deep cervical nodes at a higher level (the intermediate deep cervical nodes).

Lymph vessels drain lymph from the parotid, and mastoid nodes into superficial cervical nodes which lie slightly above the confluence of the posterior division of the retromandibular and posterior auricular veins (the origin of the external jugular vein).

The superficial cervical nodes drain into an enlarged upper deep cervical node, the jugulodigastric node, or into deep cervical nodes on a lower level.
The former lies close to the point where the internal jugular vein is crossed by the posterior belly of the digastric muscle.

The deep cervical nodes form a chain, the deep cervical chain, which lies in close relation to the internal jugular vein. The deep cervical chain on the right side joins the right lymphatic duct, and on the left side the thoracic duct.

The submental, submandibular, parotid, mastoid, and occipital nodes constitute the pericervical group which all ultimately join the deep cervical chain of nodes.

An extensive submucosal plexus interconnects all regions of the oral cavity. The oral mucosa, the teeth, alveolar processes, palate, and the tongue are drained by lymphatics which join the nodes previously mentioned.

The incisors and canines drain to submental and submandibular nodes. Lymph from the premolars and molars drains backwards via lymph vessels which pierce the superior constrictor and buccinator muscles to reach the deep parotid and deep cervical nodes. The former nodes are found deep to the parotid gland in relation to the lateral wall of the pharynx.

Lymph vessels from the anterior region of the hard palate join the anterior facial vessels in their course towards the submandibular node, while lymph from the posterior region of the palate courses towards the upper deep cervical nodes.

Lymph from the cheeks enters the buccal nodes after piercing the buccinator muscle, or may go directly to the submandibular nodes.

Lymph from the anterior part of the floor of the mouth courses to the submental nodes, or directly to lower deep cervical nodes, while the posterior part drains into the submandibular nodes.

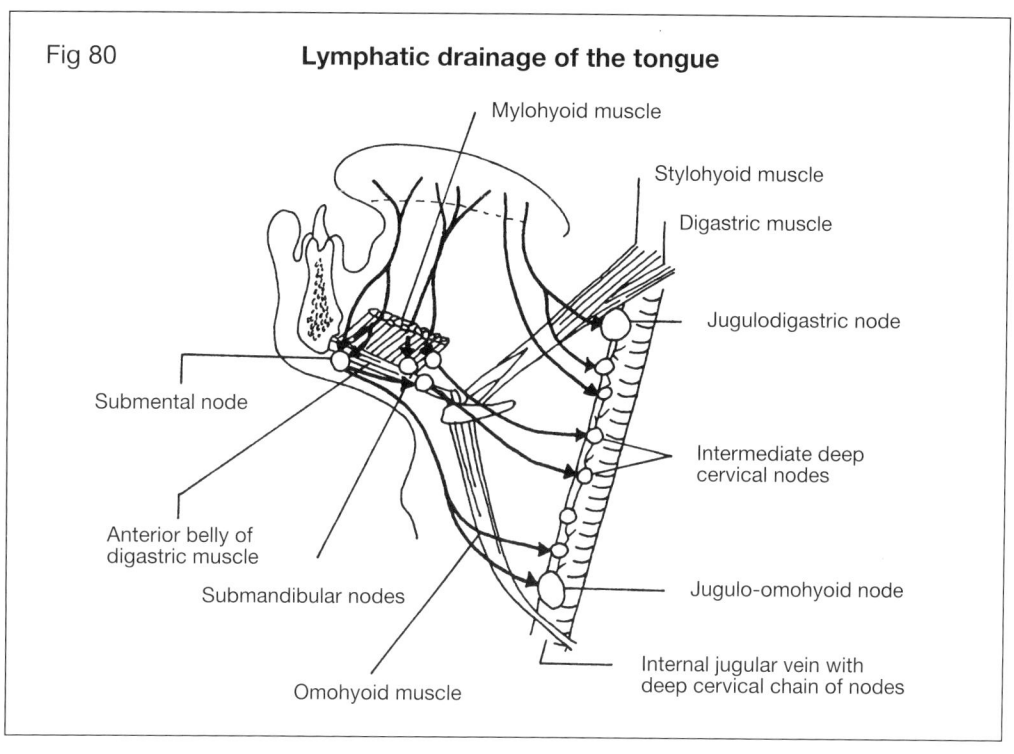

Fig 80 Lymphatic drainage of the tongue

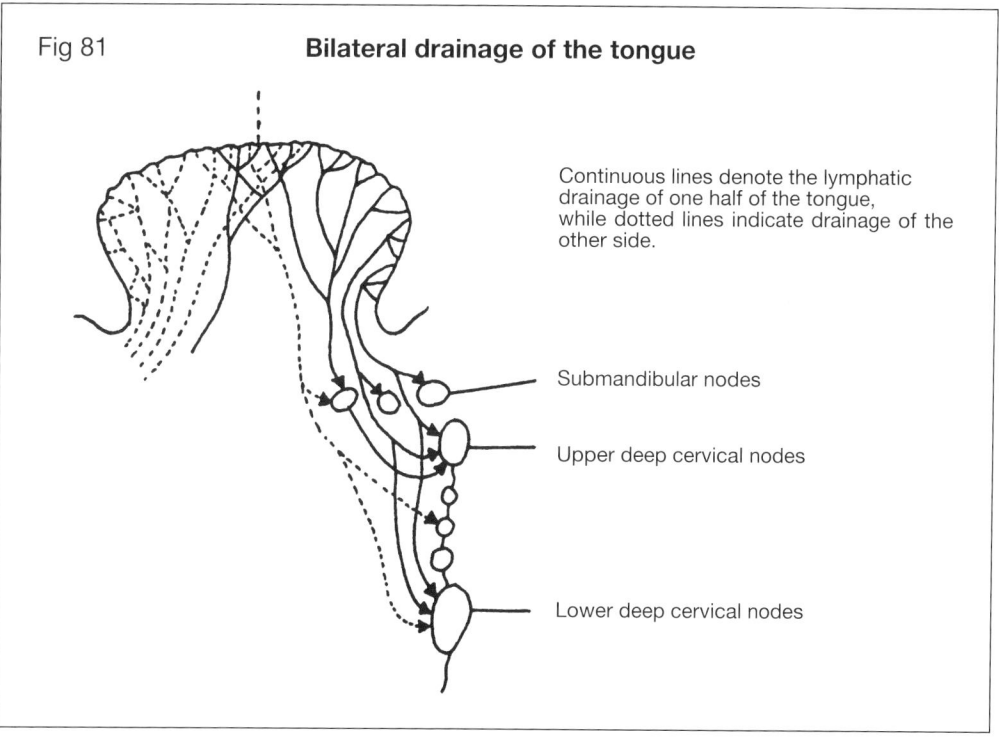

Fig 81 Bilateral drainage of the tongue

Continuous lines denote the lymphatic drainage of one half of the tongue, while dotted lines indicate drainage of the other side.

Lymph vessels from the palatine tonsil pierce the superior constrictor muscle to enter the deep parotid and upper deep cervical nodes.

The tongue has a very rich submucosal network of lymphatics which drain into the deep cervical, submandibular, and submental nodes (Fig 80). Vessels from the base of the tongue pierce the side walls of the pharynx to reach the upper deep cervical nodes. The drainage of the major central part of the tongue is to the submandibular nodes after the vessels have pierced the mylohyoid muscle, while the vessels from the anterior part of the tongue similarly pierce the mylohyoid muscle to reach the submental nodes. It is important to realize that a certain amount of bilateral drainage occurs in the tongue and it is not uncommon for cervical metastases from a tongue carcinoma to be found bilaterally or even contralaterally (Fig 81).

The lymphatic system plays a vital role in the immune reaction, and the reader is advised to consult chapters 6 (part I) and 23 (part II) in this regard.

Selected bibliography

1. Dixon, A.D. (1986) *Anatomy for Students of Dentistry*, 5th edition. Edinburgh: Churchill Livingstone.
2. Kelly, D., Wood, R.L. and Enders, A.C. (1984) *Bailey's Textbook of Microscopic Anatomy*, 18th edition. Baltimore: Williams and Wilkins.
3. Krause, W.J. and Cutts, J.H. (1986) *Concise Text of Histology*, 2nd edition. Baltimore: Williams and Wilkins.
4. McMinn, R.M.H. (ed) (1990) *Last's Anatomy*, 8th edition. Edinburgh: Churchill Livingstone.
5. West, J.B. (1991) *Best and Taylor's Physiological Basis of Medical Practice*, 12th edition. Baltimore: Williams and Wilkins.

Review questions

1. Describe the histology of a lymph node.
2. Describe the arrangement of the superficial lymph nodes of the head.
3. Describe the lymphatic drainage of the lower lip, mandible, and tongue.
4. Describe the lymphatic drainage of the maxilla and palate.

27. Charles Darwin (1809 – 1882)

Charles Darwin was born on 12 February 1809 into a wealthy family, at a time when the scientific world was still very much divided on the question of the origin and development of life forms. Some scientists were still of the opinion that life originated in a type of special creation, or spontaneous generation in which some medium, possibly menstrual blood in the case of a female mammal, was stimulated in a mysterious way to create a new being. Others believed in preformation and that a new organism was present in miniature but perfectly created form in one of the parents before mating occurred. In some way the process of mating acted as a stimulus for further development. Genetics was an unexplored science and the theory of evolution was unknown. The reader is advised to consult the chapter on the historical review of embryology (part I, chapter 2) in the above regard.

Darwin commenced his medical studies at the University of Edinburgh in 1825 but was unmotivated (he found his lectures dull), and later studied theology at the University of Cambridge. He completed his studies but did not apply for a parish because his heart was not in this profession either. He had been a lover of nature since childhood, and was shortly to receive an invitation to join H.M.S. Beagle.

When captain Fitzroy embarked on a world cruise on the survey ship, H.M.S. Beagle, in 1831 to collect biological material, Darwin was invited to join the expedition as the accompanying natural scientist. Darwin collected vast quantities of biological specimens, filled book after book with scientific notes, and came to conclusions which would stagger the scientific world, and still are the cause for many a controversy. Everywhere he went, Darwin saw evidence of variation. Members of one specific type of animal were not absolutely identical but varied in size, strength, vigour, health status, behavior, and in a host of other respects. Darwin was convinced that a form of evolution had taken place and was still in the process of taking place. At this stage he was still unable to formulate his ideas.

Darwin was a sickly person and was described by some as a gentle neurotic introvert when he returned to Great Britain on 2 October 1836. He was married on 29 January 1839 and had an extremely happy and rewarding family life. He wrote many minor articles on his findings after his return, but his ideas, which would be of major scientific importance in later years, only became formulated when he perused an article by T.R. Malthus, an English church minister, which had been published in 1798. Malthus claimed that the reproductive potential of mankind was greater than the natural resources necessary to support the growing population. In practice, the size of a population is kept in check by factors such as disease, famine, and wars, according to Malthus.

It is unfortunate that Darwin was at this stage unaware of the important work done by a contemporary, Gregor Mendel, in the field of genetics. Knowledge of this work would undoubtedly have been of very great assistance to Darwin.

Darwin and Alfred Russel Wallace, who worked in Indonesia, studied the writings of Malthus independently, unaware of one another. Darwin, in 1839, and Wallace, in 1858, noted that the answer to the evolution controversy would probably be found in the work of Malthus. They concluded that what was true for the human race was equally applicable to animals and plants. There is a struggle for survival.

In 1858 Darwin and Wallace published their theory regarding the origin of species through natural selection, and in 1859 Darwin published his well-known work *The Origin of Species by Means of Natural Selection.* This work initiated fierce differences of opinion which have persisted into modern times. Darwin hardly ever became involved in arguments.

A cornerstone of the theory of evolution is that organisms who succeed in surviving, of necessity possess characteristics which confer greater adaptability to the environment on them than are possessed by their contemporaries who do not survive. It was, furthermore, clear to Darwin and Wallace that in a population or group in which the members mate for reproduction, those properties which result in fertile and viable offspring will be maintained in a subsequent generation, while characteristics which do not, will be eliminated.

The theory, as expounded by Darwin and Wallace in 1859, can be summarized by means of four observations and three deductions:

Observation 1: Organisms produce a greater number of reproductive cells (gametes), as well as progeny, than will ever attain adulthood. A potential therefore exists to increase the size of a population progressively.

Observation 2: The number of individuals in a population tends to remain constant over a long period of time.

Deduction 1: To support the first two observations, there must be a high mortality rate amongst reproductive cells and young individuals.

Observation 3: No two members of a population are absolutely identical and there is considerable variation.

Deduction 2: Individuals who survive will, because of proven distinctively favorable characteristics, be able to bring forth a next generation. Characteristics of surviving individuals make them more adaptable to the environment.

Observation 4: Descendents mimic their parents but are not replicas of their parents.

Deduction 3: Subsequent generations retain favorable characteristics which were obtained through slow changes (genetically inherited), and improve on the degree of adaptability attained by their parents. An on-going favorable genetic change is produced in the surviving population.

According to this theory, the members of a population who are unable to adapt are eliminated, while those who can adapt are favored and survive. The theory is applicable to both the plant and animal kingdoms and is known as natural selection. This is nature's way of getting rid of incompetent or inadaptable members of a population. The ideal outcome of the process is to create a genotype which will find expression in a phenotype which is ideally adapted to its environment. If an individual (human or animal) possesses any characteristic which confers greater viability or fertility on it, then that individual will usually have a greater number of descendents. If this favorable characteristic or property is genetic in origin, the progeny will be advantaged in comparison with their contemporaries and will be in a dominant position. In due course unfavorable genes will be displaced and the phenotype of the population will change. Both dominant and recessive genes will spread if the resulting phenotypes are favorable, but the rate of change will be more rapid in the case of dominant genes since both a heterozygote and a homozygote will result in phenotypic expression.

Selected bibliography

1. Harrison, G.A., Weiner, J.S., Tanner, J.M. and Barnicott, N.A. (1977) *Human Biology,* 2nd edition. London: Oxford University Press.
2. Wichler, G. (1961) *Charles Darwin.* Oxford: Pergamon Press.

Review question

What are the broad principles on which the theory of evolution is based?

28. Man's position in the animal kingdom

General remarks on classification

Classification is the process whereby animals are orderly arranged in groups. The best-known classification currently used in the hierarchial system which was developed by Linnaeus in approximately 1758. In this system, animals which are genetically related are grouped together, and such a classification consists of a series of sets or groups which are termed categories. A category includes one or more subcategories. In the classification of man, the phylum is the highest category with several subcategories of which species is the lowest. Prefixes are used to denote finer groupings. Examples are suborder, superfamily, subfamily, subspecies etc.

Each group of organisms assigned to a category is called a taxon. In this way the phylum Chordata includes a Primate taxon, an Insectivora taxon and a Carnivora taxon. In the case of man, the designation Homo sapiens refers to a species taxon (sapiens) in the Homo genus category. Taxonomy refers to the assignment of an animal type to a taxon.

A species category includes a group of individuals who mate with one another and are reproductively isolated from all other demes (animal types). A genus category consists of groups of species which are normally reproductively isolated from one another but share important characteristics. A family category consists of a group of related genera. It is possible to classify the entire animal kingdom in the above way.

The basic classification of animals is as follows (man is taken as an example):

Kingdom:	The Animal Kingdom
Phylum:	*Chordata*
Class:	*Mammalia*
Order:	*Primates*
Family:	*Hominidae*
Genus:	*Homo*
Species:	*sapiens*

The fishes (class *Pisces*), amphibians (class *Amphibia*), reptiles (class *Reptilia*), birds (class *Aves*) and mammals (class *Mammalia*) all belong to the subphylum *Vertebrata*. This implies that all these groups possessed a notochord at some stage during their embryologic development. They are collectively grouped in the subphylum *Vertebrata*, due to the presence of a vertebral (spinal) column, of the phylum *Chordata*. The classification of man will be discussed fully later in this chapter.

The class *Mammalia* includes seven subclasses, for example *Eutheria*. Some authors regard the *Eutheria* as an infraclass in the subclass *Theria*, class *Mammalia*. The *Eutheria* are regarded as "true" mammals and are characterized by an efficient placenta between the female and the unborn young. This enables the fetus to attain a relatively advanced developmental stage before birth.

The subclass *Eutheria* comprises several orders, for example, *Primates* (lemurs, "tree shrews", tarsius, apes, and man), *Insectivora* (moles), *Chiroptera* (bats), *Carnivora* (cats, dogs, bears, seals, and walruses), *Cetacea* (whales, porpoises, and dolphins), *Perissodactyla* (horses, and all animals with a single uncloven hoof), *Artiodactyla* (pigs, sheep, cattle, antelope, camels, all animals with a cloven hoof), *Proboscidae* (elephants), and *Rodentia* (rats, mice, guinea-pigs, squirrels, porcupines). The *Insectivora* are probably the oldest order amongst the living *Eutheria* and it is likely that they gave rise to the primates.

The order *Primates* includes various animals that share many characteristics of dentition. The incisors are often spatulate, the canines are usually well developed and the molars have two or three roots and two or more cusps. The crowns of the teeth are covered by enamel, and cementum is confined to the roots of the teeth. Two dentitions, a primary (deciduous) and a secondary (permanent) dentition, develop.

Man's position

In the composition of his body, man displays such an amazing similarity to lower animals, especially to mammals, that it is surprising that doubt ever existed regarding an evolutionary relationship. His skull and skeleton consists of the same bony elements, his muscular system consists of the same muscles arranged in a similar manner, his heart and vascular system is organized in an almost identical way, and his brain, although more complex, consists of the same type of cells. Man is anatomically one of the animals and, more specifically, a mammal, since distinguishing features of this group, such as warm-bloodedness, hairiness, a heterodont dentition, and the ability to give live birth to offspring and to suckle them afterwards, are all present.

Classification of man

Man belongs to the class *Mammalia* (mammals), is naturally a member of the *animal kingdom* and is grouped into the phylum *Chordata*. Since he develops a vertebral column, man is classed in the subphylum *Vertebrata*.

Because of certain anatomical characteristics, man belongs to the order *Primates*.

It is difficult to give a satisfactory definition of the primates, since there is no single common distinguishing feature among all the members of this order. Although many other mammalian orders are characterized by strikingly distinguishing features, the primates in general developed a very unspecialized anatomy.

According to the classification of mammals by Dr. G.G. Simpson (1945, University of Harvard), man is grouped in the suborder *Anthropoidea* which also includes the anthropoid apes (man-apes) and others. The other suborder in the primates are the *Prosimii* which includes lemurs and others. Members of the *Anthropoidea* are distinguished by a human-like appearance and share a few characteristics, for example a prehensile hand, a relatively large head and flat face, the proximity of the eyes, small ears, an alert facial expression, mobility of the lips, and relatively flat nails on the hands and feet. The manner of giving birth and suckling the young are physiologic characteristics shared by this group.

The *Anthropoidea* are further divided into three superfamilies: the *Hominoidea* includes anthropoid apes and man, *Cercopithecoidea* includes the Old World Monkeys such as the common baboon an many others, and the *Ceboidea* which includes the spider monkeys, bush-babies and others.

The *Hominoidea* are divided into two families, the *Hominidae* and *Pongidae*. The *Hominidae* includes modern and extinct human forms, while the *Pongidae* includes modern and extinct forms of anthropoid apes (gorilla, chimpanzee, and orang-outang).

Man, together with extinct forms of man, belongs to the genus *Homo*, while modern man belongs to the species *sapiens*.

The face

In a typical non-primate such as a dog or sheep, the facial skeleton (the snout) projects forwards from beneath the cranium. During the evolution of the primates the face was gradually displaced backwards to a position underneath the cranium, a development which has been completed in modern man. An enlarging brain and a flexure in the axis of the cranial base in the region of the pituitary fossa were contributory factors in this development. There is probably also a connection between the upright posture and bipedal locomotion assumed by man and the above changes.

During primate evolution there was a gradual lessening of the importance of the sense of smell, and a consequent reduction in the size of the snout and nasal cavities. These changes led to a reduction in tooth space which was compensated for in anthropoid apes by a concomitant increase in size of the alveolar ridges (prognathism). The eyes moved to the front of the face in primates. The above changes resulted in a backward shift of a facial bone, the facial process of the ethmoid bone in lower animals, to the medial wall of the eye socket where it is termed the orbital plate of the ethmoid bone. In anthropoid apes, who generally favor a herbivorous diet, eating is a continuous process, resulting in a compensatory adaption of the entire masticatory apparatus (teeth, muscles, and alveolar processes). These parts are exceptionally well developed.

Mastication and heterodontism

Maintenance of a constant body temperature requires a reliable source of food that is not season-dependent. Mammals have adapted by developing a masticatory apparatus and digestive processes which allow the more efficient utilization of a wide range of available foodstuffs. This is probably a cardinal factor in the present dominant position of mammals.

The heterodont dentition of mammals, in which different teeth have different shapes and functions, evolved from a homodont dentition in which all the teeth have a similar conical form, as seen in reptiles. The jaws and teeth of most reptiles are merely traps in which the prey is caught before being swallowed. In living reptiles the teeth are often inclined backwards and this facilitates the swallowing process, while preventing escape of the prey. In extinct herbivorous reptiles the only function of the teeth was to tear leaves and other plant material into smaller portions. The masticatory apparatus of mammals can achieve much more. The jaws have been strengthened and a heterodont dentition has developed. This enables the animal to cut up the food and to reduce it to small particles by an efficient chewing process. A particularly notable distinguishing characteristic of mammals is that they can generally process a larger variety of foodstuffs. Furthermore, the efficiency of digestive juices has increased due to more efficient mastication as well as to an increased body temperature. A new enzyme, amylase, has appeared enabling the digestion of starch to commence in the mouth.

A large variety of heterodont dentitions developed in different mammalian orders, even to some extent amongst the primates, but extreme forms of specialization, for example the tusks of elephants and other examples in the animal kingdom, are not found in primates. In some mammals, however, the trend towards a heterodont dentition appears to be reversed. In dolphins, for example, the dentition is homodont, while whalebone whales have lost their teeth altogether.

A further development in mammals was the loss of a continuous succession of teeth as is seen in lower animals like fishes, reptiles, and amphibians. A mammal has only two dentitions, a primary (deciduous) and a secondary (permanent) dentition. A change of diet and of manner of use which resulted in fewer teeth being violently lost were probable factors influencing this evolutionary development.

The jaws

The concept of biologic efficiency implies that an organ, or other part of the body, will be reduced in size by genetic adaption during evolution if it is larger than is necessary. Compared to other primates, man has developed smaller jaws. This is advantageous because it ensures a better balance of the head on the vertebral column in the upright posture. A perfect balance has probably been developed in modern man.

Observations on animals reveal that the final size of the jaws and the masticatory muscles is largely dependent on the amount of work these parts are subjected to during ontogenetic growth, but that the teeth are not influenced to the same extent by environmental factors. Tooth size is more directly under genetic control.

Although the crowns of erupted teeth are subjected to environmental influences resulting in, for example, attrition (wear), the basic size and morphology of a tooth crown is predetermined. It can be said that the size of the jaws in modern man has not adapted accordingly. The phenotype (ultimate jaw size) has been altered by environmental factors but the genotype (tooth size) has not. The result of the above is that in many instances the jaws are too small for the teeth and irregularities in tooth placement occur. In this respect it can be said that orthodontic treatment neutralizes selection pressure.

The morphology of the tooth crowns and the manner of chewing, especially a forcible sideways grinding action by the molars, causes great horizontal forces on the jaws. The sideways swinging movement of the mandible is largely due to alternate contraction of the two lateral pterygoid muscles. In this way the force of the muscles is transmitted to both sides of the mandible in turn. This causes a great horizontal tension on the bone, especially in the symphyseal area. In modern man this area is strengthened by the development of a chin but in other primates, exspecially the modern large apes, this area has been reinforced on the inner inferior aspect by the development of a bony "simian" shelf which stretches from one side to the other.

As a result of the more horizontal position of the human head in relation to the vertebral column, a change which was probably made possible by a more anterior positioning of the occipital condyles, the mandible is located closer to the neck than in other animals. In the absence of any adaptation to this situation, the mandible would occlude the trachea, larynx, and the great vessels in this area when the mouth is opened. Adaptation did, however, occur and the eversion of the chin developed to create more space between the mandible and the throat.

The teeth

The general arrangement of the teeth in the jaws will be described before specific teeth are considered. In man the teeth are arranged in the U-shaped dental arches. In other primates the teeth lie in two rows which converge towards the incisors, or in the form of the three sides of a rectangle. Fossil remains tend to support the view that the human arch form probably evolved from the former arrangement.

Since the different heterodont tooth types of mammals developed from homodont dentitions of reptiles, there are often few fundamental differences between tooth types in mammals. It is, for example, not easy to distinguish between the incisors and canines, or the premolars and molars, of sheep. Many similar examples can be found. Most primates, however, possess four distinct tooth types, namely incisors, canines, premolars, and molars. The teeth are numbered from anterior to posterior. A generalized placental mammal from which man and the other mammals were probably derived, could have had the following teeth present in each half (quadrant) of both jaws: $I1$, $I2$, and $I3$ (three incisors); C (one canine); $P1$, $P2$, $P3$, and $P4$ (four premolars), and $M1$, $M2$, and $M3$ (three molars). The total of 44 teeth in this ancestral mammal has in man been reduced to 32, with loss of 12 teeth, probably $I3$, $P1$, and $P2$.

The incisors of mammals are generally adapted to cutting food into smaller pieces, although they have undergone considerable specialization in some cases, notably as the tusks of elephants. Incisors enable mammals to reduce food to smaller pieces to facilitate chewing and swallowing, a mechanism not possessed by other animals such as reptiles. The higher primates, including man, possess only four incisors, two per quadrant, in each jaw.

The canines are more interesting teeth from an evolutionary point of view. During the entire period of mammalian evolution, the canines tended to retain a pointed reptile-like form. The function of these teeth is primarily prehensile which is important for animals with a carnivorous diet. The herbivores tend to use the canine as a weapon, especially in those animals in which the claws and nails were not developed to aid in defense or offense, as in baboons who possess canines of extraordinary dimensions. In these animals the presence of large canines cannot be ascribed to the nature of their diet. The greater size of canines in the male animal is an example of sexual dimorphism, the male usually being more aggressive, threatening, and protective of the interests of his family.

Animals with large canines have a diastema, or space, in the opposing jaw into which the canines fit. The diastema for the mandibular canine is always anterior to the upper canine. The diastema is absent until the canines have erupted fully and is the result of tooth movement rather than genetic factors. Although the shape of the human canine may vary slightly from person to person, it is significant that its root is larger than that of any other tooth. This probably indicates that man's single canine in each quadrant evolved from a very large ancestral predecessor, a view supported by fossil remains.

The premolars probably had a variety of functions during mammalian evolution. In general, they gradually evolved from reptilian precursors with a prehensile function, to flatter teeth which could crush and grind food. In many cases they assumed a shape which is similar to that of molar teeth, although usually smaller. This process is referred to as molarization. Man has two premolars in each quadrant of both jaws.

The molar teeth have in the past been the topic of many discussions. It can, however, be assumed that a more or less quadrangular tooth with four cusps is found in both jaws of all higher primates, although a fifth cusp is found in the lower molars of some. In most primates there are three molars in each quadrant.

Selected bibliography

1. Le Gros Clark, W.E. (1970) *History of the Primates,* 10th edition. London: Trustees of the British Museum (Natural History).
2. Osborn, J.W. (ed) (1981) *Dental Anatomy and Embryology.* Oxford: Blackwell Scientific Publication.
3. Romer, A.S. (1965) *Man and the Vertebrates:* 1, and (1966) *Man and the Vertebrates:* 2. Harmondsworth, Middlesex, England: Penguin Books Ltd.
4. Romer, A.S. (1970) *The Vertebrate Body,* 4th edition. Philadelphia: W.B. Saunders Company.
5. Scott, J.H. and Symons, N.B.B. (1982) *Introduction to Dental Anatomy,* 9th edition. Edinburgh: Churchill Livingstone.

Review questions

1. Explain what is meant by the hierarchial system of classification.
2. Classify modern man and give a brief description of each category in which man is placed.
3. Briefly describe the evolutionary changes which the human face has undergone.
4. Discuss the advantages of a heterodont dentition in the processing of food.

29. Teeth through the ages

General review

The dentitions of modern mammals evolved from simple predecessors. Throughout this development there was in most cases a fairly close relationship between form and function, between tooth structure and diet. When the teeth of man are considered, it is similarly clear that the nature of the diet influences tooth form as well as other components of the masticatory apparatus. This aspect will be dealt with later in this chapter.

Geologic time may be divided into four major Eras. The oldest era, in which few signs of life are found, is the Precambrian Era which began with the formation of the earth's crust appromimately 4500 million years ago and ended some 570 million years ago. The earliest forms of life probably originated some 1750 million years ago and evidence indicates an aquatic habitat. These life forms were probably prokariotic (algae).

The following era was the Palaeozoic, which began at the end of the Precambrian Era and lasted until approximately 230 million years ago. During the greater part of this Era fishes were probably the only vertebrate animals, but towards the end of this era amphibians and reptiles appeared. Plant life on dry land started during this era and was probably an important factor in the development of land animals.

The Palaeozoic Era was followed by the Mesozoic Era (230 million to 65 million years ago) which is often called the Age of Reptiles, since giant reptiles ruled over all other life forms for the major part of this era and had a dominant position. The domination of the reptiles probably ceased towards the end of this era and it is probable that true mammals were by then already in existence.

Approximately 65 million years ago the Cainozoic Era began. This era is often called the Age of the Mammals, since it was dominated by mammals, and still is. The Cainozoic Era is divided into the Tertiary and Quaternary Periods. The Quaternary Period is the most recent period and began 1–3 million years ago and has lasted until the present time. Man appeared on earth in this period.

The Palaeozoic Era is divided into the following Periods (estimated time of origin before the present time in parentheses):

Period

Cambrian	(570)
Ordovician	(500)
Silurian	(450)
Devonian	(400)
Carboniferous	(350)
Permian	(280)

In the same way the Mesozoic Era is divided as follows:

Period

Triassic	(230)
Jurassic	(180)
Cretaceous	(130)

The division of the Cainozoic Era is as follows:

Period

Tertiary	(65)
Quaternary	(2+)

During the Palaeozoic Era true vertebrates appeared. They were probably eel-like fishes without scales or jaws. The mouth was funnel-shaped with many small horn-like teeth arranged on the sides. These animals led a parasitic existence by attaching to the bodies of other sea creatures and rasping off the flesh of their hosts by means of their teeth. The modern descendants of these fishes are the lampreys which are classified in the class *Agnatha* and the order *Cyclostomata*.

By the middle of the Palaeozoic Era, fishes with functional jaws appeared. The sharks were amongst these fishes. Further attention will be paid to the sharks later in this chapter but they are not very important from an evolutionary point of view since they did not show any evolutionary tendency nor any terrestrial ambition. In the class *Osteichthyes* (the higher bony fishes) which made its appearance during the Palaeozoic Era, one finds the subclasses *Actinopterygii* (ray-finned), which has little evolutionary significance, and the *Sarcopterygii* (Choanichthyes), fishes with fleshy fins and sometimes with internal nasal openings. In the latter subclass one finds two orders, the *Crossopterygii* (lobe-finned) and the *Dipnoi* (lung-fishes). The famous celacanth (Latimeria) belongs to the *Crossopterygii*.

The existence of this primeval fish (it does not possess internal nasal openings) was unknown until one was caught by a fishing trawler off the coast of East London, Republic of South Africa, and identified by J.L.B. Smith, an undisputed authority on the sea fishes of Southern Africa, in 1939. This fish had lobe-shaped fins with a skeletal support which articulated with the shoulder girdle at one end, and with a joint of which the distal element had small radiating bones, at the other end.

This arrangement was very reminiscent of the limbs of a terrestrial animal. Other specimens of this fish have been caught since.

Related forms of these fishes can utilize free air and are able to survive for long periods in the muddy beds of dry streams or ponds by respiring through small openings in the mud. Examples of these fishes, such as variants of the catfish, are found in Australia, in Africa, in South America and in the United States. It has even been reported that they can migrate by means of a crawling motion for considerable distances over dry land in search of fresh water. The African type is reportedly so dependent on free air that it will suffocate when kept submerged for too long. Contrary to expectations, the lung-fishes are not regarded as ancestral to land animals. This honour belongs to members of the *Crossopterygii.*

In the theory of evolution a critical stage has now been reached. How did the first fishes abandon the water for a terrestrial future? At this stage there were no vertebrate animals living on dry land, although plant life was plentiful. The earth was slowly drying up. Isolated pools formed, the remaining water become deprived of oxygen and stagnant. Unless fishes could overcome these obstacles they were doomed to extinction. It is believed that selection pressure caused some fishes to develop the ability to utilize surface air. Primitive lungs evolved simultaneously. Fins elongated to form short limbs and the fishes could leave stagnant pools in search of fresh water. A similar type of behavior has already been referred to in the description of the lung-fishes.

Up to this stage tooth morphology had not conformed to a consistent pattern. The teeth of the lampreys were irregular and horn-like, the celacanths had sharp pointed teeth, and the teeth of the lung-fishes were in the form of fused tooth plates. Although of little significance in evolutionary history, sharks are of interest. They belong to the class *Chondrichthyes* (cartilagenous fishes) and were already plentiful in the Palaeozoic Era. They showed little evolutionary development over millions of years. The sharks almost became extinct towards the end of this era due to the development of a shortage of other marine life, and they had to adapt to other types of food such as shellfish (crustaceans) for survival.

The Port Jackson shark *(Heterodontus japonicus)* is probably the only surviving member of a large group of sharks that became extinct. The teeth of this fish are not identical throughout the mouth and show adaptation to a shellfish diet. In the front of the jaws several rows of sharp teeth are found, while flat paving stone-like teeth are seen in the back of the mouth. The dentition is an example of a primitive heterodont type which has been adapted to prise shellfish off rocks with the anterior teeth, while crushing the shells with the posterior teeth.

Most of the present-day sharks evolved early in the Mesozoic Era. Amongst true sharks all the teeth are uniformly sharp, but is is believed that modern sharks did pass through a heterodont stage similar to that depicted by the Port Jackson shark. Sharks have a continuous succession of teeth, a characteristic called polyphyodontism. All the teeth in the mouth are similar, therefore they have a homodont dentition. The teeth may be flat and triangular with a fine serrated edge, as found in the feared maneater shark which is also known as the blue shark, blue pointer, white pointer, or great white, all being classified as *Carcharodon carcharias,* and in the Zambezi shark (*Carcharinus zambezensis,* according to J.L.B. Smith in *The Sea Fishes of Southern Africa,* 1965 and *Carcharinus leucas,* according to D.H. Davies in *About Sharks and Shark Attack,* 1964). The teeth of the ragged-tooth shark *(Carcharias taurus)* are long and cone-shaped (haplodont) with very sharp tips. If the prey is small, it is taken whole and immediately swallowed, but if it is large it is bitten and violently shaken to tear off manageable portions which are immediately swallowed. The backward inclination of the teeth assists the latter process. Mastication is impossible and no functional occlusion exists. Jaw movement is of a simple hinge-type and the upper and lower jaws are connected by fibrous tissue.

It is fitting to discuss the classification of fishes very briefly. Although all fishes are sometimes erroneously classified into one class, the great variation that exists amongst fishes merits subclassification and the following has been suggested:

Superclass: Pisces (all fishes)

Class:
(a) *Agnatha,* jawless fishes such as the lampreys and extinct forms.
(b) *Placodermi,* primitive fishes of the Palaeozoic Era, extinct.
(c) *Chondrichthyes,* cartilagenous fishes such as the sharks.
(d) *Osteichthyes,* bony fishes, presently the largest component of the fish population.

To distinguish the *Agnatha* from all other animals with functional jaws, the term "gnathostomes" is used for the latter. From an evolutionary perspective the *Agnatha* gave rise to the *Placodermi* (characterized by an armoured skin), and these in turn to both the *Chondrichthyes* and the *Osteichthyes*. The *Chondichthyes* did not develop further, and higher life forms such as amphibians, reptiles, mammals, and birds probably evolved from the bony fishes, especially as descendents of life forms which can be described as members of the *Crossopterygii.*

The development of the class *Amphibia* (a subdivision of the superclass *Tetrapoda* which includes the land animals, belonging to the classes *Reptilia,* with some exceptions, *Aves* and *Mammalia*) from fishes is based on the assumption that the ancestral fishes abandoned a strictly aquatic habitat. The ability of some fishes to achieve this has been mentioned earlier. Further developments in the fleshly bone-supported fins of members of the order *Crossopterygii* enabled them to locomote in mud and later on dry land. Primitive air-breathing lungs developed. These are the broad outlines of the evolution of the amphibians. They were initially four-limbed fishes able to utilize free air and to move in mud and on dry land. The frogs (order *Anura,* subclass *Lissamphibia*) are probably the most successful amphibians. Their teeth, when present, are arranged in a single row on the upper jaw. They are haplodont, homodont, and their dentition is polyphyodont. Their diet consists largely of insects and worms.

The amphibians are the ancestors of all land animals. The reptiles (class *Reptilia*) followed on the amphibians. Modern reptiles (lizard-like types, snakes, and tortoises) are relatively common in tropical and subtropical regions but absent in cold regions where survival is difficult for cold-blooded animals. The first reptiles were clumsy animals with limbs that projected sideways from the body. Reptiles were present towards the end of the Paleozoic Era and assumed a dominant position in the following Mesozoic Era. The dinosaurs dominated all other life forms and an outstanding example of these reptiles was *Tyrannosaurus rex*, a bipedal partially upright animal with massive hind legs and very much reduced forelimbs. From fossilized remains it is reliably estimated that this reptile was 15 m long, with a standing height of approximately 6 m and a head length of some 2 m. Its teeth were sharp and cone-shaped with crowns 125 mm long, the dentition was homopolyphyodont and consisted of 12–14 teeth in each quadrant of both jaws. *Tyrannosaurus rex* was a carnivore and its main prey was another dinosaur, the herbivorous *Triceratops*. The dinosaurs and reptiles were a very large group of animals but were subjected to catastrophically severe climatic changes in the Cretaceous Period of the Mesozoic Era. Mountains were formed, inland lakes and swamps dried up (the massive reptiles needed the support of water for buoyancy), the plants changed. The herbivores could not survive and both they and the carnivorous dinosaurs became extinct. The big dinosaurs were not responsible for further evolutionary developments which led to birds and mammals.

Birds (class *Aves*) are probably descended from primitive flying reptiles. Although certain reptiles, notably members of the order *Pterosauria* (winged reptiles) and the pterodactyls (wing-fingered) were able to glide, probably a parachute-like motion, birds are considered to have evolved from other reptiles in which flying feathers developed on the upper limbs to form the wing surfaces. The Pterosaurs "flew" by means of a membranous bat-like flap of skin which extended from the elongated fourth "finger" to the thigh region. Many modifications were, however, necessary before successful flying was possible. The skeleton had to be lightened (air sacs and hollow bones), and the birds had to develop a high rate of metabolic turnover and a constant body temperature. The exact period when flight by means of movable wings became a reality is uncertain, but was probably in the middle of the Mesozoic Era. In contrast to some extinct primitive Mesozoic forms, birds do not have teeth. A stout beak developed in place of teeth. A good example of a primitive bird with teeth is found in fossilized remains of the *Archaeopteryx* from the late Jurassic Period. This creature had a reptile-like beak and tail, with sharp claws on the wingbones.

The mammals evolved from reptiles, but the reptilian ancestors of mammals diverged from other reptilian stock at a very early stage. An evolutionary relationship between mammals and present-day reptiles is therefore so vague that it may be disregarded.

The reptilian ancestors of the mammals were small and insignificant when compared to the dinosaurs. Mammal-like reptiles are found in the Permian and Triassic Periods. They were carnivorous and agile quadriped runners in which the elbow and the knee were placed more directly under the body, unlike other early reptiles. This gave better support to the body and made greater speed of movement possible. They belonged to the order *Therapsida* (class *Reptilia*). Factors which could have contributed to the later ascendency and domination of the mammals were "intelligence", efficient movement, improvements in the blood circulation, and a high, constant body temperature (a factor linked to the development of body hair). Most of these features already developed under the domination of the big reptiles when speed and "intelligence" were essential for survival.

The teeth of a primitive Jurassic mammal-like reptile, for example Amphitherium, already show considerable variation when compared to the teeth of a typical reptile. This animal was one of the first reptiles to show signs of a heterodont dentition in which the teeth are differentiated into incisors, canines, premolars, and molars.

Many mammalian types developed to form the class *Mammalia*, for example, the subclass *Prototheria* (egg-laying mammals), and the subclass *Theria* (mammals bearing the young alive). In the latter subclass one finds the order *Marsupialia* (infraclass *Metatheria*) in which the young are born alive but in an underdeveloped state. The Australian kangaroo is an example. The infraclass *Eutheria* includes all mammals with an efficient placenta. These include the order *Insectivora* (shrews and moles), the order *Chiroptera* (bats), the order *Primates* (apes and man), the order *Carnivora* (dogs and cats), the order *Perissodactyla* (horses), the order *Artiodactyla* (cattle and sheep), the order *Proboscidae* (elephants), the order *Cetacea* (whales and dolphins), and the order *Rodentia* (rodents).

The original mammalian dentition probably consisted of three incisors, one canine, four premolars and three molars in each quadrant. The dental formula is as follows, indicating the number of teeth present in maxillary and mandibular quadrants on one side of the mouth:

$$I\frac{3}{3} \; C\frac{1}{1} \; P\frac{4}{4} \; M\frac{3}{3}$$

This was the beginning of the development of a heterodont dentition which varies in nature in different mammalian orders. In the primates extreme dental specialization, as seen in the tusks of elephants (the lateral incisors) and in the upper canine of the walrus, is not present.

A heterodont dentition made mastication possible. They could break the hard exoskeleton of insects to utilize the soft tissues within, they could crush nuts and seeds to reach carbohydrates and fats, they could liberate sugars and starches from roots and tuberous plants, and utilize proteins from plant material.

A further development was the cessation of continuous tooth succession. Mammals are characterized by one, or more commonly two, sets of teeth, a primary and a secondary dentition. Changes in the diet and the manner of tooth usage probably lessened the possibility of violent tooth loss which reduced the necessity for continuous tooth replacement.

Specializations in finer details of tooth morphology are common in different mammals. These differences are largely confined to the tooth crowns and often make it possible to deduce the nature of the animal's diet from crown morphology.

In herbivores the number of teeth varies but the morphology of the crowns is relatively constant.

Sheep have the following formula for the secondary dentition:

$$I\frac{0}{3} \ C\frac{0}{1} \ P\frac{3}{3} \ M\frac{3}{3}$$

This indicates a loss of 12 teeth when compared to the original mammalian dentition.

The horse and the pig (except the wart-hog) have the full complement of 44 teeth with the following classical formula:

$$I\frac{3}{3} \ C\frac{1}{1} \ P\frac{4}{4} \ M\frac{3}{3}$$

The canines of the horse are more prominent than those of the sheep.

The pointed teeth of early mammals were unsuited for the processing of vegetable foods and consequently the crowns of the teeth of herbivores, especially the premolars and molars, broadened with the development of vertical occlusal ridges, composed of enamel, dentine, and cementum, running in an antero-posterior direction.

The temporomandibular joints of herbivores has a flat temporal articulating surface which does not restrict the free side-to-side swinging motion of the mandible. The uneven occlusal wear of the teeth creates an ideal roughened surface for crushing plant material.

The porcupine, a large modern rodent, has the following formula for the secondary dentition:

$$I\frac{1}{1} \ C\frac{0}{0} \ P\frac{0}{0} \ M\frac{4}{4}$$

The molars of rodents are reminiscent of those of the typical herbivores, but the occlusal ridges run in a lateral direction in harmony with the movements of the mandible. The mandibular fossa on the temporal bone is a relatively deep antero-posterior groove which allows a hinge movement of the mandible, as well as an antero-posterior sliding movement. No lateral movement is possible. This ensures an optimal chewing process on the molars, while the continuously erupting incisors are ideally adapted to a gnawing function. These very long teeth are embedded in the jaws as far back as the last molar teeth, and grow in a segment of a circle.

It is interesting to note that the members of the cat family cannot chew effectively and are strictly carnivorous. Dogs, on the other hand, are not strict carnivores and do possess chewing ability. Bears, except the strictly carnivorous polar bears, are omnivores with considerable chewing ability. Their dentition is, nevertheless, not readily distinguishable from that of a typical carnivore. The carnassial element in the premolars is, however, reduced and the grinding surfaces of the molars has broadened in the herbivorous bears.

The family *Felidae* (domestic cats, lion, and leopards) are the strictest carnivores in the natural state. Their secondary dental formula is the following:

$$I\frac{3}{3} \ C\frac{1}{1} \ P\frac{3}{2} \ M\frac{1}{1}$$

The two characteristic features of their dentition are the following:

(a) The canines are robust and considerably elongated.
(b) The premolars and molars, especially the only molar (the carnassial tooth) have a blade-like appearance. The functioning molars have a scissors-like action which slices the meat.

The jaw joints of the *Felidae* permit only a hinge action since the mandibular fossae are laterally directed deep grooves. Little, if any anterior movement of the mandible is permitted. Masticatory ability is minimal and meat is merely sliced up before being swallowed.

The suborder *Anthropoidea* (apes and man) is included in the order *Primates*. In general, the primates are omnivorous and this lack of dietary specialization is reflected in poor dental specialization. On the other hand, a lack of dental specialization may be viewed as an adaptation to an omnivorous diet.

A chimpanzee has the same dental formula as a human. The formula for the secondary dentition is as follows:

$$I\frac{2}{2} \ C\frac{1}{1} \ P\frac{2}{2} \ M\frac{3}{3}$$

The incisors are relatively short and broad, similar to their human counterparts, but have an edge-to-edge contact with their opponents. The normal human incisor relationship is marked by both an overbite and an overjet. The canines, especially in males, are longer than the other teeth but not as long as in some other apes. The premolars and molars are almost identical in crown morphology to their human counterparts.

The masticatory apparatus of man with reference to evolution and the influence of diet

A total of 52 teeth develop in man. These consist of 20 teeth in the primary (deciduous) dentition and 32 in the secondary (permanent) dentition. Twenty of the secondary teeth have deciduous predecessors.

The formula for the primary dentition is as follows:

$$I\frac{2}{2} \; C\frac{1}{1} \; DM\frac{2}{2}$$

The letters DM denote the deciduous molars.
The formula for the secondary dentition is the following:

$$I\frac{2}{2} \; C\frac{1}{1} \; P\frac{2}{2} \; M\frac{3}{3}$$

Compared to the ancestral mammalian dentition, it is widely accepted that man has lost the third incisor and the first two premolars, leaving him with I1, and I2, C, P3, and P4, and M1, M2, and M3. Many authors are of the opinion that man is still in the process of tooth reduction. The root of the human canine is still disproportionately large and may be the last tooth to be lost if such an evolutionary tendency is accepted. The size of the tooth may point to a time in evolutionary history when this tooth was perhaps the greatest natural weapon of man's ancestors, and consequently has considerable genetic importance.

The most specialized members of the human dentition are possibly the incisors. In primitive man they still meet edge-to-edge as in chimpanzees. They are adapted to an omnivorous diet. They can be used to tear meat off bones, to bite into fruit, and to perform delicate sideways and nibbling movements because no restriction in the form of a large canine, or a peculiarity of temporomandibular anatomy, inhibits free movement of the mandible.

Most humans are omnivores and the average Western diet is more varied than that of any animal. Some people are vegetarian by their own choice, whilst Eskimos were in the past exclusively carnivorous in their natural state, no other foodstuffs being available. The human masticatory apparatus, consisting of the teeth, the jawbones, and the masticatory muscles, is subject to a variety of modifications depending on the physical nature of food.

The most common effect of food is attrition (wearing) of the crowns. In the past this was of fairly common occurrence in peoples who followed a hunter-gatherer lifestyle, like the Kalahari San (formerly known as Bushmen), the Australian Aborigines, and the Eskimos. The San were largely dependent on roots and tuberous plants for survival. Meat was not as freely available, and when an animal was killed most of the meat was immediately eaten. The food could not be rinsed due to the shortage of water and the attached soil particles acted as an abrasive on the teeth when chewed. The Australian Aborigines had a similar lifestyle. In these peoples, changes were confined to the tooth crowns which showed considerable attrition.

The nomadic Eskimos subsisted on meat and fat obtained mainly from whales, fish, seals, and the walrus. In summer the meat was laid on the ground to partially dry out and as a result, gathered soil particles. These had the same effect on the teeth as described for the San and the Aborigines. The dried meat (pemmican), a favorite winter food, required quite considerable force to chew and this had a stimulating effect on muscle and bone growth. Both attrition and robust masticatory muscles and alveolar processes are characteristic of the traditional Eskimos. It was reported that these changes were not as distinct in a younger generation who had had increasing contact with a Western-type diet. On the other hand, it was reported that the younger generations were more subject to crowding of the teeth.

The above observations tend to support the theory that the form and size of the basal (supporting) bone of the jaws, as well as tooth size, is under genetic control but that the form and size of the alveolar processes and the masticatory muscles can be modified by function.

Selected bibliography

1. Le Gros Clark, W.E. (1970) *History of the Primates*, 10th edition. London: Trustees of the British Museum (Natural History).
2. Osborn, J.W. (ed) (1981) *Dental Anatomy and Embryology*. Oxford: Blackwell Scientific Publications.
3. Romer, A.S. (1965) *Man and the Vertebrates:* 1, and (1966) *Man and the Vertebrates:* 2. Harmondsworth, Middlesex, England: Penguin Books Ltd.
4. Romer, A.S. (1970) *The Vertebrate Body,* 4th edition. Philadelphia: W.B. Saunders Company.
5. Scott, J.H. and Symons, N.B.B. (1982) *Introduction to Dental Anatomy,* 9th edition. Edinburgh: Churchill Livingstone.

Review questions

1. Describe the geologic time scale as set out in this chapter and refer briefly to life forms which existed in each period.
2. Describe briefly the origin of a heterodont dentition.
3. What is regarded as the ancestral mammalian dental formula?
4. What are the advantages of a heterodont dentition?
5. Describe the tooth form and mandibular fossa osteology of a typical herbivore, carnivore, and omnivore.
6. Describe briefly the influence of diet on the masticatory apparatus of man.

Part II

1. A tooth and its surroundings

The development of a tooth

The development of the teeth commences during the 7th week of intra-uterine life when the embryo is approximately 15 mm long. The epithelium of the embryonic mouth thickens in a U-shaped band in the area of the future alveolar processes and this primary epithelial thickening, or band, rapidly differentiates on its deep aspect into a buccal vestibular lamina and a lingual dental lamina. The vestibular lamina later undergoes certain changes which transform it into the vestibule of the mouth, while the more important dental lamina will subsequently give origin to the teeth. Both laminae are composed of surface oral epithelium and are separated from the underlying mesenchyme by a basal (germinal) cell layer and a basal lamina.

The development of ten small localized thickenings of the dental lamina, the tooth buds of the primary teeth, leads to the next stage of tooth development. At this stage there is a marked increase in mitotic activity in the bud and in the adjacent mesenchyme. Bone formation in the future alveolar process begins at the same time. The tooth bud undergoes peripheral proliferation into the mesenchyme, leaving a central mesenchyme-filled depression, to form the tooth cap which consists of the same components as the tooth bud.

With further peripheral proliferation of the cap epithelium into the mesenchyme, the depression on its deep aspect deepens. The epithelial organ resembles a bell and this stage of development is referred to as the bell stage. At the same time, histodifferentiation commences. The basal cells which line the mesenchyme-filled depression become the inner, or internal, enamel epithelium, while the basal cells which cover the outer surface of the epithelial organ become known as the outer, or external, enamel epithelium. Intercellular spaces between the internal cells of the "bell" widen considerably, but they retain desmosomal contact. They assume a star-shaped appearance and consequently this region is known as the stellate reticulum. A few layers of flattened squamous-type cells form between the inner enamel epithelium and the stellate reticulum. This is the stratum intermedium or intermediate cell layer.

The mesenchyme which is included in the invagination of the deep surface of the epithelial organ, is the dental (tooth, or dentine) papilla, while the mesenchyme surrounding the outer enamel epithelium is the dental follicle.

The epithelial organ, composed of the outer and inner enamel epithelia, the stellate reticulum and the stratum intermedium, constitutes the enamel organ which initially remains connected to the overlying oral epithelium by a thinned dental lamina. The enamel organ, together with the dental papilla and the dental follicle, is the tooth germ. The enamel organ will give rise to the enamel, while the dental papilla will give rise to the pulp and dentine, and the follicle will give rise to cementum, the periodontal ligament and alveolar bone. The root of the tooth will arise where the outer and inner enamel epithelia meet (the cervical loop), from where a combined epithelial membrane, from which the stellate reticulum and the intermediate cells are excluded, proliferates into the underlying mesenchyme. This membrane is the root sheath of Hertwig.

A tooth is formed by a complex cellular and biochemical process. Cells on the periphery of the dental papilla, adjacent to the inner enamel epithelium, are induced by the overlying inner enamel epithelium to differentiate into odontoblasts which are involved in dentine formation. When dentine starts forming, the name of the papilla changes to dental pulp. The earliest dentine (predentine) induces the cells of the inner enamel epithelium to become the ameloblasts which form the enamel. The future neck of the tooth, or amelocemental (enamel-cement) junction, will form at the point where the root sheath arises. Dentine forms on the papillary aspect of the root sheath in the same way as described above, while cementum forms in association with the outer aspect of the root sheath. With the formation of cementum, the root sheath cells lose continuity but persist as a network of cells, the cell rests of Malassez, amongst the fibers of the periodontal ligament. The periodontal ligament fibers form in the dental follicle and attach the cementum to the alveolar bone.

Tooth eruption

Eruption of a tooth commences soon after completion of the crown. Various theories regarding the mechanism of tooth eruption exist and these will be discussed later. They include root development, changes in the dental pulp, changes in the alveolar bone, changes in the periapical tissues and changes in the periodontal ligament. The last-mentioned theory currently enjoys the most support.

After the tooth crown erupts into the mouth through the oral mucosa, the epithelial coverings of the crown rapidly disintegrate and become incorporated into the gingiva to form the junctional, or attachment, epithelium, which attaches to the enamel, and the sulcular (crevicular) epithelium which lines the gingival sulcus. The position of the junctional epithelium is not static. It is initially attached to enamel as far cervically as the amelocemental junction. Under the influence of factors such as aging, poor oral hygiene, or poor brushing techniques, the attachment may migrate in an apical direction till it is found entirely on cementum. The junctional and sulcular epithelia are not keratinized.

The components of a tooth

A tooth consists of three mineralized tissues, enamel, dentine, and cementum, and a dental pulp. The enamel is of ectodermal origin while the other components are of mesenchymal origin. The enamel consists of millions of rods, or prisms, which extend from the amelodentinal (enamel-dentine) junction to the surface of the crown. Each rod is composed of apatite crystals, mainly hydroxyapatite, and is optically separated from adjacent rods by a rod sheath, or cortex. The existence and nature of the rod sheath is controversial. Some believe that the appearance is due to the presence of a higher content of organic material, while others believe that differences in crystal orientation of adjacent rods in this zone, with resultant differences in light refraction, cause an optical illusion seen as a sheath. These views are probably reconcilable.

The chemical composition of enamel is important. It consists of 96-97% inorganic material (chiefly hydroxyapatite-$Ca_{10}(PO_4)_6.(OH)_2$, while other elements are also present), 1% organic material and 3-4% water by weight. It is the hardest and most highly calcified biological tissue known.

Dentine consists of mineralized tissue (chiefly hydroxyapatite) in which microscopic tubules are found extending from the pulp to very close to, or even through, the amelodentinal junction. These dentinal tubules partly contain the cytoplasmic processes of the odontoblasts which are found on the pulpal aspect of the dentine, as well as some nerves. Dentine is composed of 70% inorganic material, 18% organic material (collagen) and 12% water by weight. It is the second most highly mineralized tissue of the body.

Cementum is similar to bone in many respects. The main difference is the vascularity of bone compared to cementum. Parts of the cementum contain enclosed cells, the cementocytes, which are morphologically identical to the osteocytes of bone. Cementum also provides attachment to the fibers of the periodental ligament. On a wet weight basis, cementum consists of 65% inorganic material (hydroxyapatite), 23% organic material and 12% water.

The dental pulp is the central hollow of a tooth. It contains delicate connective tissue, nerves and both blood and lymph vessels and is lined peripherally by odontoblasts. It communicates by means of one or more openings (the apical foramina) with the periapical tissues.

The periodontium

The periodontium enables a tooth to maintain a functional position and consists of those tissues which surround and support the tooth. These are the gingiva and the junctional epithelium with the associated fibers, the cementum, the periodontal ligament, and the alveolar bone.

That part of the maxilla and mandible which houses the teeth, is the alveolar process. This contains the tooth sockets (alveoli) which are lined by a thin layer of compact bone, the alveolar bone. This contains very many small perforations through which blood vessels and nerves reach the periodontal space, and is also known as the cribriform plate. The dense alveolar bone is continuous over the rim of the socket (the alveolar crest) with the outer cortical plate of the alveolar process. Cancellous, or spongy, bone is present between the alveolar bone and the outer cortical bone of the alveolar process.

Periodontal ligament fibers, and the fibers associated with the gingiva, not only attach a tooth to the surrounding alveolar bone but also extend between adjacent teeth, as well as into the gingiva to maintain a close-fitting gingival seal around the neck of the tooth, the so-called gingival collar. The latter is most important since it protects the integrity of the junctional epithelium and the more superficial periodontal fibers. The width of the periodontal space varies from person to person, and from tooth to tooth, but is usually 0.1-0.4 mm.

Selected bibliography

1. Bhaskar, S.N. (1991) editor. *Orban's Oral Histology and Embryology,* 11th edition. St. Louis: Mosby-Year Book, Inc.
2. Mjör, I.A. and Fejerskov, O. (1986) editors. *Human Oral Embryology and Histology.* Copenhagen: Munksgaard.
3. Osborn, J.W. (1981) editor. *Dental Anatomy and Embryology.* Oxford: Blackwell Scientific Publications.
4. Osborn, J.W. and Ten Cate. A.R. (1983) *Advanced Dental Histology,* 4th edition. Bristol: Wright PSG.
5. Scott, J.H. and Symons, N.B.B. (1982) *Introduction to Dental Anatomy,* 9th edition. Edinburgh: Churchill Livingstone.
6. Ten Cate, A.R. (1989) *Oral Histology: Development, Structure and Function,* 3rd edition. St. Louis: C.V. Mosby Company.

(This chapter must be regarded as an introduction to, and a summary of, subsequent chapters on subjects mentioned. It does not contain sufficient information for test or examination purposes).

A sketch of a longitudinal section of a tooth and its immediate surroundings (Fig 82) appears below. The salient features are indicated by numbers and are briefly explained on the following page.

Fig 82 **Longitudinal section through a tooth and its periodontium**

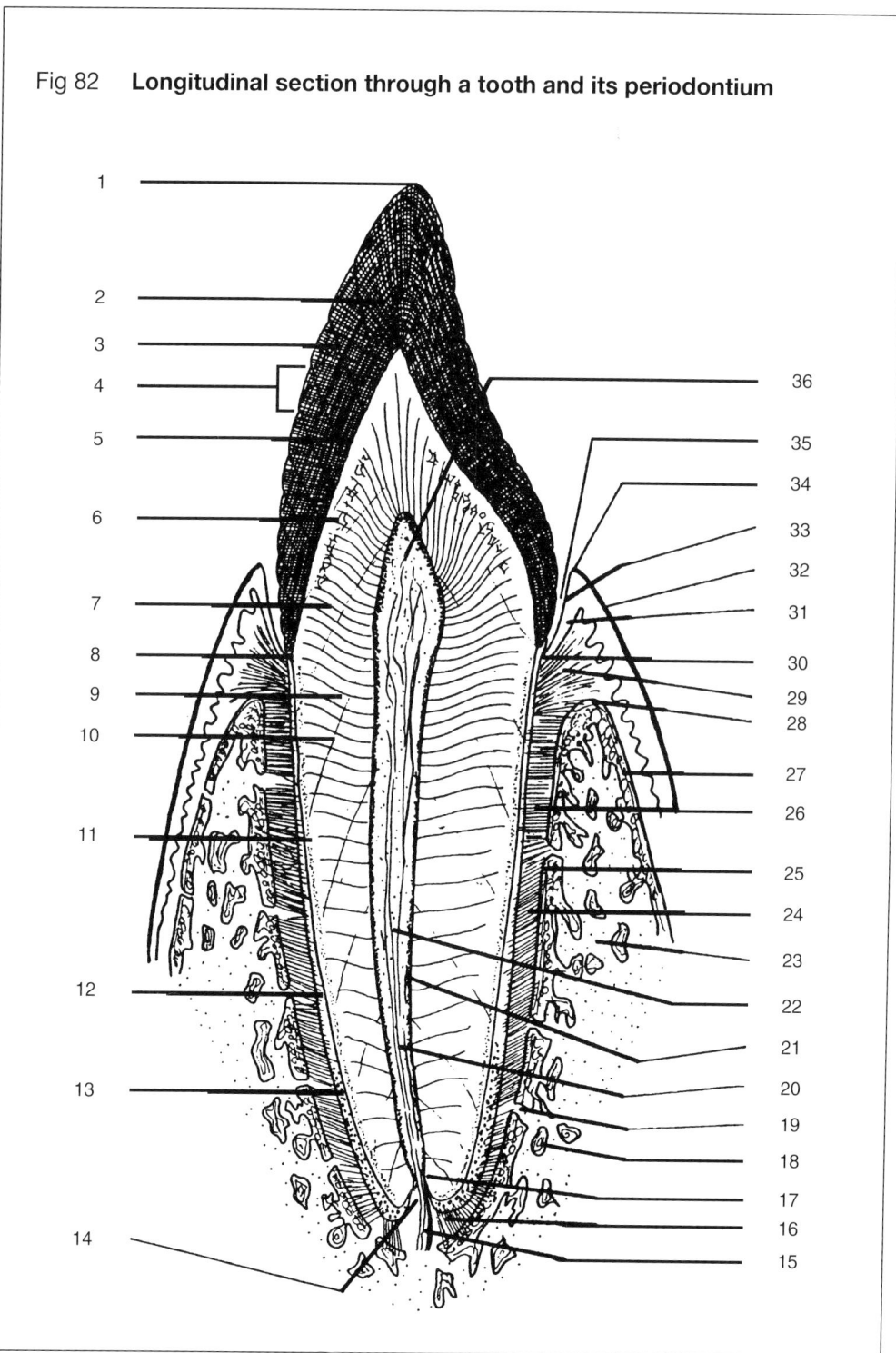

Description of longitudinal section

1. Incisive edge. This is morphologically similar to the tip of a cusp in a premolar or molar tooth.
2. Enamel. The enamel consists of many enamel rods which leave the amelodentinal junction at right angles and run to the surface of the crown. They, in turn, are crossed by cross-striations which run parallel to the amelodentinal junction.
3. A line of Retzius. Both the Retzius lines and the cross-striations are indications of the incremental manner of formation of enamel.
4. Perikymata. A newly erupted tooth has low tranverse ridges, or perikymata, where the lines of Retzius reach the surface of the crown. They are later worn away.
5. Amelodentinal junction. It has a finely scalloped appearance.
6. Interglobular dentine. This represents areas of defective mineralization of dentine and is found under enamel.
7. Dentinal tubules. Dentinal tubules partly contain the processes of odontoblasts and follow an S-shaped course (the primary curvature) from the pulp to the periphery of the dentine.
8. Amelocemental junction. Enamel and cementum meet at this junction where the cementum commonly overlaps the enamel slightly.
9. Dentine.
10. Incremental lines in the dentine.
11. The granular layer of Tomes. This layer in the periphery of root dentine has the appearance of small granules.
12. Cementum (acellular).
13. Cementum (cellular). The cells inside cementum are cementocytes and represent cementoblasts which were included in the cementum during its formation.
14. Apical foramen. The pulp communicates with the periapical tissues through the apical foramen.
15. Apical blood vessels and nerves.
16. Apical periodontal fibers.
17. Constriction in the apical part of the root canal.
18. Cancellous (spongy) bone.
19. Volkmann's canals. These are small perforations in the alveolar bone of the tooth socket.
20. Root canal (radicular pulp).
21. Odontoblasts. The odontoblasts line the entire pulp cavity. Their processes are found inside the dentinal tubules. Odontoblasts remain active thoughout the life of a vital tooth.
22. Blood vessels and nerves in the root canal.
23. Bone marrow.

24. Oblique fibers of the periodontal ligament.
25. Alveolar bone.
26. Horizontal fibers of the periodontal ligament.
27. Cortical bone on the outside of the alveolar process. It is covered by a periosteum.
28. Alveolar crest. Alveolar crest fibers of the periodontal ligament attach to the bony alveolar crest.
29. Gingival fibers. These extend from the cementum into the gingiva.
30. Junctional epithelium. This part of the gingival epithelium is attached to the tooth surface by means of hemidesmosomes.
31. Crevicular epithelium. This forms the outer lining of the gingival sulcus.
32. Free (marginal) gingiva. The free gingiva is keratinized and not directly attached to the underlying bone.
33. Non-keratinized epithelium.
34. Gingival crest.
35. Gingival sulcus.
36. Pulp chamber (coronal pulp).

2. Development of the teeth - a general review

The role of ectomesenchyme in tooth development is discussed in the next chapter. This chapter provides a general review of the whole process of tooth development, while specific topics such as amelogenesis, dentinogenesis, and cementogenesis are dealt with in subsequent chapters.

The origin of the dental lamina

At the start of tooth development, the epithelial lining of the mouth comprises a layer of low columnar cells. In some areas, especially in those regions where the teeth will form, the superfine layers of the epithelium are arranged in the form of two to three layers of flattened cells covering a basal cell layer which is separated from the underlying mesenchyme by a basal lamina. At this stage there is still no sign of a lip or a tooth-bearing alveolar ridge (Figs 83, 84, I).

The first indication of tooth formation is the appearance, between the 6th and 7th weeks of intra-uterine life, of a continuous U-shaped band of proliferating epithelium in the future tooth-bearing area of both the maxilla and mandible (Fig 84, II). The embryo is then approximately 15 mm long. This proliferation, which is initially confined to the basal cells, projects into the underlying mesenchyme and is termed the primary epithelial thickening.

During the 7th week each primary epithelial thickening divides to form an outer (buccal) vestibular lamina and an inner (lingual) dental lamina (Fig 84, III). The vestibular lamina indicates the future division between the lips (and the cheeks) and the tooth-bearing areas of the mouth. It thickens and develops a sulcus which is destined to become the vestibule of the mouth. Already at this stage the first signs of ossification appear in the jaws.

The dental lamina is intimately concerned with tooth formation and proliferates deeper into the tissues of each jaw. Further localized areas of proliferation occur along the length of each dental lamina and these tooth buds (Fig 84, IV) are the future enamel organs of the primary teeth (Fig 84, V, VI). Later, as the dental lamina proliferates distally, tooth buds for the three secondary molar teeth form. The basal cells of the tooth buds are continuous with the basal cell layer of the oral epithelium, and the more central cells show few differences when compared to more superficial epithelial cells at this stage. Already at this stage there is a fairly close relationshaip between nerves and the tooth buds.

The dental papilla

The mesenchymal tissue surrounding the tooth buds shows an increased mitotic activity and cellularity, and later gives rise to a dental papilla and dental follicle (Fig 85, II) for each of the developing teeth. The deep (papillary) surface of each bud is initially bulb-shaped but rapidly becomes concave (the cap stage of development) due to an increased peripheral mitotic activity. The rim of the epithelial organ grows deeper into the underlying tissue and partly encloses mesenchymal tissue (the dental papilla). Due to continued proliferation the cap stage changes shape and becomes transformed into a bell-shaped structure, the bell stage of the enamel organ.

Changes in the dental lamina

The above changes are accompanied by an elongation of the dental lamina. On its lingual aspect, the dental lamina of the primary teeth show the development of a second tooth bud which will undergo the same changes as the initial tooth bud. These are the enamel organs of the permanent successors. The dental laminae fragment quite early in development and may persist for some time as isolated groups of cells, the so-called cell rests, or epithelial pearls, of Serres (Fig 85, III, IV).

The origin of the enamel organ

The cells forming the outer layer of the epithelial downgrowth are short columnar cells and very similar to the basal cells of the oral epithelium. The deeper cells are rounder in shape. As the bud becomes transformed into an early cap stage, the cells forming the papillary surface of the future enamel organ elongate and become known as the inner enamel epithelium (Fig 85, I), while the cells on its follicular surface are known as the outer enamel epithelium. The inner and outer enamel epithelia meet at the rim of the epithelial organ where marked mitotic activity takes place to form the cervical loop. The tooth is now in an early bell stage of development.

Glycosaminoglycans are secreted by the internal cells and cause a measure of intercellular separation, probably as a result of osmotic changes (glycosaminoglycans are associated with a comparatively large water component). The cells become compressed but retain desmosomal contacts at the ends of cell processes. This gives the cells a star-shaped appearance and this zone is known as the stellate reticulum. These cells contain the enzyme alkaline phosphatase but only small amounts of RNA and glycogen. The fourth zone to differentiate in the enamel organ is the intermediate cell layer, or stratum intermedium. It consists of a few layers of flattened cells between the inner enamel epithelium and the stellate reticulum. These cells are rich in alkaline phosphatase and are probably involved in protein synthesis and the transport of substances to and from the future ameloblasts.

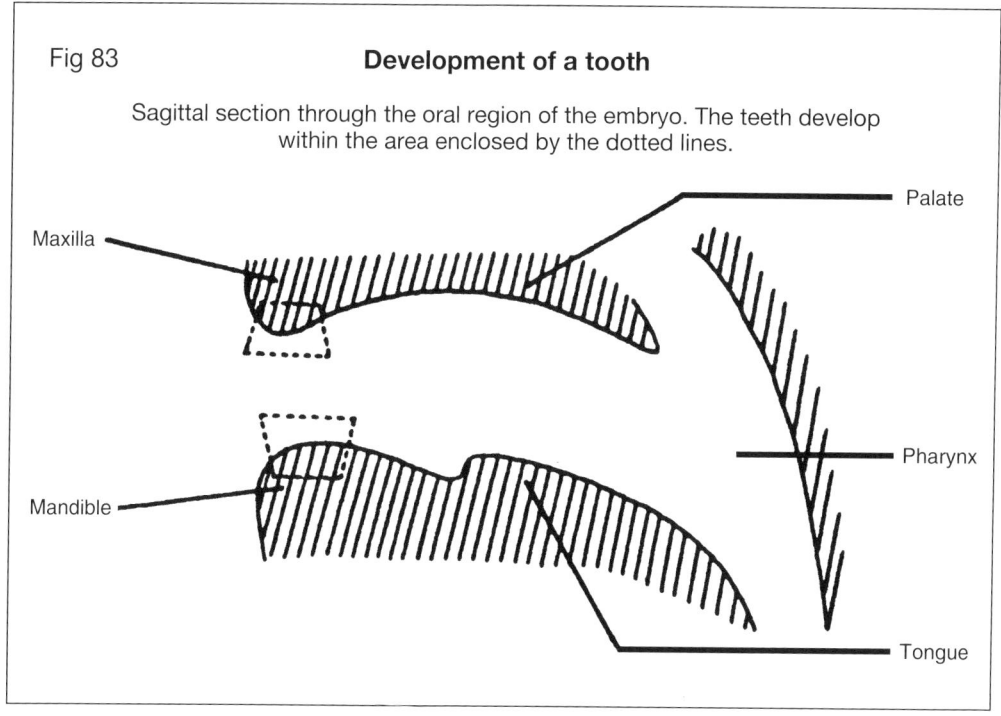

Fig 83 **Development of a tooth**

Sagittal section through the oral region of the embryo. The teeth develop within the area enclosed by the dotted lines.

Fig 84 — Early stages in tooth development
(a mandibular incisor is taken as an example)

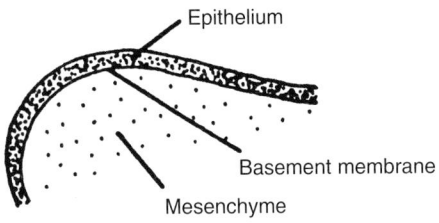

I. The epithelial lining of the oral cavity before development commences

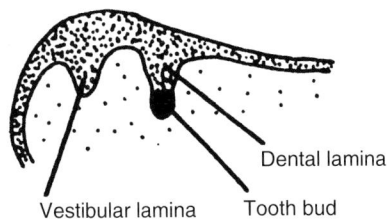

IV. Tooth bud

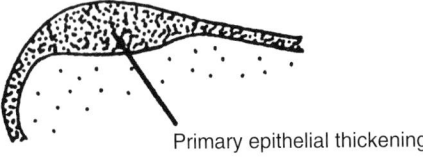

II. The primary epithelial thickening

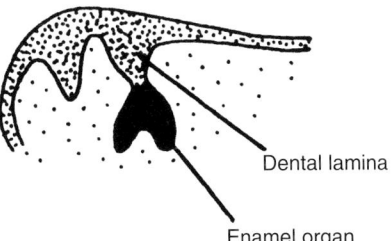

V. Cap stage of development

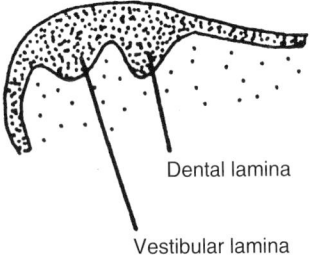

III. Vestibular and dental laminae

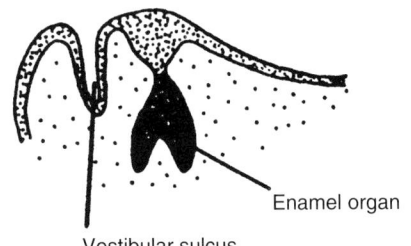

VI. Vestibular sulcus (lip cleft)

Fig 85 **Later development of a tooth**

I. Enamel organ

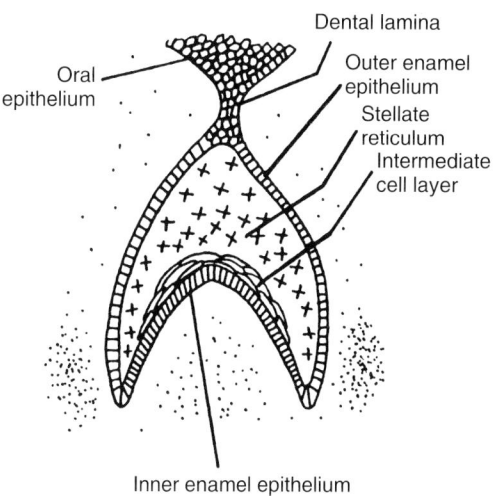

The Enamel organ develops from oral ectoderm.

II. Tooth germ

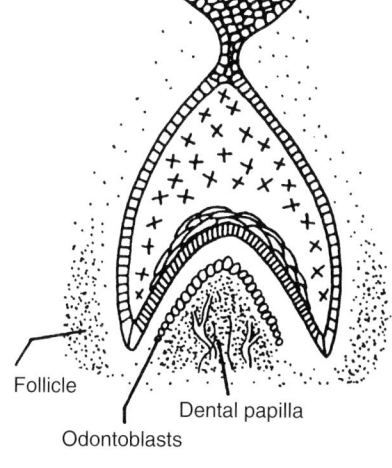

The tooth germ consists of the enamel organ, the follicle and the dental papilla.

III. Formation of dentine and enamel

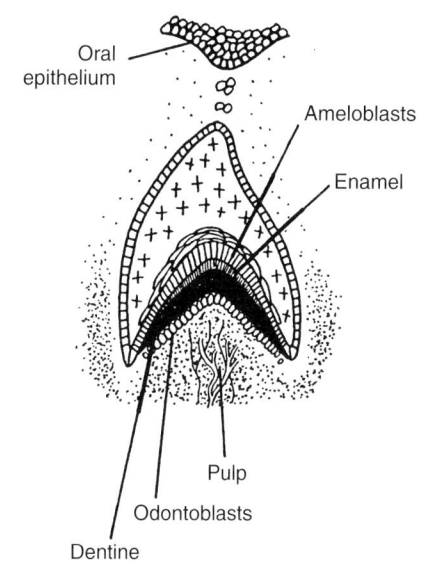

IV. Root sheath of Hertwig

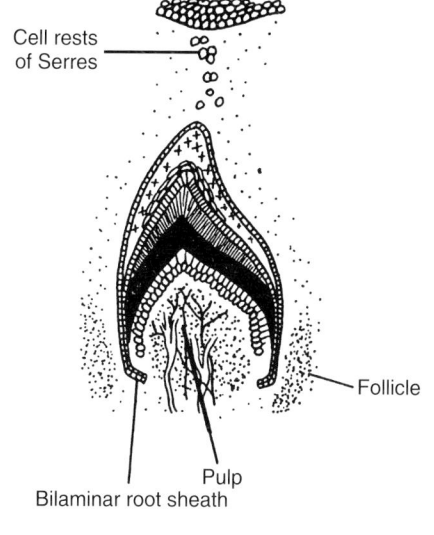

The process of histodifferentiation in the enamel organ is completed with the appearance of the intermediate cells. Blood vessels and nerves can be observed in the dental papilla of the enamel organ during the bell stage of development.

After differentiation of all four cellular constituents, the epithelial organ is known as the enamel organ, since it possesses the ability to form enamel in conjunction with the papillary mesenchyme. The internal enamel epithelium becomes the ameloblast layer when enamel formation commences.

The epithelial sheath of Hertwig

After the enamel organ has determined the morphology of the tooth crown by means of differential mitotic activity in the inner enamel epithelium (see Part II, Chapter 4: Determination of crown form), and dentine and enamel formation commences, the epithelial root sheath of Hertwig forms. This sheath arises in the area where the outer and inner enamel epithelia meet and is the result of mitotic activity in the cervical loop area. The root sheath extends into the underlying mesenchyme as a double-layered tubular membrane without any intermediate cells or stellate reticulum between the two layers. It determines root form. In the case of a tooth with one root, it retains a simple tubular form, but in multirooted teeth it becomes subdivided into two or more tubes, depending on the number of roots to be formed.

Changes in the papilla

The dental papilla initially comprises densely packed mesenchymal tissues, but when the bell stage of development is reached, delicate fibers appear on its periphery. Blood capillaries appear at the same time and are indicative of metabolic activity which will eventually result in dentine formation.

Nerves appear prior to the differentiation of odontoblasts but the nerve supply of the pulp and dentine is not well developed before birth.

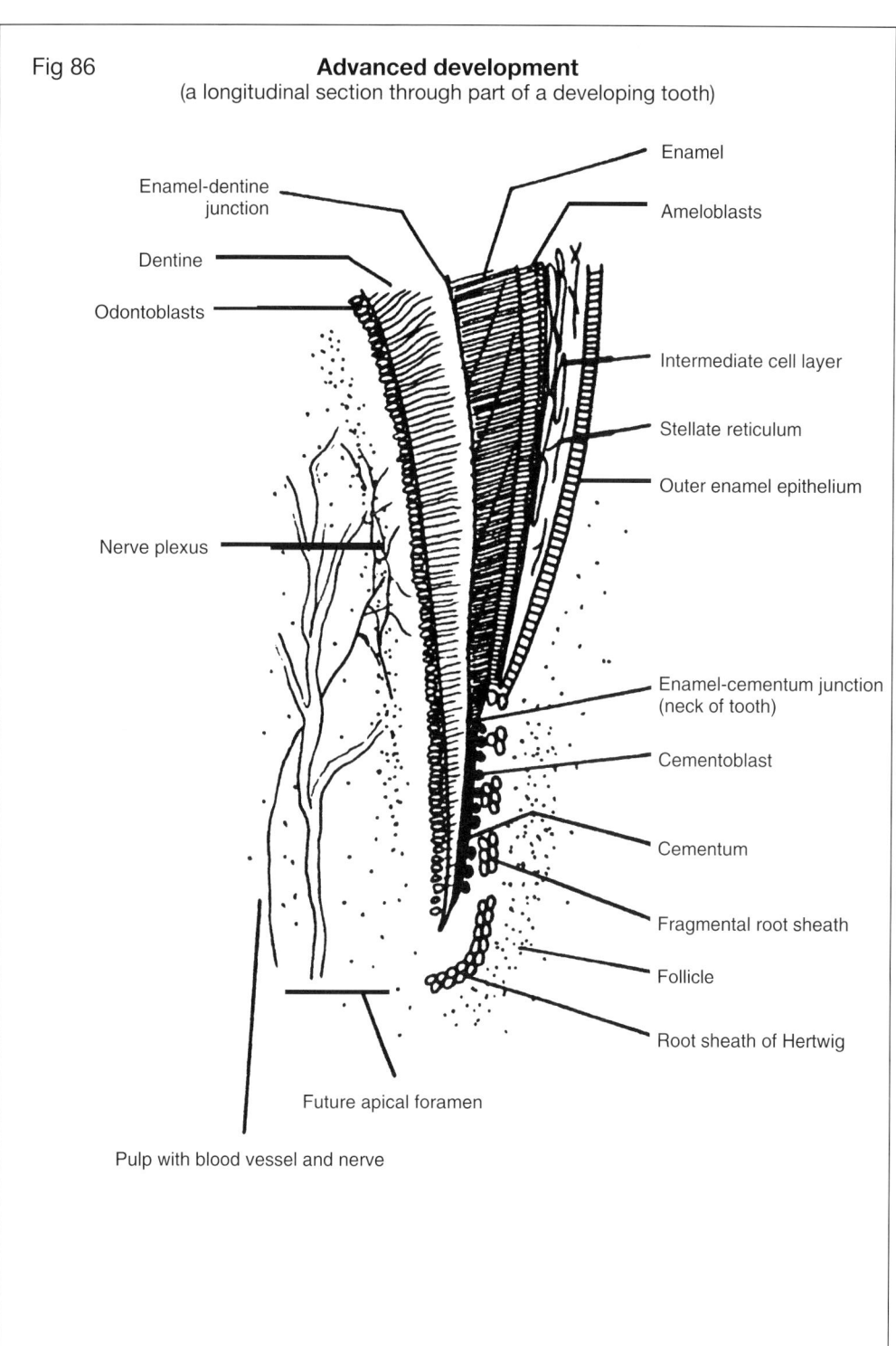

The dental follicle

In the crown area of a developing tooth, the dental follicle forms a fibrovascular covering around the developing enamel. It is responsible for the nutrition of the forming enamel, acts as a protective buffer for the crown, and determines the size of the crypt in which the tooth develops. In the root area the follicle gives origin to the cementum, periodontal ligament, and alveolar bone.

The tooth germ

A tooth germ is composed of the enamel organ, the dental papilla, and the follicle and therefore has the potential to form an entire tooth and most of the periodontium.

Enamel, dentine, and cementum

The basal lamina separating the inner enamel epithelium from the dental papilla represents the amelodentinal junction. Enamel, of ectodermal origin, is laid down by the ameloblasts (former inner enamel epithelium cells) which move away from the dentine surface and leave enamel behind. Dentine, of mesenchymal origin, is formed by odontoblasts (formerly the peripheral cells of the dental papilla) in an inward direction away from the amelodentinal junction. The ameloblasts and odontoblasts therefore move in opposite directions. Cementum forms on the surface of the dentine in the root of the tooth after the epithelial root sheath loses its continuity (Fig 86).

Selected bibliography

1. Bhaskar, S.N. (1991) editor. *Orban's Oral Histology and Embryology,* 11th edition. St. Louis: Mosby-Year Book, Inc.
2. Mjör, I.A. and Fejerskov, O. (1986) editors. *Human Oral Embryology and Histology.* Copenhagen: Munksgaard.
3. Osborn, J.W. (1981) editor. *Dental Anatomy and Embryology.* Oxford: Blackwell Scientific Publications.
4. Osborn, J.W. and Ten Cate, A.R. (1983) *Advanced Dental Histology,* 4th edition. Bristol: Wright PSG.
5. Scott, J.H. and Dixon, N.B.B. (1982) *Introduction to Dental Anatomy,* 9th edition. Edinburgh: Churchill Livingstone.
6. Ten Cate, A.R. (1989) *Oral Histology: Development, Structure and Function,* 3rd edition. St. Louis: C.V. Mosby Company.

Review questions

1. Discuss the development of an enamel organ.
2. Describe a tooth germ and enumerate all the tissues and structures which arise from each of its components.

3. Epithelium-ectomesenchyme interaction in tooth development

General remarks

Tooth development involves a continuous series of complex co-ordinated events, namely formation of the dental lamina, histogenesis of the enamel organ, crown morphogenesis, and terminal differentiation of odontoblasts, ameloblasts, and cementoblasts. Progressive developments which lead to the formation of the tooth germ and the tooth are controlled by a multiphasic process of epithelium-mesenchyme interaction. In this process the extracellular matrix, including the basal lamina, initially plays an essential roll as an inductive mediator of communication between epithelium and ectomesenchyme. Explanations for the controlling mechanisms for tooth formation are largely speculative at this stage and are based on experimental and electronmicroscopic observations.

The origin of ectomesenchyme

With the development of the neural tube in a human embryo certain cells, the neural crest cells (neuroectoderm), which are initially located on the periphery of the neural plate and later on the crests of the neural groove, are not incorporated into the tube. They migrate to form a sagittally elongated cord of cells which lie dorsolateral to the neural tube under cover of the surface ectoderm. The cord of neural crest cells undergoes segmentation to form the ganglia of the autonomic nervous system, as well as the ganglia of the sensory spinal and cranial nerves (cranial nerves V, VII, IX and X). In addition, neural crest tissue differentiates into Schwann cells and pigment cells of the skin. They contribute to the meninges and much of the bone (osteoblasts) and cartilage (chondroblasts) of the facial skeleton, as well as to the odontoblasts.

Mesenchyme, or embryonic connective tissue which initially fulfils a filling role, is composed of a loose network of stellate and spindle-shaped cells in an amorphous ground substance and is found between the three primary germ layers. It is derived mainly from mesoderm. Mesenchyme gives rise to a wide range of cells and tissues such as hemopoietic tissue, tendons, fibrous tissue, and muscle.

Some neural crest cells migrate into the mesenchyme and lose all contact with the neural tube and its derivatives. It must be borne in mind that neural crest cells initially cannot, by ordinary microscopy, be clearly identified, especially in mammals. The paths of migration and the ultimate destinations of these cells have been traced by labelling them with radioactive isotopes, such as tritiated thymidine. Refined biochemical methods, based on the higher RNA content of these cells compared to the mesenchyme into which they migrate, have similarly been employed.

The combined tissue type (neural crest cells and mesenchyme) is known as ectomesenchyme and is rapidly found in the deeper regions of the embryo. In the head region it can be assumed that undifferentiated ectomesenchyme is the tissue of origin of general connective tissue, cartilage, and bone. In tooth development an interaction occurs between oral ectoderm and ectomesenchyme to initiate the process.

The mechanism of epithelium - ectomesenchyme interaction in tooth development

Introduction

Initiation of tooth development involves the establishment of a dentally active ectomesenchyme and the subsequent changes in the overlying ectoderm. This leads to the formation of the primary epithelial thickening, followed by the dental lamina and the tooth bud. Ectomesenchyme is primarily involved in the above initiation process, while the role of the oral ectoderm is uncertain in this phase of tooth development.

Interaction, or communication, between heterotypic cell populations occurs mainly by means of cell surface receptors possessed by one cell responding to extracellular matrix macromolecules secreted by another cell (the inducer cell), or responding to physical cell membrane contact with the inducer cell. A cell which is able to react in the above way is said to be competent. The basal lamina probably plays a major role in the above mechanisms, especially the former. Inductor substances include collagen formed by epithelia (type IV), proteoglycans and laminin, collagen formed by ectomesenchyme (types I and III), and fibronectin (all are part of the basal lamina complex), as well as phosphoproteins, glycoproteins, and mRNA.

A cytoplasmic response is initiated by the activated receptor. The response is carried over to the nucleus in which new activities, such as mitosis, histodifferentiation, and functional changes are generated. The cell is said to be differentiating.

Cellular interactions of the above nature between epithelium and ectomesenchyme give rise to the formation of the tooth bud, tooth germ, and subsequent formation of dentine, enamel, and cementum, as well as to crown morphology.

The relative importance of epithelium (in the form of the enamel organ) and ectomesenchyme (in the form of the dental papilla) in determining crown form is illustrated by the following experiment:

Experimental recombination of an incisor enamel organ with a molar dental papilla of a mouse, after separation from the respective incisor dental papilla and molar enamel organ before hard tissue genesis has started, results in a molariform tooth, and vice versa. The conclusion which may be drawn is that the dental papilla is the determinant regarding crown form. The inductive influence of ectomesenchyme is further illustrated by the fact that dental papilla cells are experimentally able to induce the development of an enamel organ from epithelium of the sole of the foot of the mouse.

Interaction in the initiation phase of tooth development (phase I)

The formation of the primary epithelial thickening, dental lamina, tooth bud, and the subsequent cap and bell stages of tooth development are autonomic events which can be attributed to genetically predetermined processes. These events cannot take place in the absence of dentally active ectomesenchyme which becomes concentrated in the regions of presumptive tooth development before any histological signs of such events are observed.

The undifferentiated inner enamel epithelium of the enamel organ induces the ectomesenchyme of the dental papilla to become orientated towards the intervening basal lamina and to differentiate into pre-odontoblasts.

Interaction in the differentiation of odontoblasts and the initiation of ameloblasts (phase II)

Continued induction by cells of the inner enamel epithelium results in completion of the differentiation of pre-odontoblasts. These cells enter a postmitotic phase and rapidly become odontoblasts which form predentine.

Early predentine collagen plays an important role in the basal lamina-mediated initiation of pre-ameloblast differentiation from the inner enamel epithelium. This involves a lengthening of the cells, polarization, and a specific arrangement of cytoplasmic organelles.

Differentiation of ameloblasts (phase III)

The differentiation of pre-ameloblasts and cessation of mitotic activity in these cells is accompanied by a degradation of the adjacent basal lamina by collagenases. This allows cell membrane contact between the above cells and the odontoblasts (and predentine) in the papilla. Cell-mediated interaction between these cells results in progressive maturation and terminal differentiation of the pre-ameloblasts to become functional ameloblasts. Enamel formation commences.

Enamel and dentine formation (phase IV)

The differentiation of both odontoblasts and ameloblasts commences in the tips of the future cusps, or incisal edges, after crown form has been determined. This process gradually extends towards the future amelocemental junction where differentiation of ameloblasts and enamel formation ceases. The inner enamel epithelium component of the root sheath of Hertwig, however, retains the ability to induce papillary mesenchyme to differentiate into odontoblasts and dentinogenesis continues in the root of the tooth. Reciprocal induction of ameloblast differentiation does not occur due to the absence of intermediate cells in the root sheath.

Differentiation of cementoblasts and formation of cementum (phase V)

Before the root sheath becomes discontinuous, the root sheath cells secrete a thin hyaline layer, consisting of epithelium-produced amino acids and collagen, on the surface of the dentine. After the root sheath subsequently fragments, this hyaline layer (ectodermal in orgin) induces the undifferentiated follicular ectomesenchyme to form cementoblasts which commence cementum formation.

The histogenesis and morphogenesis of teeth are more fully discussed in those chapters which deal specifically with development of dental hard tissues, while other aspects of ectomesenchyme are dealt with in Chapters 3, 11, and 13 of part I.

Selected bibliography

1. Berkovitz, B.K.B. and Moxham, B. (1981) Development of dentition: early stages of tooth development. In *Dental Anatomy and Embryology*. Osborn, J.W. (editor) Oxford: Blackwell Scientific Publications.
2. Fejerskov, O. and Josephson, K. (1986) Odontogenesis. In *Human Oral Embryology and Histology*. Mjör, I.A. and Fejerskov, O. (editors) Copenhagen: Munksgaard.

3. Karcher-Djuricic, V., Staubli, A., Meyer, J-M. and Ruch, J-V. (1985) Acellular dental matrices promote functional differentiation of ameloblasts. *Differentiation, 29,* 169-175.
4. Osborn, J.W. and Ten Cate, A.R. (1983) The role of ectomesenchyme in tooth formation and induction. In *Advanced Dental Histology,* 4th edition. Bristol: Wright PSG.
5. Ruch, J-V, Lesot, H., Karcher-Djuricic, V., Meyer, J-M. and Olive, M. (1982) Facts and hypotheses concerning the control of odontoblast differentiation *Differentiation, 21,* 7-12.
6. Ruch, J-V., Lesot, H., Karcher-Djuricic, V., Meyer, J-M. and Mark, M. (1983) Epithelial-mesenchymal interactions in tooth germs: mechanism of differentiation. *Journal de Biologie Buccale, 11,* 173-193.
7. Sperber, G.H. (1989) Development of the dentition (odontogenesis). In *Craniofacial Embryology,* 4th edition. London: Wright.

Review questions

1. Describe the origin of ectomesenchyme.
2. Discuss the broad principles of interaction between heterotypic cell populations.
3. Describe the different phases of tooth development in which epithlium-ectomesenchyme interaction plays a role.

4. Development of dentine

General remarks

Initially all cells of the dental papilla look alike. In due course odontoblasts are formed by induction of papillary mesenchyme (Fig 87,I) by the inner enamel epithelium. Before odontoblasts differentiate, the peripheral cells of the papilla are irregularly arranged with a fairly disorganized appearance of cytoplasmic organelles. The first relatively regular layer of cells to appear are pre-odontoblasts (Fig 87, II), a layer of short columnar cells which immediately start producing dentine matrix. Fibers produced in the peripheral parts of the papilla undergo marked changes at this time. They are initially arranged in a complex network but soon groups of fibers become orientated with their long axes perpendicular to the basement membrane of the inner enamel epithelium. These fibers are often demonstrated by means of silver impregnation techniques and appear as coarse black fibers which follow a spiral course between the odontoblasts and fan out next to the basement membrane. These von Korff fibers are often referred to as argyrophilic fibers (because of their staining characteristics), or "corkscrew fibers" because of the course they follow. They mingle with irregular finer subodontoblastic fibers.

The von Korff fibers form the organic matrix of the outer, or mantle, dentine.

Differentiation of odontoblasts

Pre-odontoblasts undergo differentiation to form odontoblasts (Fig 87, III) which show many differences when compared to their predecessors, and have lost the ability to undergo mitotic cell division.

An odontoblast is a columnar cell with a proximally (basally) situated nucleus. Rough endoplasmic reticulum surrounds the nucleus and is also present in the distal, or formative, end of the cell and around the Golgi apparatus which lies distal to the nucleus. Mitochondria and free ribosomes are found throughout the cell. Histochemically, an odontoblast contains a high concentration of RNA and is characterized by a high oxidative and hydrolytic enzyme activity. A newly differentiated odontoblast does not possess an alkaline phosphatase activity of sufficient magnitude to be detected by light microscopic methods. This activity is, however, clearly detectable in the subodontoblastic layer. Vesicles, or secretory granules, are initially only present in the region of the nucleus but move to the distal end of the cell as activity increases. The distal end of the cell shows infoldings of the cell membrane in the pre-odontoblast stage, but a single cytoplasmic process develops with full differentiation of the odontoblast. This process is sometimes referred to as the Tomes process but this term should not, in the author's opinion, be used for odontoblasts but be reserved for a short cytoplasmic extension of an ameloblast. The odontoblast process does not contain all the organelles of the cell body but does contain vesicles in which a finely stippled material is seen, microfilaments, microtubules, some mitochondria, and granular material. The vesicles, which are formed in the Golgi apparatus, move to the cell membrane of the odontoblast, or its process, and release their contents extracellularly by exocytosis. Histochemical studies have demonstrated the presence of hydrolytic enzymes and lipids in particulate form in the odontoblast process, which should therefore not be regarded as a metabolically inactive part of the cell.

Formation of the matrix and mineralization

The first stage in the formation of dentine (mantle dentine) is the deposition of matrix which is composed of collagen fibers in a ground substance rich in glycosaminoglycans (Figs 87, III, IV). Although the fibers are initially arranged in a network, they soon group together to become orientated with their long axes perpendicular to the basement membrane separating the inner enamel epithelium from the papilla. These von Korff fibers follow a spiral course between the early odontoblasts to fan out next to the basement membrane. They are often referred to as argyrophilic because of their affinity for silver in silver impregnation techniques, possibly due to the presence of reducing sugars in the surrounding ground substance. The same argyrophilic property is shared by reticulin fibers.

While both odontoblasts and subodontoblast cells may be responsible for forming the matrix of mantle dentine, results of electronmicroscopic studies, and of studies based on the presence of alkaline phosphatase in these cells, suggest that while subodontoblast cells may be responsible for a proportion of the very first matrix, including the von Korff fibers, odontoblasts form the major part of mantle dentine matrix. The direction followed by the fibers in mantle dentine is quite different from the orientation of the fibers in later additions to the dentine (circumpulpal dentine) (Fig 87, IV).

Fig 87   **Early development of dentine**

I Undifferentiated mesenchymal cells of dental papilla

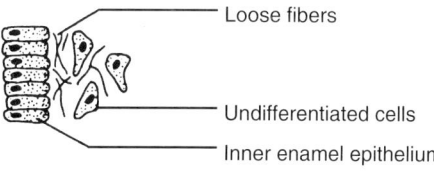
- Loose fibers
- Undifferentiated cells
- Inner enamel epithelium

II Pre-odontoblasts

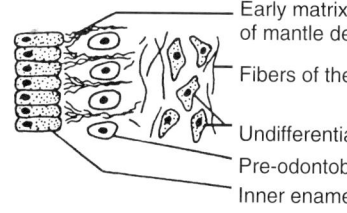
- Early matrix fibers of mantle dentine
- Fibers of the pulp
- Undifferentiated cells
- Pre-odontoblasts
- Inner enamel epithelium

III Odontoblasts

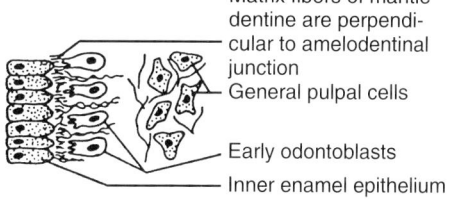
- Matrix fibers of mantle dentine are perpendicular to amelodentinal junction
- General pulpal cells
- Early odontoblasts
- Inner enamel epithelium

IV Matrix formation

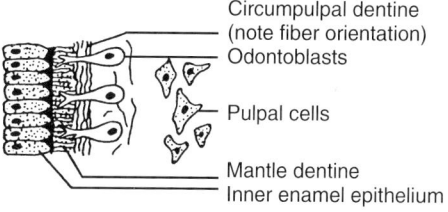
- Circumpulpal dentine (note fiber orientation)
- Odontoblasts
- Pulpal cells
- Mantle dentine
- Inner enamel epithelium

Odontoblasts move away from the amelodentinal junction and leave their processes in the matrix. Early mineralization commences.

Fig 88 — Later development of dentine

I Matrix formation and mineralization

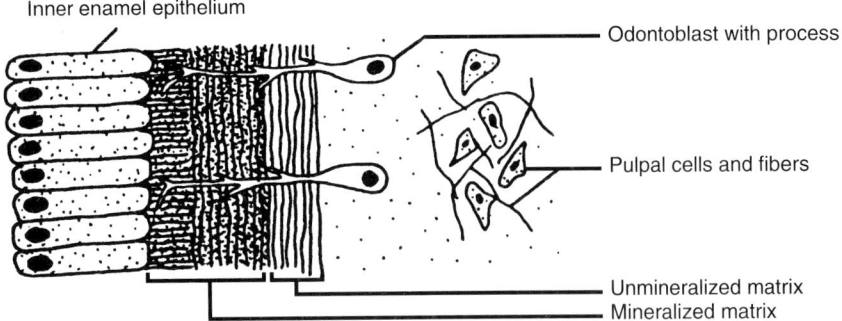

In the above sketch stippling indicates mineralization in both mantle and some circumpulpal dentine. Predentine is not mineralized.

II Mineralization

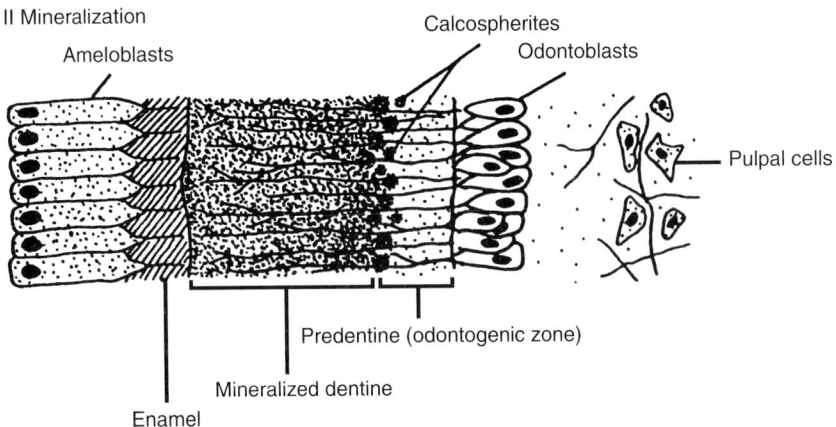

III Intratubular (peritubular) and intertubular dentine (cross-section)

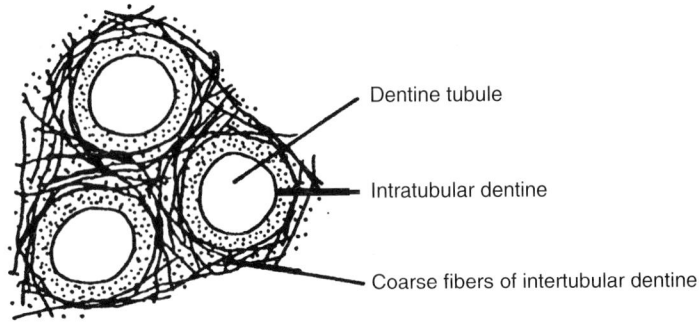

The cell processes of the odontoblasts are embedded in the matrix and lengthen as the matrix thickens and the cell bodies of the odontoblasts recede (Fig 88, I).

The name of the papilla changes to dental pulp at the commencement of dentine formation. Dentine is initially organic, inorganic material only being deposited in the first-formed matrix (close to the amelodentinal junction) when it has gained in thickness.

The layer of dentine adjacent to the pulp remains unmineralized throughout the life of a vital tooth (Figs 88, I, II) and is known as the odontogenic zone, or predentine. As the dentine thickens, the initially fine matrix fibers gradually become orientated with their long axes more transversely arranged and group together to form thicker collagen bundles. This later dentine is referred to as circumpulpal dentine.

When mineralization commences, dentinal tubules form around the lengthening odontoblast processes. The tubules and processes extend from the odontoblast cell bodies in the peripheral pulp to close to the amelodentinal junction. In their course through the mineralized dentine the tubules show numerous lateral branches (caused by mineralization around lateral cytoplasmic extensions of odontoblast processes), as well as quite extensive terminal branching under the amelodentinal junction. More numerous, and finer, lateral branches are present in root dentine than in crown dentine. A difference of opinion exists regarding the depth of penetration of the processes into the tubules in adult dentine, but it can safely be assumed that they do not necessarily course the whole distance to the junction of enamel and dentine.

Soon after dentine starts forming, the odontoblast cell bodies become pear-shaped and assume a pseudostratified arrangement (Fig 88, II). This appearance can probably be attributed to two factors. Firstly, the cell bodies are not of equal lengths and the basal nuclei are thus found at different levels and, secondly, a certain amount of crowding of the cell bodies is caused by a gradually decreasing pulp volume as additional dentine is formed.

Mineralization is somewhat retarded compared to matrix formation and results in the presence of an unmineralized zone, the predentine, between the odontoblast cell bodies and the fully mineralized dentine. Mineralization is histologically observed as the formation of spheres, or calcospherites, of inorganic material which increase in size and coalesce to form fully mineralized dentine (Fig 88, II). The calcospherites form around small seed molecules (hydroxyapatite) which are secreted by the odontoblasts and are dispersed within the matrix around the collagen fibers. Mineralization commences in the matrix of the cuspal part of the dentine and is continued by conical additions of mineral down the length of the crown and into the root. This rhythmic process results in the formation of incremental lines. The calcospherites usually coalesce to form a homogeneous mass of mineralized dentine but fusion between them is sometimes incomplete and results in interglobular spaces (interglobular dentine). The reason for this regional defective mineralization is unknown.

The majority of apatite crystals in dentine are arranged lengthwise along the collagen fibers although some are found within the fibers. In the light of the complex arrangement of the fibers, it is obvious that the crystal pattern is very complex. The crystals not only follow a more complex pattern than in enamel but are also smaller. Odontoblasts and subodontoblast cells contain phosphatase and some glucagon in early stages of dentine formation. Furthermore, odontoblasts, as well as their processes, are rich in minerals throughout dentinogenesis and this indicates that the cell is actively involved in mineralization. If minerals are present in the odontoblast, the following questions arise: How are the calcium and phosphates transported through the predentine to deposit in the mineralizing dentine, and why are they not immediately precipitated in the predentine? No definite answers are available. The inorganic material is probably transported in dentine fluid to deeper (older) parts of the dentine matrix and accumulate around previously secreted seed molecules. Predentine must undergo certain biochemical changes to make it receptive for mineral deposition, but the nature of these changes is largely unknown.

Intratubular (peritubular) and intertubular dentine

Dentine is deposited throughout the life of a vital tooth. Not only is the total thickness of dentine increased, but additional dentine is deposited in due course on the inner walls of the tubules, thereby decreasing their diameter. The term intratubular dentine (Fig 88, III) is preferred for this dentine, although the traditional term peritubular dentine is widely used. The older dentine which lies between adjacent tubules is termed intertubular dentine. Intratubular dentine contains less matrix and is of a more homogeneous nature with fewer delicate collagen fibers. The latter may be random inclusions. The mineral phase does not show a typical apatite crystalline morphology but consists of small (25 nm diameter) densely packed polygonol or round particles of amorphous calcium phosphate.

Intratubular dentine becomes visible 60-100 μm from the predentine-dentine junction and gradually thickens towards the periphery of the dentine. It is therefore not present in predentine and, for obvious reasons, not in interglobular dentine.

Selected bibliography

1. Bhaskar, S. N. (1991) editor. *Orban's Oral Histology and Embryology,* 11th edition. St. Louis: Mosby-Year Book, Inc.

2. Cole, A.S. and Eastoe, J.E. (1988) *Biochemistry and Oral Biology,* 2nd edition. London: Wright.
3. Jenkins, G.N. (1978) *The Physiology and Biochemistry of the Mouth,* 4th edition. Oxford: Blackwell Scientific Publications.
4. Mjör, I.A. and Fejerskov, O. (1986) editors. *Human Oral Embryology and Histology.* Copenhagen: Munksgaard.
5. Osborn, J.W. (1981) editor. *Dental Anatomy and Embryology.* Oxford: Blackwell Scientific Publications.
6. Osborn, J.W. and Ten Cate, A.R. (1983) *Advanced Dental Histology,* 4th edition. Bristol: Wright PSG.
7. Scott, J.H. and Symons, N.B.B. (1982) *Introduction to Dental Anatomy,* 9th edition. Edinburgh: Churchill Livingstone.
8. Ten Cate, A.R. (1989) *Oral Histology: Development, Structure, and Function,* 3rd edition. St. Louis: C.V. Mosby Company.

Review questions

1. Discuss the differentiation of an odontoblast and describe its electron-microscopic appearance after formation of the odontoblast process.
2. What is the origin of the fibrous component of dentine matrix?
3. What is the main difference between mantle and circumpulpal dentine?
4. Write brief notes on dentine mineralization.
5. Discuss briefly the differences between intratubular and intertubular dentine.

5. Dentine

General remarks

Dentine forms the major portion of a tooth and surrounds the pulp cavity. It is covered in the crown by enamel and in the root by cementum. It has a pale yellow color, possesses a high degree of elasticity, and is harder than bone or cementum but less hard compared to enamel.

Chemical composition

The organic content of dentine (18% of the mass) is much higher than in enamel. It consists mainly of 93% collagen, 0.9% citric acid, 0.2% insoluble protein, and 0.2% each of glycosaminoglycans and lipids. Inorganic constituents account for 70-75% by weight, and the rest of the mass is water. The inorganic phase of dentine consists almost entirely of apatite, mainly hydroxyapatite, although amorphous calcium phosphate is also present. The apatite crystals of dentine are smaller than those of enamel. The reader is advised to consult the bibliography for detailed discussions of the chemical composition of dentine, since values provided by different authors vary.

General structure

Although it is often stated that dentine is composed of the odontoblasts and intercellular substance which lies between the protoplasmic processes of these cells it is also true that the odontoblasts form an integral part of the pulp. Some authors refer to this area as the pulp-dentine complex. The metabolism and survival of the odontoblasts depend on the state of vitality of the dental pulp. Dentine is crossed in almost its entire thickness by microscopic dentinal tubules which arise in the pulp and extend to the vicinity of the amelodentinal junction (Fig 89). The processes of the odontoblasts extend for varying distances into the tubules which contain dentinal fluid in those sections not filled by the processes. Adjacent tubules run a parallel course. It is calculated that there are approximately 65 000 tubules/mm^2 of dentine close to the pulp, while approximately 15 000 tubules/mm^2 are found peripherally due to the slight fanning out of the tubules. An average number of 35 000/mm^2 may be found in the middle of the dentine.

The intercellular substance consists of collagen fibers embedded in a mineralized ground substance. The matrix fibers in the dentine layer close to the enamel (mantle dentine) are relatively coarse and arranged at right angles to the amelodentinal junction while the finer fibers in the bulk of the dentine (circumpulpal dentine) course in lattice-like fashion between the tubules, parallel to the pulpal surface of the dentine.

Odontoblasts and odontoblast processes

In a vital tooth the odontoblasts form a closely arranged pseudostratified layer in the periphery of the pulp. They are elongated cells with an expanded basal end, in which the nuclei are situated, and a protoplasmic process which extends for varying distances into the tubules. Each process has very many fine branches along its entire length and these lie in corresponding lateral branches of the tubules (Fig 89). Those tubules not occupied in their full length by processes show numerous lateral extensions which indicate that mineralization occurred around processes and their lateral branches present at the time of initial matrix formation and mineralization. The process, especially that section found in predentine, contains the same cytoplasmic organelles as the cell body, namely ribosomes, endoplasmic reticulum, and mitochondria. A fine network of filaments as well as vesicles with a finely stippled content are present. The vesicles appear to secrete their contents into the space surrounding the process and confirms the view that the odontoblast process has a secretory function. Intratubular dentine is almost certainly formed in this way.

The depth to which the odontoblast processes extend into the dentine has been the subject of much research (Fig 90). the older view was that processes extend the full length of the tubules. Later, it was stated that the processes are confined to approximately the pulpal one third of the tubules for a distance not exceeding 0.7 mm. More recent electronmicroscopic studies are reputed to have shown that the processes penetrate deeper into the tubules and, as a rule, end closer to the amelodentinal junction. It is also stated that nerves are more commonly found inside the tubules than previously believed.

In the light of present knowledge it can, however, be assumed that neither the odontoblast process nor nerves are found within the tubules according to a regular pattern. In certain parts of the tooth, notably the neck area, the processes are said to extend further towards the periphery than in other areas. Nerves are assumed to be randomly present in the tubules. Those peripheral sections of the tubules not occupied by processes contain dentine fluid. It can further be assumed that the processes do not entirely fill the lumen of the tubule and that a narrow periodontoblastic space is present. The above discussion is of importance when sensitivity of teeth is considered (Part II, Chapter 19).

Predentine

Predentine is a layer of dentine, 10-20 µm thick, which is not completely mineralized and lies adjacent to the pulpal surface of the dentine. In a decalcified section of dentine stained with hematoxylin and eosin, predentine appears pale in color compared to the darker pink color of fully mineralized dentine. The transition zone is marked by the presence of discrete calcospherites, or calcospherites in the process of fusion. This is the mineralizing front of the dentine and is present throughout the life of a vital tooth.

Dentinal tubules

The diameter of the tubules is 1-5 µm with an average of 1.5 µm. On their pulpal aspect the tubules are closer together than peripherally and there is correspondingly more intertubular dentine in the peripheral regions.

Tubules do not follow a perfectly straight course but describe two curvatures, a primary and secondary curvature (Fig 91). By means of fairly low magnification of a longitudinal section of a tooth, it can be seen that the tubules follow a shallow S-shaped curve in their course to the periphery. The first convexity of this primary curvature is in an apical direction, and the second coronal. The peripheral ends of the tubules finish more coronally than their commencement pulpally. Beneath the incisal edges and cusps, as well as in the roots of teeth, the tubules follow a straighter course. By means of higher magnification the tubules are seen to follow a fine wavy course, the secondary curvatures, which are the result of a spiral movement of the odontoblasts during dentine formation.

Close to the amelodentinal junction, the majority of tubules divide into two or more terminal branches which unite with terminal branches of other tubules to form a plexus. In the crown, some tubules are found within the mineralized enamel spindles, while in peripheral root dentine, adjacent to the cementum, the plexus is regarded as giving rise to the so-called granular layer (of Tomes). This layer is best seen in a ground section of a tooth.

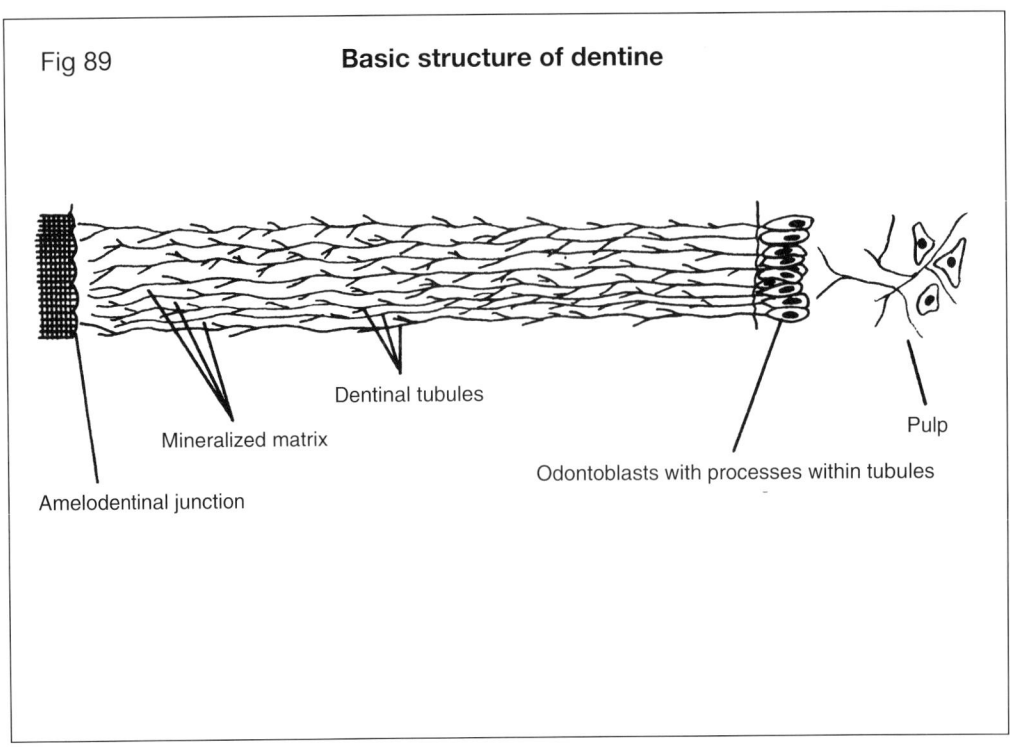

Fig 89 Basic structure of dentine

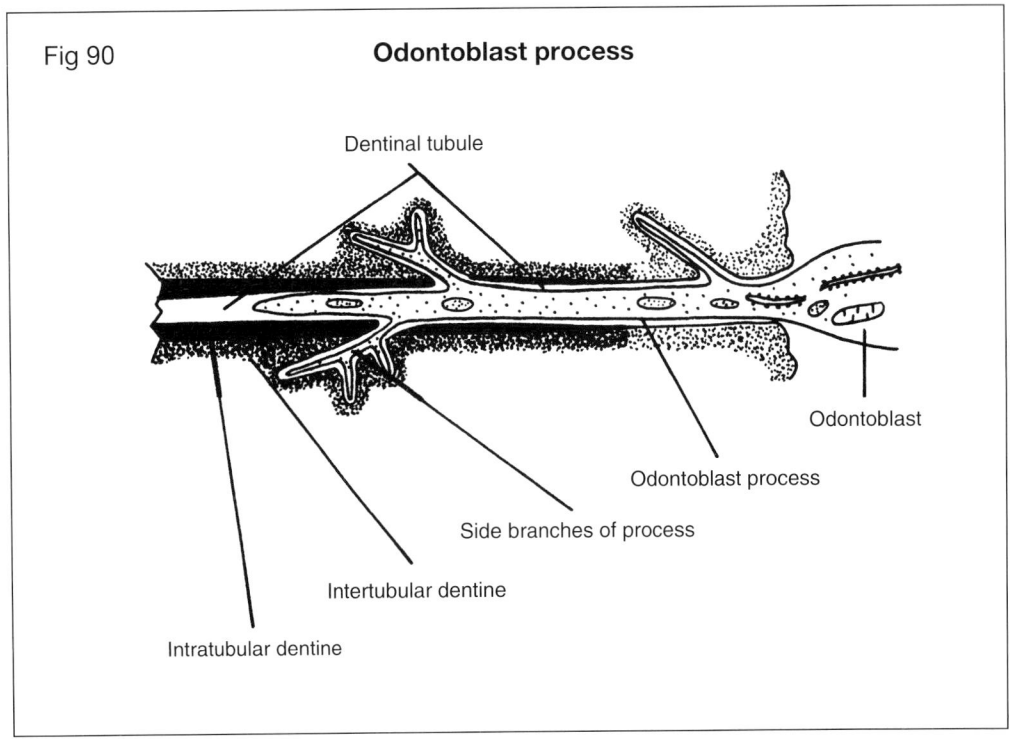

Fig 90 Odontoblast process

Intratubular (peritubular) dentine

In ground sections prepared at right angles to the dentinal tubules and viewed microscopically by means of transmitted light, a clear transparent ring is seen around each tubule lumen. It was previously pointed out that this intratubular, or peritubular, dentine is a secondary deposit of a denser, more homogeneously mineralized, dentine on the inner aspect of the tubule wall. This fact accounts for the above appearance, since less light diffraction occurs in this dentine. By means of microradiography, the intratubular dentine is more radiopaque than the intertubular dentine for the same reasons.

An intratubular zone is not present along the entire length of a tubule, being absent in predentine. When a tubule is followed peripherally the diameter of the tubule lumen becomes narrower with a corresponding increase in the width of the intratubular zone.

Increased deposition of intratubular dentine eventually leads to a complete obliteration of the lumens of tubules. This results in a transparent appearance of the dentine, a condition known as transparent, or sclerotic, dentine. This is an aging phenomenon, which commences apically and extends coronally.

Interglobular dentine

Dentine mineralized with the coalescence of adjacent calcospherites in the matrix to form a homogeneously calcified mass. These calcospherites sometimes remain discrete in certain areas of the dentine, resulting in small areas of matrix which are not fully mineralized. These areas of interglobular dentine are outlined by the curved outlines of unfused calcospherites (Fig 92). Areas of interglobular dentine are normally found adjacent to the crown enamel. Tubules passing through these interglobular areas do not contain intratubular dentine.

Incremental lines

Dentine matrix is formed rhythmically by adjacent odontoblasts acting in concert at a rate of approximately 4 µm per day. Between these daily increments slight changes in the orientation of collagen fibers occur. These changes are more exaggerated in a 5-day cycle when they are more easily demonstrated histologically as the incremental lines of von Ebner (Fig 93) at right angles to the tubules. These lines are not parallel to the outer surface of the dentine and indicate the inner or

Fig 91 — Curvatures of tubules

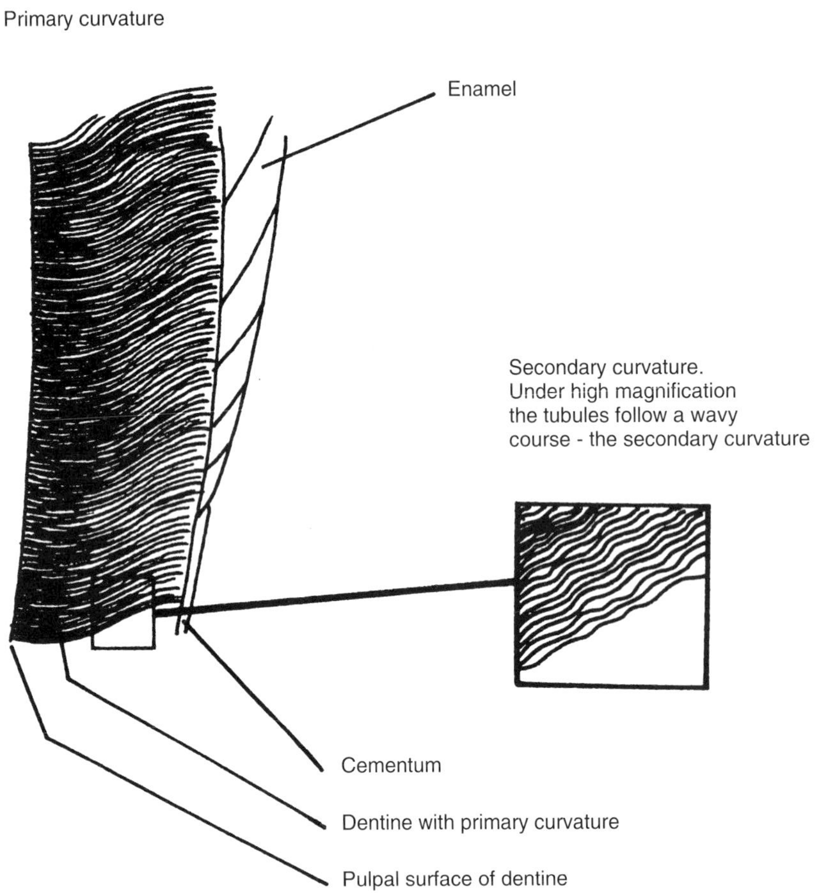

Primary curvature

- Enamel
- Secondary curvature. Under high magnification the tubules follow a wavy course - the secondary curvature
- Cementum
- Dentine with primary curvature
- Pulpal surface of dentine

Fig 92 — Interglobular area

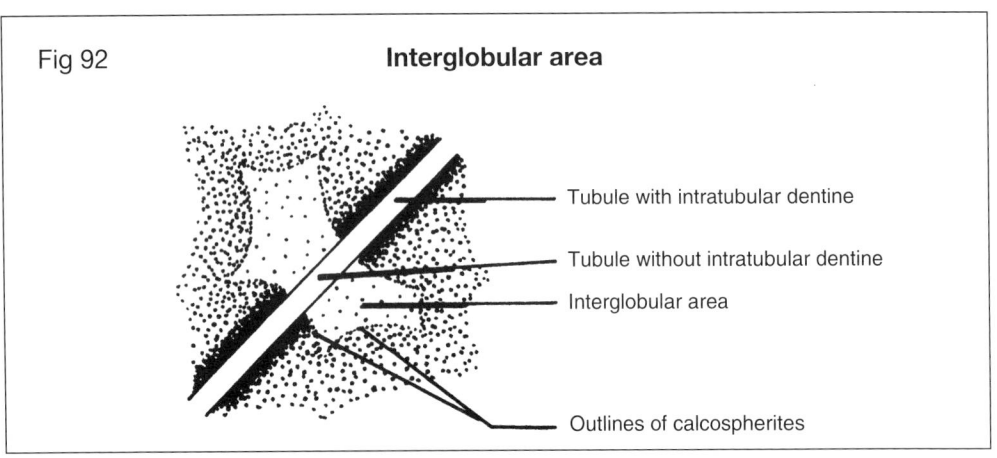

- Tubule with intratubular dentine
- Tubule without intratubular dentine
- Interglobular area
- Outlines of calcospherites

pulpal surface of each conical addition of successive layers of dentine. An accentuated line of von Ebner, or neonatal line, may separate prenatal from postnatal dentine but is never as clearly seen as a corresponding line in the enamel. Occasionally lines in dentine are seen which represent a coincidence of secondary curvatures of the tubules. These contour lines of Owen (Fig 93) are irregularly arranged at an angle to the direction followed by the tubules.

There appear to be differences of opinion and terminology in the literature regarding these incremental lines in dentine.

Granular dentine (granular layer of Tomes)

In a dry ground section of a tooth a "granular" layer is seen in peripheral dentine adjacent to the cementum covering the root, and is known as granular dentine, or granular layer of Tomes (Fig 94). A traditional view was that this layer is composed of many minute areas of interglobular dentine caused by some interference with mineralization in this area. A more recent and well-documented view is that these "granules" represent true spaces in extensively and irregularly looped terminal portions of dentinal tubules. These spaces are exposed during the preparation of the section and irregular diffraction of light away from the microscope lens by the air within the cut tubules gives a darkly granular appearance when viewed with transmitted light.

Amelodentinal junction (enamel-dentine junction)

The junction between enamel and dentine has a scalloped appearance with the convexities on the side of the dentine. On the enamel side of the junction, enamel spindles and enamel tufts are found. These are fully discussed in the chapter on enamel (Part II, Chapter 7).

Three-dimensionally, the junction, as seen from the enamel side, consists of interlacing ridges between hollows into which domes of enamel fit. This pattern is best developed in crown dentine where stresses on tooth structure are the greatest, and supports the view that the junction plays an important supporting role.

Fig 93

Incremental lines

Fig 94

Granular dentine

(a longitudinal section)

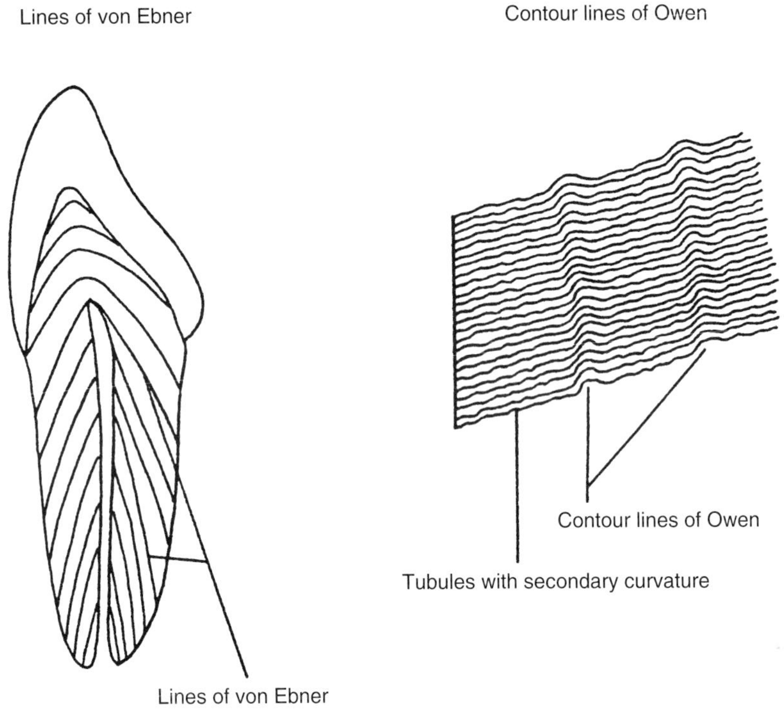

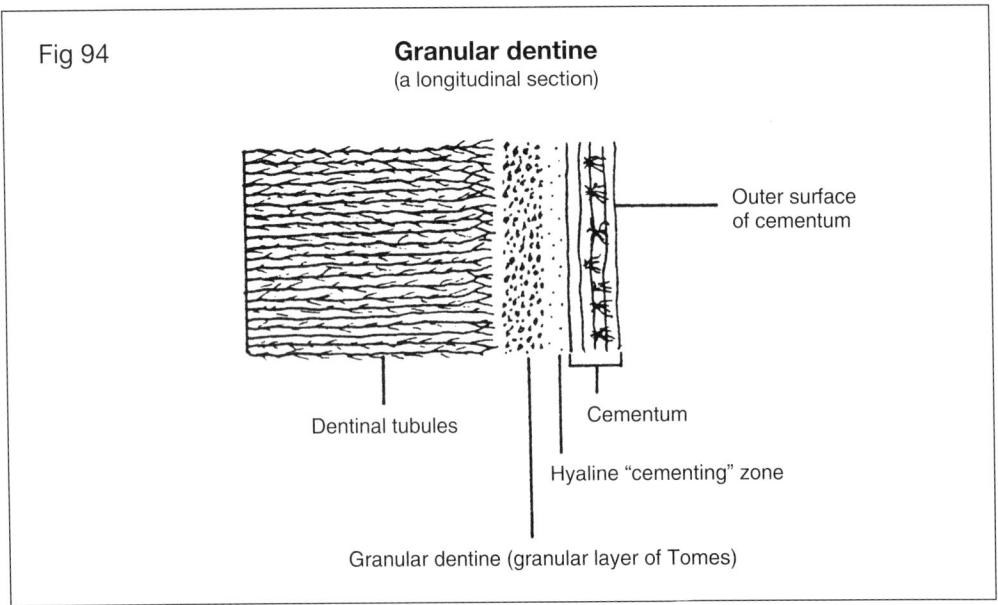

Dentinocemental junction

The boundary between dentine and cementum is often difficult to see. This is particularly so in those areas where acellular cementum covers the dentine. When present, the boundary is seen as a narrow zone with an amorphous and hyaline appearance external to the granular layer (Fig 94). Various views regarding the origin and nature of this hyaline "cementing" layer exist. One view, not yet fully investigated, is that this layer is secreted by root sheath cells before the sheath becomes discontinuous, and consists of mineralized amino acids and collagen of epithelial origin. This layer is then responsible for inducing cementoblast differentiation from follicular mesenchyme. Such a train of thought seems logical in the light of views on epithelium-mesenchyme interaction in tooth development. Both dentine and cementum are of mesenchymal origin and interactions of this nature between homotypic cells or tissues are unknown.

The nerve supply of dentine is discussed fully in the chapter on the nerve supply and sensitivity of teeth (Part II, Chapter 19), while secondary dentine and other changes in morphology of dentine are dealt with in the chapter on permeability and age changes of teeth (Part II, Chapter 13).

Selected bibliography

1. Bhaskar, S.N. (1991) editor. *Orban's Oral Histology and Embryology,* 11th edition. St. Louis: Mosby-Year Book, Inc.
2. Franck, R.M. and Steuer, P. (1988) Transmission electron microscopy of the human odontoblast process in peripheral root dentine. *Archives of Oral Biology, 33,* 91-98.
3. Holland, G.R. (1985) The odontoblast process: form and function. *Journal of Dental Research, 64,* 499-514.
4. Mjör, I.A. and Fejerskov, O. (1986) editors, *Human Oral Embryology and Histology.* Copenhagen: Munksgaard.
5. Osborn, J.W. (1981) editor. *Dental Anatomy and Embryology.* Oxford: Blackwell Scientific Publications.
6. Osborn, J.W. and Ten Cate, A.R. (1983) *Advanced Dental Histology,* 4th edition. Bristol: Wright PSG.
7. Schroeder, H.E. (1991) *Oral Structural Biology.* New York: Thieme Medical Publishers, Inc.
8. Scott, J.H. and Symons, N.B.B. (1982) *Introduction to Dental Anatomy,* 9th edition. Edinburgh: Churchill Livingstone.
9. Sigal, M.J., Aubin, J.E. and Ten Cate, A.R. (1985) An immunocytochemical study of the human odontoblast process using antibodies against tubulin, actin and vimentin. *Journal of Dental Research, 64,* 1348-1355.

10. Ten Cate, A.R. (1989) *Oral Histology: Development, Structure, and Function*, 3rd edition. St. Louis: C.V. Mosby Company.

Review question

Describe dentine with reference to the following:

(a) location;
(b) dentinal tubules;
(c) odontoblasts and odontoblast processes;
(d) dentine matrix;
(e) predentine;
(f) intratubular (peritubular) dentine;
(g) interglobular dentine;
(h) incremental lines;
(i) granular dentine;
(j) amelodentinal junction;
(k) dentinocemental junction.

6. Development of enamel

General remarks

Development of enamel, or amelogenesis, involves the formation of an organic matrix and subsequent mineralization of the matrix. The cells of the inner enamel epithelium differentiate to form ameloblasts. During amelogenesis the ameloblasts have the appearance and properties of secretory cells. Towards the end of the process they influence the withdrawel of organic material from the matrix and finally form part of the reduced enamel epithelium which plays an important role in the process of eruption and establishment of the dentogingival attachment apparatus.

Although enamel is of ectodermal origin in contrast to other mineralized tissues of the body, it possesses the same basic properties, namely a matrix which is produced by a cell layer and which is secondarily mineralized by chiefly hydroxyapatite.

Determination of crown form

Crown form, and especially the morphology of the incisal edges and cusps, is determined by the inner enamel epithelium during the bell stage of development. It has been suggested that this epithelium ceases mitotic activity in the prospective incisal edge or cusp regions, but that this area where mitosis has ceased buckles outwards in the direction of the outer enamel epithelium as a result of pressure created by continued mitotic activity in the adjacent inner enamel epithelium cells. The valley (future cusp slopes and fissures) between the presumptive cusps hereby increases in depth and appears to deepen into the papilla. This process takes place in the region of each presumptive cusp.

The inner enamel epithelium cells in the future cusp tips are the first to induce adjacent mesenchyme to form odontoblasts. The predentine which is subsequently formed has a reverse inductive influence on these enamel epithelium cells which differentiate into ameloblasts and start forming enamel (Fig 95).

The result of the above changes is that the incisal edges and cusp tips are the first parts of the tooth crown in which dentine and enamel form. The process described above spreads down the cusp slopes and results in the eventual formation of the entire crown.

Differentiation of the ameloblasts

During the differentiation of the inner enamel epithelium to form pre-ameloblasts, the precursors of ameloblasts, several morphological changes take place in these cells. Cell height increases to about 40 µm while the cell width decreases to approximately 7 µm. The cells are regularly arranged with oval nuclei occupying the basal parts of the cells next to the intermediate cell layer (stratum intermedium) (Fig 96).

The intermediate cell layer and the ameloblasts must be regarded as a functional unit in enamel formation, since alkaline phosphatase is initially found almost exclusively in the intermediate cells. Alkaline phosphatase is associated with biological transport mechanisms and is instrumental in facilitating the transport of selected substances to ameloblasts. Ameloblasts are rich in RNA and possess a high oxidative enzyme activity. In this respect they are similar to cells responsible for other mineralized tissues.

Between the basal part of the cell and the nucleus, a dense concentration of mitochondria is found but these organelles are also dispersed throughout the entire cell. A Golgi apparatus is present on the distal, or formative, side of the nucleus while the central parts of the cell contain both smooth and rough endoplasmic reticulum. Free ribosomes are dispersed throughout the cell. Soon after a thin layer of enamel has formed, a short pointed cytoplasmic extension, the Tomes process (Fig 97, II), develops and projects into the newly formed matrix. These processes are the secretory surfaces of an ameloblast.

Fig 95 Development of enamel

Tooth germ

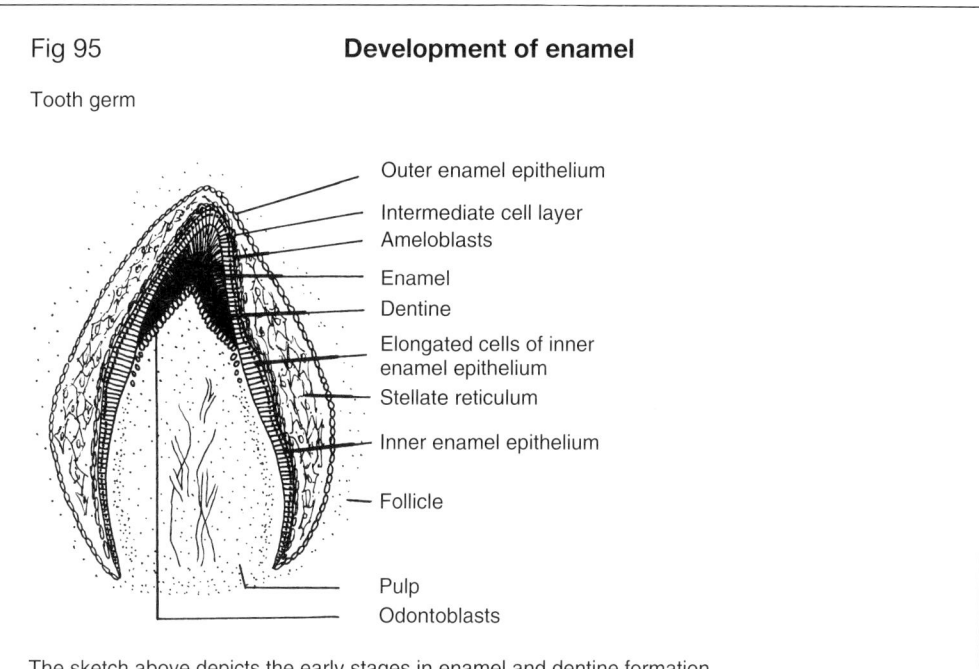

The sketch above depicts the early stages in enamel and dentine formation.
Note that the ameloblasts are elongated in the area of odontoblast differentiation.

Fig 96 A pre-ameloblast shortly before amelogenesis begins

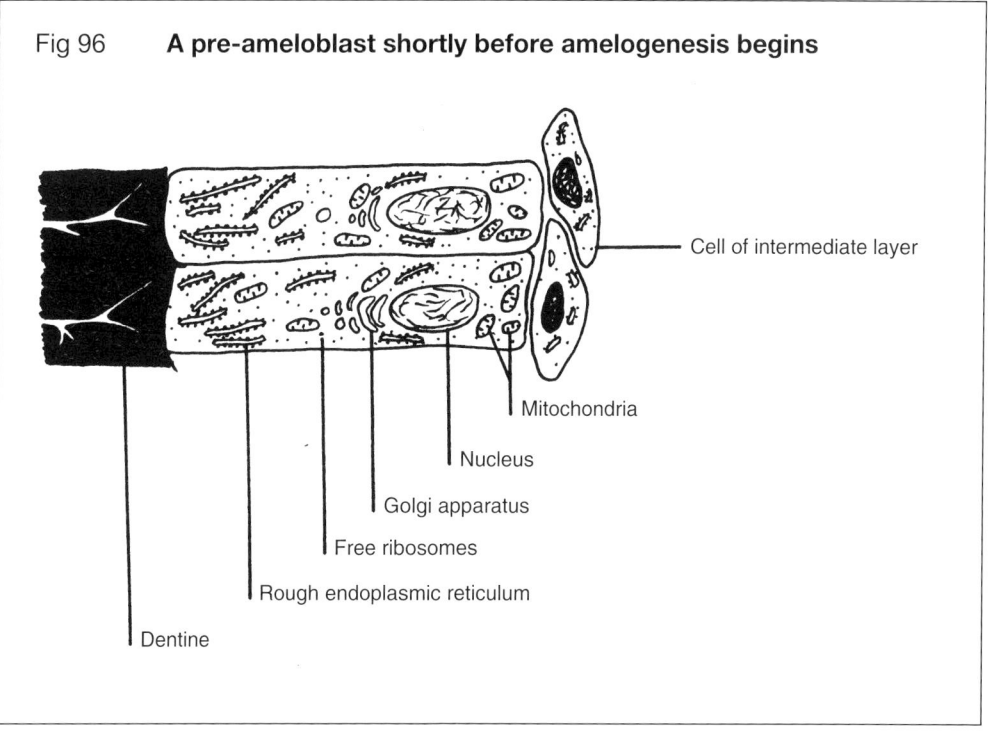

Fig 97 **Amelogenesis**

I. Initial matrix formation

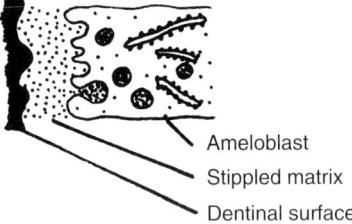

- Ameloblast
- Stippled matrix
- Dentinal surface

The pre-ameloblasts move from the surface of the dentine and secrete matrix via secretory vesicles.

II. Formation of the Tomes process and early crystallization

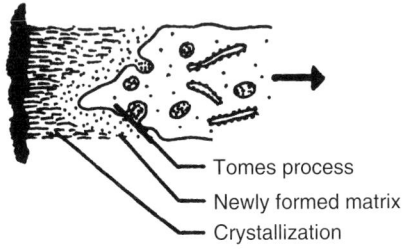

- Tomes process
- Newly formed matrix
- Crystallization

After a thin layer of matrix has been secreted, the Tomes process forms and crystallization commences. The ameloblasts move away from the dentine.

III. The enamel cuticle

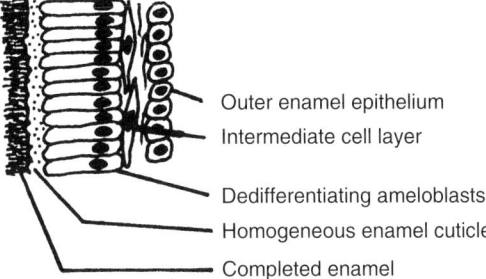

- Outer enamel epithelium
- Intermediate cell layer
- Dedifferentiating ameloblasts
- Homogeneous enamel cuticle
- Completed enamel

IV. The reduced enamel epithelium

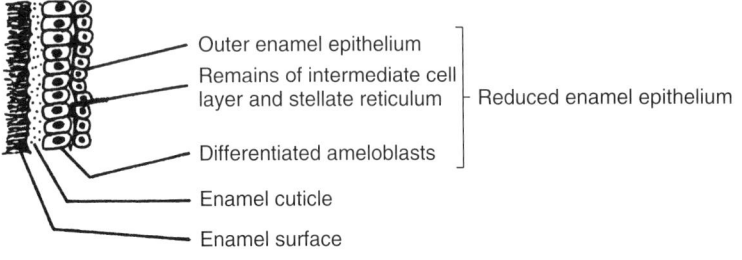

- Outer enamel epithelium ⎫
- Remains of intermediate cell layer and stellate reticulum ⎬ Reduced enamel epithelium
- Differentiated ameloblasts ⎭
- Enamel cuticle
- Enamel surface

Formation of the matrix and early mineralization

As in other secretory cells, vesicles appear first in the Golgi apparatus. These secretory vesicles contain an electron-dense stippled material and eventually move to the distal formative end of the cell where their contents are discharged by exocytosis (Fig 97, I). The above mechanism forms the basis for secretion of both the organic matrix and the mineral component of enamel (Fig 97, II). Enamel formation is thus an extracellular (secretory) process. Enamel crystals (hydroxyapatite) appear almost immediately after the initial matrix is deposited. A very thin layer of unmineralized matrix with a finely stippled appearance surrounds the Tomes processes for the duration of active enamel formation, but a clear layer of unmineralized matrix, similar to predentine, is never present.

The thickness of the enamel increases with the deposition and mineralization of additional matrix and the ameloblasts move away from the amelodentinal junction. Their path of movement is, however, not in a direct line with the long axes of the cells and is later reflected in the arrangement of enamel crystals in the rods.

The nature of enamel matrix

Early enamel matrix consists of 65% water, small amounts of proteoglycans, glycosaminoglycans, lipid, and citrate, and a small percentage of inorganic ions. The organic (proteinaceous) material forms a heterogeneous mixture, constituting about 20% of the matrix, and can be divided into two main groups, the amelogenins, and the enamelins. Amelogenins initially form the bulk of the matrix but are withdrawn in large quantities during maturation of the enamel, accompanied by a relative increase in the enamelin component.

Amelogenins are characterized by the presence of large amounts of proline, leucine, histidine, and glutamic acid, while hydroxyproline, hydroxylysine, and cystine are absent. Glycine is present only in minute quantities. The enamelins contain less proline, glutamic acid, and histidine than the amelogenins but greater quantities of glycine. It is therefore obvious that enamel matrix is devoid of both collagen and keratin.

Enamel matrix is present in very small quantities and is therefore difficult to investigate. It has been described as a gel-like material in which crystals are deposited. In some carefully demineralized preparations it appears as a delicate network surrounding each apatite crystal. This has in the past led some researchers to believe that it contains a keratin-like fibrous protein but there is little basis for this belief. It is generally accepted that enamel matrix is composed of a large variety of amino acids present as a gel in a mature tooth. It has been suggested that enamel matrix can flow under presure from the growing crystals, a view which is strengthened by the fact that administration of a single dose of radioactive amino acid spreads throughout the matrix and is subsequently found in matrix formed both before and after administration.

Mineralization and maturation of enamel

A partial mineralization occurs almost immediately after deposition of the first enamel matrix. Needle-shaped crystals appear after the deposition of a thickness of approximately 50 nm of matrix and are initially thin and widely dispersed. They rapidly increase in size and become hexagonal with a concomitant reduction in the quantity of matrix separating them. This is speedily followed by a secondary final mineralization which commences at the amelodentinal junction and is a rapid process. This secondary mineralization cannot be easily distinguished from the initial phase. The enamel is hereby transformed from a relatively soft material into a hard substance.

After formation of the full thickness of the enamel and during the process of final mineralization referred to above, in which large quantities of inorganic material are deposited in the matrix, the ameloblasts undergo various changes. The appearance of large numbers of lysosomes in the Tomes process raises the possibility that hydrolytic enzymes may be responsible for a reduction in the organic component of the matrix which occurs at this time. This reduction is both quantitative and qualitative, since the composition of the matrix changes. Maturation of enamel is also marked by a loss of water from the matrix.

The formative ends of the ameloblasts develop a brush border with loss of the Tomes processes during the final stages of maturation. This increases the contact area with the enamel. Mitochondria aggregate in this part of the cell and there is an increase in acid phosphatase activity. The above changes are seen as an indication of catabolic activity, since this enzyme is found in large quantities in osteoclasts. It is therefore assumed that the ameloblasts degrade material which is selectively withdrawn from the maturing enamel. Alkaline phosphatase activity in the ameloblasts is seen as an indication of transport of mineral salts across the cell membrane as part of the final maturation of enamel.

The reduced enamel epithelium and enamel cuticle

The last phase in the life history of an ameloblast may be regarded as dedifferentiation (Fig 97, III, IV). A homogeneous layer of approximately 1 µm is found between the dedifferentiated ameloblasts and the enamel surface. This layer has in the past erroneously been described as a thin surface layer of unmineralized enamel matrix. This layer has also been known as the enamel cuticle, or primary enamel cuticle. In the light of confusion which exists regarding the existence of a so-called secondary enamel cuticle, it is suggested that before eruption of the tooth crown this homogeneous layer be known only as the enamel cuticle. This "cuticle" is, in fact, a basal lamina-type junctional region between the inactive ameloblasts and the enamel surface. After eruption of a tooth this "cuticle" is found between the junctional epithelium and the enamel surface and is known as the internal basal lamina or attachment lamina. This is fully discussed in the chapter on the periodontium (Part II, Chapter 18).

After completion of enamel formation, but before the tooth erupts, the ameloblasts shorten to cuboidal cells and form the inner component of the reduced enamel epithelium (Fig 97, IV). The remainder of the enamel organ (stratum intermedium, outer enamel epithelium, and stellate reticulum) forms the outer parts of the reduced enamel epithelium and possesses the ability to divide mitotically. In due course the different cell types in the reduced enamel epithelium become indistinguishable. The reduced enamel epithelium, which is only a few cells thick, and the enamel "cuticle" have in the past collectively been known as Nasmyth's membrane.

During eruption of the tooth the outer cells of the reduced epithelium proliferate to unite with oral epithelium to form a united enamel, or junctional, epithelium. These developments are fully discussed in the chapter on the periodontium (Part II, Chapter 18).

Selected bibliography

1. Bhaskar, S.N. (1991) editor. *Orban's Oral Histology and Embryology,* 11th edition. St. Louis: Mosby-Year Book, Inc.
2. Cole, A.S. and Eastoe, J.E. (1988) *Biochemistry and Oral Biology,* 2nd edition. London: Wright.
3. Osborn, J.W. (1981) editor. *Dental Anatomy and Embryology.* Oxford: Blackwell Scientific Publications.
4. Osborn, J.W. and Ten Cate, A.R. (1983) *Advanced Dental Histology,* 4th edition. Bristol: Wright PSG.
5. Schroeder, H.E. (1991) *Oral Structural Biology.* New York: Thieme Medical Publishers, Inc.
6. Scott, J.H. and Symons, N.B.B. (1982) *Introduction to Dental Anatomy,* 9th edition. Edinburgh: Churchill Livingstone.
7. Ten Cate, A.R. (1989) *Oral Histology: Development, Structure, and Function,* 3rd edition. St. Louis: C.V. Mosby Company.
8. Williams, R.A.D. and Elliott, J.C. (1979) *Basic and Applied Dental Biochemistry.* Edinburgh: Churchill Livingstone.

Review questions

1. Discuss amelogenesis with reference to the following:

    (a) differentiation of ameloblasts;
    (b) formation of enamel matrix and early mineralization;
    (c) the nature of enamel matrix; and
    (d) processes associated with maturation of enamel.

2. Write brief notes on the structures which cover the enamel after completion of amelogenesis but before tooth eruption takes place.

7. Enamel

General remarks

The crowns of teeth are covered by enamel which is the hardest and most highly mineralized tissue of the body. It varies in thickness from 2 to 2.5 mm in the tips of cusps to a very fine edge at the amelocemental junction. Healthy young enamel is translucent and the color of a tooth depends to a large extent on this translucency. The underlying dentine normally has a yellow color which bestows on a tooth a light yellow color. It is common to observe a bluish tint in the enamel of the incisive edges of anterior teeth. This appearance is due to the fact that dentine does not extend into the incisive edges and the relative darkness of the oral cavity is seen through the translucent enamel.

Chemical composition

Enamel consists of 96–97 % inorganic material by mass, less than 1 % organic material, the remainder being water. By volume, the inorganic component accounts for 86 %, the organic component for 2 %, and water for 12 %.

The inorganic component consists almost exclusively of hydroxyapatite ($Ca_{10}(PO_4)_6.(OH)_2$) but fluoride ions may play an important role in replacing the hydroxyl ions and thereby converting the hydroxyapatite into fluorapatite. Fluorapatite crystals are less acid soluble than hydroxyapatite and their presence bestows on enamel a greater resistance to caries. This process is discussed in the chapter on fluoride (Part II, Chapter 24).

Organic matrix is sparsely distributed in enamel and occupies the submicroscopic spaces between hydroxyapatite crystals. It is difficult to demonstrate. The nature of the matrix is discussed in the previous chapter.

The structure of enamel

The basic structural components of enamel are the enamel rods, previously known as prisms, which extend from the amelodentinal junction to the surface of the crown (Fig 98). The rods, of which each tooth has some millions, are about 3 µm wide at the amelodentinal junction and widen to approximately 6 µm close to the surface. In cross-section (Fig 98) the rods have a fish-scale or keyhole appearance and it has been customary to describe a rounded "head" and a narrower "tail" region in each rod. The "tails" fit into the area created by the convexities of adjacent rounded "heads". The rounded "heads" are commonly directed towards the incisive edges or cusps, while the "tails" are directed towards the necks of the teeth. It must be emphasized that neither of the above appearances is found exclusively in cross-sections of enamel and adjacent rods may be differently shaped.

Enamel rods do not follow a straight course throughout their whole length and are sometimes intertwined, especially close to the dentine and in the tips of incisal edges and cusps. Enamel with this appearance is known as gnarled enamel. It is common for the rods to follow a horizontal course close to the cervical edges of primary teeth and of the majority of secondary teeth, but in approximately 25% of premolars they show an apical inclination in this region. In primary teeth and in approximately 70% of secondary teeth, no rod structure is seen on the surface of the enamel. In this surface layer the crystals in the rods are uniformly arranged with their long axes perpendicular to the surface.

Enamel crystals and rod sheaths

Mature enamel is basically composed of needle-shaped hexagonal hydroxyapatite crystals which are about 160 nm long, 40 nm wide and 25 nm thick. The crystals in the heads of the enamel rods are arranged with their long axes parallel to the long axes of the rods. They change direction towards the "tail" and become arranged perpendicular to the long axes of the rods.

The long axis of an ameloblast forms a considerable angle with the direction followed by the enamel rod during amelogenesis (Fig 99). Each pointed Tomes process occupies a hollow in the forming enamel. Enamel crystals are deposited with their long axes perpendicular to that surface of the Tomes process from which they arise. This results in considerable differences in crystal orientation in a rod, and this difference is most marked in a line corresponding to the tip of the Tomes process as it recedes. This line represents the area of the so-called rod sheath. It is apparent that not less than three or four ameloblasts participate in the formation of a single enamel rod. It is clear that rod sheaths are absent in surface enamel of those primary and secondary teeth in which no clear rod structure is present.

Fig 98 **Enamel**

The relationship between enamel and the other mineralized dental tissues

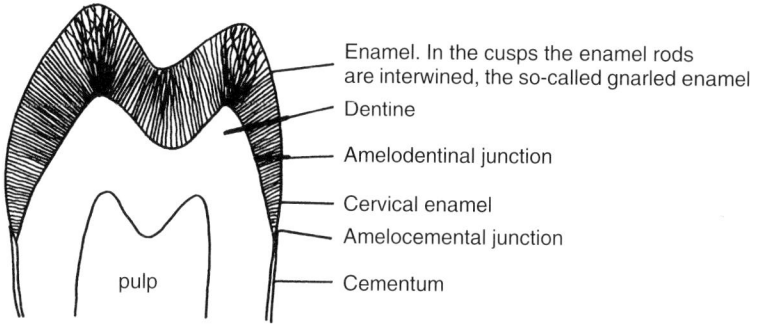

- Enamel. In the cusps the enamel rods are interwined, the so-called gnarled enamel
- Dentine
- Amelodentinal junction
- Cervical enamel
- Amelocemental junction
- Cementum

pulp

Enamel rods in longitudinal section

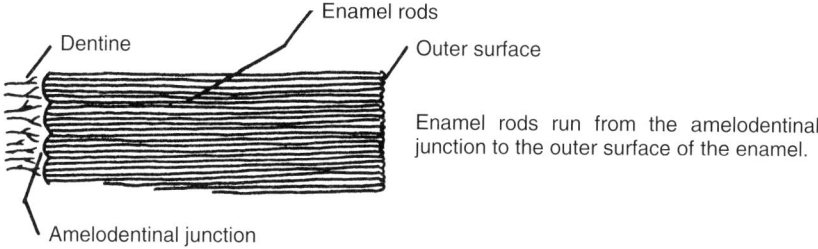

- Dentine
- Enamel rods
- Outer surface
- Amelodentinal junction

Enamel rods run from the amelodentinal junction to the outer surface of the enamel.

Enamel rods in cross-section

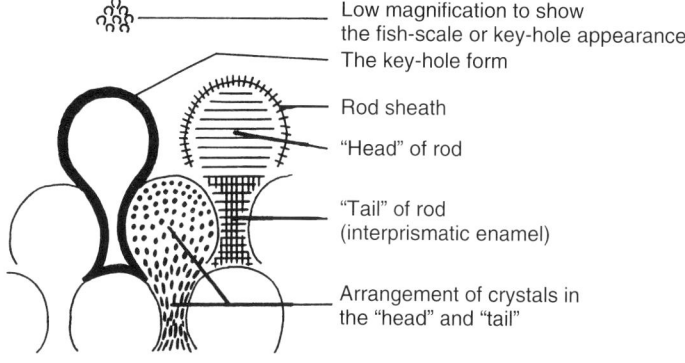

- Low magnification to show the fish-scale or key-hole appearance
- The key-hole form
- Rod sheath
- "Head" of rod
- "Tail" of rod (interprismatic enamel)
- Arrangement of crystals in the "head" and "tail"

The convex head of a rod is capped by a rod sheath which appears different from the rest of the rod when viewed both lightmicroscopically and electronmicroscopically. The existence of this structure has in the past been the subject of much research and speculation. According to some researchers the thickness of the sheath is less than 0.5 µm. Others believe that it is slightly thicker. It is occasionally indistinct or absent. It is commonly believed that the appearance of a sheath is due to differences in crystal orientation in this region (Fig 99). As described previously, a fairly sharp difference is found in crystal orientation in the "head" and the "tail" region of a rod. Where the "head" of one rod abuts against the "tail" of an adjacent rod, the irregular interface causes differences in light diffraction which is seen as a sheath. Another view is that the sheath represents an area with a higher organic content. The term inter-rod, or interprismatic, enamel is used to describe the rod substance between two adjacent heads.

Cross striations

Periodic bands, or cross striations (Fig 99), are seen at intervals of 4–6 µm across the rods. It has in the past been stated that these striations reflect variations in the density of mineralization occurring in a diurnal pattern. Scanning electronmicroscopy reveals that constrictions along the length of a rod are responsible for the striations, while another view holds that the constrictions represent areas of higher organic, and less inorganic, content.

A more recent view is that the enamel rods follow a rhythmically undulating course with a periodicity of about 4 µm and that this is responsible for the optical appearance of the narrowings as seen in a longitudinal ground section. It is widely held that the cross striations represent daily increments of growth.

Lines of Retzius

The incremental lines, or striae, of Retzius (Fig 100) are brown lines which are irregularly arranged in enamel. These lines vary in width from 150 µm to the width of a cross striation. In a ground longitudinal section of enamel it is seen that these lines commence at the amelodentinal junction and run obliquely outward in the direction of the cusps or incisive edges. They reach the surface of the enamel on the buccal and lingual surfaces of the tooth, but do not reach the surface in the cusps or incisive edges, where they surround the dentine tip to end at the amelodentinal junction on the opposite side of the tooth. Each line represents the out-

Fig 99   Movement of ameloblasts and crystallization

Direction of movement of ameloblasts

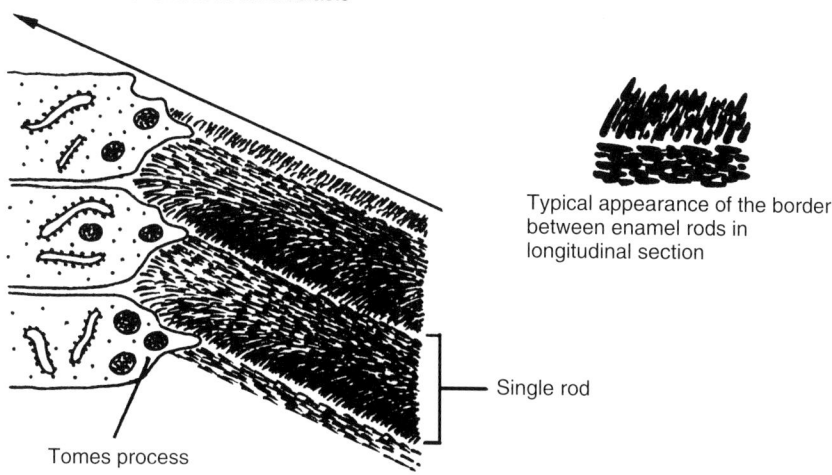

Typical appearance of the border between enamel rods in longitudinal section

Single rod

Tomes process

This sketch shows the relationship between the area of secretion of the ameloblasts (hexagonal areas), enamel sheaths (horse-shoe areas) and a single rod (stippled).

Cross striations of enamel rods

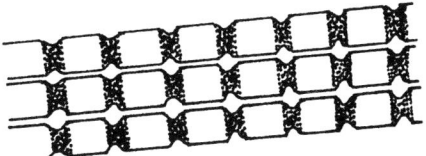

Cross striations (stippled) are 4-6 μm apart and are thought to represent daily increments of growth.

Fig 100 **Formation of the lines of Retzius**

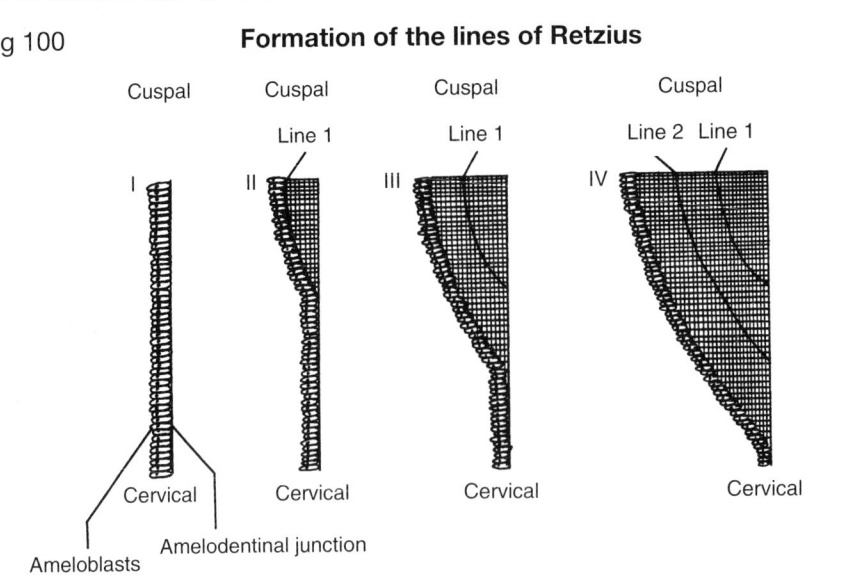

Amelogenesis takes place in a cervical direction. Cuspally-situated ameloblasts are the first to become active (sketch II). The process is then interrupted and a line of Retzius (line 1) forms. The former ameloblasts then resume their function and are joined by a more cervical group. When a second interruption occurs, a further line of Retzius forms (line 2, sketches III and IV). In this manner further lines of Retzius are added.

Fig 101 **Formation of perikymata**

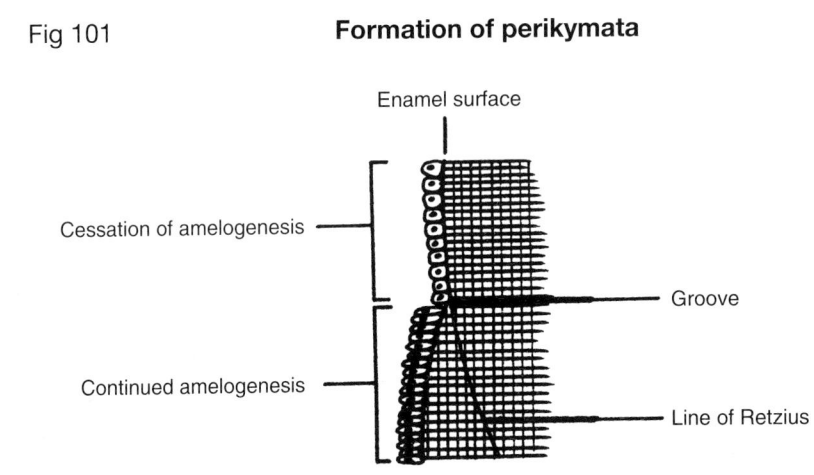

In the later stages of amelogenesis, cuspally-situated ameloblasts stop functioning, while the process is continued cervically. A surface groove represents the termination of a line of Retzius and indicates a point at which a group of ameloblasts have stopped functioning. The grooves and ridges thus formed are the perikymata.

line of enamel at succeeding stages in amelogenesis. The first line to form surrounds the dentine tip and indicates that this portion of enamel was formed first. As more ameloblasts become active, and more enamel forms, additional "domes" of enamel are added peripherally. In this way the pattern of lines is established.

By means of high magnification, it is obvious that the lines of Retzius cross adjacent enamel rods in step-wise fashion on the cross striations. This emphasizes the periodic nature of enamel deposition, and the direction followed by the lines is a further indication that groups of ameloblasts become active at different times. It is commonly accepted that the lines of Retzius are a result of disturbed mineralization and therefore have a systemic basis. The neonatal line, an accentuated line of Retzius, forms in all those teeth which are in the process of mineralization at the time of birth and reflects the marked physiologic disturbance occurring at this time. Corresponding lines form in all the teeth mineralizing when disturbance occurs. They are rare in prenatal enamel and most prominent in secondary teeth.

Where the lines of Retzius meet the surface of the enamel shallow transverse grooves, separated by low ridges, are seen in a newly-erupted crown (Fig 101). These structures are circumferentially arranged around the crown in a horizontal plane. Scanning electronmicroscopy reveals very small irregular hollows on the ridges. These represent identations produced by the Tomes processes of ameloblasts at the termination of enamel formation.

Terminology in the literature regarding these grooves and ridges is confusing. It is recommended that the term perikymata be used to collectively describe the grooves and ridges which follow a wave-like pattern on the enamel surface. The enamel surface is, however, rapidly worn smooth during function.

Bands of Hunter-Schreger

The bands, or lines, of Hunter-Schreger (Fig 102) are best observed in a ground longitudinal section of enamel viewed by reflected light. They appear as alternating light and dark horizontal bands which run in the same general direction as the enamel rods and are found in the inner part of the enamel. They are caused by successive changes in direction followed by groups of enamel rods in a horizontal plane, and the dark and light appearance can be reversed by changing the incident angle of the light.

Enamel rods leave the amelodentinal junction in a relatively perpendicular direction. After a short distance, horizontally arranged successive groups of rods deviate to right or to left and run obliquely for some distance before again coursing in a straight line to the surface of the enamel. If a tooth crown is seen as a successive series of horizontally stacked discs, the deviation of the group of rods in each disc varies from the direction followed by rods in adjacent discs.

When the incident light falls on the edge of a group of rods, it is reflected to the eye and results in a light appearance. Should the incident light fall on the cut edges of a group of rods, light is absorbed by the rods and the appearance is dark. This phenomenon is responsible for the alternating light and dark bands of Hunter-Schreger.

The amelodentinal junction

The amelodentinal junction (Fig 103) has a scalloped appearance with the hollows on the side of the enamel and the convexities on the side of the dentine. This pattern is occasionally indistinct, or may be absent. This junction is more fully dealt with in the chapter on dentine (Part II, Chapter 5).

Enamel tufts

Enamel tufts (Fig 103) arise at the amelodentinal junction and extend for a short distance into the enamel. They resemble tufts of grass and occur developmentally due to abrupt changes in the direction followed by groups of rods after leaving the amelodentinal junction. The tufts are best observed in a ground transverse section of a tooth and are probably due to the same directional changes of rods responsible for the bands of Hunter-Schreger. These ribbon-like structures appear to contain greater concentrations of enamel protein and are hypomineralized. They may represent wide interrod areas or prism sheaths.

Enamel spindles

When dentinal tubules are found within enamel, they usually appear as dark club-shaped structures which project into the enamel for varying distances, perpendicular to the amelodentinal junction. They are independent of the course followed by the enamel rods.

These enamel spindles (Fig 103) are more abundant in the cusps of teeth and arise when a single odontoblast process invades the pre-ameloblast layer at the commencement of dentinogenesis. This may be due to a relative crowding of odontoblasts in this area and formation of a process before meaningful movement of the cell body away from the junction occurs. The process penetrates between pre-ameloblasts and is subsequently embedded in enamel.

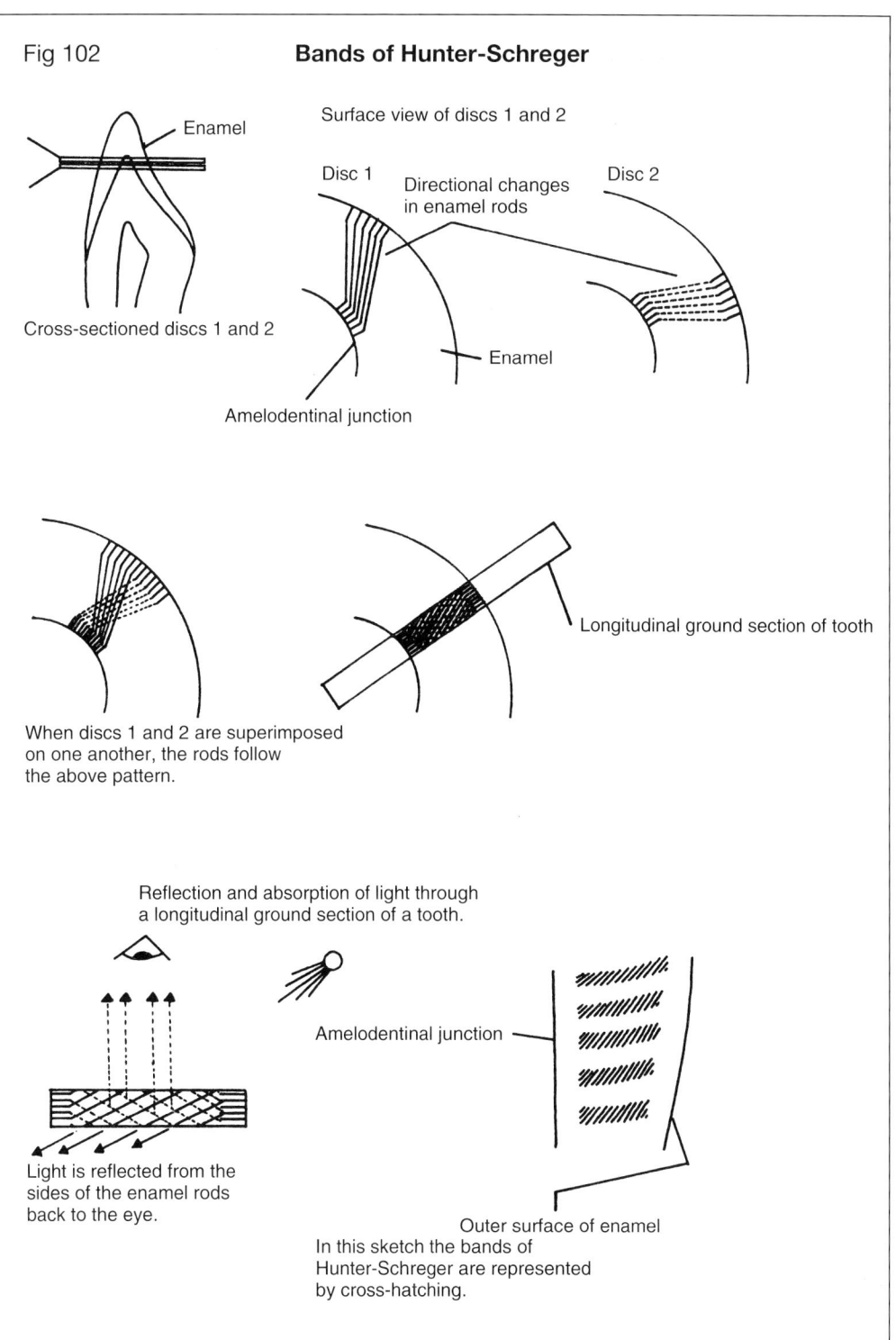

Fig 103   **Structures found at the enamel-dentine junction**

Amelodentinal junction

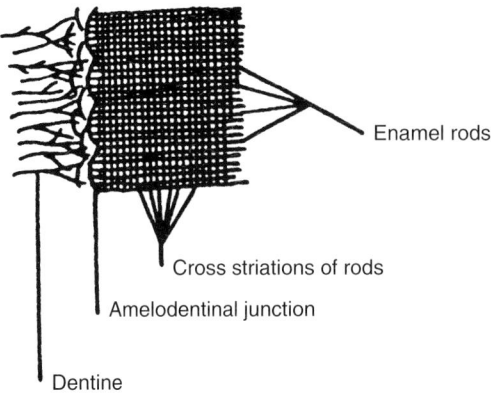

Enamel rods
Cross striations of rods
Amelodentinal junction
Dentine

Enamel tufts

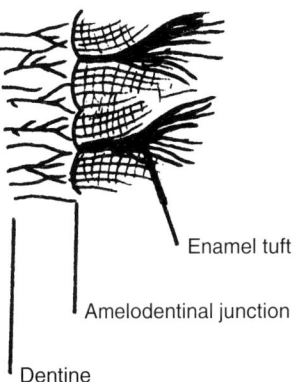

Enamel tuft
Amelodentinal junction
Dentine

Enamel spindles

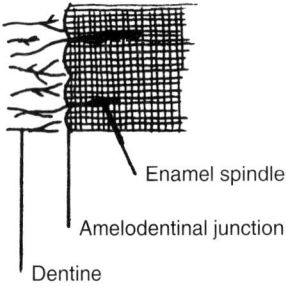

Enamel spindle
Amelodentinal junction
Dentine

Enamel lamellae

Enamel lamellae are thin leaf-like unmineralized structures arranged vertically in the enamel and extend for varying depths from the surface. In unerupted teeth, lamellae consist of unmineralized enamel matrix. These primary lamellae may be due to shrinkage of enamel during the final stages of amelogenesis. The small cracks thus formed may be secondarily filled with enamel protein.

Secondary lamellae are found in the enamel in the post-eruptive phase. They have the same appearance as primary lamellae but appear to be the result of occlusal trauma. They are filled with organic debris from the oral cavity.

Enamel permeability

Various experiments have shown that enamel is permeable, both from the surface and from the pulp through the dentine. This phenomenon is discussed in more detail in the chapter on permeability and age changes of teeth (Part II, Chapter 13).

Selected bibliography

1. Bhaskar, S.N. (1991) editor. *Orban's Oral Histology and Embryology,* 11th edition. St. Louis: Mosby-Year Book, Inc.
2. Melcher, A.H. and Zarb, G.A. (1973) editors. *Oral Sciences Reviews,* Volume III: *Dental Enamel.* Copenhagen: Munksgaard.
3. Mjör, I.A. and Fejerskov, O. (1986) *Human Oral Embryology and Histology.* Copenhagen: Munksgaard.
4. Osborn, J.W. (1981) editor. *Dental Anatomy and Embryology.* Oxford: Blackwell Scientific Publications.
5. Osborn, J.W. and Ten Cate, A.R. (1983) *Advanced Dental Histology,* 4th edition. Bristol: Wright PSG.
6. Schroeder, H.E. (1991) *Oral Structural Biology.* New York: Thieme Medical Publishers, Inc.
7. Scott, J.H. and Symons, N.B.B. (1982) *Introduction to Dental Anatomy,* 9th edition. Edinburgh: Churchill Livingstone.
8. Ten Cate, A.R. (1989) *Oral Histology: Development, Structure, and Function,* 3rd edition. St. Louis: C.V. Mosby Company.

Review questions

Describe enamel with reference to the following:

(a) enamel rods and sheaths;
(b) incremental lines in enamel;
(c) enamel tufts and spindles; and
(d) enamel lamellae.

8. Development of cementum

General remarks

Development of cementum and the root of a tooth commences when enamel formation is completed. The outer and inner enamel epithelia together form the epithelial root sheath of Hertwig which is responsible for determining the shape of the root (Fig 104, I). These two epithelial layers, which are separated in the enamel organ by the stratum intermedium and stellate reticulum, become continuous in the area of the future junction of the enamel and cementum to form a two-layered sheath which grows into the underlying mesenchyme. It is currently accepted that growth of the sheath does not occur downwards into the jaws, but that proliferation in the cells of the sheath causes an upward movement of the developing tooth. Thus the apical portion of the sheath remains relatively static, while the more coronal portion, which is associated with dentine and cementum formation, moves in the direction of the oral cavity. It is only after opposing teeth meet during eruption that a downward growth of the apical portion of the root sheath is described. The apical portion of the root sheath forms a partial diaphragm which separates the dental papilla from the surrounding mesenchyme.

Cells of the inner epithelial layer of the root sheath do not differentiate into ameloblasts due to the absence of other components of the enamel organ, but retain the ability to induce adjacent papillary undifferentiated mesenchymal cells to differentiate into odontoblasts and to proceed with the formation of predentine and dentine (Fig 104, II).

Changes in the root sheath

After dentine formation has commenced, changes occur in the root sheath. There are indications that the surface of root dentine is covered by a layer which is the product of the epithelial cells of the sheath before it becomes discontinuous. This layer is approximately 10 µm thick with a hyaline appearance and contains very fine granules and fibrils. This layer is described as being very similar to enameloid which covers the teeth of many fish. Enameloid resembles enamel but is formed by both the dental papilla and the inner enamel epithelium. The latter contributes significantly to enameloid. This layer, of ectodermal origin, later probably induces follicular mesenchyme to differentiate into cementoblasts. While this layer is forming, the cells of the root sheath develop increasing quantities of rough endoplasmic reticulum which indicates secretory activity. The hyaline layer contains epithelium-formed collagen and amino acids, according to some reports, and is often referred to as a cementing zone between dentine and cementum.

In due course the root sheath becomes discontinuous (Fig 104, III) and this enables the surrounding follicular mesenchyme to come into contact with the surface of the dentine. Changes in the root sheath have not been studied in human material but is has been shown in experimental animals that the first change occurs in the basal lamina. It becomes indistinct and disappears. Groups of epithelial cells then lose contact and fine fibers from the follicle are observed between these groups which once again develops a basal lamina. No degenerative changes develop in these epithelial cells which remain in the form of a network of groups of cells in the periodontal space for many years as the cell rests of Malassez.

Cementoblasts (Fig 104, IV, V) differentiate from follicular undifferentiated mesenchyme through an intermediate precementoblast stage. The cementoblasts develop all the cytoplasmic organelles which characterize protein-synthesizing and protein-secreting cells, such as numerous mitochondria, a well-developed Golgi apparatus and much rough endoplasmic reticulum, in addition to a well-marked alkaline phosphatase activity.

Formation of cementum

Cementoblasts (Fig 105) are cuboidal cells and arrange on the outer surface of the hyaline layer which covers the dentine. They are responsible for the deposition of the organic matrix of cementum, which consists of a proteoglycan ground substance and intrinsic collagen fibers, as well as being responsible for subsequent mineralization of the organic matrix (cementoid or precementum) in relation to the collagen fibrils.

Before a tooth erupts, fibers from the follicle (extrinsic fibers) are incorporated in the cementum and lie parallel to the root surface. When eruption commences, the inclination of these follicular fibers changes. They become oblique and can at this stage already be regarded as the precursors of periodontal ligament fibers.

Mineralization (Fig 105) commences when a thin layer of cementoid has formed. Mineral salts are derived from tissue fluid containing calcium and phosphate ions and are deposited as hydroxyapatite crystals along the long axes of collagen fibers. As cementum continues to increase in thickness, more follicular (extrinsic) fibers become included in the cementum and will eventually be known as Sharpey's fibers when the periodontal ligament becomes established (Fig 106).

Fig 104   **Early development of cementum**

I. Root sheath of Hertwig

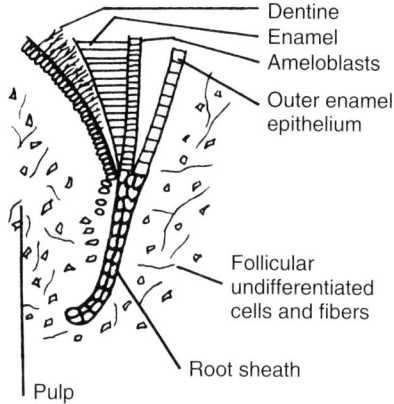

II. Proliferation of root sheath and further dentine formation in an apical direction

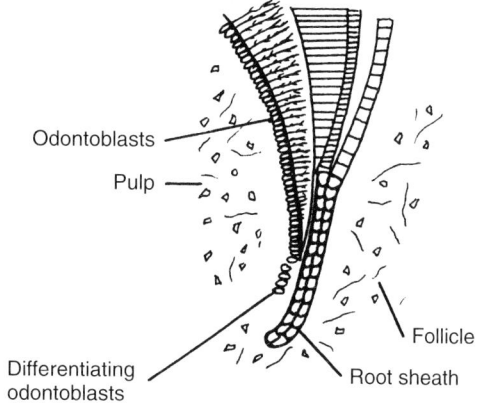

III. Fragmentation of root sheath

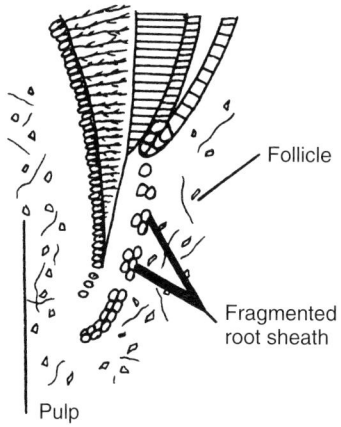

IV. Follicular cells and fibers contact dentine surface

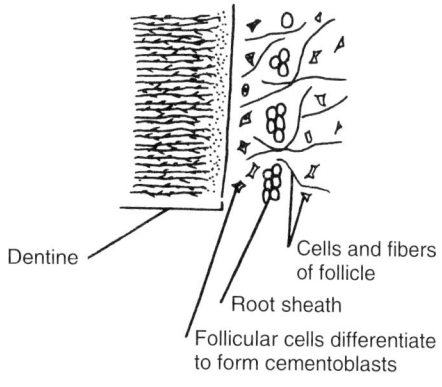

V. Differentiation of cementoblasts

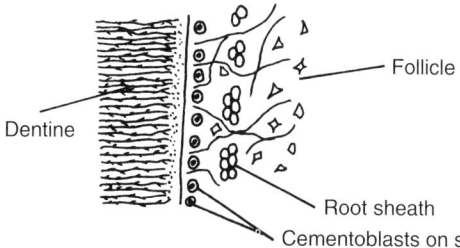

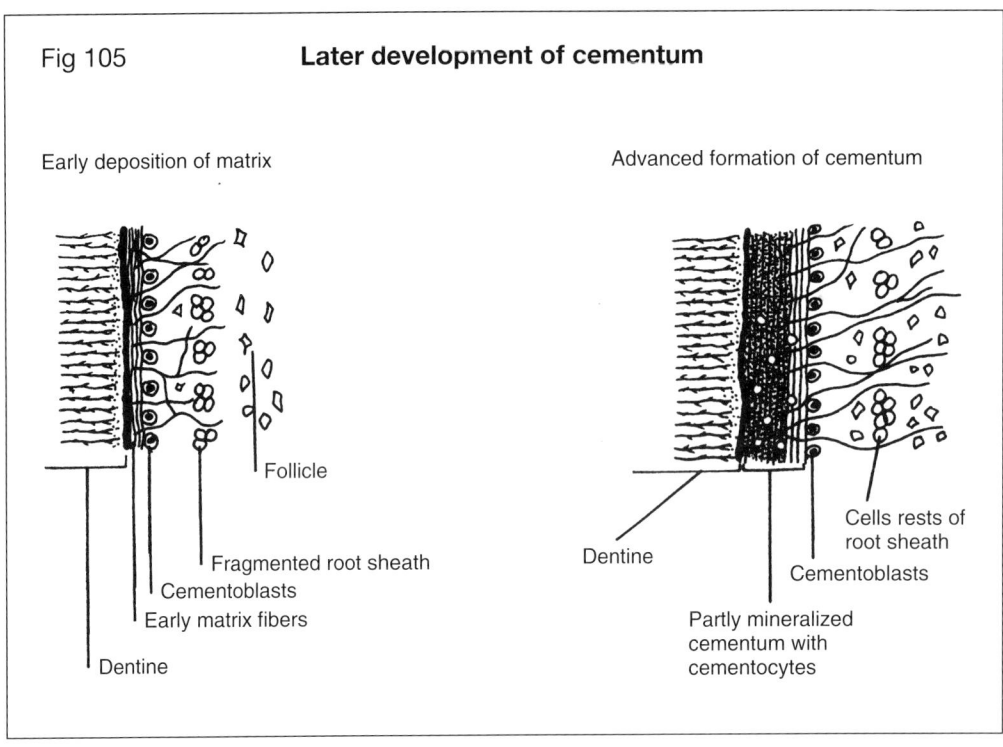

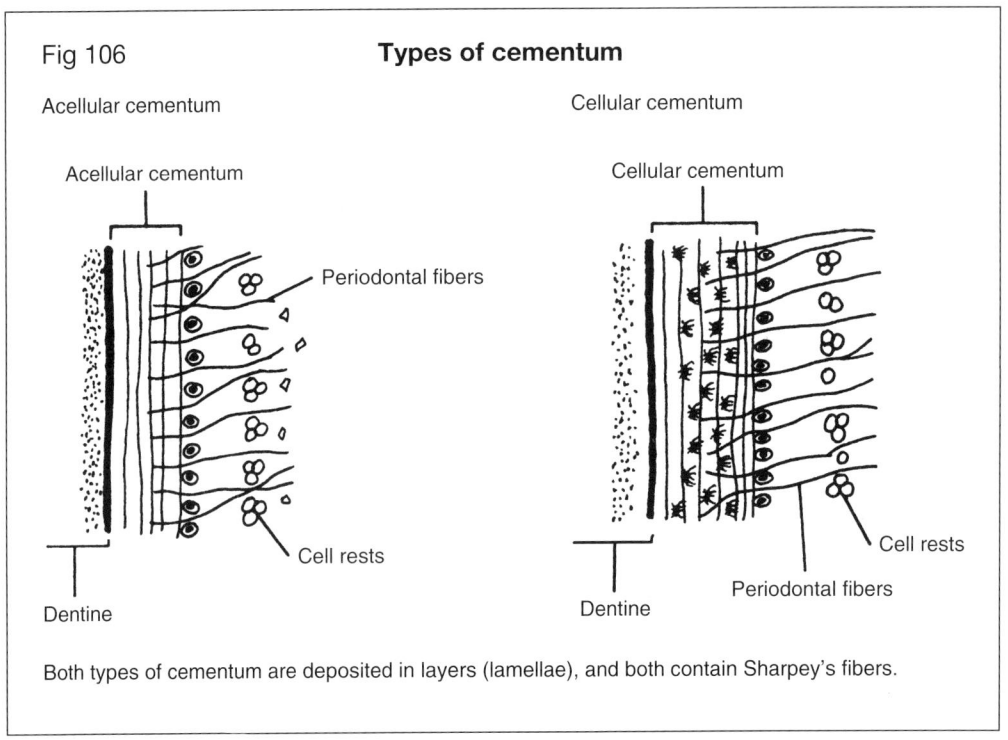

Proliferation of cells of the epithelial root sheath of Hertwig, dentine formation, fragmentation of the older parts of the sheath, differentiation of new cementoblasts and formation of cementum continues in an apical direction for the duration of root development. Cementum which is formed first does not contain any enclosed cells, but cementoblasts are included in later cementum. These enclosed cementoblasts are known as cementocytes and are found in lacunae.

In teeth with more than one root, the initial single primary apical foramen (Fig 107, I), formed by the epithelial diaphragm of the root sheath of Hertwig, becomes divided into two or more secondary apical foramina by tongues of epithelial tissue from the diaphragm (Fig 107, II). These fuse in the future furcation area of the roots. The number of secondary apical foramina is determined by the presence of groups of blood vessels which enter the dental papilla. These groups determine the number and location of roots and are surrounded by the ingrowing epithelial tongues from the diaphragm. Differential mitotic activity in the region of the diaphragm after formation of the secondary apical foramina (Fig 107, III) causes the roots to separate and to lengthen by continued deposition of dentine and cementum.

Types of cementum

Two distinct types of cementum are usually described, depending upon the absence or presence of cementocytes: acellular and cellular (Fig 106). Both forms are incrementally deposited as lamellae.

Acellular, or primary, cementum

Acellular cementum is usually found as a thin layer covering the dentine from the enamel-cementum junction to close to the root apex. It does not contain any cementocytes but usually a clear lamellar pattern.

Cellular, or secondary, cementum

Cellular cementum usually covers the acellular cementum in the apical parts of the root. It is also found in the furcation area of the roots and is deposited throughout life. It contains cementocytes in lacunae with radiating canaliculi in which fine cell processes of the cementocytes are found. Cellular cementum appears more irregular than acellular cementum and the lamellar structure is not as obvious. This is probably due to alternating phases of resorption and apposition of cementum to adapt to factors such as physiologic tooth movement. This is the main type to form after eruption of a tooth commences. Sharpey's fibers are found in both types.

Fig 107 — Formation of more than one root

These sketches represent three-dimensional views of a developing tooth sectioned longitudinally

I. A single apical foramen at the start of root development

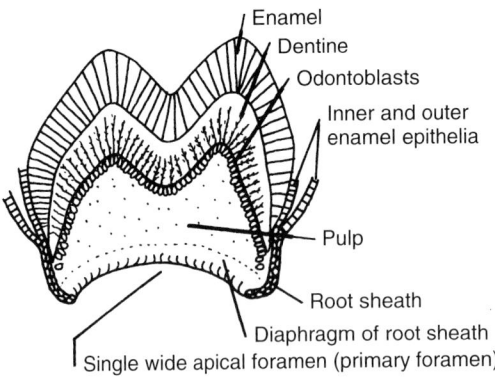

Labels: Enamel; Dentine; Odontoblasts; Inner and outer enamel epithelia; Pulp; Root sheath; Diaphragm of root sheath; Single wide apical foramen (primary foramen)

II. Division of the single apical foramen by ingrowing epithelial tongues from diaphragm of root sheath

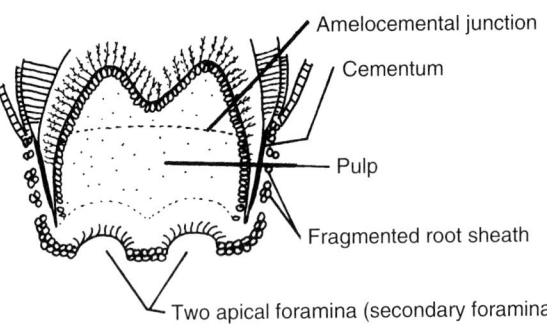

Labels: Amelocemental junction; Cementum; Pulp; Fragmented root sheath; Two apical foramina (secondary foramina)

III. Elongation of the roots

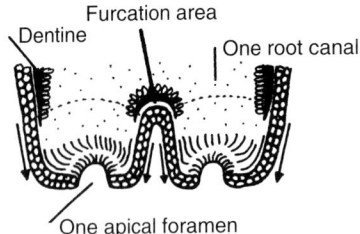

Labels: Furcation area; Dentine; One root canal; One apical foramen

Increased mitotic activity (arrows in sketch) on the periphery of each wide foramen gives origin to two tubes of root sheath cells. Mitosis is static in the furcation area. Deposition of dentine and cementum leads to the formation of two roots.

Cementum is formed throughout life and may reach a considerable thickness in the apical area of a tooth. Cementum is avascular and dependent only on periodontal tissue fluid for nutrition.

Disturbed development

Enamel pearls

Enamel "pearls" may form when a localized area of the epithelial root sheath shows full development of all the cell types of the enamel organ. The same inductive processes as found in crown development result in a circumscribed area of enamel, based on dentine and surrounded by cementum, on the root surface. These "pearls" are found most commonly in the furcation area of molar roots.

Accessory root canals and apical foramina

These may form when the continuity of the epithelial root sheat becomes interrupted before dentine formation commences, or if odontoblasts fail to differentiate. In these cases a corresponding break is found in the dentine and is converted into accessory root canals or apical foramina when the surrounding dentine continues to form normally.

Furthermore, the presence of supernumerary groups of blood vessels and nerves entering the dental papilla through the primary apical foramen may influence the development of the diaphragmatic "tongues" (described earlier) and lead to disturbed apical development.

Hypercementosis

This is a condition marked by excessive formation of cellular cementum in the apical region of the root. It may form in response to localized chronic inflammation and be confined to a single tooth, or be generalized, when it may be traced to genetic factors, or be due to systemic factors.

Selected bibliography

1. Bhaskar, S.N. (1991) editor. *Orban's Oral Histology and Embryology,* 11th edition. St. Louis: C.V. Mosby Company.
2. Osborn, J.W. (1981) editor. *Dental Anatomy and Embryology.* Oxford: Blackwell Scientific Publications.
3. Osborn, J.W. and Ten Cate, A.R. (1983) *Advanced Dental Histology,* 4th edition. Bristol: Wright PSG.
4. Schroeder, H.E. (1991) *Oral Structural Biology.* New York: Thieme Medical Publishers, Inc.
5. Scott, J.H. and Symons, N.B.B. (1982) *Introduction to Dental Anatomy,* 9th edition. Edinburgh: Churchill Livingstone.
6. Ten Cate, A.R. (1989) *Oral Histology: Development, Structure, and Function,* 3rd edition. St. Louis: C.V. Mosby Company.

Review questions

1. Describe the formation of the epithelial root sheath of Hertwig and its role in the development of a tooth root.
2. Discuss the formation of acellular (primary) cementum.
3. What are the morphological differences between acellular and cellular (secondary) cementum?
4. Describe how two roots form on a tooth.
5. Describe the formation of an enamel pearl.

9. Cementum

General remarks and chemical composition

Cementum covers the entire root of a tooth. It is similar to bone in many respects including structure, composition and behavior but is avascular. It is light yellow in color and is the softest of the dental mineralized tissues.

Cementum is composed of an organic matrix and inorganic mineral salts. The matrix consists of collagen fibers embedded in a proteoglycan ground substance and constitutes 23% of the total mass of cementum, while submicroscopic crystallites of mainly hydroxyapatite contribute 65% to the total mass. The rest is water.

Structure and distribution

Two types of cementum are traditionally described: acellular (primary) and cellular (secondary) cementum (Fig 108). Cementum formation continues throughout life and is normally characterized by life-long alternating phases of resorption and deposition to enable the tooth to adapt to periodontal conditions and function.

Collagen found in cementum is of two types. The first type consists of intrinsic matrix fibers formed by cementoblasts during cementogenesis, and arranged parallel to the root surface, while fibers of the second type are synthetized by periodontal (follicular) fibroblasts and become secondarily incorporated into the cementum. These extrinsic fibers are known as Sharpey's fibers when they form part of the periodontal ligament.

Acellular (primary) cementum is deposited first and forms a thin, fairly homogeneous, layer over the entire root surface. Cellular (secondary) cementum only starts to form at the commencement of tooth eruption, and is usually confined to the apical half of the root where it forms a layer of variable thickness external to the acellular cementum. Its thickness increases in an apical direction in a fully formed root in which it usually surrounds the apical foramen, or foramina, through which the pulp communicates with the periapical tissues (Fig 109). In a newly completed root the apical foramen is the most constricted part of the root canal and indicates the apical termination of dentine. With later formation of additional cellular cementum around the apical foramen, a funnel shaped cementum-lined apical foramen is formed peripheral to the constriction which now constitutes a dentine-cementum junction within the apical root canal.

Cementocytes

Cementocytes (Fig 110) are irregularly scattered in cellular cementum. They represent cementoblasts which were included and embedded in cementum during cementogenesis. They are found in lacunae while their radiating cytoplasmic processes fill fine canaliculi which are directed mainly in the direction of the periodontal ligament. The cementocytes lying some distance from the surface have less cytoplasm and associated organelles than those lying closer to the periodontal space, but all cementocytes have the same basic cytoplasmic structure as cementoblasts and cannot be regarded as dormant cells. The cementum immediately surrounding each lacuna and its canaliculi is especially dense, an indication of continued secretory activity by the cells.

Both types of cementum are deposited in lamellae, or layers, which are separated by incremental lines similar to growth rings in a tree. These lines are easily observed in acellular cementum but somewhat obscured in cellular cementum and have been used with varying degrees of success in human age determination studies.

Sharpey's fibers

Those parts of the periodontal ligament fibers embedded in cementum are termed Sharpey's fibers (Fig 111), while the same terminology is applied to the ends of the fibers embedded in alveolar bone. The attachment proper of the periodontal fibers is, however, in the most recently formed layer of cementum. These peripheral parts of the Sharpey's fibers are in a direct line with their direction in the periodontal ligament, and consequently stresses are always applied along their long axes, mechanically the most favorable way. In the deeper layers of cementum the Sharpey's fibers may run in different directions, an indication of previous tooth movements and different orientations of periodontal fibers.

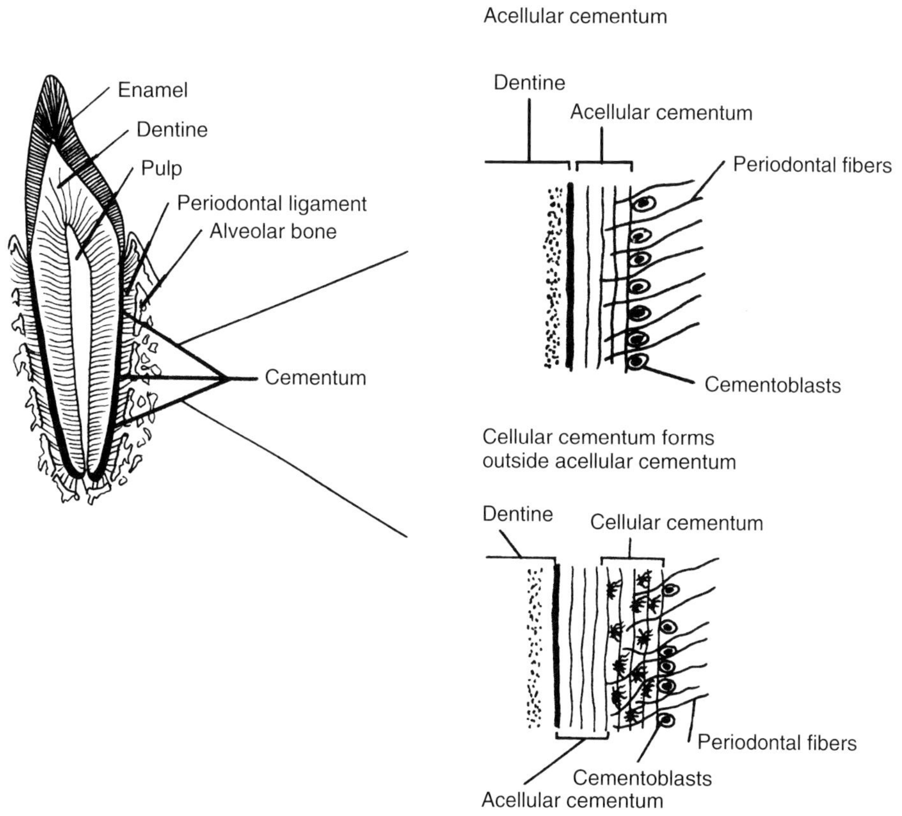

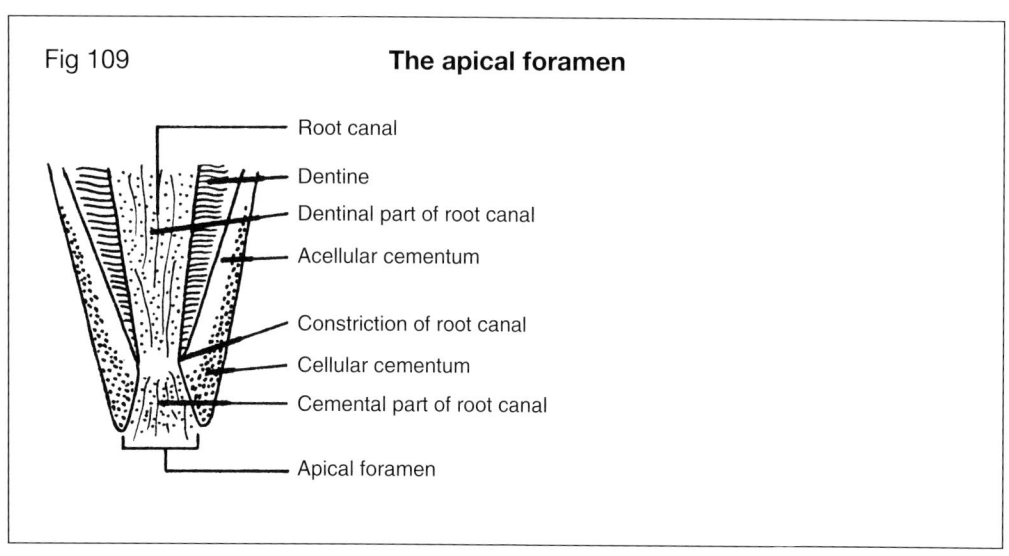

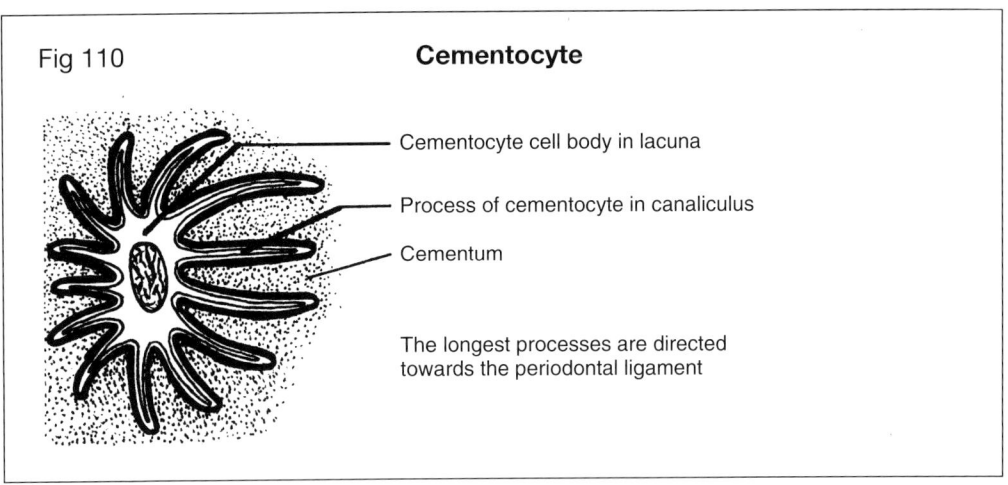

Fig 110 **Cementocyte**

— Cementocyte cell body in lacuna
— Process of cementocyte in canaliculus
— Cementum

The longest processes are directed towards the periodontal ligament

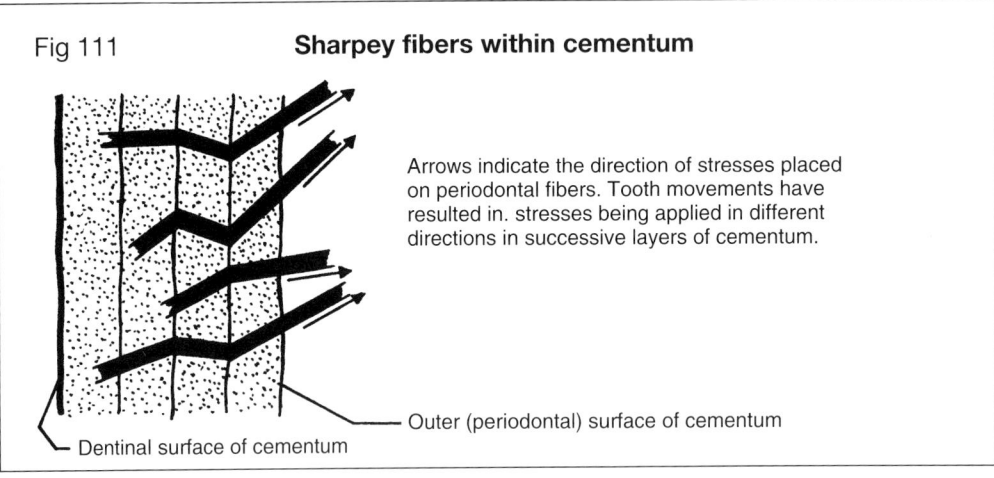

Fig 111 **Sharpey fibers within cementum**

Arrows indicate the direction of stresses placed on periodontal fibers. Tooth movements have resulted in. stresses being applied in different directions in successive layers of cementum.

— Outer (periodontal) surface of cementum
— Dentinal surface of cementum

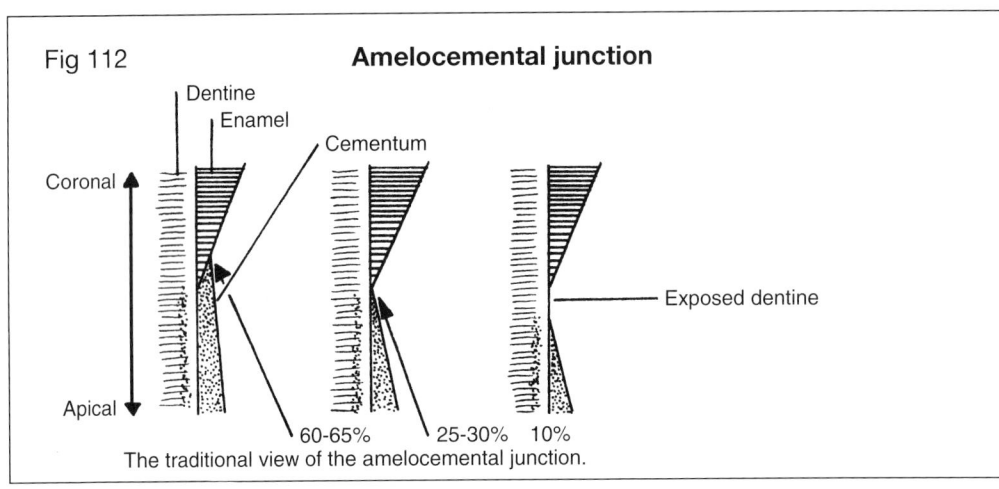

Fig 112 **Amelocemental junction**

Dentine
Enamel
Cementum
Coronal
Apical

— Exposed dentine

60-65%   25-30%   10%

The traditional view of the amelocemental junction.

The amelocemental (enamel-cementum) junction

The relation between enamel and cementum at the necks of teeth varies considerably (Fig 112). It has been maintained for very many years that the cementum overlaps the enamel at the amelocemental junction in about 60% of teeth; that in about 30% of teeth enamel and cementum meet edge to edge, and that in about 10% of teeth a gap is present between enamel and cementum, thus leaving some root dentine exposed. In these cases sensitivity may develop at the necks of the teeth when a certain degree of gingival recession occurs. Overlapping of enamel by cementum is probably due to disintegration of the reduced enamel epithelium close to the necks of the teeth after completion of enamel formation. This allows follicular mesenchyme to differentiate into cementoblasts under inductive influences from the cervical enamel and to produce an overlapping spur of acellular cementum.

More recent well-documented investigations have cast doubts on the validity of the percentages quoted above. A fairly consistent finding has been that the enamel and cementum meet as a butt joint in the majority of junctions studied, followed by cementum overlapping enamel. All agree that a minority of junctions showed a gap between the two tissues but emphasize that all three junctional relationships may occur side by side along the circumference of a single tooth.

In newly erupted teeth the dentogingival attachment ends at the amelocemental junction but it commonly migrates on to cementum as part of the aging process, and as a result of inadequate oral hygiene measures or faulty tooth brushing techniques.

Functions and clinical behavior

The primary function of cementum is to act as a medium for the attachment of periodontal ligament fibers and thus to ensure stability of the teeth in their sockets. Since cementum is formed throughout life, the attachment of ligament fibers is constantly altered due to factors such as tooth movement during the different phases of eruption (pre-eruptive, prefunctional and functional), or as a result of orthodontic treatment procedures. When a tooth changes its position, the periodontal fibers no longer leave the cemental surface in a straight line but at a new angle. Further deposition of cementum, however, encloses the fibers in their new inclination to once again ensure that the fibers exit from the cementum in a straight line.

When light pressure is applied to the side of a tooth, alveolar bone is resorbed on the side where pressure is applied and deposited on the side of tension. This is the basis for orthodontic tooth movement. Cementum is not normally influenced to the same degree as alveolar bone, but excessive pressure may cause necrosis of periodontal tissues (alveolar bone, periodontal ligament fibers, and cementum). Cementum has a remarkable regenerative capacity.

The roots of primary teeth are resorbed to enable their permanent successors to erupt in a normal position. This process is not continuous but is alternated by periods of recovery and cementum deposition. However, resorption remains the dominant process.

Resorption of primary incisor and canine roots commences in their linguo-apical parts, since these areas are closer to the erupting secondary successors. In primary molars, however, resorption commences between the roots in close proximity to the succeeding developing premolars. Resorption of the hard tissues of primary teeth is achieved by odontoclasts which are histologically identical to osteoclasts, being large multinucleated cells containing many lysosomes and mitochondria, and with a ruffled border in contact with the hard tissue to be resorbed. Both odontoclasts and osteoclasts are derived from mononuclear hematopoietic progenitor cells but their exact mode of differentiation remains obscure. The connective tissues of the periodontium are resorbed by fibroblasts and macrophages.

Selected bibliography

1. Bhaskar, S.N. (1991) editor. *Orban's Oral Histology and Embryology*, 11th edition. St. Louis: Mosby-Year Book, Inc.
2. Mjör, I.A. and Fejerskov, O. (1986) *Human Oral Embryology and Histology*. Copenhagen: Munksgaard.
3. Muller, C.J.F. and Van Wyk, C.W. (1984) The amelo-cemental junction. *Journal of the Dental Association of South Africa*, 39, 799-803.
4. Osborn, J.W. (1981) editor. *Dental Anatomy and Embryology*. Oxford: Blackwell Scientific Publications.
5. Osborn, J.W. and Ten Cate, A.R. (1983) *Advanced Dental Histology*, 4th edition. Bristol: Wright PSG.
6. Provenza, D.V. (1988) *Fundamentals of Oral Histology and Embryology*, 2nd edition. Philadelphia: Lea and Febiger.
7. Schroeder, H.E. (1991) *Oral Structural Biology*. New York: Thieme Medical Publishers.
8. Scott, J.H. and Symons, N.B.B. (1982) *Introduction to Dental Anatomy*, 9th edition. Edinburgh: Churchill Livingstone.
9. Ten Cate, A.R. (1989) *Oral Histology: Development, Structure, and Function*, 3rd edition. St. Louis: C.V. Mosby Company.

Review questions

1. Discuss cementum with reference to the following:

   (a) structure and distribution;
   (b) periodontal fiber attachment;
   (c) amelocemental junction; and
   (d) functions and clinical behavior.

10. The dental pulp

General remarks

The dental pulp consists of delicate vascular connective tissue components with specialized cells, the odontoblasts, arranged peripherally in close relation to the dentine. The pulp is intimately concerned with reactions of the tooth to stimuli reaching it as a result of loss of tooth material or damage of any nature.

The pulp fulfils various functions:

Inductive and formative functions

The primary function of the papilla (the later dental pulp) is to interact with the inner enamel epithelium of the enamel organ. In this way crown form is determined. A further result of this interaction is that odontoblasts differentiate and dentine production commences. This leads to differentiation of ameloblasts and the start of enamel formation. Odontoblasts continue to form dentine throughout the life of a vital pulp.

Reparative function

The pulp responds to any form of low-grade irritation, whether bacterial, mechanical, chemical or thermal, by producing secondary dentine in an attempt to isolate the pulp from the source of irritation (see Part II, Chapter 13).

Defensive, or protective, function

Acute irritation of the pulp, such as caused by bacteria, deep cavity preparation or an irritating filling, causes an inflammatory response in which mast and plasma cells, monocytes, macrophages, lymphocytes, and neutrophils are involved. Hyperemia and edema may, however, lead to the accumulation of excessive extravascular fluid which is confined by the rigid dentinal walls. The pressure of this fluid may lead to ischemia with resultant pulp necrosis.

Nutritive function

The blood vascular system of the pulp nourishes the dentine and enamel through the odontoblasts and the odontoblast processes. Dentinal tubules are bathed in pulpal fluid which also serves a moistening function in the enamel.

Sensory function

All stimuli reaching the vital pulp cause pain and, in a sense, this may also be regarded as a protective function. The sensitivity of teeth is dealt with in Part II, Chapter 19.

General appearance and structure

The central, or pulp, cavity of a tooth is occupied by the dental pulp (Fig 113). The pulp cavity is surrounded by dentine, except in its extreme apical part where the apical foramen is surrounded by cellular cementum in an older tooth. It can be divided into a pulp chamber, or coronal pulp, in the crown and neck area of a tooth, and the root canal, or radicular pulp, in the roots. The pulp communicates with the periapical tissues through a foramen, or several foramina, through which nerves, blood vessels, and lymphatics enter. Considerable variation occurs in the number and shape of root canals in a tooth. Accessory root canals may be present anywhere along the length of a root, while lateral root canals may connect any part of the root canal with the periodontal tissues.

It is, furthermore, unusual for a single apical foramen to be present on a root apex. Any number of small accessory foramina may connect the root canal to the periapical tissues in this region.

The shape of the dental pulp is roughly similar to the external shape of the tooth. The pulp chamber is prolonged in the form of small projections, the pulp horns or cornua, under each cusp. With continued dentine formation in older teeth these pulp horns become less prominent.

The dental pulp is a delicate connective tissue and is composed of cells, blood and lymph vessels, nerves and thin collagenous fibers in an abundant gelatinous ground substance. The cellular elements are prominent in a young pulp but the number of cells diminishes in an older pulp with a concomitant increase in the fibrous elements.

Fig 113 **The dental pulp**

General morphology

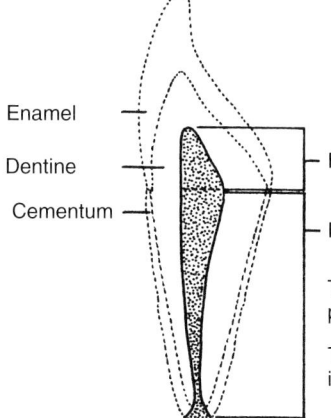

- Enamel
- Dentine
- Cementum
- Pulp chamber (coronal pulp)
- Root canal (radicular pulp)

The pulp cavity consists of the pulp chamber plus the root canal.

The apical part of the root canal is discussed in the previous chapter.

The pulp horns (cornua)

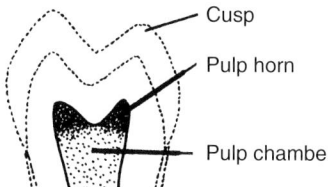

- Cusp
- Pulp horn
- Pulp chamber

Variations of the apical part of the root canal

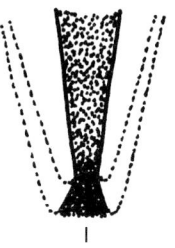 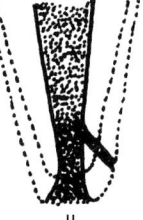

I II

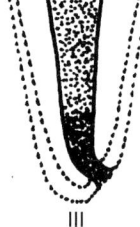

III

I. One foramen on the anatomical apex.

II. Accessory lateral canal.

III. A foramen on the side of the anatomical apex.

Regions of the dental pulp

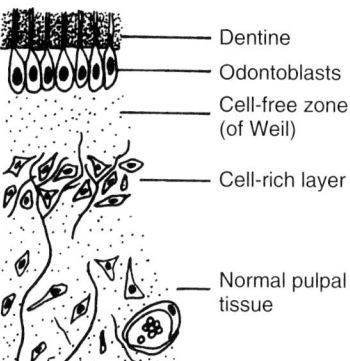

- Dentine
- Odontoblasts
- Cell-free zone (of Weil)
- Cell-rich layer
- Normal pulpal tissue

Cells of the pulp

On the periphery of the pulp the odontoblasts are found in close relationship to the predentine. A relatively cell-free zone, the basal layer of Weil, can often be seen immediately beneath the odontoblasts and this zone is separated from the normal pulpal tissue by a relatively cell-rich subodontoblastic zone with a rich capillary and nerve network. The basal layer of Weil is not seen in a developing tooth, while the subodontoblastic layer is best observed during active dentine formation.

Fibroblasts are the most numerous cells of the pulp and are flattened spindle-shaped cells with oval nuclei and long cytoplasmic processes which may have desmosomal contact with processes of adjacent fibroblasts. These cells contain the cytoplasmic organelles concerned with protein synthesis, namely ribosomes, endoplasmic reticulum, mitochondria, and a Golgi apparatus. Undifferentiated mesenchymal cells are smaller than fibroblasts and are usually arranged perivascularly. They cannot otherwise be readily distinguished. Macrophages, lymphocytes, plasma cells, and eosinophil granulocytes are occasionally seen in healthy pulps.

Fibers of the pulp

Young pulps contain few fibers and, when present, they occur singly or grouped in delicate bundles which are irregularly arranged. The fibers are collagen, in which the typical cross-banding pattern can be observed by means of the electron microscope, as well as reticulin fibers, especially in relation to blood vessels. These fibers are demonstrated by means of silver stains. Some von Korff fibers are revealed by the same method. The amount of collagen increases with age. Elastic fibers are found in the walls of blood vessels.

Blood vessels of the pulp

The pulp has a very rich blood supply. The vessels enter and leave the pulp through the apical foramen and accessory and lateral foramina. Small arterioles enter the pulp in small bundles which branch out and run in the periphery of the root canals towards the pulp chamber. The branches anastomose with branches of neighboring vessels. In the pulp chamber, capillaries course towards the odontoblasts where a rich subodontoblastic capillary network forms. Loops from this network are found between the odontoblasts but no capillaries enter the dentinal tubules. Venules returning to the foramina run more centrally in the pulp. Blood vessels and nerves in the pulp are usually grouped in neurovascular bundles.

Lymphatics of the pulp

Although doubts have been expressed regarding the presence of lymphatics in the dental pulp, studies have shown that they are present. They are often perivascular but may be present as discrete vessels which communicate with lymphatics of the periodontal ligament.

Nerve supply

Both myelinated and unmyelinated nerve fibers are found in the dental pulp.

1. *Myelinated fibers* are terminal branches of the trigeminal nerve. They are afferent somatosensory nerves with free nerve endings and are able to register only pain, independent of the nature of the stimulus to the pulp. These nerves enter the root canals in the form of two or three large trunks and course to the pulp chamber where they branch considerably. In the relatively cell-rich layer and in the cell-free zone the sensory fibers lose their myelin sheaths and form a plexus (of Raschkow) from where unmyelinated fibers course towards the odontoblasts. These fine branches are covered only by a Schwann cell and cannot be distinguished from fibers of the autonomic nervous system. They may form a second marginal plexus on the surface of the predentine. Terminal fibers may end free in the vicinity of the odontoblasts, may form short loops into the predentine or enter some tubules for varying distances. All the terminal branches form free nerve endings.
2. *Unmyelinated fibers* belong to the autonomic nervous system and course in the walls of blood vessels and innervate the smooth muscle fibers to regulate vasoconstriction.

A detailed discussion of the nerve supply of the dental pulp can be found in Part II, Chapter 19.

Selected bibliography

1. Bhaskar, S.N. (1991) editor. *Orban's Oral Histology and Embryology*, 11th edition. St. Louis: Mosby-Year Book.
2. Mjör, I.A. and Fejerskov, O. (1986) *Human Oral Embryology and Histology*. Copenhagen: Munksgaard.
3. Osborn, J.W. (1981) editor. *Dental Anatomy and Embryology*. Oxford: Blackwell Scientific Publications.
4. Osborn, J.W. and Ten Cate, A.R. (1983) *Advanced Dental Histology*, 4th edition. Bristol: Wright PSG.
5. Provenza, D.V. (1988) *Fundamentals of Oral Histology and Embryology*, 2nd edition. Philadelphia: Lea and Febiger.
6. Schroeder, H.E. (1991) *Oral Structural Biology*. New York: Thieme Medical Publishers.
7. Scott, J.H. and Symons, N.B.B. (1982) *Introduction to Dental Anatomy*, 9th edition. Edinburgh: Churchill Livingstone.
8. Ten Cate, A.R. (1989) *Oral Histology: Development, Structure, and Function*, 3rd edition. St. Louis: C.V. Mosby Company.

Review questions

1. Write brief notes on the functions of the dental pulp.
2. Describe the morphology of the dental pulp.
3. Discuss the histology of the dental pulp with reference to the following:

    (a) cell population;
    (b) fibers present;
    (c) blood supply and lymphatic drainage; and
    (d) nerve supply.

11. Form, arrangement and chronology of teeth

General remarks

Man has two dentitions, a primary, or deciduous dentition, and a secondary, or permanent dentition. The primary dentition starts to appear about 6 months after birth and eruption is completed at about 2 $\frac{1}{2}$ years of age. The secondary teeth begin to erupt after about 6 years with accompanying shedding of the deciduous teeth. At about the age of 13 years, all primary teeth have been replaced by secondary teeth and the secondary dentition is usually complete at 20 years of age.

The developmental dates (chronology) of teeth are dealt with later in this chapter. The sequence of eruption may be summarized as follows:

Primary teeth

1. Mandibular central incisor
2. Mandibular lateral incisor/maxillary central incisor
3. Maxillary lateral incisor
4. Mandibular first molar
5. Maxillary first molar
6. Mandibular canine
7. Maxillary canine
8. Mandibular second molar
9. Maxillary second molar

Secondary teeth

1. Mandibular first molar
2. Maxillary first molar
3. Mandibular incisors
4. Maxillary incisors
5. Mandibular canines
6. Mandibular first premolar

7. Maxillary first premolar
8. Mandibular second premolar/Maxillary second premolar
9. Maxillary canine
10. Second molars
11. Third molars

The teeth of both jaws are arranged in symmetrical dental arches (Fig 114) with the same number of teeth on both sides of the midline which is represented by an imaginary point between the two central incisors in each arch.

In each half (quadrant) of each jaw the following teeth are located from the midline backwards. Teeth are numbered according to the system proposed by the Féderation Dentaire Internationale (FDI):

Deciduous teeth (a total of 20 teeth)

Central incisors (maxilla, 51 and 61; mandible, 71 and 81)
Lateral incisors (maxilla, 52 and 62; mandible, 72 and 82)
Canines (maxilla, 53 and 63; mandible, 73 and 83)
First molars (maxilla, 54 and 64; mandible, 74 and 84)
Second molars (maxilla, 55 and 65; mandible, 75 and 85)

Secondary teeth (a total of 32 teeth)

Central incisors (maxilla, 11 and 21; mandible, 31 and 41)
Lateral incisors (maxilla, 12 and 22; mandible 32 and 42)
Canines (maxilla, 13 and 23; mandible, 33 and 43)
First premolars (maxilla, 14 and 24; mandible, 34 and 44)
Second premolars (maxilla, 15 and 25; mandible, 35 and 45)
First molars (maxilla, 16 and 26; mandible, 36 and 46)
Second molars (maxilla, 17 and 27; mandible, 37 and 47)
Third molars (maxilla, 18 and 28; mandible, 38 and 48)

The formula for the primary dentition is as follows:

$$I\frac{2}{2}\,C\frac{1}{1}\,M\frac{2}{2}$$

The formula for the secondary dentition is as follows:

$$I\frac{2}{2}\,C\frac{1}{1}\,P\frac{2}{2}\,M\frac{3}{3}$$

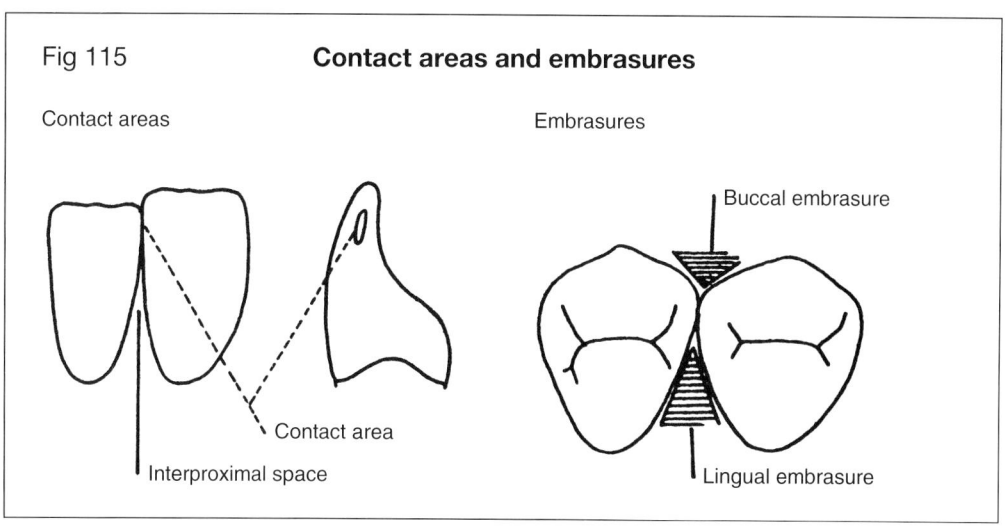

Fig 114 **The dental arches**

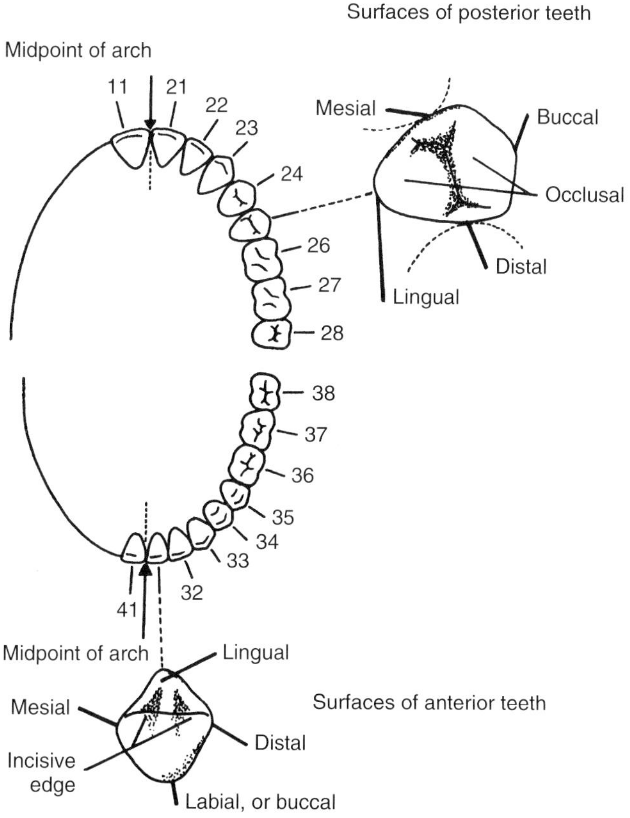

Fig 115 **Contact areas and embrasures**

325

The crowns of the incisors and canines (anterior teeth) possess four surfaces and a cutting margin, the incisive edge. The surfaces are as follows (Fig 114):

mesial = nearest to the midline of the arch
distal = furthest from the midline of the arch
buccal (also termed labial or vestibular) = the surface that makes contact with the cheeks and lips at rest
lingual (also termed palatal) = the surface that makes contact with the tongue at rest

The crowns of the premolars and molars (cheek teeth) possess the abovementioned four surfaces (no incisive edges), plus an occlusal, or crushing surface on which cusps are found. The adjacent surfaces of two teeth in the same dental arch are termed the proximal surfaces, represented by the distal surface of one tooth and the mesial surface of the next tooth.

The area of contact between two teeth is termed the contact area, or the contact point (Fig 115). These contact areas are round or oval and are of different sizes. The rounded surfaces of the crowns tend to diverge in all directions from the contact area, thus forming spaces around such an area. Buccally, lingually and incisally, or occlusally these spaces are known as embrasures (Fig 115).

The contact areas are located near the incisive edges of the incisors, but shift progressively in a cervical direction posteriorly. The interdental space, normally almost completely occupied by a gingival papilla buccally and lingually, is situated between the contact area and the bony alveolar crest. The buccal and lingual papillae are connected below the contact area by a thinned septum of gingiva, the col.

The mandibular dental arch is slightly smaller than the maxillary dental arch, resulting in the incisive edges of the upper incisors and canines, and the buccal cusps of the premolars and molars, biting just outside the corresponding parts of the lower teeth. According to many authors, the incisors and canines form part of the anterior, or labial segment, while the premolars and molars form part of the posterior, or buccal segment. The anterior segment may be arched or relatively straight, while the posterior segment may be straight or convex buccally. The canines may appear to be part of one or the other segment and are regarded as constituting a prominent corner-stone between the two segments.

In both arches, the incisive edges of the incisors and canines are aligned with the buccal cusps of the molars. The lingual cusps of the molars tend to decrease in size from the third molar to the first, and are represented in the anterior teeth by a cingulum, a lingual bulge of the cervical third of the teeth. In the dentitions of man, the teeth form an uninterrupted series without any natural spaces, or diastemata, between them.

The long axes of mandibular molars show a slight lingual inclination, while the maxillary premolars and molars show a slight buccal inclination. In both jaws, most of the teeth show a slight mesial inclination, which is more clearly seen in the premolars and molars where the roots are frequently distally inclined. As a result of the mesial inclination of the molars, they are driven in a mesial direction during function. This forms, according to some investigators, part of the mechanism of mesial drift in teeth.

The maxillary incisors show a slight mesial and a labial inclination, and are consequently inclined to shift in these directions. A mesial drift of the teeth of one side is, however prevented by the corresponding teeth of the other side in an intact arch. During function, great force is exerted by both segments on the upper canine, and a tendency for this tooth to be displaced labially could be expected. The importance of the buccinator muscle of the cheeks, and the orbicularis oris muscle of the lips, to prevent such a labial displacement of especially the upper canines, cannot be overemphasized. The lower incisors show a natural labial inclination, but bite against the lingual surfaces of the upper incisors. As long as the upper teeth are stable, no tendency towards a labial displacement of mandibular incisors will occur. The upper arch is supported lingually by the lower teeth and buccally by the cheeks and the lips, while the lower arch is supported buccally by the teeth in the upper arch and lingually by the tongue.

Dental articulation, or occlusion, is the relationship between the teeth of opposing dental arches with the jaws closed and with the mandibular condyle in a rest position in the glenoid fossa of the temporal bone. The following are characteristics of normal dental articulation:

1. The buccal cusps and incisive edges of the maxillary teeth bite outside the corresponding parts of the mandibular teeth.
2. The lower canines always bite on the mesial side of the upper canines.
3. The mesiolingual cusps of the upper molars bite into the central fossae of the corresponding lower teeth. The tips of the mesiobuccal cusps of the upper molars bite on the outside of, but opposite, the buccal groove which separates the mesiobuccal and distobuccal cusps of the lower molars.
4. Each tooth in the upper arch is opposed by not only the corresponding tooth in the lower arch, but also by the tooth distal to it. The only exception is the upper third molar which makes contact only with the lower third molar.

Terminology

When studying individual teeth, it is necessary to define the following terms:

cusp
cusp ridge
incisive, or incisal, edge/margin
cervical margin/edge/line
tubercle
cingulum
ridge
marginal ridge
triangular ridge

oblique ridge
fossa
triangular fossa
pit
groove/fissure
developmental groove
supplemental groove
sulcus

A *cusp* is a major elevation on the occlusal surface of a premolar or molar tooth.

A *cusp ridge* is the mesial or distal slope of a cusp as seen from the buccal or lingual side of the tooth.

An *incisive edge*, or *incisive margin*, is the cutting edge of anterior teeth.

A *cervical margin*, also known as a *cervical edge* or *line* is the area where the crown of a tooth meets the root at the neck of the tooth, and is represented by the enamel-cementum junction.

A *tubercle* is a small elevation on the crown of a tooth and is not a typical feature of that tooth.

A *cingulum* is an elevation on the lingual surface of the crowns of anterior teeth (incisors and canines). It usually makes up the bulk of the cervical third of the tooth.

A *ridge* is a linear elevation on the surface of a tooth.

A *marginal ridge* is a rounded border of enamel found on the mesial and distal margins of the occlusal surface of a tooth.

A *triangular ridge* is a ridge descending from the apex of a cusp towards the central part of the occlusal surface and is so named due to the triangular shape of the ridge in cross-section.

An *oblique ridge* is found on maxillary molar teeth and is formed by the union of the triangular ridges of the mesiolingual and distobuccal cusp. It crosses the occlusal surface obliquely.

A *fossa* is a major irregular concavity on a tooth surface. It usually has a pit near its centre.

A *triangular fossa* is a fossa on the occlusal surface of premolars and molars, mesial or distal to marginal ridges.

A *pit* is a small pin-point depression found at the junction of grooves.

A *groove*, or *fissure* is a sharply-defined linear depression on a tooth surface found mainly on occlusal surfaces and separating cuspal areas.

A *developmental groove* is a fissure between the primary parts of a crown and is a typical feature of a tooth.

A *supplemental groove* is less strongly marked than a fissure and may not be a characteristic feature of a tooth.

A *sulcus* is a valley-like depression between ridges and cusps. It usually has a developmental groove in its deepest part.

The primary dentition

Maxillary central incisor (51, 61) (Fig 116)

Buccal (labial) aspect. The crown is longer mesiodistally than in a cervico-incisal direction. The opposite holds true for the corresponding secondary tooth. The slightly convex labial surface is smooth and the incisive edge is straight. The incisive angle formed by the incisive edge and the distal surface is more rounded than the mesial incisive angle. Vertical developmental grooves are seldom seen. The root is long in comparison with the length of the crown.
Lingual aspect. Well-developed marginal ridges are present and the cingulum is prominent, virtually separating the lingual surface into a mesial and a distal fossa. The root narrows lingually from a relatively flat labial surface, and a cross-section through the cervical part of the root has a triangular outline, presenting distolingual, mesiolingual, and labial surfaces.
Mesial and distal aspects. The mesial and distal aspects of the tooth are similar, the buccolingual width at the cervical third of the crown being only about 1 mm less than its cervico-incisal height. The mesial and distal outlines are wedge-shaped. The curvature of the cervical margin is prominently convex incisally, but not as prominent as in the secondary tooth.

Maxillary lateral incisor (52, 62) (Fig 117)

The outline of this tooth is similar to the outline of the central incisor, except for a smaller crown. The mesiodistal length is also less than the cervico-incisal height and the distal incisive angle of the crown is more rounded. Although the root has a similar outline, it is longer in relation to the crown size when compared to the central incisor.

Maxillary canine (53, 63) (Fig 118)

Buccal (labial) aspect. This is not comparable with that of the central or lateral incisors, but the root shape is similar.

The crown is narrowed at the neck and the mesial and distal surfaces are prominently convex. The incisive edge is not straight but has a long, well-developed cusp. The mesial slope, or cusp ridge of the incisive edge is longer than the distal ridge and is not as acutely slanted as is the case in the secondary canine. The root is long.
Lingual aspect. Prominent vertical enamel ridges are seen on the lingual aspect, in the form of the central cingulum and mesial and distal marginal ridges. A lingual tubercle on the cusp tip is an elongation of the lingual ridge connecting the cingulum with the cusp. This lingual ridge divides the lingual fossa into mesiolingual and distolingual halves.

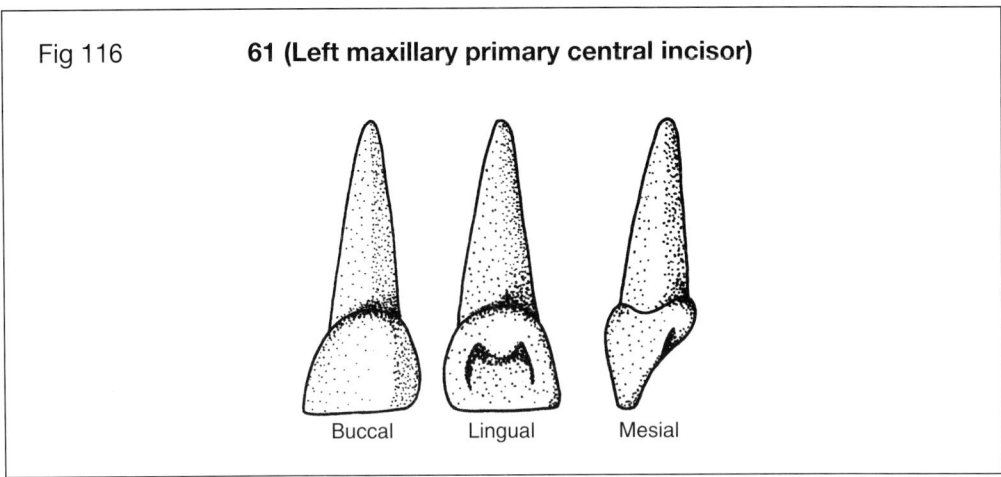

Fig 116 — **61 (Left maxillary primary central incisor)**
Buccal   Lingual   Mesial

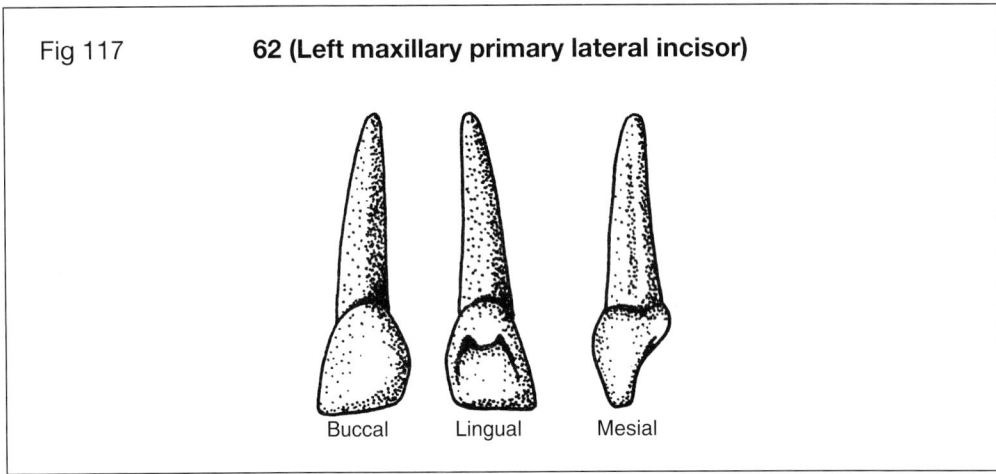

Fig 117 — **62 (Left maxillary primary lateral incisor)**
Buccal   Lingual   Mesial

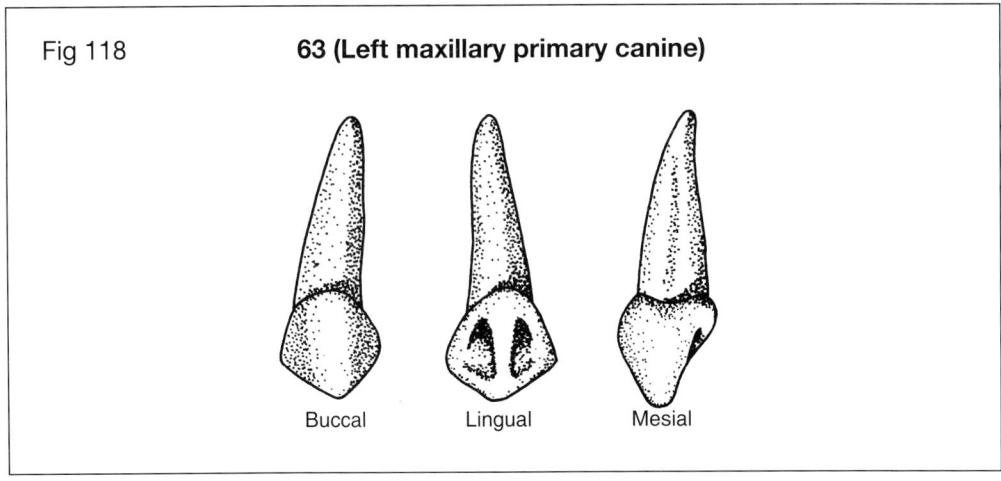

Fig 118 — **63 (Left maxillary primary canine)**
Buccal   Lingual   Mesial

Mesial and distal aspects. The outlines of the tooth are similar to those of the incisors, but the crown is more robust in its cervical third.

Mandibular central incisor (71, 81) (Fig 119)

Buccal (labial) aspect. This is relatively flat without any developmental grooves. The mesial and distal sides become narrower in a cervical direction from the contact area and the crown is wide in relation to its cervico-incisal height. The root is long, thin, and gradually narrows to a sharp apex. The incisive edge is straight.
Lingual aspect. A cingulum and marginal ridges are present on the lingual surface. In the middle and incisal third, the lingual surface may either be flat, or display a lingual fossa. Both the crown and root narrow lingually.
Mesial aspect. The mesial surface shows the typical outline of an incisor, but is small. The cervical convexities of the labial and lingual surfaces are just as prominent as in other primary incisors. Although a smaller tooth, its labiolingual measurement is only about a millimeter less than the corresponding measurement of the upper central incisor.

The mesial surface of the root is nearly flat, and the root narrows gradually in an apical direction, although the apical area appears more blunt. The cervical margin is similar to that of the other incisors.
Distal aspect. The outlines of the tooth are similar to those of the mesial aspect, but the cervical margin shows a less prominent convexity in an incisal direction.

Mandibular lateral incisor (72, 82) (Fig 120)

This tooth is similar to the central incisor but larger in all dimensions except labiolingually, where they have more or less the same measurements. The cingulum may be somewhat larger and the lingual surface more concave than in the maxillary lateral incisor and the incisive edge shows a greater tendency to descend distally.

Mandibular canine (73, 83) (Fig 121)

When the whole form of this tooth is considered, few differences are seen between it and the maxillary canine, except for differences in dimensions. The tooth is smaller, especially labiolingually, and the cervical convexities (labially and lingually) are less prominent. A striking difference is, however, the longer distal cusp ridge, compared to the mesial cusp ridge, in contrast to the relative lengths of these ridges in the upper canines.

Maxillary first molar (54, 64) (Fig 122)

The greatest length of the crown is between the mesial and distal contact areas from where the crown narrows cervically. The roots are thin, relatively long, and

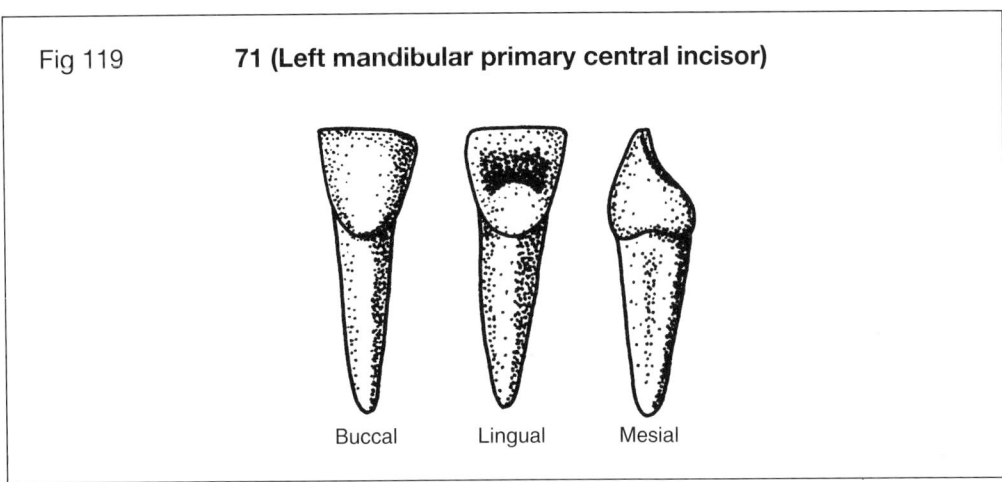

Fig 119 — 71 (Left mandibular primary central incisor) — Buccal, Lingual, Mesial

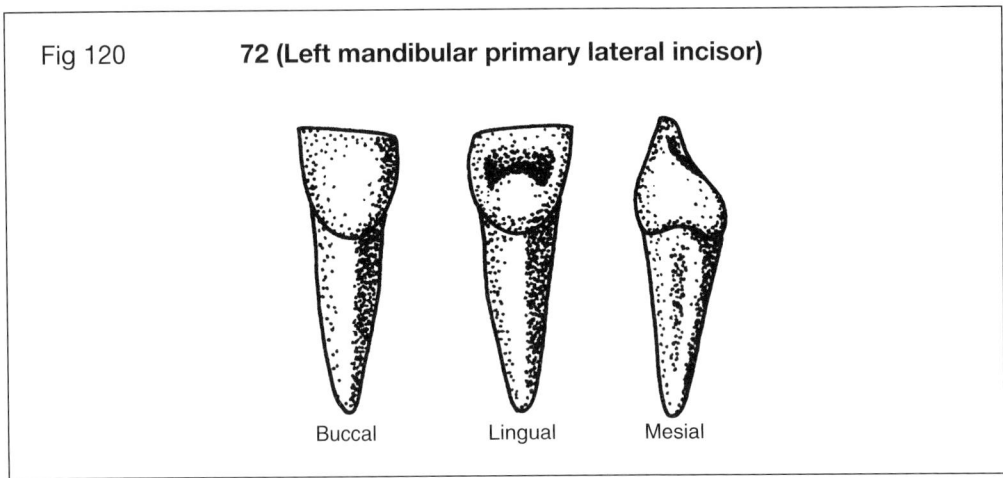

Fig 120 — 72 (Left mandibular primary lateral incisor) — Buccal, Lingual, Mesial

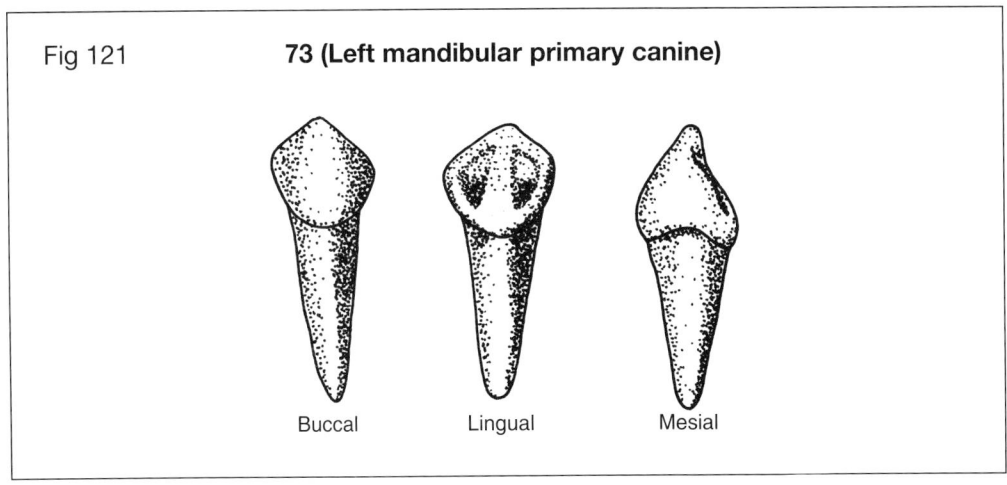

Fig 121 — 73 (Left mandibular primary canine) — Buccal, Lingual, Mesial

Fig 122   **64 (Left maxillary primary first molar)**

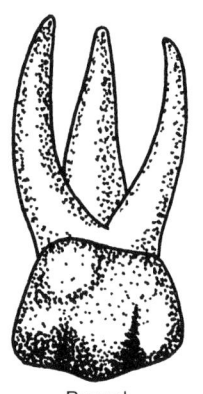
Buccal

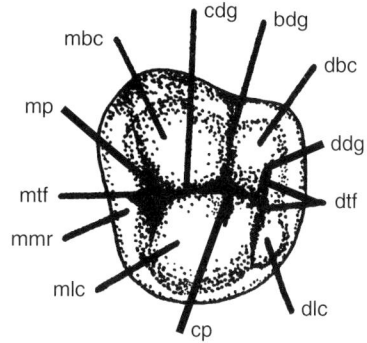
Occlusal

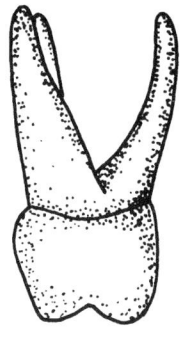
Mesial

bdg = buccal developmental groove
dbc = distobuccal cusp
dtf = distal triangular fossa
dlc = distolingual cusp
ddg = distal developmental groove
ldg = lingual developmental groove
mbc = mesiobuccal cusp
mtf = mesial triangular fossa
mlc = mesiolingual cusp
mmr = mesial marginal ridge
mp = mesial pit
cdg = central developmental groove
cp = central pit
or = oblique ridge

65 (Left maxillary primary second molar)

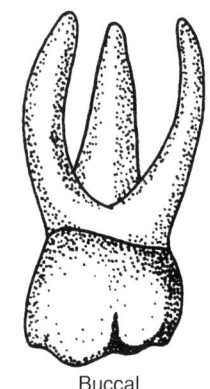
Buccal

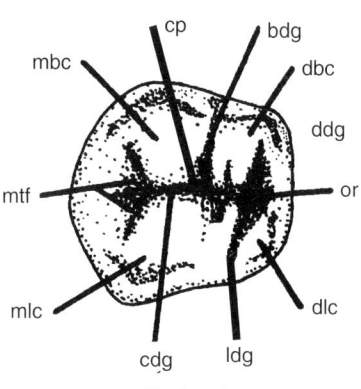
Occlusal

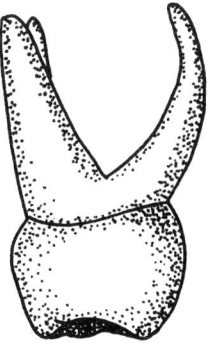
Mesial

widely divergent. The tooth has three roots, and in this respect corresponds to the root pattern of the secondary molar, two being situated bucally and one lingually. The distal root is much shorter than the mesial root and the division of the roots, or furcation area, is close to the cervical margin. The lingual root is the thickest. From an occlusal aspect, the buccal surface of the crown is longer than the lingual surface.

There are two main cusps, a buccal and a lingual cusp. The buccal cusp is an elongated ridge which may be partially divided into two, sometimes three smaller cusps by developmental grooves. There is usually one prominent buccal developmental groove. These grooves arise in the broad, but relatively shallow central fossa from a central developmental groove. The lingual cusp is usually divided into a large mesiolingual and a small distolingual cusp by means of a distal developmental groove. The distal marginal ridge is small compared to the mesial marginal ridge. When a low oblique ridge connects the large mesiolingual cusp with the small distobuccal cusp, the connection between the central pit and the distal developmental groove is absent.

The buccal surface of the crown shows a prominent bulge close to the cervical margin and opposite the mesiobuccal root. The cervical margin slopes in a gingival direction mesially on the buccal side, is relatively straight lingually, and shows a slight mesial and distal occlusal convexity.

Maxillary second molar (55, 65) (Fig 122)

Except for features such as the marked divergence of the roots, the bulbous shape of the crown, a narrowed cervical area, and a short root stem, the primary second molar resembles the secondary second molar. The cusps and roots have a similar arrangement. The mesiolingual cusp is the largest and is connected to the distobuccal cusp by a low oblique ridge. This ridge divides the occlusal surface into a large mesial and a small distal fossa. The tubercle of Carabelli is often present on the lingual surface of the mesiolingual cusp. The buccal cusps are more or less of equal size and are separated by a buccal developmental groove. The occlusal surface has a central fossa with a central pit, a well developed triangular mesial fossa with a mesial pit, and a central developmental groove which connects the mesial and central pits. The buccal developmental groove arises in the central pit. An oblique ridge is present and the distal triangular fossa, with its distal developmental groove, is found distal to the oblique ridge. This distal groove separates the mesiolingual and distolingual cusps and extends onto the lingual surface. The two marginal ridges are well developed and the cervical line resembles that of the first molar.

Fig 123   **74 (Left mandibular primary first molar)**

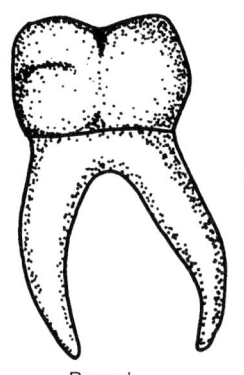
Buccal

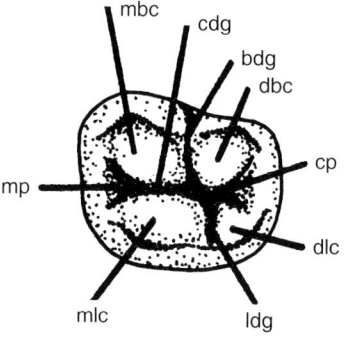
Occlusal

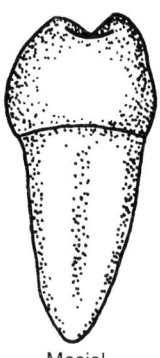
Mesial

```
bdg   = buccal developmental groove
dbc   = distobuccal cusp
dbdg  = distobuccal developmental groove
dc    = distal cusp
dlc   = distolingual cusp
dp    = distal pit
ldg   = lingual developmental groove
mbc   = mesiobuccal cusp
mbdg  = mesiobuccal developmental groove
mlc   = mesiolingual cusp
mp    = mesial pit
cdg   = central developmental groove
cp    = central pit
```

75 (Left mandibular primary second molar)

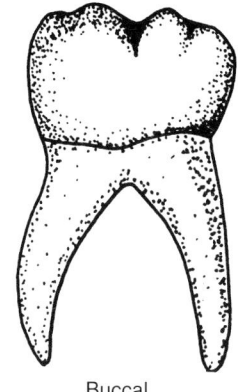
Buccal

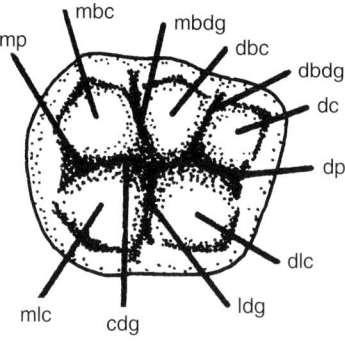
Occlusal

Mesial

Mandibular first molar (74, 84) (Fig 123)

The occlusal surface is elongated mesiodistally and supports four cusps which are often relatively indistinct. This tooth does not resemble any other primary or secondary molar.

The mesial surface descends nearly vertically from the contact area to the cervical line, and consequently the tooth shows very little cervical constriction mesially. The distal convexity of the crown is, however, similar to that of the other deciduous molars. The cervical margin opposite the mesial root descends in an apical direction, with the result that the mesial part of the crown appears to be higher than the distal part. The cervical margin is relatively straight on the lingual surface while exhibiting a slight convexity in an occlusal direction mesially and distally. The buccal surface of the crown shows a very prominent bulge near the cervical line, opposite the mesial root. The mesial and distal roots are thin and diverge sharply. Occlusally, the mesiolingual cusp is the largest and is partly separated from a small distolingual cusp by an indistinct lingual developmental groove. The buccal developmental groove separates the two buccal cusps and arises in a central pit in a large distal fossa. The central developmental groove starts here, courses mesially between the mesiobuccal and mesiolingual cusps, and ends in a mesial pit in the smaller mesial triangular fossa.

Mandibular second molar (75, 85) (Fig 123)

Except for general differences between the primary and secondary teeth, this tooth resembles the secondary lower first molar. It has the same number and arrangement of cusps and roots.

The crown, compared to the deciduous first molar, shows a more pronounced cervical constriction, but is as a whole much larger. The mesial and distal roots are long, thin, and divergent and bifurcate close to the cervical margin.

The occlusal surface carries five cusps, three buccally and two lingually. One buccal cusp is in a buccodistal position and is frequently called the distal cusp. The three buccal cusps are more or less of the same size, as are the two lingual cusps. The total mesiodistal length of the buccal surface of the crown is greater than the length of the lingual surface.

An irregular central developmental groove runs from a mesial pit in the mesial triangular fossa to a distal pit in the distal triangular fossa. Transverse developmental grooves arise from the central groove and separate the three buccal cusps and the two lingual cusps. The mesial and distal buccal grooves are continued as developmental grooves onto the buccal surface, while the lingual developmental groove is continued onto the lingual surface of the crown to separate the two lingual cusps. The central pit is located at the point where the lingual and mesiobuccal grooves meet the central groove.

Articulation (occlusion) of the deciduous teeth

The deciduous dentition is arranged in the form of two arches. A line connecting the labial and buccal surfaces of the upper teeth describes part of an oval which is wider than a corresponding line around the lower teeth. The teeth articulate in such a way that each tooth contacts two opposing teeth, except the lower central incisors and the upper second molars. The deciduous dentition should be in normal alignment soon after the age of 2 $\frac{1}{2}$ years.

A year or two after eruption of the deciduous dentition, jaw growth accelerates to such an extent that spaces develop between some teeth. Spacing usually occurs between the anterior teeth and begins to develop between the 4th and 5th years. The canines and molars usually maintain contact with each other but move relative to one another during the growth process. It seems highly likely that these constant changes in position of the deciduous teeth contribute to the excessive attrition usually observed on these teeth. Eruption of the secondary first molar distal to the second deciduous molar at the age of about 6 years is considered to be a stabilizing factor.

Articulation of the deciduous dentition may be summarized as follows:

1. The mesial surfaces of the upper and lower central incisors correspond to the midline of the dental arch.
2. The upper central incisor articulates with the lower central incisor as well as with the mesial third of the lower lateral incisor. The incisive edges of the lower anterior teeth make contact with the lingual surfaces of the upper incisors close to their incisive edges.
3. The upper lateral incisor articulates with the distal two thirds of the lower lateral incisor, as well as with that part of the lower canine situated mesial to the tip of its cusp.
4. The upper canine articulates with the remaining distal part of the lower canine crown, as well as with the mesial part of the first lower molar on the mesial side of its mesiobuccal cusp.
5. The upper first molar articulates with the distal two thirds of the lower first molar as well as with the mesial part of the second molar, involving its mesial marginal ridge and triangular fossa.
6. The upper second molar articulates with the remaining part of the lower second molar and extends distally, slightly beyond the distal surface of the lower tooth.

Mineralization and eruption

Mineralization of deciduous teeth begins after about 4 months of intra-uterine life, and all teeth are actively mineralizing by the 6th month. There are usually no teeth present in the mouth at birth. It must be emphasized that tables in which mineralization and eruption times are given, reflect approximate values since no two individuals will chronologically develop in precisely the same way. Tables are, nevertheless, of value in the diagnosis of abnormal development. In this regard the following words of Schour are significant: *"It must be pointed out that the tooth is more than an organ of mastication. During the development of its enamel and dentine the tooth is also a biologic recorder of health and disease, especially of alterations in mineral metabolism. The incremental layers of enamel and dentine reflect its life history (weather, nutrition etc)."*

One must guard against an attitude of temporariness with regard to the primary teeth and that taking care of these teeth is not as important as the care of secondary teeth. All the primary teeth are in use from about the age of 2 years until the age of 6 or 7 years, a total of up to 5 years. Furthermore, the deciduous dentition plays an important role for a total of 11-12 years, or even longer, and must be regarded as a significant factor in the maintenance of the child's welfare during his first years of growth and development, both physically and mentally.

The developmental dates of the deciduous teeth are as follows:

Tooth	**First signs of mineralization**	**Crown completed**	**Appearance in mouth**	**Root completed**
Maxilla				
central incisor	3-4 m i.u.	4 m	7.5 m	1.5-2 a
lateral incisor	4.5 m i.u.	5 m	8 m	1.5-2 a
canine	5.5 m i.u.	9 m	16-20 m	2.5-3 a
first molar	5 m i.u.	6 m	12-16 m	2-2.5 a
second molar	6 m i.u.	10-12 m	20-30 m	3 a
Mandible				
central incisor	4.5 m i.u.	4 m	6.5 m	1.5-2 a
lateral incisor	4.5 m i.u.	4.5 m	7 m	1.5-2 a
canine	5 m i.u.	9 m	16-20 m	2.5-3 a
first molar	5 m i.u.	6 m	12-16 m	2-2.5 a
second molar	6 m i.u.	10-12 m	20-30 m	3 a

The secondary dentition

The incisors

In reproducing the sketches of secondary teeth on the following pages, tribute is paid to the memory of J. G. de Boer, former professor of Cariology at the University of Groningen, the Netherlands.

Maxillary central incisor (11, 21) (Fig 124)

The labial surface is convex but often becomes straight as the incisive edge is approached. Two small vertical grooves which reflect a phase in tooth development may be present on the labial surface near the incisive edge. Viewed from the front, the mesial surface appears more vertical than the slightly rounded distal surface. The lingual surface of the crown is concave, except in the area of the cervical margin where a prominent bulge, the cingulum, is present. Marginal ridges extend mesially and distally from the incisive edge into the cingulum. The lingual fossa is situated between the marginal ridges and is sometimes divided into two halves by a slight central vertical ridge.

Viewed from the mesial or distal aspect, the crown is wedge-shaped. The mesial incisive angle is virtually a rectangle, while the distal incisive angle is more rounded.

The single root is oval to triangular in cross-section with rounded angles separating labial, mesiolingual, and distolingual surfaces which are not clearly demarcated. The root gradually becomes thinner towards the apex. The triangular pulp chamber has two small cornua, or pulp horns directed towards the distal and mesial incisive angles. The root canal is round. The cervical margin is convex in an incisal direction on the mesial and distal surfaces, but relatively straight on the labial and lingual surfaces.

Maxillary lateral incisor (12, 22) (Fig 125)

The general shape of the lateral incisor resembles that of the central incisor. The crown is, however, smaller while the roots may be of nearly equal length. The pulp is relatively large. The distal incisive angle is more rounded than in the central incisor, and a deep pit or fossa is not uncommon on the lingual surface, incisal to the cingulum. The cervical margin is similar to that of the central incisor. It is not uncommon for the lateral incisors to be congenitally absent or otherwise deformed. The most common deformity is a peg-shaped or conical tooth.

Mandibular central incisor (31, 41) (Fig 126)

It is the smallest tooth in the secondary dentition. It is long, narrow and chisel-shaped. The labial surface of the crown is convex, but straightens towards the incisive

Fig 124 **21 (Left maxillary secondary central incisor)**

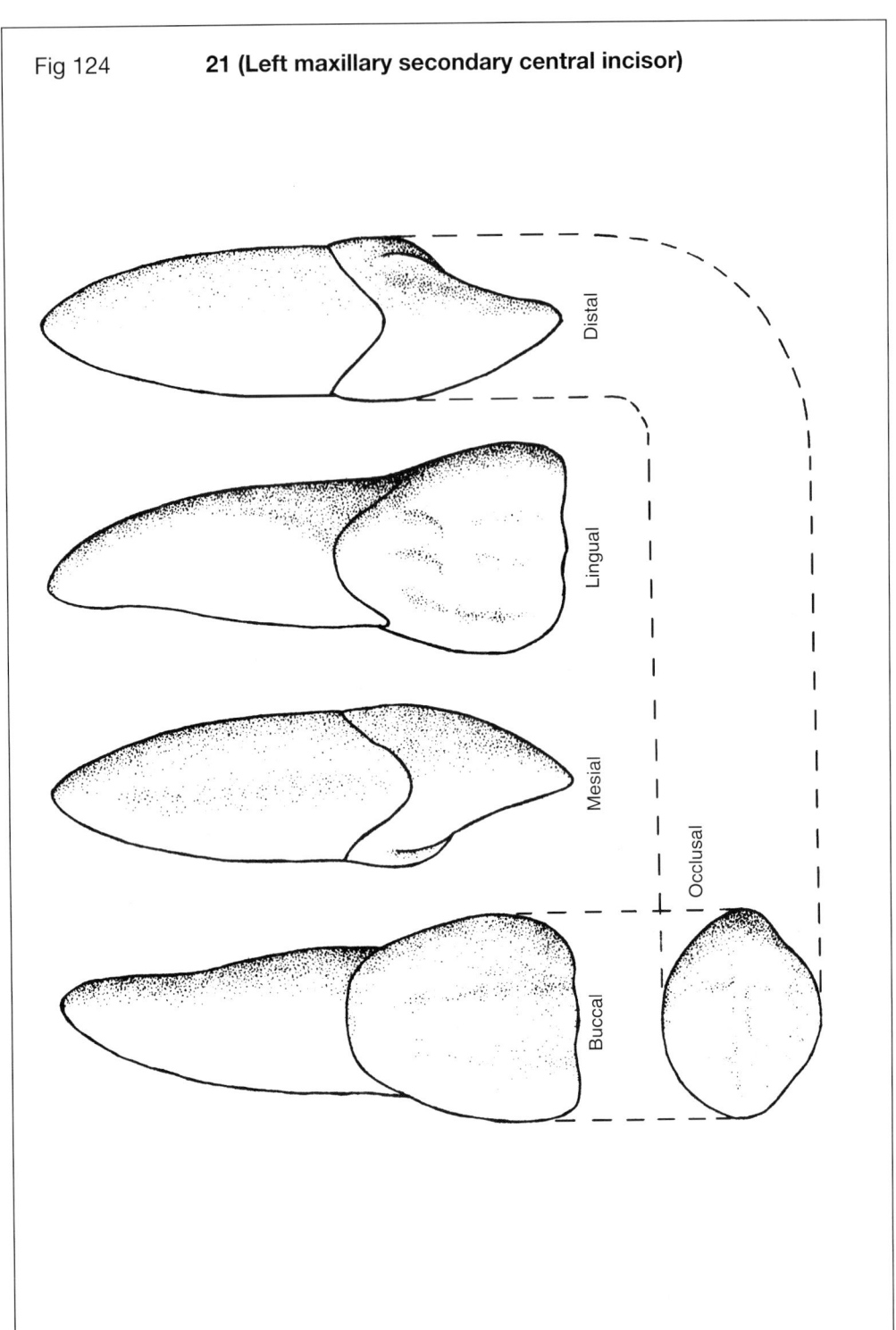

Fig 125 **22 (Left maxillary secondary lateral incisor)**

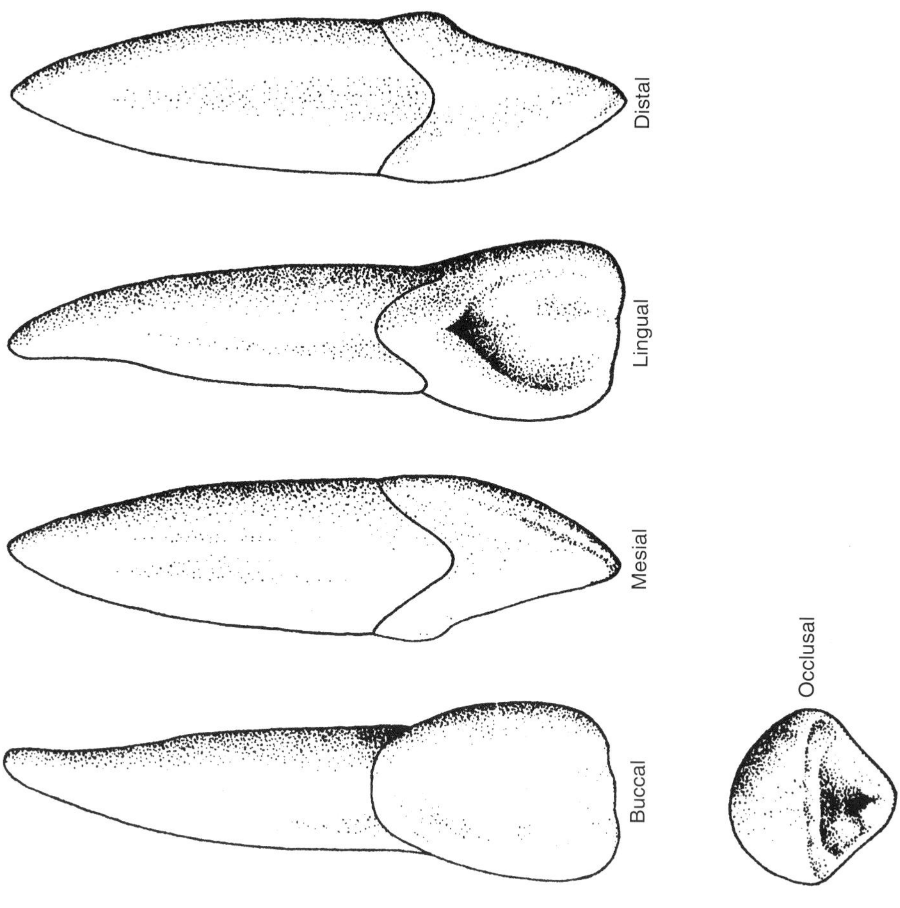

Distal

Lingual

Mesial

Buccal

Occlusal

edge. The lingual surface of the crown is moderately concave, except close to the cervical margin where it is slightly convex, although a distinct cingulum is not often present. Marginal ridges are not present lingually. The crown appears wedge-shaped from the mesial or distal aspect.

The single oval root is slightly flattened mesially and distally with shallow vertical grooves on both these surfaces. The distal groove is deeper.

As in the maxillary central incisor, the pulp has cornua directed towards the mesial and distal incisive angles.

The cervical margin is convex in an incisal direction on the mesial and distal surfaces of the tooth, but relatively straight labially and lingually.

Mandibular lateral incisor (32, 42) (Fig 127)

This tooth is slightly larger than the central incisor. The incisive edge is also longer mesiodistally, and usually the mesial and distal marginal ridges on the lingual side are more noticeable than in the central incisor. The teeth are very similar in other respects.

The canines

Although the roots of human canines are larger and more robust than any other single root in the human dentition, the crowns do not extend markedly above the level of the other teeth. This is one of the factors allowing a greater degree of sideways movement of the human mandible during mastication. There are no natural spaces, or diastemata between the canines and adjacent teeth as in many other animal species.

Maxillary canine (13, 23) (Fig 128)

The incisive edge is characterized by a pointed cusp. The labial surface is convex and shows an indistinct vertical ridge running from the cusp to the cervical margin. The lingual surface nearly always shows a prominent cingulum and may be convex, flat, or slightly concave above the cingulum. A prominent vertical ridge runs from the cingulum to the tip of the cusp and is flanked by depressions on either side.

The lingual surface is bordered by marginal ridges. The distal incline, or slope (cusp ridge), of the incisive edge is shorter than the distal incline. The junction of the distal surface and the incisive edge is rounded while the mesial junction is more angular. The cervical margin is convex in an incisal direction of the mesial and distal surfaces, but straight or convex in an apical direction on the labial and lingual surfaces.

The single root is more or less triangular in cross-section and shows labial, distolingual, and mesiolingual surfaces. A vertical groove is often present on both the mesiolingual and distolingual root surfaces. The large pulp cavity is oval in cross-section without cornua, but narrows to a point within the crown.

Fig 126   **31 (Left mandibular secondary central incisor)**

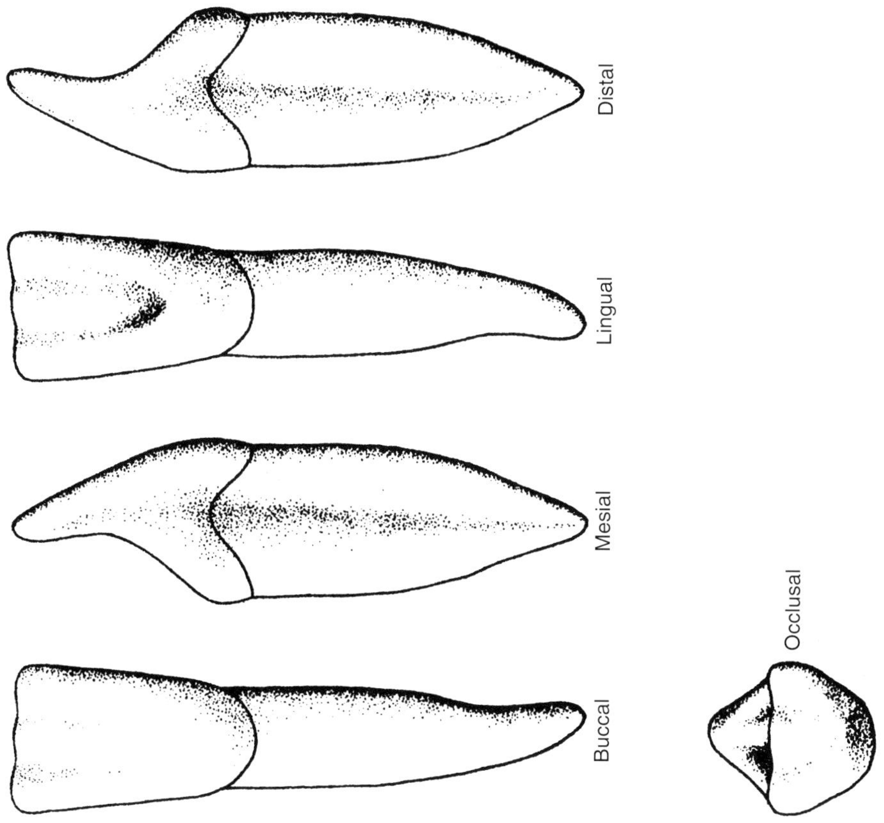

Fig 127   **32 (Left mandibular secondary lateral incisor)**

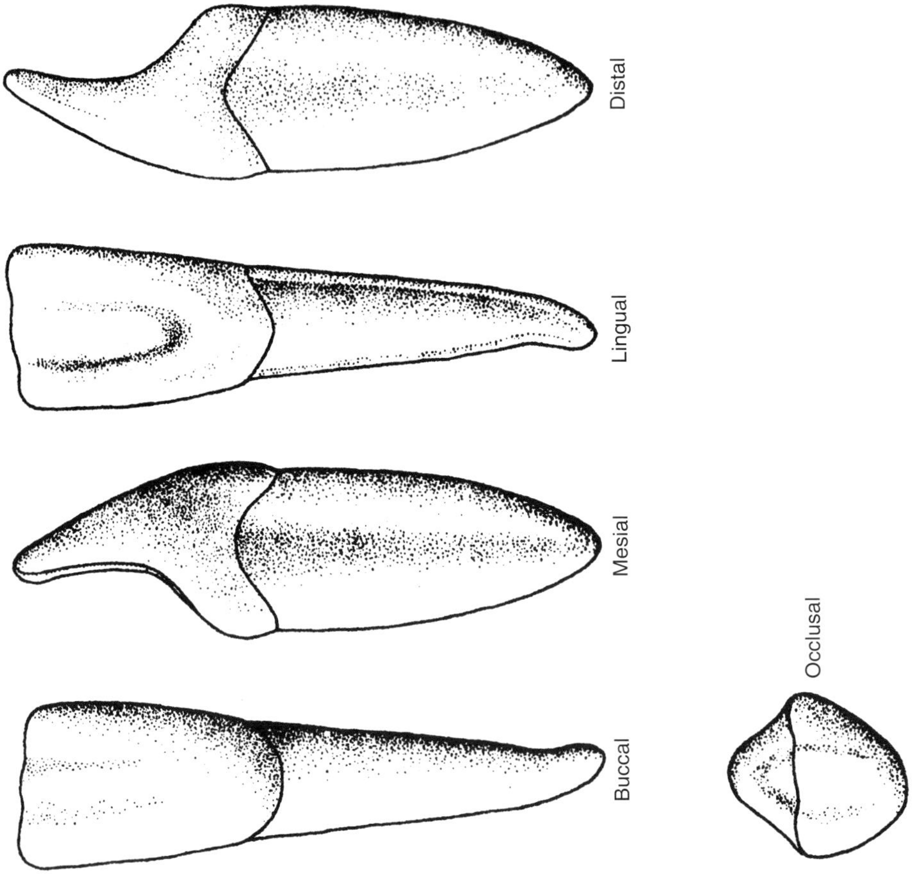

Fig 128   **23 (Left maxillary secondary canine)**

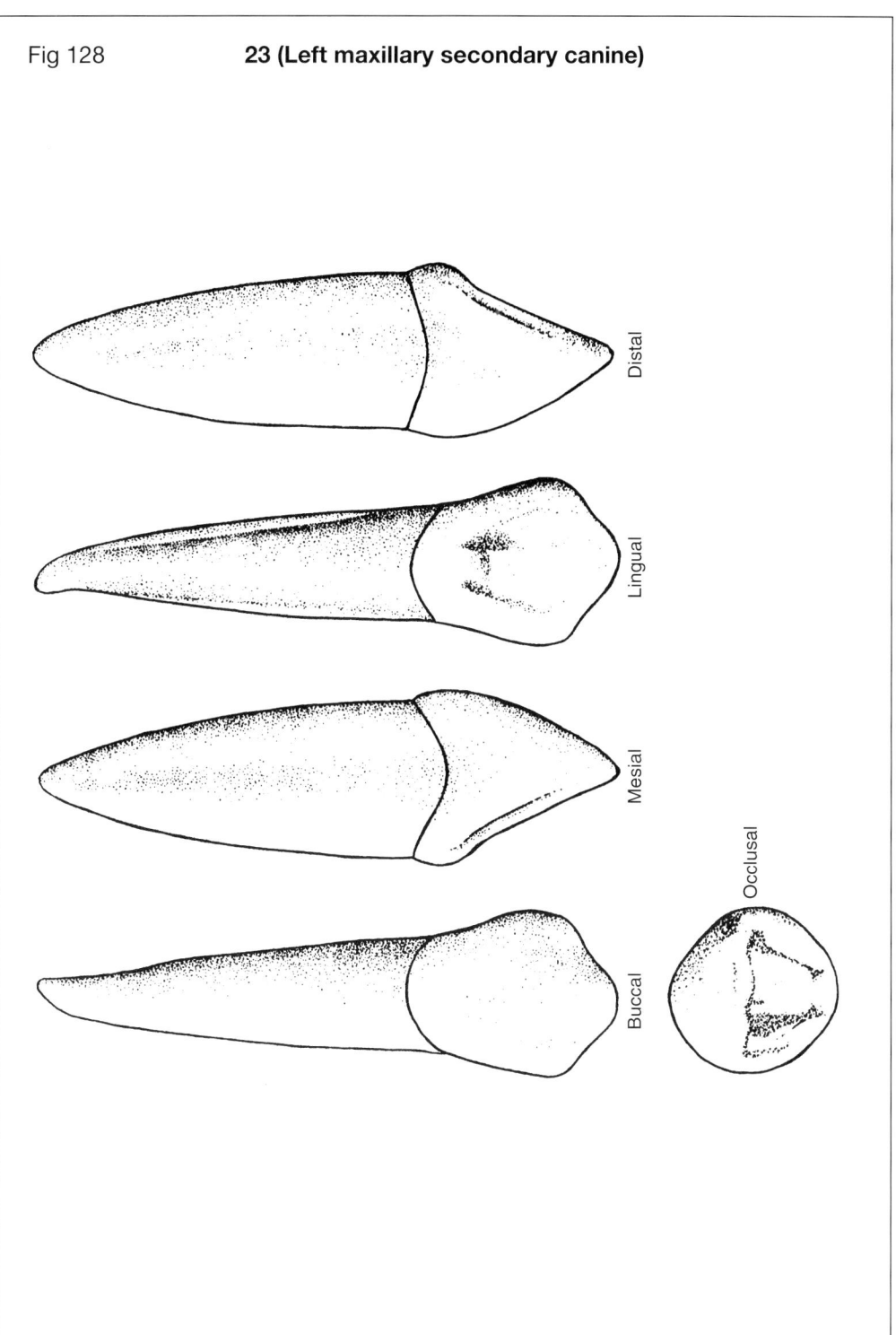

Fig 129   **33 (Left mandibular secondary canine)**

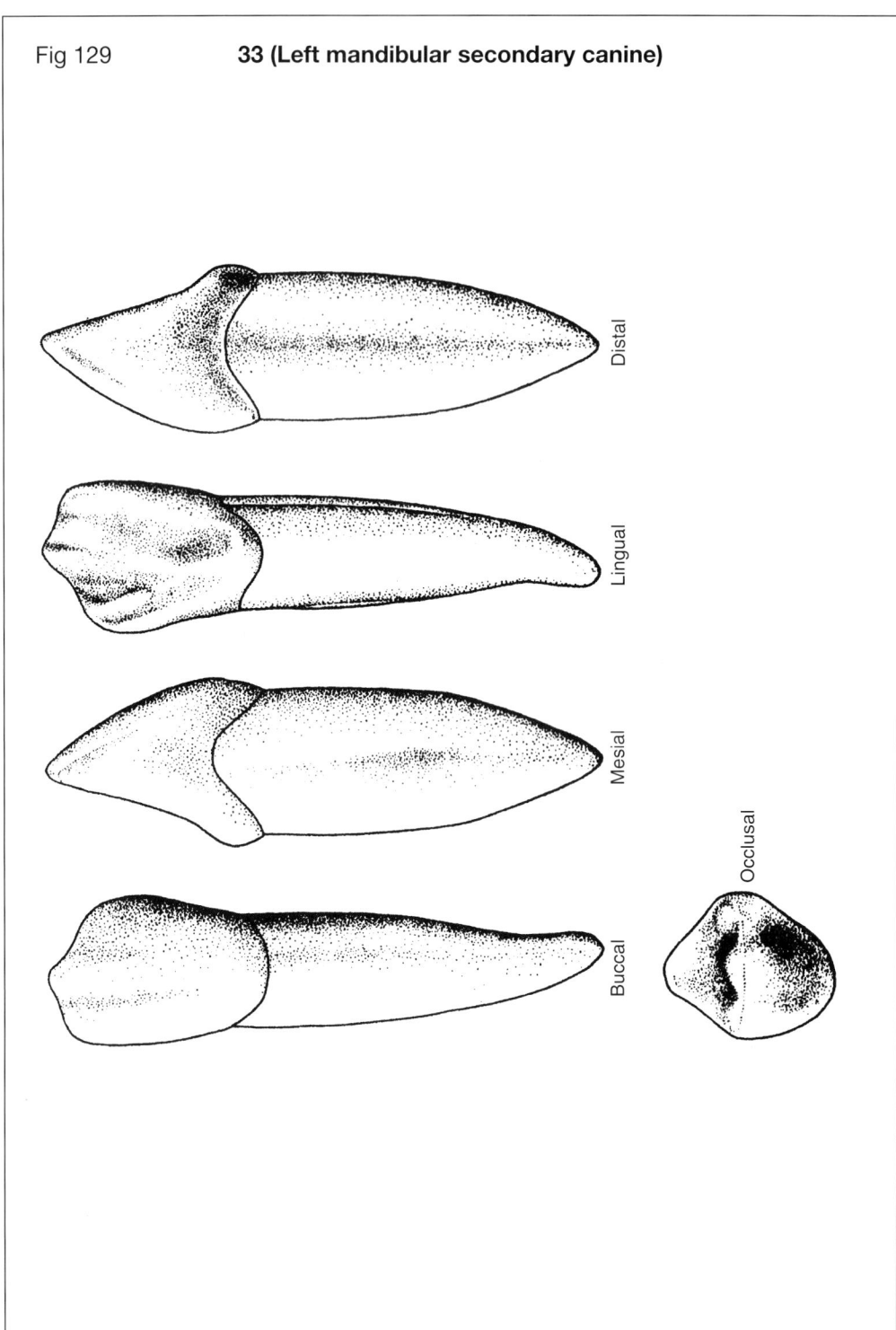

Mandibular canine (33, 43) (Fig 129)

Compared to the upper canine, the crown is narrower mesiodistally and the cusp on the incisive edge is less pointed. The distal incisive angle is rounded. The labial surface is convex and may show the presence of a vertical ridge, similar to that of the upper canine.

A well-marked cingulum is present on the lingual surface and this surface is similar to the lingual surface of the lower incisors except for a slight vertical enamel ridge. This runs from the tip of the cusp to the cervical margin. Depressions, bordered mesially and distally by marginal ridges, are present on both sides of the ridge. The mesial and distal surfaces of the crown are wedge-shaped. The cervical margin is convex in an incisal direction on the mesial and distal surfaces. The oval root is usually single, although two roots are sometimes found. It is slightly flattened mesially and distally and usually shows vertical mesial and distal grooves on these surfaces. The pulp cavity is smaller than that of the upper canine but otherwise similar in all respects.

The premolars

The premolars are only found in the secondary dentition. They are the successors to the deciduous molar teeth and are situated anterior to the secondary molars. Their occlusal surfaces, although simpler than those of the molars, are also used for crushing food. Usually only a buccal and a lingual cusp are present, the buccal cusp being the larger. The lingual cusp is a development of the cingulum of the anterior teeth. In the mandibular dentition there is a gradual transition from the incisors without marked cingula, to the canine with a small cingulum, then to the first premolar with a small lingual cusp, and finally the second premolar with a relatively large lingual cusp.

The cervical margin is less sinuous than in the canines or incisors. On the buccal and lingual surfaces the cervical margin is slightly convex apically, while being slightly convex in an occlusal direction on the mesial and distal surfaces.

Maxillary first premolar (14, 24) (Fig 130)

The occlusal surface shows the following characteristics:
The outline is egg-shaped, and broader on the buccal than on the lingual side. The mesiodistal length of the tooth is considerably less than the buccolingual width. The buccal, lingual, and distal surfaces of the crown are rounded but a distinguishing feature of this tooth is a depression in the mesial cervical region of the crown, extending onto the root. This cavity is sometimes known as the canine fossa, or the mesial developmental depression. The buccal cusp is the larger of the two cusps. The tip of the buccal cusp is closer to the central point of the root stem than is the lingual cusp.

In most cases the occlusal surface of this tooth shows no supplemental grooves.

A distinct central developmental groove divides the occlusal surface into buccal and lingual halves. The central developmental groove extends from a point situated just mesial to the distal marginal ridge and extends over this ridge as a mesial marginal developmental groove to end on the mesial surface of the crown.

Two collateral developmental grooves unite with the central groove on the inside of the marginal ridges. These collateral grooves are known as the mesiobuccal and distobuccal collateral developmental grooves. The connection between the collateral grooves and the central groove forms the mesial and distal developmental pits.

The hollow in the occlusal surface, mesial to the distal marginal ridge, which houses the distal collateral developmental groove and pit, is called the distal triangular fossa. The mesial triangular fossa takes a corresponding position in the mesial part of the occlusal surface.

The triangular ridge of the buccal cusp is prominent and arises near the midpoint of the central groove, extending to the tip of the buccal cusp. The lingual triangular ridge is less prominent. It also arises near the midpoint of the central fossa and extends to the lingual cusp tip.

The tooth usually has a buccal and a lingual root. When a single root is present, it is vertically grooved mesially and distally, the mesial groove usually being deeper. A small percentage of upper first premolars have three roots, one being lingual and two on the buccal side. The distance from the cervical margin to the division of the roots may vary.

In cross-section, the pulp cavity has an elongated oval form with the long axis directed buccolingually. Buccal and lingual cornua of the pulp chamber are directed towards the corresponding cusps.

In a double-rooted tooth two separate root canals are present, one in each root. In a single-rooted tooth, two canals leave the pulp chamber but join to form a single canal close to the apex of the root. A tooth with three roots has a separate canal in each root.

Maxillary second premolar (15, 25) (Fig 131)

In its general morphology, the crown of the upper second premolar is similar to that of the first but the occlusal surface is more oval in outline and slightly smaller. The two cusps are more equal in size, but shorter and more blunt. The distance between the tips of the cusps may be greater than in the first premolar, giving rise to a greater width of the occlusal surface in a buccolingual plane.

The central developmental groove is more irregular than in the first premolar and supplemental grooves are often present, radiating from the central groove. Due to these grooves, the occlusal surface may appear wrinkled. The mesial marginal developmental groove, a feature of the first premolar, is usually absent and the mesial and distal marginal ridges are uninterrupted. Distinct collateral developmental grooves are not an outstanding feature of this tooth.

Fig 130   **24 (Left maxillary secondary first premolar)**

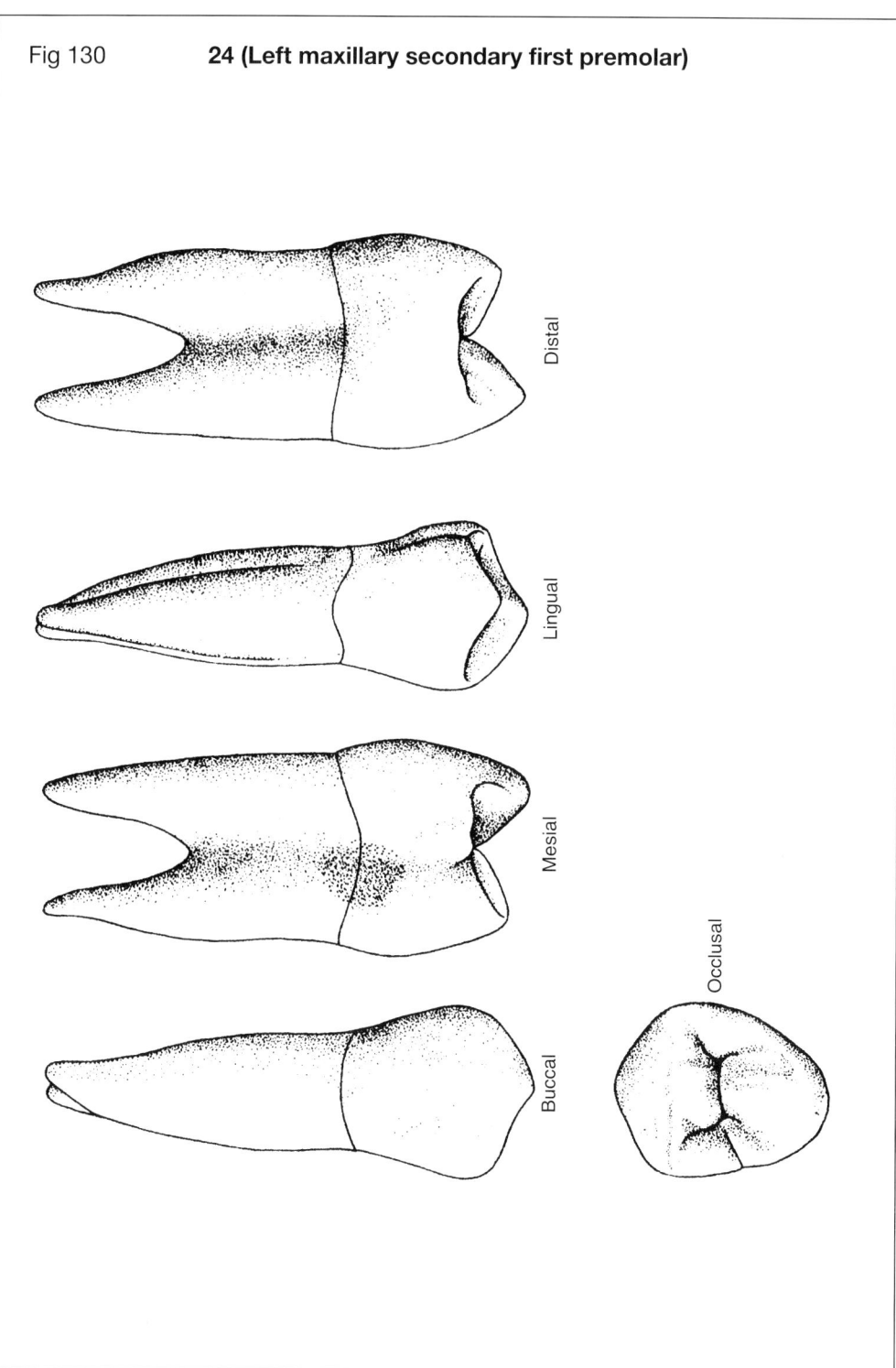

Fig 131   **25 (left maxillary secondary second premolar)**

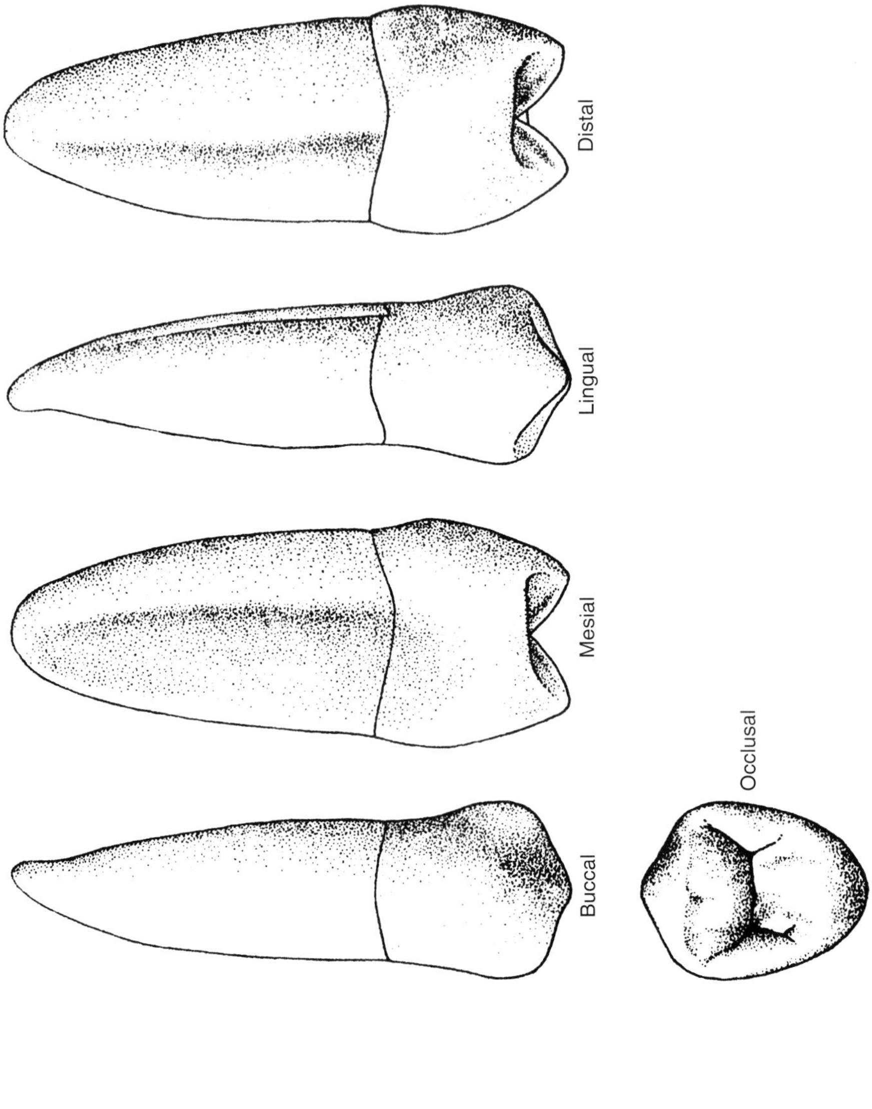

Distal

Lingual

Mesial

Buccal

Occlusal

Similarly, the mesial developmental depression, or canine fossa is not present in the second premolar and the entire mesial surface is convex.

There is usually only one root showing mesial and distal vertical grooves. The pulp cavity is similar to that of the first premolar with two root canals. Sometimes only one root canal is present.

The mandibular premolars

The first premolar has a large, well-formed buccal cusp and a small, non-functioning lingual cusp. This is, in many teeth, not larger than the cingulum of some lower canines.

The second premolar usually has two cusps, a large cusp buccally and a smaller cusp lingually, but it is not uncommon for the lingual cusp to be divided into two subsidiary cusps, the mesial being the larger.

The first premolar has many of the characteristics of a small canine since the pointed buccal cusp is the only part of the tooth making contact with opposing teeth. It is usually smaller than the second, in contrast to the relative sizes of the corresponding maxillary teeth.

The second premolar has many features of a small molar tooth since the lingual cusps are in most cases well-formed. The marginal ridges are consequently at such a level that efficient occlusal contact with opposing teeth is possible.

Mandibular first premolar (34, 44) (Fig 132)

This tooth is situated between the canine and the second premolar and has charcteristics of both. The features that are similar to those of the canine are as follows:

1. The buccal cusp is long, sharply pointed, and is the only articulating cusp.
2. Buccolingual measurements are similar.
3. The occlusal surface slopes lingually in the direction of the lingual cervical margin.
4. The mesiobuccal cusp ridge is slightly shorter than the distobuccal cusp ridge.

Features similar to those of the second premolar are the following:

1. The profile of the tooth, viewed from the buccal side, is the same.
2. The contact areas, both mesial and distal, are at the same vertical height.
3. The curvature of the cervical margin is similar.
4. The tooth has more than one cusp, although the lingual cusp is much smaller.

Although the root is shorter than the root of the second premolar, its length is closer to that of the second premolar than to that of the canine.

Fig 132   **34 (Left mandibular secondary first premolar)**

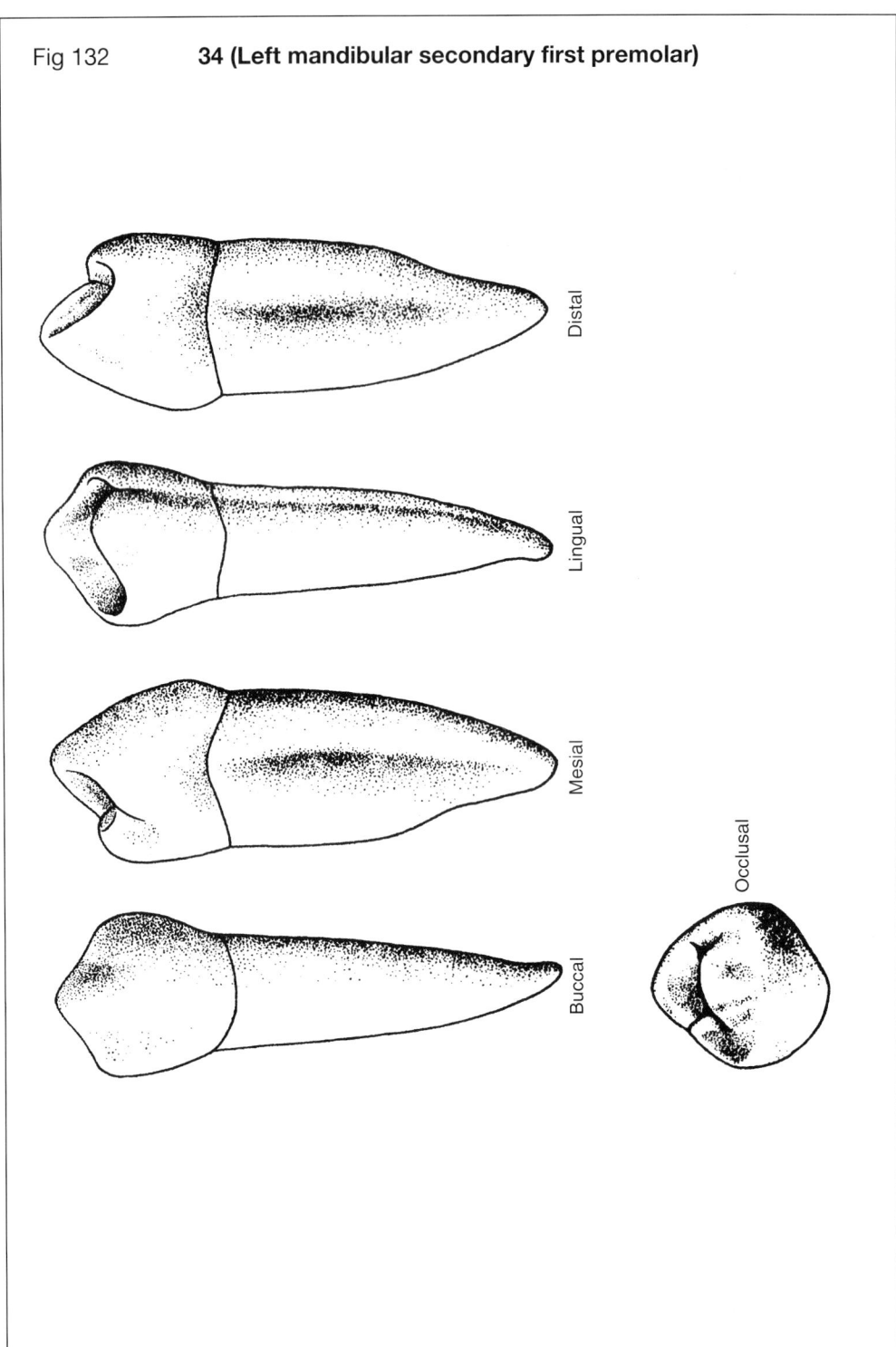

The tooth has two cusps, one arranged buccally and one lingually. The buccal cusp is much larger than the lingual cusp and has well developed mesial and distal ridges. A central developmental groove separates the large buccal cusp with its prominent buccal triangular ridge from the small lingual cusp which does not have a triangular ridge. Marginal ridges are comparatively well-developed.

A mesial developmental groove, running in a buccolingual direction, arises in the central developmental groove in the mesial fossa and is continued over the mesiolingual part of the mesial marginal ridge to form the mesiolingual developmental groove on this surface of the crown.

The distal fossa may contain a small distal developmental groove, as well as a shallow developmental pit with radiating supplemental grooves.

The tip of the lingual cusp is on a much lower level than the tip of the buccal cusp. The single root is oval, and slightly flattened mesially and distally. Vertical grooves are present on these surfaces, the mesial groove being the more marked. The coronal pulp usually has a single cornu directed in the direction of the buccal cusp. There is usually only one root canal.

Mandibular second premolar (35, 45) (Fig 133)

When viewed from the buccal side, the second premolar is similar to the first. The two cusps are nearly equal in size, the buccal cusp being larger but less pointed than in the first premolar. It has well-developed mesial and distal cusp ridges.

There are two common occlusal forms, a two-cusped form and three-cusped form: In the two-cusped form, the single lingual cusp is opposite the buccal cusp. The central developmental groove runs in a mesiodistal direction. It may be straight but is usually curved around the buccal cusp. The ends of the central groove are situated in the mesial and distal fossae from where supplemental grooves radiate, as well as from the central groove itself. Some teeth show mesial developmental pits at the ends of the central groove in the mesial and distal fossae. Most of the lower second premolars with two cusps have a developmental groove that crosses the distolingual cusp ridge from the central groove.

The three-cusped form has three distinct cusps arranged as follows: a large buccal cusp, a main mesiolingual cusp and a small distolingual cusp. Each of the three cusps has a well developed triangular ridge, separated by relatively deep developmental grooves which converge in a central pit on the occlusal surface in the shape of a Y. A mesial developmental groove runs from the central pit in a slightly mesiobuccal direction and ends in the mesial triangular fossa. The shorter distal developmental groove runs in a distobuccal direction and ends in the distal triangular fossa, mesial to the distal marginal ridge which is at a lower level than the mesial ridge. The lingual developmental groove courses between the lingual cusps. The mesiolingual cusp is longer in a mesiodistal direction than the distolingual cusp, resulting in the lingual developmental groove being located on the distal side of the middle of the crown. Supplemental grooves and hollows often radiate from the developmental grooves and may cross either of the marginal ridges.

Fig 133   **35 (Left mandibular secondary second premolar)**

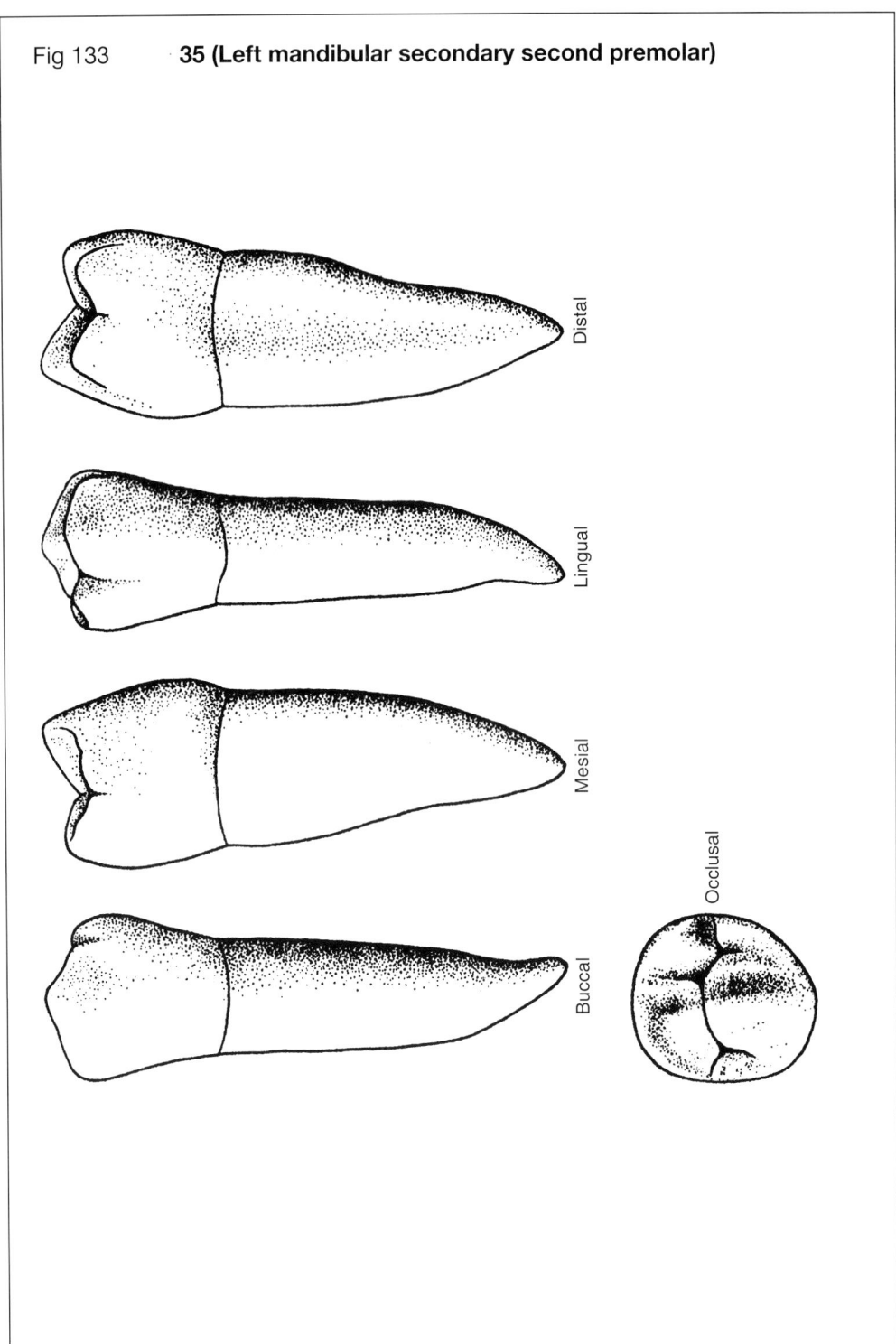

The single oval root is often grooved mesially and distally and is curved in a distal direction, often close to the root apex. The pulp chamber has two or three cornua, depending on the number of cusps present, and there is usually only one root canal.

The molars

The molar teeth are adapted to crushing and grinding food. Molars are multicusped and usually have more than one root. The occlusal surfaces are large and surmounted by three to five cusps with a complex arrangement of grooves.

The molars in the upper dentition decrease in size from before backwards. In the mandibular dentition, the first molar is the largest tooth, while the second molar may be larger or smaller than the third molar.

The cervical margin of these teeth is less sinuous than that of all the teeth previously mentioned. The upper molars usually have three roots, while the lower molars have two. The roots of the molars become shorter and show a greater tendency to fuse towards the rear of the series, while they may develop a progressively more marked distal curvature.

The maxillary molar teeth

In general, the upper molars have large crowns with four well-developed cusps, two being found buccally and two lingually. There are two buccal roots and a single large lingual root.

Maxillary first molar (16, 26) (Fig 134)

The crown is somewhat wider buccolingually than its mesiodistal measurement, and slightly rhombic in occlusal outline. It is usually the largest maxillary tooth and has four well-developed functioning cusps. A non-functioning supplemental fifth cusp, the cusp or tubercle of Carabelli, is present in about 60% of these teeth.

The four large cusps are situated mesiobuccally, mesiolingually, distobuccally, and distolingually. The fifth cusp is located on the lingual side of the mesiolingual cusp which is the largest of the four functioning cusps. This fifth cusp is situated about one third the distance from the mesiolingual cusp tip to the cervical margin. Quite often it is so undeveloped that it is hardly noticeable, but yet a small developmental groove may still be discernable. The buccal cusps are more pointed than the lingual cusps. The mesiobuccal cusp is the second largest, followed by the distolingual, the distobuccal, and lastly the fifth cusp.

All the cusps show mesial and distal cusp ridges, although these ridges sometimes appear indistinct on the smaller and rounder distolingual cusp. Marginal ridges connect the two mesial cusps and also the distal cusps with each other. The

Fig 134   **26 (Left maxillary secondary first molar)**

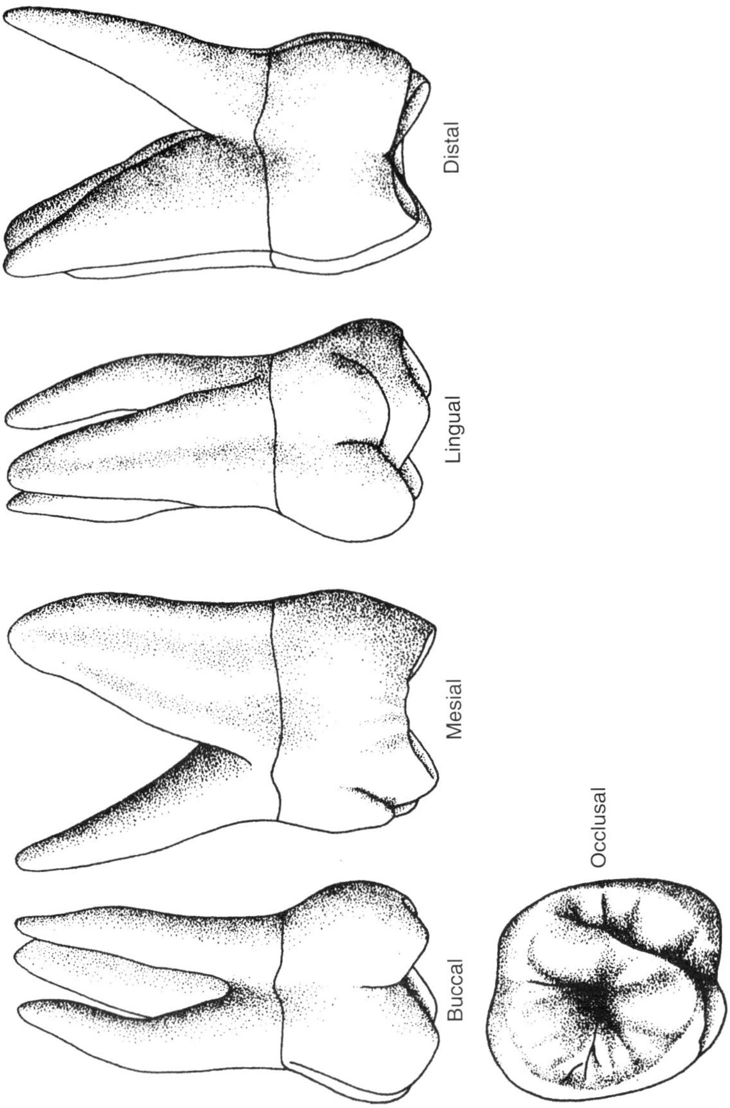

Distal

Lingual

Mesial

Occlusal

Buccal

mesiolingual cusp is connected to the distobuccal cusp by means of an oblique ridge. The height of the oblique ridge decreases towards the middle of the occlusal surface where it is about equal to that of the marginal ridges. Sometimes the two large occlusal fossae are connected by a shallow groove. The oblique ridge divides the occlusal surface into two parts, the mesial being the largest.

The larger and more central mesial fossa on the occlusal surface is bordered by the distal cusp ridge of the mesiobuccal cusp, the mesial cusp ridge of the distobuccal cusp, The crest of the oblique ridge and the crests of the triangular ridges of the mesiobuccal and mesiolingual cusps. This fossa usually has a central pit. From the central fossa, buccal and mesial developmental grooves, each contained in a sulcus, pass between the buccal and mesial cusps. The buccal groove runs onto the buccal surface of the tooth. The mesial marginal ridge forms the base of the mesial triangular fossa. The apex of the fossa is more or less at the point where short collateral supplemental grooves join the end of the mesial developmental groove. The distal fossa is situated distal to the oblique ridge. An irregular developmental groove (the distal oblique groove) crosses the distal fossa in its deepest part and separates the distobuccal cusp and the buccal part of the oblique ridge from the smaller distolingual cusp. It also separates the lingual part of the oblique ridge from the distolingual cusp. It joins the lingual developmental groove at the junction of the cusp ridges of the mesiolingual and distolingual cusps, and ends on the lingual surface of the crown. If the fifth cusp is well developed, its developmental groove joins the lingual groove.

The distal oblique groove usually has a considerable number of supplemental grooves, with two distinct short grooves, one on the buccal and one on the lingual side of the distal fossa. These two sides, together with the mesial incline of the distal marginal ridge, form the distal triangular fossa.

There are three roots: one mesiobuccal, one distobuccal, and one lingual. They do not divide close to the cervical margin but are connected to the crown by means of an undivided stem.

The buccal roots are usually slightly flattened on their mesial and distal surfaces. The mesiobuccal root is usually slightly thicker and longer than the distobuccal root, while the lingual root is longer and thicker than either of the buccal roots. The lingual root is cone-shaped and develops in a lingual direction away from the two buccal roots.

The cervical margin is relatively straight around the tooth but may show an occlusal convexity on the mesial side. The cornua of the pulp chamber are directed in the direction of the cusps. There is one root canal in each root.

Maxillary second molar (17, 27)

The upper second molar can best be described by way of comparison with the first molar, considering the overall resemblance.

The roots are the same length as the roots of the first molar, if not slightly longer. The mesiobuccal and mesiolingual cusps are just as well developed as those of the first molar. The distobuccal cusp and the distolingual cusp are smaller and the fifth cusp is absent. The crown is slightly shorter in a cervico-occlusal direction than that of the first molar, but their buccolingual widths are more or less identical.

Viewed occlusally, two types of upper second molars are identified:
1. The most common type has an occlusal surface similar to that of the first molar except that the mesiodistal length is less.
2. The second type has a crown more similar to the third molar and shows a very poorly developed distolingual cusp, emphasizing the size of the other three cusps. This type of crown has a heart-shaped outline.

The buccal roots are about the same length, nearly parallel to each other, and both following a distal curvature, placing the apex of the mesiobuccal root opposite the buccal groove of the crown, and not in line with the mesiobuccal cusp tip as in the first molar. The apex of the lingual root is in line with the distolingual cusp tip, and not with the lingual groove as in the first molar. The lingual root does not diverge as far from the direction of the buccal roots as it does in the first molar.

There are more supplemental grooves and pits on the occlusal surface of this tooth than on the first molar.

Maxillary third molar (18, 28)

The upper third molar is often deformed and very irregular in appearance. The occlusal outline is often heart-shaped with three cusps, namely two buccal cusps and a single large lingual cusp. The distolingual cusp and distal fossa is absent in such a tooth, or an extremely small distolingual cusp may be present. When the distolingual cusp is absent, the distal marginal ridge is formed by the oblique ridge between the distobuccal and mesiolingual cusps, no lingual groove being present. The crown is shorter in both a cervico-occlusal and a mesiodistal direction than the crown of the second molar.

The roots are relatively short, are usually partially united and function as one large single root.

The mandibular molar teeth

The lower molars are the largest teeth in the mandibular arch. There are three molars in each quadrant of the mandible and, although they resemble each other in functional morphology, differences exist in the number of cusps, general size, the placing and size of the roots, and occlusal pattern.

The outlines of the mandibular molar crowns are similar and each has one mesial and one distal root. The roots of the third molars are quite often fused. The crowns of all lower molars are roughly quadrilateral and slightly longer in a mesiodistal direction than the buccolingual width. The crowns of lower molars are relatively short cervico-occlusally compared to the anterior teeth.

The roots of the molars are individually possibly not as large as those of some of the other lower teeth, but the combined surface area of the roots results in a stronger periodontal attachment, or anchorage, and great masticatory efficiency. The total mesiodistal length of the molars is usually greater than the mesiodistal length of all the more anteriorly positioned teeth.

Mandibular first molar (36, 46) (Fig 135)

Viewed from an oculusal aspect, the crown appears somewhat assymmetrical and oblong with the buccal side longer than the lingual side, and the mesial part slightly broader than the distal part. There are five cusps, but some anatomists consider four, namely the mesiobuccal, mesiolingual, distobuccal, and distolingual as primary or principal cusps and the smaller distal cusp as secondary. This fifth, or distal cusp is situated between the primary distal cusps and is slightly buccally placed. The mesiobuccal cusp is the largest, followed by the two lingual cusps which are of about equal size, then the smaller distobuccal cusp, and finally the smallest distal cusp. The buccal cusps and the distal cusp are relatively flat, while the lingual cusps are more pointed.

A main fossa and two smaller fossae are found on the occlusal surface, the main fossa being centrally situated. The two smaller fossae are the mesial triangular fossa situated distal to the mesial marginal ridge, and the distal triangular fossa situated on the mesial side of the distal marginal ridge. The developmental grooves are the central developmental groove and the lingual and distobuccal developmental grooves.

The central developmental groove courses in an irregular manner in a mesial direction from the central pit and ends in the mesial triangular fossa. In its course it joins the mesiobuccal developmental groove which runs mesiobuccally in the deepest part of the sulcus separating the mesiobuccal and distobuccal cusps.

The mesiobuccal groove of the occlusal surface is continued onto the buccal surface of the tooth as the mesiobuccal groove of the crown. The lingual developmental groove of the occlusal surface runs irregularly in the sulcus between the triangular ridges of the lingual cusps to where the cusp ridges of the lingual cusps meet. Here the lingual developmental groove is continued as the lingual groove on the lingual surface of the crown.

The central developmental groove runs slightly in a distobuccal direction from the central pit to meet the distobuccal developmental groove of the occlusal surface. From this point the groove runs slightly distolingually to the distal triangular fossa.

The distobuccal developmental groove of the occlusal surface runs in a distobuccal direction until it joins the distobuccal groove of the buccal surface of the crown at the point where the cusp ridges of the distal and distobuccal cusps meet. Supplemental grooves appear as sidebranches of developmental grooves.

Fig 135 **36 (Left mandibular secondary first molar)**

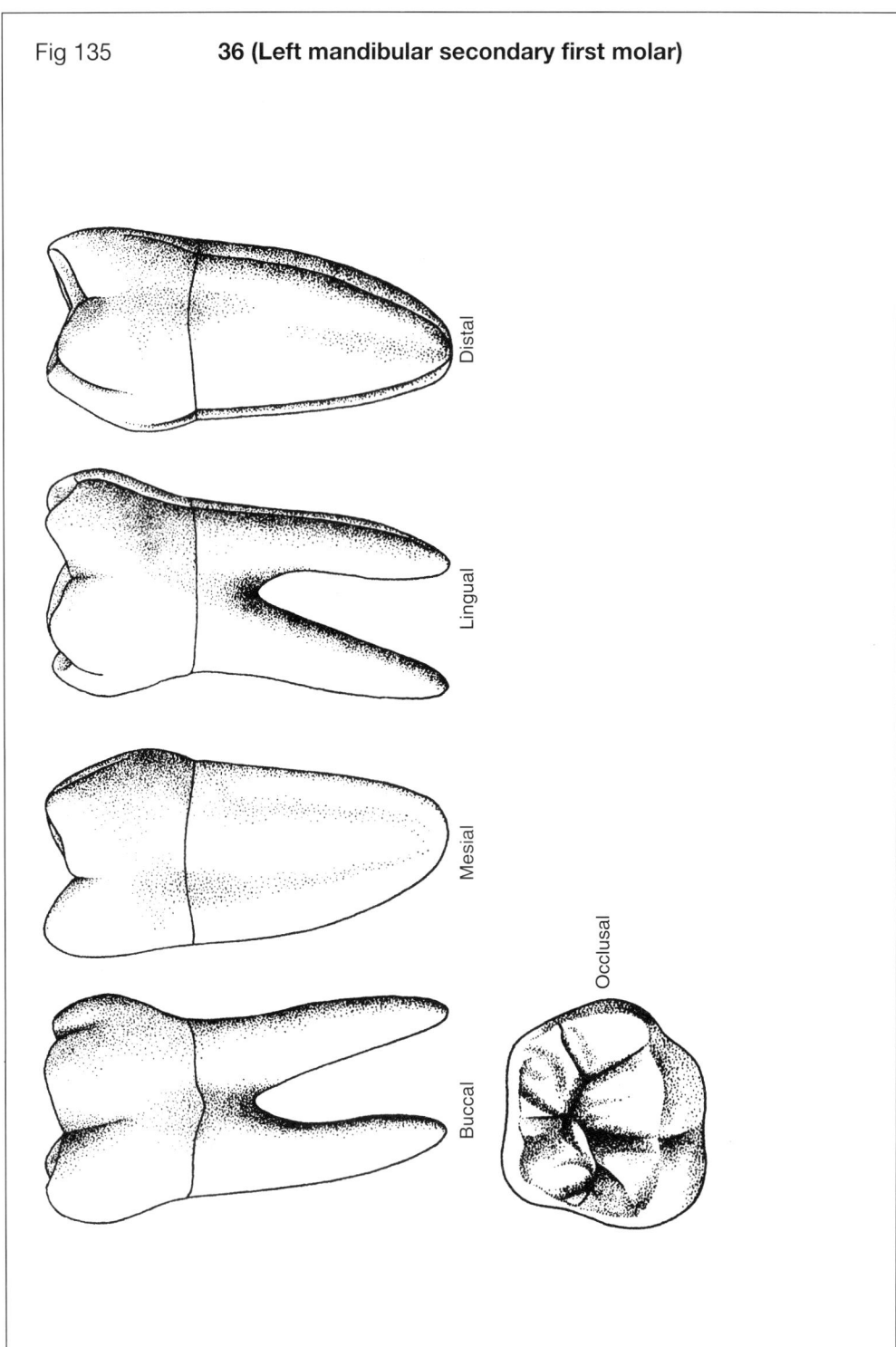

The cervical margin is slightly convex cervically on the buccal and lingual sides of the tooth. Mesially, the cervical margin appears convex in an occlusal direction while it is straighter on the distal surface. In all teeth, the cervical margin is at a slightly higher level on the lingual than on the buccal side.

The tooth has two roots, one mesial and one distal. The roots do not divide close to the neck of the tooth but are connected to the crown by a short common root stem. The broad roots are flattened mesially and distally and the mesial root is the largest. It usually shows quite a sharp distal curvature. On the distal surface of the mesial root there is usually a distinct vertical groove which is not as well developed as on the mesial side of the distal root. The distal root does not usually show a sharp distal curvature.

The coronal pulp contains cornua, directed to the principal cusps. Two root canals are usually present in the mesial root, while the distal root has a single canal.

Mandibular second molar (37, 47) (Fig 136)

The second molar is usually fractionally smaller than the first molar. It is, however, not uncommon for the second molar crowns to be larger than the first molar crowns and, although the roots are usually not as robust and well-developed, they may be longer.

There are four well developed cusps, two on the buccal and two on the lingual side. The distobuccal cusp is larger than that of the first molar.

The occlusal surface shows distinct differences when compared with the first molar. The small distal cusp is absent and there is no sign of the distobuccal developmental groove. The buccal and lingual developmental grooves meet the central developmental groove at a central pit in the form of a cross that divides the occlusal surface into four parts, each of about equal size. Cusp ridges are not as smooth as those of the first molar due to the presence of many supplemental grooves that radiate from the developmental grooves. Only one developmental groove is present on the buccal surface, namely the buccal developmental groove between the two buccal cusps which are of equal height. In many instances the cervical margin is sharply convex cervically on the buccal side of the tooth. The roots are irregular in length and in arrangement. They are not as broad as the roots of the first molar and are usually fairly close together or fused. They may, however, diverge considerably. They usually show a sharper distal inclination than the roots of the first molar.

Mandibular third molar (38, 48)

The morphology of the third molar varies considerably and shows many anomalies. Although conforming generally to the design of lower molar teeth, its occlusal pattern is unpredictable and the roots underdeveloped and deformed. There are usually two roots, one mesial and one distal, both being acutely inclined di-

Fig 136   **37 (Left mandibular secondary second molar)**

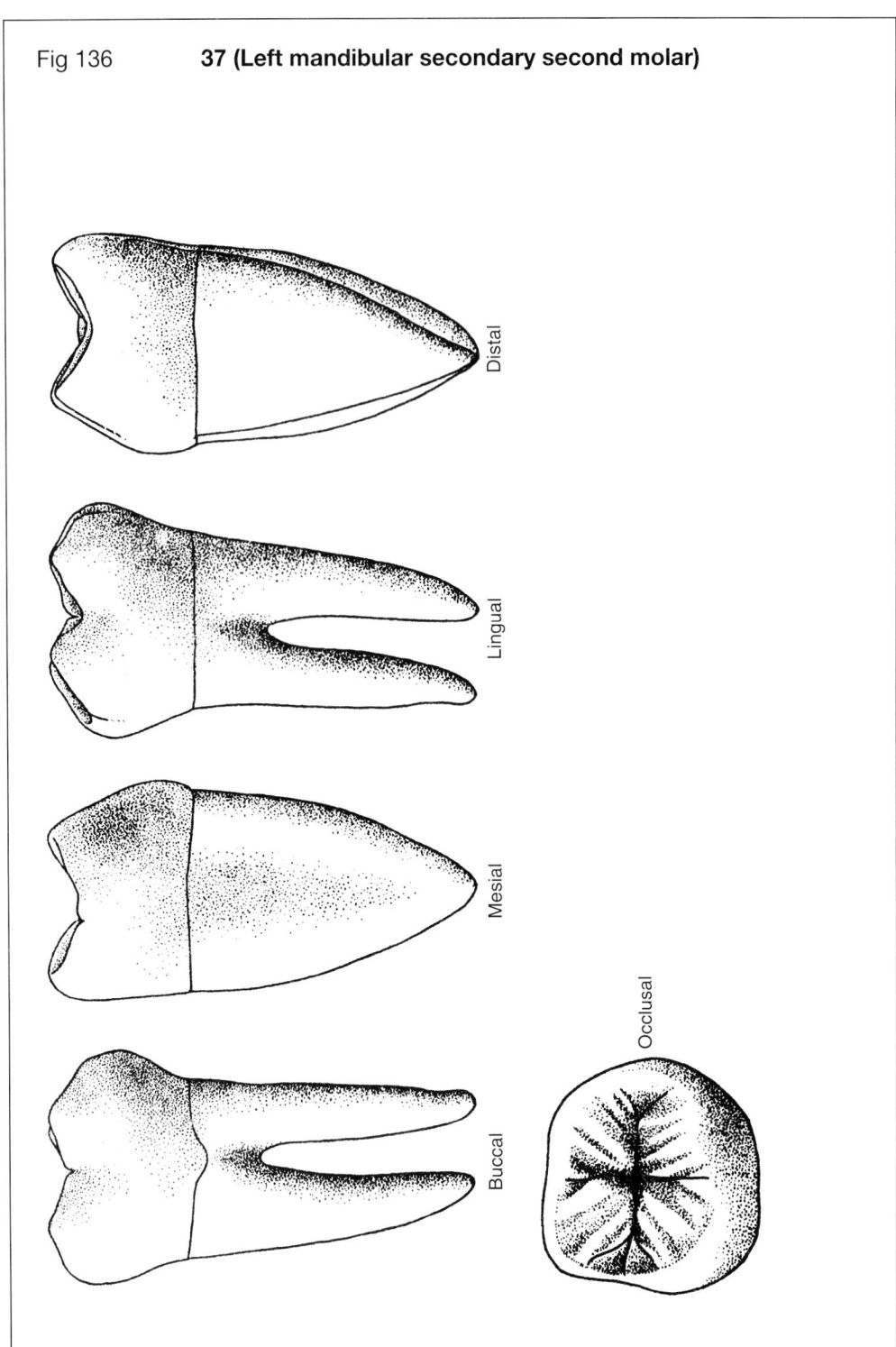

stally. They are usually short, thick, close together and often fused. There are usually two buccal and two lingual cusps. All the cusps are short and rounded.

The developmental dates of the secondary teeth are as follows:

Tooth	Crown completed	Appearance in the mouth	Root completed
Incisors	4-5 a	upper: 7-9 a lower: 6-8 a	
Canines	6-7 a	upper: 11-12 a lower: 9-10 a	2-3 a after appearance in the mouth
Premolars	5-7 a	10-12 a	
1st Molars	2.5-3 a	6-7 a	
2nd Molars	7-8 a	11-13 a	
3rd Molars	12-16 a	17-21 a	

Differences between primary and secondary teeth

1. The crowns of the primary teeth are longer in a mesiodistal direction, relative to their crown height, than the crowns of secondary teeth and have a rounder profile.
2. The roots of the primary teeth are thinner, and longer, relative to their crown lengths, than the roots of secondary teeth. The thin roots and wide crowns accentuate the bulbous shape of the crowns and differ markedly from the appearance of the cervical parts of the secondary teeth.
3. The roots of primary molars are more divergent and divide closer to the necks of the teeth, probably because the secondary premolar tooth germs develop between the roots.
4. The primary incisor crowns are smaller in all dimensions than their successors, and the primary molars are smaller than the secondary molars but longer mesiodistally than the premolars.
5. The cusps of primary teeth are more pointed when they erupt, and the contact areas are smaller due to the bulbous shape of the crowns.
6. The cervical margins of primary teeth are less sinuous than the cervical margins of the secondary teeth, and the crowns of the first primary molars show a prominent bulge close to the cervical margin mesiobuccally.
7. Primary tooth crowns are whiter.
8. Primary teeth undergo attrition to a greater extent.
9. The enamel of primary molars is thinner and of a more uniform thickness.

10. A relatively greater thickness of dentine is found over the pulp chamber in the occlusal fossae of the primary molars.
11. The pulp chambers of primary teeth are generally larger, and the pulp cornua of primary molars higher and more pointed than in secondary teeth.
12. The enamel rods close to the necks of primary teeth are often arranged horizontally instead of inclining cervically as in the secondary teeth.

Selected bibliography

1. Scott, J.H. and Symons, N.B.B. (1982) *Introduction to Dental Anatomy*, 9th edition. Edinburgh: Churchill Livingstone.
2. Ash, M. (1984) *Wheeler's Dental Anatomy, Physiology, and Occlusion*, 6th edition. Philadelphia: W.B. Saunders.

Review questions

1. Name the teeth that are located
   (a) in a quadrant of the primary dentition; and
   (b) in a quadrant of the secondary dentition.
2. Give the formulae for both the primary and secondary dentitions.
3. Describe the surfaces of the teeth.
4. Write short notes on:
   (a) contact areas (contact points);
   (b) embrasures; and
   (c) the axial inclination of teeth individually.
5. Describe the primary teeth individually.
6. Describe the primary dentition with reference to:
   (a) the nature of the articulation (occlusion); and
   (b) chronology of tooth development.
7. Define the following terms in relation to the morphology of teeth:
   (a) developmental groove
   (b) supplemental groove
   (c) cingulum
   (d) fossa
   (e) canine fossa
   (f) central fossa

(g) triangular fossa
(h) cusp
(i) ridge
(j) cusp ridge
(k) marginal ridge
(l) oblique ridge
(m) triangular ridge
(n) cervical margin
(o) incisive edge
(p) sulcus
(q) pit

8. Compile a list to indicate the following dates with reference to the secondary teeth:
   (a) completion of the crowns;
   (b) appearance in the mouth (eruption); and
   (c) completion of the roots.
9. Compare the occlusal morphology of the secondary maxillary first and second molar teeth.
10. Compare the occlusal morphology of the secondary mandibular first and second molar teeth.
11. Compare the occlusal morphology of the mandibular first and second premolars.
12. Describe the differences between primary and secondary teeth.

12. Composition of teeth

Major differences exist in the composition of enamel and dentine and it is essential to separate them carefully before any studies on composition are undertaken. Difficulties are encountered when pure samples of cementum are required, and careful shaving of a root surface may give the best results. Various methods have been used to obtain samples of tooth material.

Methods to obtain samples of tooth material

Mechanical methods

The oldest method is to splinter fragments of enamel off a tooth with a chisel, or to grind a tooth crown until only dentine remains, or to drill away the dentine core of a crown until only an enamel shell is left. These methods are time-consuming and samples are never absolutely pure.

Flotation method

This method entails grinding a whole tooth to a fine powder which is then placed in an open-ended pointed tube and introduced into a centrifuge tube containing a solution of bromoform (91%) and acetone (9%). The tubes are then centrifuged. The enamel component of the powder will pass through the open tip of the inner tube and collect at the base of the outer tube, while dentine and cementum layers will remain in the inner tube. This separation of tooth material is based on the following facts: when enamel, dentine, and cementum are centrifuged in a solution with a specific gravity of 2.70, such as bromoform and acetone described above, the enamel portion (specific gravity 2.9-3.0) will sink to the bottom, while dentine (specific gravity 2.14) and cementum (specific gravity 2.03) will remain supernatant in the solution. A finger is then placed over the top opening of the pointed tube to prevent loss of material when it is lifted out, and the tube is removed. Relatively pure samples of especially enamel can be obtained in this way. The density of cementum (2.03) is so similar to that of dentine that a clear separation is difficult but some results may be obtained if a mixture of dentine and cementum is treated with fluid of density of 2.07.

Chemical methods

Acids are used to etch off portions of enamel and dentine to obtain solutions of dissolved tooth material or residues of insoluble components. Etching can be done to a depth of 10 µm.

Hardness

Some authors maintain that the hardness of enamel gradually diminishes from the incisal edge (or cusp tip) to the cervical margin and, furthermore, that enamel is hardest at the surface, diminishes in hardness immediately below the surface, and then remains at a constant hardness adjacent to the amelodentinal junction where the hardness diminishes rapidly. The enamel of primary teeth is softer than that of secondary teeth.

According to other authors, the hardness of enamel remains fairly constant throughout its entire thickness.

Hardness can be used as an indication of calcium content and consequently of the state of mineralization. The softness of carious enamel is due to a loss of calcium (demineralization).

Dentine is softer than enamel and variations in hardness at different depths have not been described. Cementum is the softest of the mineralized dental tissues and its hardness is comparable to that of bone.

General remarks on chemical composition

One of the first descriptions of the composition of mineralized tissues was provided by Empedocles (492-432 B.C.), founder of the Sicilian Medical School, who stated that mineralized tissues are composed of 2 parts water, 4 parts fire and 2 (or 8) parts soil. Aristotle (384-322 B.C.) later found that bone is composed of 3 parts fire and 2 parts soil. In 1803 W.H. Pepys probably did the first scientific chemical analysis of enamel and found that it contains 78 % calcium phosphate, 6% calcium carbonate and 16% water. By the middle of the nineteenth century, scientists were convinced that a tooth is composed mainly of calcium phosphate with lesser quantities of calcium carbonate, magnesium phosphate, other mineral salts, water, and organic substances. Towards the end of the nineteenth century, Tomes maintained that dentine contains 72.5% calcium salts, that carious teeth contain less calcium than healthy teeth, and that posterior teeth are more densely calcified than anteriors. Many investigations were carried out with dissimilar results. It was found that enamel and dentine differ in composition, that carious and healthy teeth have the same mineral composition, that variations in the composition of the teeth of one person are just as great as between the teeth of different persons, that aging does not influence chemical composition of teeth, that gingivitis causes a lessening of carbonate content, that no relation exists between the chemical composition of teeth and caries susceptibility, and that primary teeth contain more water and less inorganic material than secondary teeth.

Enamel

Mature enamel contains 96-97% inorganic material by mass, while the rest is composed of organic material (0.4-0.8% in secondary enamel and 0.5-0.9% in primary enamel) and water.

A major part of the organic content of enamel is found in the lamellae and enamel tufts in the inner parts of the enamel. In newly formed enamel a larger proportion of matrix is present (as high as 19%) but this percentage is reduced in the maturation process by active withdrawal. The nature of the enamel matrix is discussed in detail in the chapter on amelogenesis (Part II, Chapter 6). The inorganic portion of enamel has the following approximate composition, according to mass:

Ca 37% Na 0.5%
Mg 0.5% PO_4 55.5%
CO_3 3.5%

Mineral salts are present chiefly in the form of hydroxyapatite with an empiric formula of $Ca_{10}(PO_4)_6 \cdot (OH)_2$. It is present in the form of submicroscopic hexagonal crystallites which are larger than the crystallites of dentine or bone.

Dentine

Dentine has a markedly higher organic content (18%) by mass than enamel. It consists largely of collagen (93%). Approximately 70 % of dentine is inorganic and the rest is water. The apatite of dentine is also present as submicroscopic crystallites but they are smaller than those of enamel and similar to crystallites of cementum and bone.

Cementum

Cementum contains a high proportion of organic material (23%) by mass. This consists largely of collagen and ground substance. It is further composed of inorganic substances (65%) and the rest is water. Calcium salts are present as apatites in crystallites which are found inside and around collagen fibers which are embedded in the cementing ground substance.

Trace elements

The following elements have been found in enamel and dentine:

Ag (silver)	Al (aluminum)	Ba (barium)
Cu (copper)	Fe (iron)	Mg (manganese)
Ni (nickel)	Pb (lead)	Si (silicon)
Sr (strontium)	Ti (titanium)	V (vanadium)
Zn (zinc)	Cr (chrome)	K (potassium)
Li (lithium)	Mn (manganese)	Sn (tin)
Na (sodium)	P (phosphorus)	Co (cobalt)
Sb (antimony)	Pt (platinum)	Rb (rubidium)
Ca (calcium)	F (fluorine)	

Little is known of the significance of most of these elements, many of which may be fortuitously present as contaminants. Only the presence of fluorides will be briefly discussed.

The fluoride content of teeth has enjoyed much attention since the discovery of a relationship between fluoride intake and dental caries. The fluoride content of teeth is largely determined by the intake of fluoride in food and drinking water during the time that mineralization of teeth is taking place.

Fluoride is found in greatest concentrations in the outer layers of enamel (up to ten times more than in enamel as a whole), even in areas where fluoride is not present in drinking water. This is due to topical deposition of fluoride from foods and beverages, such as tea, in the enamel surfaces and most researchers agree that the caries- inhibiting action of fluoride can be ascribed to this surface concentration. Unerupted teeth similarly have a high surface concentration of fluoride and this is due to absorption of the ions from tissue fluid. The surface fluoride content is, however, never as high as in erupted teeth. It is therefore obvious that fluoride is taken up in enamel both before and after eruption. The main age difference in this regard is an increased depth of fluoride concentration in enamel of erupted teeth.

The fluoride concentration in dentine is higher than in enamel and increases gradually from the amelodentinal junction to the predentine. The concentration on the dentinal side of the junction is, in fact, three to four times greater than on the enamel side. The effects of fluoride on the teeth are discussed fully in Part II, Chapter 24.

Selected bibliography

1. Jenkins, G.N. (1978) *The Physiology and Biochemistry of the Mouth*, 4th edition. Oxford: Blackwell Scientific Publications.
2. Cole, A.S. and Eastoe, J.E. (1988) *Biochemistry and Oral Biology*, 2nd edition. London: Wright.
3. Mjör, I.A. and Fejerskov, O. (1986) *Human Oral Embryology and Histology*. Copenhagen: Munksgaard.
4. Schroeder, H.E. (1991) *Oral Structural Biology*. New York: Thieme Medical Publishers.
5. Williams, R.A.D. & Elliott, J.C. (1979) *Basic and Applied Dental Biochemistry*. Edinburgh: Churchill Livingstone.

Review questions

1. Describe three methods to obtain samples of mineralized dental tissues.
2. Write brief notes on the hardness of enamel, dentine, and cementum.
3. Describe the chemical composition of enamel, dentine, and cementum.
4. Discuss the fluoride content of teeth.

13. Permeability and age changes of teeth

General remarks

Much speculation has existed in the past regarding the vitality of mineralized dental tissues and age changes which a tooth undergoes, and to what extent dental tissues undergo metabolic change or are influenced by changes in the body as a whole. Some authors have in the past regarded a tooth as being outside the sphere of general body changes, while others recognized the presence of submicroscopic circulatory channels in enamel and dentine through which odontoblasts and their processes exert some metabolic control. The permeability of enamel and dentine and age changes in dental tissues will subsequently be discussed.

Permeability of enamel

There is presently no doubt that enamel is permeable, but to a lesser extent than dentine. In one experiment, the pulp cavities of extracted teeth were connected to a capillary tube containing water. The surface of the enamel was covered with immersion oil and studied microscopically. Although no pressure was applied to the water, small droplets of water collected spontaneously under the oil on the surface after 2-3 hours. Although the largest droplets were found over defects in the enamel, such as lamellae, droplets were generally distributed. It is probable that physical forces, such as capillarity through the submicroscopic spaces between the crystallites, are responsible for this phenomenon.

When teeth are treated with radioactive ions, the ions are absorbed over the entire surface. Obviously, greater quantities are absorbed by defects such as carious lesions than by intact surfaces. The loss of certain constituents of the enamel during the carious process, and the resulting greater exposure of crystals and organic material, are responsible for this finding. For this reason it has been shown that enamel in early carious lesions contains a higher concentration of fluoride than intact enamel, fluoride being absorbed from the oral contents.

Permeability of dentine

When soluble or insoluble stains have been placed in the pulps of teeth, or sealed on the enamel or into the dentine, the following have been demonstrated:

1. Enamel is less permeable than dentine;
2. Advancing age tends to reduce permeability; and
3. Dentine is permeable from both the amelodentinal junction and the pulp.

In experiments in which the rate of movement of stains in dentine was investigated, argyrol (a silver-containing stain) was introduced into the dental pulps of living dogs through small holes in the necks of the teeth. The teeth were removed at varying periods after introduction of the stain and sectioned for histology. Half the thickness of the dentine was penetrated by the stain after 13 minutes, while the entire thickness of the dentine was stained after 17 minutes.

Transverse sections of the dentine showed the presence of the stain in the odontoblast processes, showing that this was a major route of diffusion. When radioactive solutions of urea were used, similar results were obtained.

Diffusion of stains takes place in all directions, as shown by experiments in which stains are introduced into the dentine through a small hole in the enamel. The stain spreads to the amelodentinal junction, as well as into the enamel to a limited extent, to the pulp cavity and from here to other parts of the dentine.

The to and fro movement, or microcirculation, of fluid in dentine enjoyed much attention in the past. To substantiate such a possibility, a space must be present between the tubule wall and an odontoblast process through which pulpal fluid can circulate. A widely held view is that an odontoblast process completely fills the lumen of the tubule in that part of the dentine in which it is present, and that the space is a potential one. It is concluded that any circulation takes place via the interior of the process and peripheral dentine fluid in those parts of the tubule in which a process is absent.

Permeability of cementum

It has been shown experimentally that primary cementum in older animals is impermeable to stains from both its outer and inner surfaces. Secondary (cellular) cementum in older animals is permeable from the outer surface, indicating that intercommunicating processes of cementocytes offer a route for penetration of dyes. The use of radioactive substances has shown that cementum is somewhat more permeable than described above, probably due to smaller ions in these substances and the greater sensitivity of detection methods.

Withdrawal of minerals from enamel and dentine

Since both enamel and dentine are permeable to a lesser or greater degree, the question arises whether mineral salts can normally be withdrawn from these tissues to the same extent as from bone. This is improbable in the case of enamel since fluid movement is very slight and cells are absent. Dentine is, however, more permeable, is intimately associated with odontoblasts and their processes, and is constantly being formed both pulpally and in the intratubular zone. Mineral salt exchange in dentine is therefore a distinct possibility.

Most experimental systemic conditions in which withdrawal of mineral salts could be expected, such as in conditions of hyperparathyroidism or mineral deficiencies did, however, not lead to demineralization, although minute mineral withdrawals remain a possibility. It is wrong to state that disturbances in calcium metabolism will necessarily be accompanied by changes in the mineral composition of teeth.

Age changes in teeth and their reaction to irritation

Secondary dentine

It is clear that continuous formation of dentine will reduce the size of the pulp cavity. Some authors distinguish between dentine formed before the roots are completed (primary dentine) and dentine formed later in the absence of obvious trauma to the tooth, such as caused by mechanical wear (attrition), erosion (chemical factors), or dental caries. This second type of dentine is termed secondary dentine, or physiologic secondary dentine and can be attributed to factors such as a progressive crowding of odontoblasts in a reduced pulp cavity which leads to irregularities in the course followed by tubules, or to normal aging processes.

A third type of dentine, or tertiary dentine, is described by some authors to result from rapid loss of tooth material by attrition or erosion, or due to caries of dentine. In tertiary dentine the tubules follow irregular patterns, are reduced in number, or cellular inclusions may be present. The above changes are, according to these authors, the results of damage to, or death of, odontoblasts.

Dentine tends to be deposited in greater amounts in certain areas of the pulp cavity with aging, for example, in the floor of the pulp chamber (furcation area of the roots).

It is considered unnecessary to recognize more than two types of dentine. It is suggested that the term primary dentine be applied to all dentine with a normal appearance with respect to the number of tubules and the course followed by the tubules, irrespective of the age of the tooth. This type of dentine is found even in old teeth not subjected to attrition or caries. The term secondary dentine should be used to describe dentine which has been formed following damage to dentine or the odontoblasts and differs from primary dentine. Secondary dentine is subdivided into relatively regular secondary dentine, as described above under physiologic secondary dentine, and into irregular secondary dentine which shows the features described above under tertiary dentine. The demarcation line between primary and secondary dentine is usually very conspicuous.

Transparent or sclerotic, dentine

Transparent or sclerotic, dentine is another type of secondary dentine. When loss of coronal tooth material is slow, large quantities of mineral salts are deposited in the dentinal tubules, and intratubular dentine may be deposited in such quantities as to occlude the tubules. This may create a relatively impermeable zone of dentine beneath the area of tooth loss. Pulpal to this barrier, the odontoblast and their processes may be quite normal. Sclerotic dentine is often present in such quantities in older teeth that all the dentine appears translucent, even in teeth showing no visible damage. Transparent dentine usually forms first midway between the surface of the pulp and the amelodentinal junction, and apically. Progressive coronal spread of transparent dentine from the apex with aging is one of the criteria used in age determination.

Dead tracts

When irritation of odontoblasts is more severe, as in the case of rapid dental caries, these cells may be severely damaged or destroyed. The related parts of the dentine are then known as dead tracts and extend from the pulp to the damaged surface. The affected tubules appear empty and are often covered at their pulpal ends by a barrier of irregular secondary dentine.

Dead tracts are, however, also found in teeth which never erupted and in teeth with little or no visible damage. They can therefore also be regarded as an aging phenomenon.

Changes in enamel

Enamel becomes less permeable with age, possibly as a result of surface consolidation of crystals, formation of fluorapatite and a reduction of matrix between individual crystals.

A further age change is attrition (wear resulting from masticatory movements of teeth and friction of food particles).

The surface of enamel is normally worn smooth with loss of perikymata soon after eruption.

Changes in the dental pulp

The dental pulp shows a relative loss of cells and an increase in its fibrous component with age. Odontoblasts may atrophy and collagen may increase to such an extent that a form of fibrosis results. Abnormal calcifications in the form of pulp stones (denticles) are fairly common. The blood supply of the pulp gradually diminishes, possibly due to a partial obliteration of the apical foramen by deposits of cellular cementum, and the pulp may become less sensitive. There is a tendency for dentine and enamel to dehydrate and to form minute cracks which make the tooth increasingly unsuitable for the placement of restorations.

Changes in cementum

Since cementum forms throughout life, it gradually increases in thickness, especially in the region of the tooth apex. Irregular deposits of cementum in this region may also lead to the formation of additional foramina. Areas of resorption of cementum and deposition of new cementum can usually be seen in all parts of a tooth root.

Selected bibliography

1. Bhaskar, S.N. (1991) editor. *Orban's Oral Histology and Embryology*, 11th edition. St. Louis: Mosby-Year Book, Inc.
2. Jenkins, G.N. (1978) *The Physiology and Biochemistry of the Mouth*, 4th edition. Oxford: Blackwell Scientific Publications.
3. Mjör, I.A. and Fejerskov, O. (1986) editors, *Human Oral Embryology and Histology*. Copenhagen: Munksgaard.
4. Osborn, J.W. and Ten Cate, A.R. (1983) *Advanced Dental Histology*, 4th edition. Bristol: Wright PSG.
5. Scott, J.H. and Symons, N.B.B. (1982) *Introduction to Dental Anatomy*, 9th edition. Edinburgh: Churchill Livingstone.
6. Ten Cate, A.R. (1989) *Oral Histology: Development, Structure, and Function*, 3rd edition. St. Louis: C.V. Mosby Company.

Review questions

1. Discuss the permeability of enamel.
2. Discuss the permeability of dentine.
3. What is your opinion regarding the withdrawal of mineral salts from the tooth?
4. Describe age changes in enamel.
5. Describe age changes in dentine.
6. Describe age changes in cementum.
7. What are the possible functional implications, if any, of the changes mentioned in your answers to questions 4, 5 and 6?

14. Oral mucosa

General remarks

The oral cavity is the first part of the intestinal tract. It communicates posteriorly with the oropharynx through the oropharyngeal isthmus, or fauces, and is bounded by the hard and soft palate above, the floor of the mouth below, and the cheeks and lips at the sides and front. It contains the teeth and the tongue and is lined by a mucous membrane, or mucosa.

The oral mucosa serves a variety of functions. A multitude of minor salivary glands in the oral mucosa, in addition to major glands which are situated at varying distances from the oral cavity, moisten the surfaces of the mouth and facilitate the processes of chewing, tasting, and swallowing. It provides a vast amount of information regarding events inside the oral cavity by means of taste buds and receptors that responds to pain, temperature, and touch. Furthermore, the oral mucosa protects the deeper tissues from the oral environment. It has a limited digestive function through the α-amylase which hydrolyses cooked starch to maltose and which is secreted by serous salivary glands.

The oral mucosa resembles the skin in many ways. It is composed of a surface epithelium and a deeper layer of connective tissue, the lamina propria. The epithelium is of either endodermal or ectodermal origin, depending on its relation to the embryonic buccopharyngeal membrane. The major part is, however, of ectodermal origin. The lamina propria is mesodermal in origin. Irregular connective tissue papillae of the lamina propria interdigitate with epithelial ridges, or rete ridges. These connective tissue papillae constitute the papillary layer of the lamina propria, while the deeper part of the lamina propria is known as the reticular layer. The oral mucosa is separated from underlying structures, such as muscle and bone, by a submucosa of varying thickness.

Classification and nature of oral mucosa

The morphology of the oral mucosa varies in different parts of the oral cavity (Fig 137). Around the teeth and in the hard palate, the mucosa is subjected to considerable friction during mastication and is known as masticatory mucosa. The mucosa of the lips, cheeks, soft palate, and the floor of the mouth is not subjected to trauma during mastication and is a more simple lining mucosa, while the mucous membrane on the dorsum of the tongue is a specialized mucosa since it shares the characteristics of both a masticatory and a gustatory mucosa.

The oral epithelium is a stratified squamous epithelium (Fig 138) and is separated from the lamina propria by a basal lamina. As in skin, several cell layers can be recognized. The cells which lie on the basal lamina are low columnar or cuboidal cells. Since much of the mitotic division responsible for the more superficial cells of the epithelium takes place in this layer, it is often known as the stratum germinativum. Those cells in this basal layer destined to form keratin are distinguished by the presence of keratin filaments and are known as keratinocytes. The volume of the cells increases during their upward migration till they form the stratum granulosum, but the volume of keratinized surface cells is less, although they cover a larger surface area.

Superficial to the single layer of basal cells, several layers of polyhedral cells are found. Some mitotic division still takes place in these cells. Clear intercellular spaces with numerous desmosomes separate these cells which are known as the stratum spinosum, or prickle-cell layer, because of their "prickly" appearance. The cells of this layer are slightly basophilic and contain bundles of tonofibrils. The stratum germinativum and the stratum spinosum are often collectively known as the stratum Malpighii.

The stratum granulosum overlies the stratum spinosum in orthokeratinized epithelium, and consists of slightly flattened cells arranged in two to three layers. These cells contain conspicuous keratohyalin granules associated with large bundles of tonofibrils. These granules stain intensely basophilic with acid dyes such as hematoxylin and are intimately associated with orthokeratin formation. Their nuclei show some signs of degeneration and are pyknotic.

The stratum lucidum is absent in oral epithelium or is very poorly developed.

In the orthokeratin layer, or stratum corneum, the cells have flattened considerably and lost their nuclei, keratohyalin granules, and intracellular cytoplasmic organelles such as ribosomes and mitochondria. The cells are filled completely with keratin and stain eosinophilic. Keratin is a tough insoluble protein which more or less completely fills the interior of the shrunken cells and contains a high proportion of disulfide linkages, probably derived from cystine, joining different polypeptide chains together. Desmosomes become indistinct. The process of keratinization commences before birth on the lingual and buccal aspects of the alveolar ridges, an indication that the process is genetically predetermined.

Fig 137 **Regional classification of oral mucosa**

Regional classification of oral mucosa as seen in a coronal section through the molar region of the oral cavity.

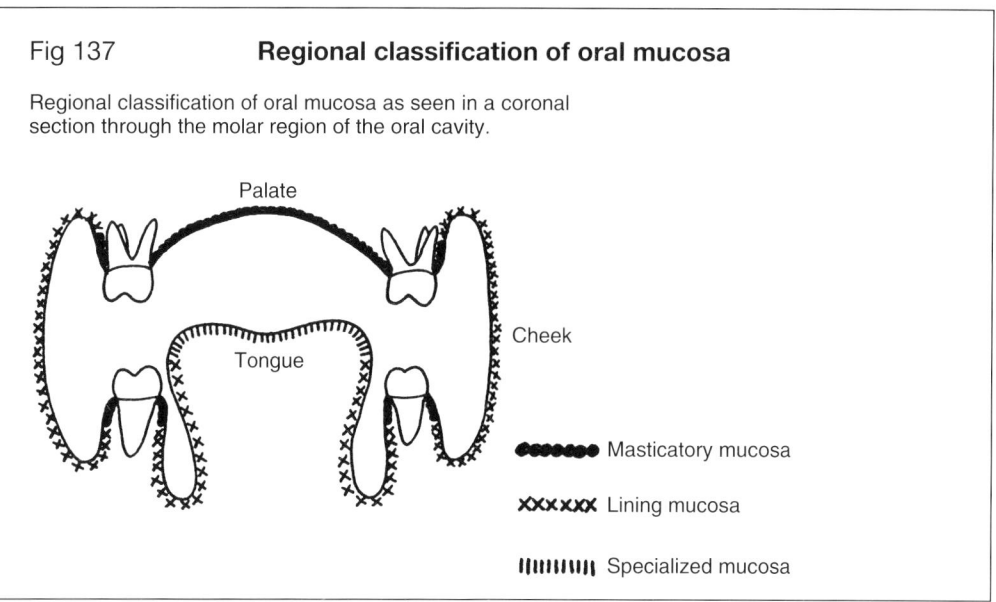

Palate
Cheek
Tongue
Masticatory mucosa
Lining mucosa
Specialized mucosa

Fig 138 **Oral mucosa**

General histologic structure

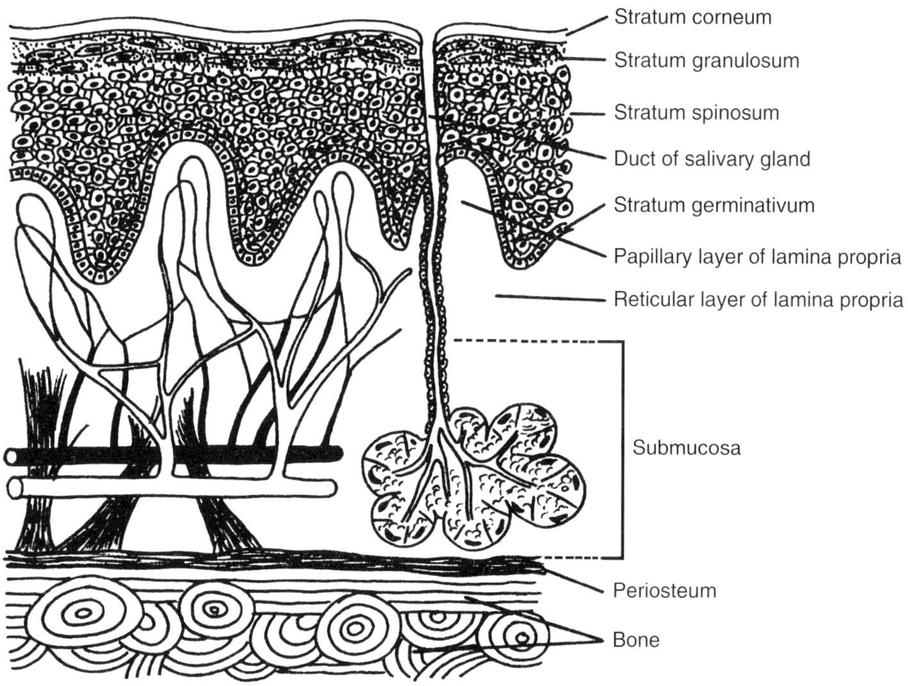

Stratum corneum
Stratum granulosum
Stratum spinosum
Duct of salivary gland
Stratum germinativum
Papillary layer of lamina propria
Reticular layer of lamina propria
Submucosa
Periosteum
Bone

As indicated earlier, the above description applies to an orthokeratinized epithelium. A variety of keratinization, termed parakeratinization, or partial keratinization, is widely encountered in the epithelium of masticatory mucosa. In a parakeratinized epithelium the surface layer (stratum corneum) stains for keratin but retains flattened pyknotic nuclei in many of the squames. Keratohyalin granules are very often completely absent and therefore a clear granular layer is frequently difficult to recognize.

An admixture of partially keratinized and non-keratinized epithelium is usually found in the lining mucosa. A non-keratinized epithelium consists of a basal cell layer, or stratum germinativum, similar to other oral epithelia, and superficial cells which are arbitrarily divided into an intermediate cell layer, or stratum intermedium, and a superficial cell layer, or stratum superficiale. The cells of the intermediate layer are slightly larger than those of the stratum spinosum and desmosomes are much less obvious. This layer unobtrusively becomes a superficial layer in which the cells are not flattened and contain plump nuclei. This layer desquamates in the same way as parakeratin or orthokeratin.

Endogenous pigmentation of the oral mucosa occurs most frequently in the attached gingiva below the interdental papillae, in the hard palate, buccal mucosa, and tongue. It is caused by melanin and varies from light brown to almost black. Melanin is produced by melanocytes, which are situated amongst the basal cells. These cells are derived from neural crest cells and migrate into the epithelium at an early embryonic age. They form a self-producing population of cells with long dendritic processes which pass between the cells of the adjacent layers. Melanosomes, melanin-containing oval or round granules, pass along the dendritic processes and into the cytoplasm of adjacent epithelial cells.

Two other types of cells encountered in the oral epithelium, are Langerhans cells and Merkel cells. The cells of Langerhans are also dendritic cells and are found in the upper layers of the stratum spinosum. They are involved in the immune response by recognizing and processing antigenic material entering the epithelium and presenting it to cells of the lymphoid system. They can probably migrate from the epithelium to regional lymph nodes.

Merkel cells do not possess any long cytoplasmic processes and are found in the basal cell layer, especially along the epithelial rete ridges. They form desmosomes with neighboring epithelial cells and are often closely associated with free intra-epithelial nerve endings. It is commonly accepted that these cells are sensory receptors, responding to pressure and touch. Sensory nerves in the oral mucosa terminate as both free and organized nerve endings perceiving sensations such as cold, heat, touch, pain, and taste.

The lamina propria consists of a connective tissue rich in cells and collagen and containing a network of young collagen fibers, bundles of collagen, numerous associated fibroblasts, macrophages (histiocytes), undifferentiated mesenchyme, mast cells and other cell types, some elastic fibers, blood vessels, lymphatics, and nerves. The papillary layer has a more delicate fiber arrangement than the reticular layer and contains capillary loops and nerves.

The submucosa (Fig 138) is more loosely constructed than the lamina propria and contains the larger arteries, veins, and nerves. The minor salivary glands lie in the submucosa and their secretory ducts pass through the submucosa and lamina propria to penetrate the epithelium. The distinction between the lamina propria and the submucosa is arbitrary.

Specific parts of the oral mucosa will subsequently be discussed:

Masticatory mucosa is found on the gingiva and hard palate. It is subjected to considerable friction during mastication. The gingiva is separated from the general lining alveolar mucosa on its vestibular (buccal) side by an irregular scalloped mucogingival line. The gingiva normally has a light pink color but may have a greyish hue in the presence of orthokeratin. It is currently accepted that masticatory epithelium, especially the gingiva, is predominantly parakeratinized, while a small minority of surfaces are covered by orthokeratin, and very much fewer surfaces by an incompletely parakeratinized and non-keratinized layer. The gingiva is histologically characterized by deep and narrow epithelial ridges of epithelium into the lamina propria, in contrast to shallow or absent ridges in lining mucosa. A detailed description of the gingiva will be found in the chapter on the periodontium (Part II, Chapter 18).

The mucosa of the hard palate (Fig 139) is similar to that of the gingiva and is firmly attached to the underlying periosteum. The following regions are described:

1. A gingival zone, adjacent to the teeth.
2. A median midpalatine raphe extending backwards from the incisive papilla.
3. An anterolateral area between the raphe and the gingiva, containing much fatty tissue in the submucosa.
4. A posterolateral area between the raphe and the gingiva, containing many minor mucous glands in the submucosa.

A distinct submucous layer is present, except for specific and narrow zones, namely the extreme peripheral gingival part and in the area of the midpalatine raphe.

Palatal rugae, or ridges, are irregular sidways elevations in the anterior part of the hard palate. They consist of an elevated core of connective tissue covered by epithelium. Individual ruga patterns are unique, like finger-prints, and have been used for post-mortem identification purposes.

The major part of the oral surface of the soft palate is covered by a stratified squamous epithelium similar to lining mucosa, and is continuous with that of the

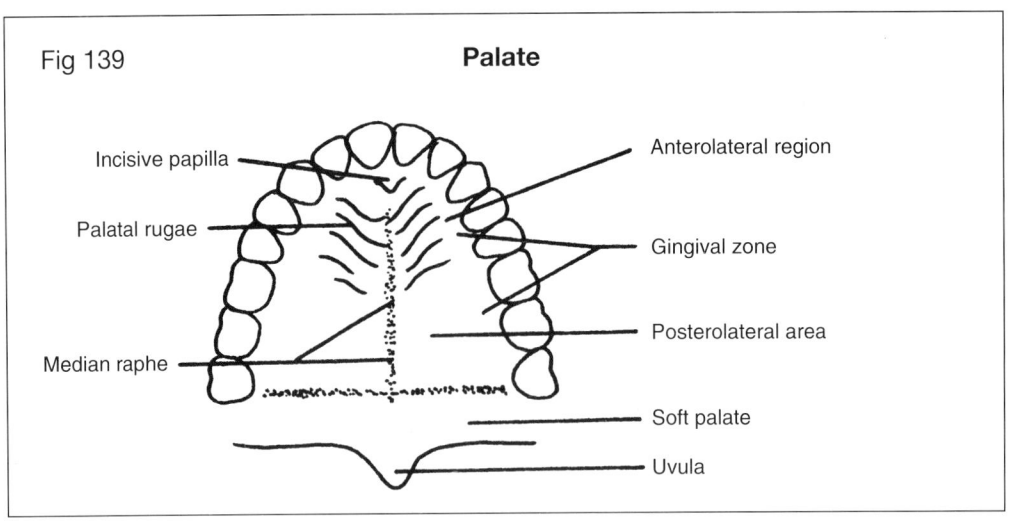

Fig 139 Palate

Fig 140 Tongue

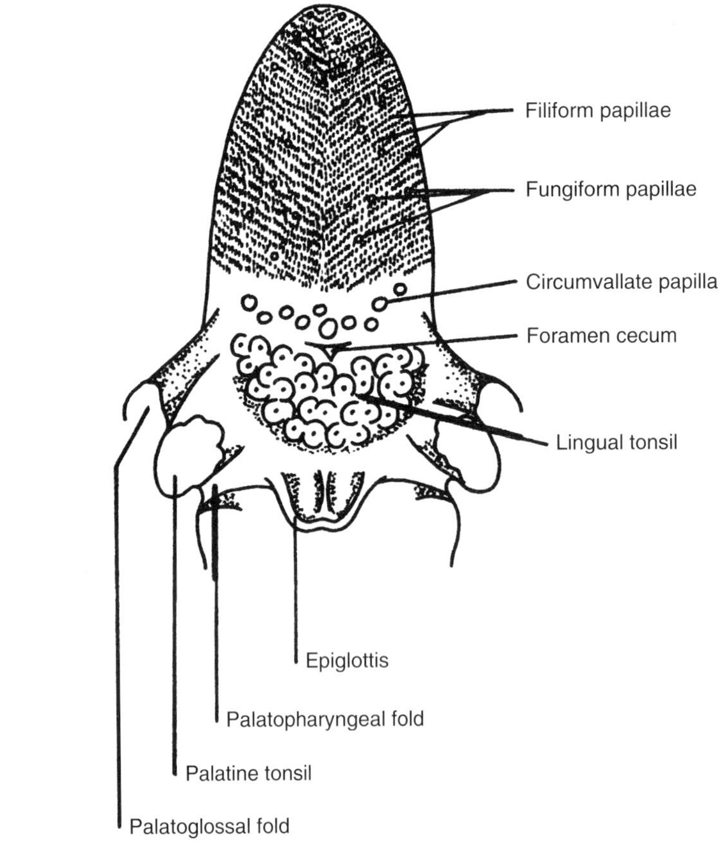

hard palate in such a way that a transition is scarcely noticeable. Respiratory epithelium covers the nasal surface of the soft palate. The submucosa of the soft palate contains many minor salivary (mucous) glands.

Specialized oral mucosa is found on the dorsal surface of the tongue (Fig 140). The ventral surface and sides of the tongue are smooth and covered by lining mucosa. The anterior part (body) of the dorsal surface is covered by many papillae, while the smaller posterior part (base) is smoother. Many small masses of lymphoid tissue, the lingual follicles or tonsillar tissue, are present on the posterior part and produce oval or round swellings. A small opening on each swelling leads into a relatively deep tonsillar crypt. The anterior and posterior parts are separated by a V-shaped shallow sulcus, the terminal sulcus, with the apex directed posteriorly.

Three types of papillae are found on the tongue (Fig 141): filiform (thread-shaped), fungiform (mushroom-shaped), and circumvallate (rampart-shaped) papillae. A fourth type, the foliate (leaf-shaped) papillae, are well-developed in rabbits but are considered rudimentary in humans. When present, they form indistinct mucosal folds on the lateral part of the dorsum of the tongue close to the circumvallate papillae. Taste buds are present in the epithelium on the lateral slopes of these folds. Filiform papillae are the most numerous and are found in the mucosa of the tongue anterior to the row of circumvallate papillae. They are described as forming oblique rows which radiate from the median line of the tongue, but this pattern is never well established. Fungiform papillae are scattered among the filiform papillae and are found in the greatest numbers close to the tip and sides of the tongue. Circumvallate papillae are present in an irregular row anterior to the terminal sulcus. They are the largest papillae and their number may vary from 8 to 12. Taste buds are primarily associated with these papillae.

Filiform papillae (Fig 141,I)

Filiform papillae consist of a central core of connective tissue with many secondary papillary projections. Irregular keratinized epithelium (both orthokeratin and parakeratin are present) covers the connective tissue projections. The filiform papillae do not prossess any taste buds and are frequently inclined in a backward direction.

Fungiform papillae (Fig 141,II)

Fungiform papillae are much less numerous than the filiform papillae. They have a short stem and a flattened hemispherical surface. The connective tissue core has a few low secondary papillae and the smooth surface is covered by a relatively thin parakeratinized epithelium. The lateral surfaces are non-keratinized. A few taste buds may be found on the dorsal surface. These papillae project slightly above the general surface of the tongue and usually have a light red color due to a rich capillary network visible through the relatively thin epithelium.

Fig 141 — Papillae of tongue

I Filiform papillae

II Fungiform papilla

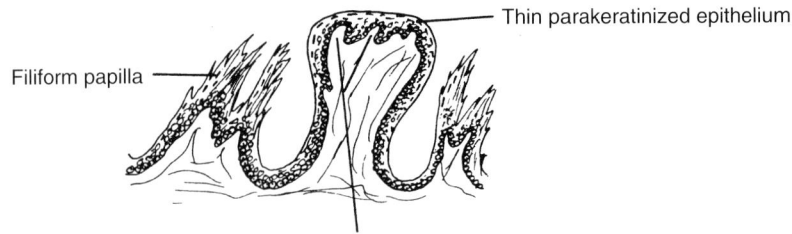

III Circumvallate papilla

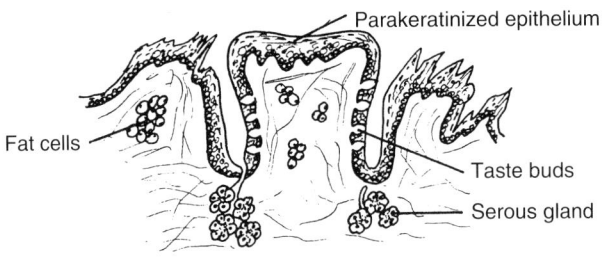

IV Taste bud

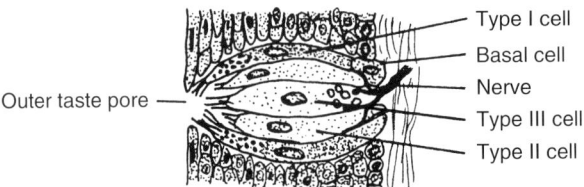

Circumvallate papillae (Fig 141,III)

The circumvallate papillae are partly submerged beneath the surface of the tongue and are surrounded by a circular groove. These large papillae contain a central connective tissue core with shallow secondary papillae below the dorsal surface. The surface is smooth and covered by orthokeratinized epithelium. Numerous taste buds are found on the lateral margins of these papillae. The ducts of small serous glands (of von Ebner), present in the lamina propria, open into the depths of the moat-like groove surrounding each papilla. Their surfaces appear light red.

Taste buds (Fig 141, IV)

Taste buds appear as pale oval structures, 30-50 µm in diameter, in the surrounding darker epithelium and consist of 20-50 cells (numbers in the literature vary from 20 to 100). They extend from the basal lamina of the epithelium through its full thickness to the surface where they communicate with the oral cavity by means of an outer taste pore with a diameter of approximately 2.5 µm. Three cell types are generally recognized in a taste bud by means of a light microscope, namely a few small irregular cells close to the base of the taste bud, columnar cells with darkly stained round or oval nuclei and a granular cytoplasm (dark cells), and columnar cells with oval nuclei and a lightly stained cytoplasm (light cells). In the past the lightly stained columnar cells have been described as taste cells, while the darkly stained cells were considered supporting cells.

By means of the electron microscope three cell populations can be clearly recognized:

1. Dark cells (type I) are long slender cells and possess many microvilli that project into the outer taste pore. Their nuclei contain much heterochromatin (dark) and the cytoplasm contains closely-packed free ribosomes, rough endoplasmic reticulum, filaments, and dark membrane-bound granules in the apical parts. The function of the dark cells is probably the secretion of pore substance between the tips of the cells inside the pore. These cells constitute up to 80% of the total cell population.
2. Light cells (type II) are less numerous (15-30% of the total) with short cytoplasmic processes and microvilli: The nuclei are light (less heterochromatin) and the cytoplasm contains fewer organelles. The morphology of these cells varies considerably and their function is unknown.
3. Intermediate cells (type III) are the least numerous of the cells of a taste bud (percentages of the total cell population vary from 5 to 30%, according to the literature), and are always surrounded by type I cells. They have short microvilli and groups of cytoplasmic vesicles of which some

are found peripherally in contact with nerve endings. These are regarded as synaptic vesicles and these cells are considered to be responsible for taste perception, although nerve fibers from a plexus in the lamina propria are also found in association with the other cell types.

The basal cells seen by light microscopy are sometimes classified as type IV cells and are possibly undifferentiated precursors of the main types, although some authors consider renewal to occur from peripheral cells. The cell population of a taste bud undergoes constant renewal over a period of about 11 days.

Selected bibliography

1. Bhaskar, S.N. (1991) editor. *Orban's Oral Histology and Embryology*, 11th edition. St. Louis: Mosby-Year Book, Inc.
2. Lavelle, C.L.B. (1988) *Applied Oral Physiology*, 2nd edition. London: Wright.
3. Mjör, I.A. and Fejerskov, O. (1986) editors, *Human Oral Embryology and Histology*. Copenhagen: Munksgaard.
4. Roth, G.I. and Calmes, R. (1981) *Oral Biology*. St. Louis: C.V. Mosby Company.
5. Schroeder, H.E. (1991) *Oral Structural Biology*. New York: Thieme Medical Publishers.
6. Scott, J.H. and Symons, N.B.B. (1982) *Introduction to Dental Anatomy*, 9th edition. Edinburgh: Churchill Livingstone.
7. Ten Cate, A.R. (1989) *Oral Histology: Development, Structure, and Function*, 3rd edition. St. Louis: C.V. Mosby Company.

Review questions

1. Write short notes on the functions of the oral mucosa.
2. Classify the oral mucosa on a regional basis.
3. Give a detailed description of an orthokeratinized epithelium.
4. Enumerate the differences between an orthokeratinized epithelium and a parakeratinized epithelium.
5. Briefly describe the morphology and function of melanocytes, Langerhans cells and Merkel cells.
6. Briefly describe the morphology of the hard palate.
7. Give a detailed description of the papillae found on the dorsum of the tongue.
8. Give a detailed description of a taste bud as seen with the electron microscope.

15. Calcium metabolism and bone mineralization

General remarks

Calcium is of such major biological importance that at least two hormones and one vitamin are specifically concerned with the regulation of its metabolism. Calcium plays an indispensable part in the following physiological processes:

1. Mineralization of the skeleton;
2. Mineralization of teeth;
3. Control of excitability of membranes of nerves and transmission of impulses at synapses:
4. Control of impulse transmission at myoneural junctions;
5. Stabilization of cell membranes and adhesion between cells;
6. Activation of enzymes involved in inflammation and in the clotting mechanism;
7. Secretion of hormones;
8. Synthesis of DNA in nuclei and of proteins in microsomes;
9. Cell division; and
10. Milk production.

The adult human body contains approximately 1.1 kg calcium (1.5% of total body mass) of which 98-99% is in the skeleton and the teeth. Although little calcium is therefore present in the general cells and body fluids, this ion plays a major role in many cell functions. The plama level is normally maintained within close limits and ranges between 2.25 and 2.75 mmol/l (9.2-10.8 mg/100ml), comprising 40% bound to protein (mainly albumin, but also globulin), 50% free or ionized, and 10% complexed with citrate, phosphate, and bicarbonate. The ionized and complexed calcium (60%) represents the diffusable fraction, while the protein-bound fraction (40%) is non-diffusable.

Calcium absorption

Calcium is relatively poorly absorbed from the gut and 70% of ingested calcium is excreted in the feces. The mechanism whereby calcium absorption from the intestinal tract, chiefly in the duodenum, is regulated, is not completely understood. Absorption is retarded by substances which form insoluble salts with calcium, for example, phosphates and oxalates, or by alkalis. A high protein diet, on the other hand, promotes absorption of calcium.

Calcium is actively absorbed from the intestinal tract by epithelial cells. The active transport of calcium across the intestinal wall is enhanced by the presence of vitamin D which acts directly on the mucosal epithelium of the duodenum and jejunum. The term "vitamin D" refers to a group of related sterols formed by the action of ultraviolet light on certain provitamins. In the skin of mammals cholecalciferol (vitamin D_3) is formed by the action of sunlight on 7-dehydrocholesterol (provitamin D_3). Vitamin D_3 is transported to the liver where it becomes altered to a metabolite of the vitamin, 25-hydroxycholecalciferol. This substance is hydroxylated further in the kidney to produce 1.25-dihydroxycholecalciferol (a potent antirachitic substance), which acts on the nuclei of intestinal epithelial cells to form the appropriate *m*RNA which dictates the sythesis of calcium-binding protein. This protein (Ca BP) causes calcium to pass across the wall of the intestine. Vitamin D_3 is the natural vitamin D and is synthesized in the body, while vitamin D_2 (ergocalciferol) is synthetic and is not manufactured in the body. A Vitamin D deficiency causes poor calcium absorption from the gut, resulting in hypocalcemia. In infants and young children rickets is caused, a condition characterized by abnormal endochondral ossification and the formation of deforming cartilagenous swellings at the ends of the long bones. The legs cannot support the body and become bowed.

Hormonal influences

Parathyroid hormone (parathormone, PTH) increases plasma calcium levels, mobilizes calcium from the skeleton, and raises the urinary excretion of phosphate. The degree to which PTH is secreted varies inversely with the level of ionized calcium in the plasma, and the main function of PTH is the maintenance of a constant Ca^{++} level in extracellular fluid. Calcitonin, a hormone produced by the parafollicular C cells of the thyroid gland in mammals, is a hypocalcemic hormone which is secreted when plasma calcium ion levels are elevated. These two hormones are responsible for maintaining calcium homeostasis.

It was earlier believed that an increase in plasma calcium levels, caused by PTH secretion, was secondary to a lowering of plasma phosphate through an increased urinary excretion of the latter, and that the action of the hormone was based on its phosphaturic effect.

An inversely proportional ratio exists between calcium and phosphorus in the plasma, and the product of the concentrations of Ca and P (in g/l) is normally 40-55 in children, and 30-40 in adults. Parathormone has a direct mobilizing effect on calcium in the skeleton whilst promoting renal reabsorption of calcium. The latter effect of PTH is, however, not of sufficient magnitude to prevent increased urinary loss of calcium in states of hyperparathyroidism. Parathormone also increases the absorption of calcium from the intestinal tract in the presence of vitamin D.

Circulating ionized calcium has a direct action on the parathyroid gland to regulate secretion of PTH. When the calcium level is high, secretion is inhibited and calcium is deposited in the skeleton. Low levels, on the other hand, stimulate secretion. Raised plasma phosphate levels stimulate PTH secretion as a result of the lowered calcium level and not as the result of a direct action of phosphate on the parathyroid gland.

Calcitonin is a hormone which lowers both the plasma calcium and phosphate levels by directly inhibiting bone resorption. Calcitonin is relatively inactive in normal adults but may play an active role in skeletal development in children. In birds and fish it is secreted by the ultimobranchial body which is embryologically derived from the fourth pharyngeal pouch. In mammals, these calcitonin-producing cells have migrated into the thyroid gland where they form the parafollicular, or C, cells. Calcitonin is also secreted by the thymus gland. Although this hormone is sometimes known as thyrocalcitonin, the term calcitonin is preferred.

Vitamin D is necessary for the action of both calcitonin, which inhibits bone resorption, and parathormone, which promotes resorption. A complex mechanism involving all three these substances ensures calcium homeostasis in the body and the maintenance of calcified tissues.

Osteogenesis (chemical aspects)

The mineral component of bone is hydroxyapatite, with the empiric formula $Ca_{10}(PO_4)_6.(OH)_2$. It is largely made up of calcium and phosphate, while ions of magnesium, sodium, potassium, and carbonate are also present, mainly in adsorbed form.

Bone is composed of an organic matrix and mineral salts, mainly calcium. Ordinary compact bone contains approximately 25% organic matrix by mass, consisting largely of collagen (97%), 3% ground substance, and 75% minerals.

The exact mechanism by which new bone matrix calcifies is not clear. Although bone can form by intramembranous or endochondral ossification, the basic process is the same: mineral salts are deposited in an organic matrix. The first sign of bone formation is an increased local vascularity of the mesenchyme. The local mesenchymal cells differentiate into osteoblasts. These cells possess the characteristics of secretory cells (RNA associated with a well-developed rough ER, and a Golgi apparatus), as well as an alkaline phosphatase activity. The osteoblasts synthesize the ground substance and the collagen, constituting the organic matrix. Inorganic salts reach the bone-forming site by way of the blood vascular system and tissue fluid. For the mineral salts to reach the bone matrix, they must pass through the osteoblasts. The salts are transferred across the cell membranes of osteoblasts into their interior through the action of alkaline phosphatase and, once inside, mitochondria may be involved in storing the minerals. Protoplasmic buds of the osteoblast cell membranes contain mineral salts and these form matrix vesicles which pass into the matrix of collagen and ground substance. The first needle-like apatite crystals appear in these vesicles but the vesicles rupture in due course with continued growth of the crystals and the crystals are liberated to form crystal clusters. Adjacent clusters then fuse to form a mass of apatite crystals within and around collagen fibers, initially according to a fiber-controlled pattern since the minerals are deposited at definite points within the 64 nm periodicity of the fibers. In the above way the organic matrix is mineralized. When the mineralized mass encloses the osteoblasts during continued new bone formation, these cells are known as osteocytes.

The above description of the mineralization process is a synthesis of current views on the subject and the reader is referred to the bibliography for detailed discussions.

Osteoclasts are bone-resorbing cells. They are large multinucleated cells with a well-developed acid phosphatase activity and contain many lysosomes, mitochondria, and a Golgi apparatus. The mechanism whereby bone is resorbed by osteoclasts is not clearly understood. Their action may be preceded by surface fibroblasts liberating collagenase and other proteolytic enzymes under certain conditions which may be hormone or pressure-related, thereby resorbing the osteoid layer and exposing the mineralized surface of the bone. The osteoblasts may similarly be influenced and, by exposing the minerals, exert a stimulating and chemotactic effect on osteoclasts. Proteolytic enzymes of these cells degrade the organic matrix further and an ideal acid pH is created for dissolution of apatite and uptake by pinocytosis.

Under normal circumstances a perfect balance is maintained between the rate of bone resorption and bone formation, but resorption predominates under conditions of localized pressure, and deposition under conditions of direct tension. Immobilization of a limb leads to demineralization of bone.

Osteogenesis (morphological aspects)

Osteogenesis takes place by intramembranous or endochondral ossification.

Intramembranous osteogenesis (Fig 142)

The term "intramembranous" was first used by early anatomists who described new bone formation in the connective tissue membrane which spans the fontanelles in fetal skulls. It may be regarded as a simpler form of bone formation. The first signs of new bone is a membrane-like mass composed of small spindle-shaped mesenchymal cells arranged parallel to one another to form smooth layers, as in the flat bones of the skull, or arranged in the form of irregular condensations, as in the maxilla.

In further development, these mesenchymal cells are responsible for the synthesis of collagen. (According to some authors the mesenchymal cells differentiate into fibroblasts which form the collagen.) The mesenchymal cells then differentiate into osteoblasts which are arranged on the surface of bundles of collagen fibers and continue the process of collagen synthesis. These cells increase in size and become cuboidal with cytoplasmic processes which may form desmosomal attachments with processes of adjacent cells.

Formation of the bone matrix, or osteoid, is completed by accumulation of ground substance between the collagen fibers. Osteoid has a high content of proteoglycans and is a region where polymerization of collagen molecules to form fibers is enhanced. The appearance of osteoid is rapidly followed by deposition of mineral salts in the form of needle-shaped crystals by the osteoblasts inside the osteoid. There is a concomitant increase in capillary blood supply to the area of new bone formation. This will in due course supply the bone marrow.

As the area of ossification increases, greater numbers of mesenchymal cells differentiate into osteoblasts, collagen and ground substance is formed in greater quantities, mineralization continues, and the mass of bone grows dynamically. As more and more bone matrix is formed and mineralization continues, the osteoblasts and their processes become trapped in the bone and are then called osteocytes. They lie in lacunae. Intramembranous bone forms in several foci resulting in the formation of interlinked spicules and trabeculae of irregular sizes. These grow by apposition of new bone at their surfaces, resulting in a mass of bone of a spongy (cancellous) nature. On the outer surfaces of the bone, compact cortical bone forms when surface trabeculae increase in size and the spaces between them become obliterated, resulting in the formation of, for example, the inner and outer tables of the skull. The outer surface of the cortical bone is covered by a membrane composed of an inner layer of osteoblasts and their mesenchymal precursors, or osteoprogenitor cells, and outer fibrous tissue. These together form the periosteum which has a rich blood supply. The inner cellular layer of the periosteum is responsible for the continued differentiation of new osteoblasts and the bone grows by surface apposition.

Fig 142 — Intramembranous osteogenesis

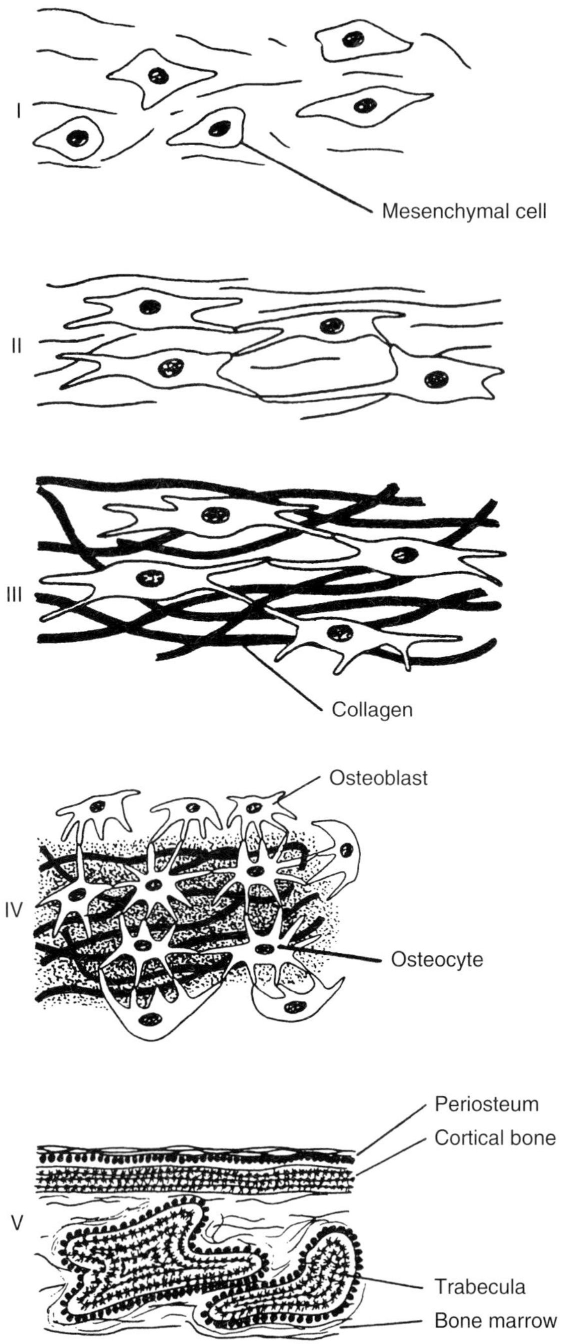

I — Loose membranous tissue composed mainly of undifferentiated mesenchymal cells.

Mesenchymal cell

II — The cells arrange into layers with their long axes parallel to each other.

III — Collagen synthesis starts and the mesenchymal cells differentiate into osteoblasts.

Collagen

IV — Mineral salts (stippled) are deposited in the osteoid. Osteoblasts are found on the surface of the new bone and have also been incorporated inside the bone as osteocytes.

Osteoblast
Osteocyte

V — Bony trabecula with surface osteoblasts and enclosed osteocytes are covered on the surface by cortical bone and periosteum.

Periosteum
Cortical bone
Trabecula
Bone marrow

Where inner spongy bone persists, trabeculae stop increasing in size and the connective tissue separating them becomes the marrow. The osteoblasts and the condensed connective tissue surrounding the bone trabeculae becomes part of the endosteum.

Bone which forms intramembranously basically consists of an inner mass of spongy bone enclosed by a surface layer of compact bone of periosteal origin.

Endochondral osteogenesis (Fig 143)

This type of osteogenesis can best be explained by a description of the formation of a long bone. Such a bone is preceded by a cartilagenous model which mimics the shape of the future bone fairly accurately. It is surrounded by a perichondrium which consists of well-vascularized connective tissue and perichondrial cells which retain the ability to form new cartilage. The cartilagenous model must be removed before bone can form.

Osteogenesis is initiated when a primary ossification center appears in the middle of the model in the region of the future shaft, or diaphysis. This involves the following processes. The cartilage cells, or chondrocytes, which are arranged in columns, enlarge together with their lacunae, and their cytoplasm becomes vacuolated. The intervening matrix becomes calcified, thus reducing the flow of nutrients to the cells which degenerate and die. While the above events are taking place, the perichondrium surrounding the center of the shaft assumes osteogenic properties and becomes known as the periosteum. Osteoblasts differentiate in the inner layer of the periosteum and a periosteal collar of compact, or cortical, bone forms around the center of the shaft. This collar is therefore of intramembranous origin.

A vascular connective tissue bud, containing osteogenic cells and chondroclasts, invades the central part of the shaft (future diaphysis) and reaches the calcified cartilage matrix. The thinned calcified lacunar partitions between the degenerated chondrocytes are removed by chondroclasts, resulting in narrow parallel tunnels between the remaining columns of calcified matrix. The tunnels are occupied by the blood vessels and osteogenic cells which rapidly line the surface of the remaining cartilage matrix and differentiate into osteoblasts. They begin to form osteoid which is soon mineralized, resulting in bone. Early trabeculae unite with one another (and with the bony periosteal collar) and are composed of cores of cartilage surrounded by bone. Remodeling of these trabeculae results in removal of all the remaining calcified cartilage matrix and establishment of the diaphysis.

In the meantime, chondrogenesis continues on each side of the diaphyseal ossification center to replenish cartilage which progressively becomes ossified. In this way the bony shaft elongates in both directions.

The ends, or epiphyses, of a long bone usually ossify after birth with the appearance of epiphyseal, or secondary, centers of ossification. These centers arise in the same way as the diaphyseal center by proliferation of cartilage, death of chondrocytes, and subsequent spread of ossification in all directions. A mass of cancellous bone results.

Fig 143 — Endochondral osteogenesis

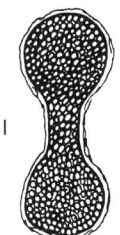

I — A cartilagenous model, surrounded by a perichondrium, exists before ossification of a long bone commences.

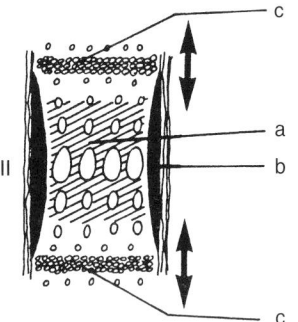

II — Sketch II represents mineralization of the cartilage matrix (a) and formation of a collar of periosteal bone (b) around the diaphysis. Active chondrogenesis (c) continues on both sides of the mineralized cartilage and is responsible for elongation of the model.

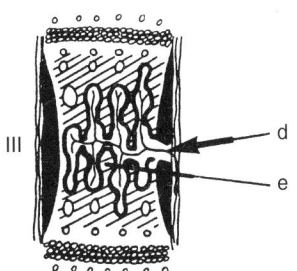

III — Sketch III shows the ingrowing vascular connective tissue bud (d) to give rise to the primary ossification center (e) by deposition of bone (thick black lines) on the remaining mineralized cartilage matrix. This is followed by remodelling and establishment of cancellous bone.

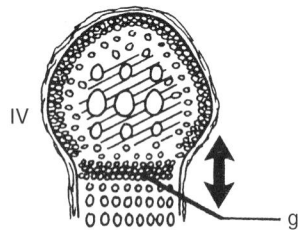
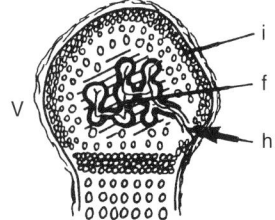

IV, V — Sketches IV and V show the appearance of an epiphyseal ossification center (f) in the same way as in the diaphyses. An epiphyseal plate (g) is responsible for continued chondrogenesis and the arrows in sketch IV indicate the directions of proliferation of cartilage. A vascular bud (h) is responsible for the ossification centre. The epiphysis remains surrounded by articular cartilage (i) until the growth process terminates.

All the cartilage of the epiphysis ultimately becomes replaced by bone except for a zone between the diaphysis and the epiphysis. This zone is known as the epiphyseal plate, and continues to proliferate to permit growth in length of the bone. Cartilage also persists on the periphery of the epiphysis as the articular cartilage.

The end of growth of the bone in length is heralded when chondrogenesis in the epiphyseal plate ceases and it becomes completely replaced by bone to unite the diaphysis and the epiphysis. The entire bone, except for the articular surfaces, is surrounded by a periosteum which is responsible for cortical bone growth. Cortical bone consists of surface lamellar bone and deeper Haversian systems which are continuous with the more centrally situated cancellous bone.

The reader is referred to the bibliography for a detailed discussion on remodeling processes in bone.

Important information

Recommended daily intake of calcium

1. Under 1 year: 0.4-0.6 g/day
2. 1-18 years: 0.7-1.4 g/day
3. Above 18 years: 0.75 g/day
4. During pregnancy and lactation: 1.5-2.0 g/day

The above figures were accepted for many years but it is reported that the World Health Organisation has, on evidence supplied by global surveys, reduced the recommended daily intake to 0.5-0.6 g for infants, older children and adults, and 1.0-1.2 g/day during pregnancy and lactation.

Dietary sources of calcium

The calcium content in mg/100 g of a variety of common (uncooked) foods are given in the table below:

Hard cheese	800
Green vegetables	25-250
Shelled nuts	13-250
Dried pulses	40-200
Fish	20-120
Root vegetables	20-100
Cow's milk	120

White bread (70% extraction fortified)	100
Soft cheese	80
Eggs	56
Oatmeal	55
Human milk	30
Wholemeal bread (100% extraction)	25
Maize	12
Potatoes	8
Rice	6

Dietary sources of phosphorus

Meat, dairy products, eggs and cereals are all good sources of phorphorus which occurs abundantly in plant and animal tissues.

Selected bibliography

1. Cole, A.S. and Eastoe, J.E. (1988) *Biochemistry and Oral Biology,* 2nd edition. London: Wright.
2. Krause, W.J. and Cutts, J.H. (1986) *Concise Text of Histology,* 2nd edition. Baltimore: Williams and Wilkins.
3. Lavelle, C.L.B. (1988) *Applied Oral Physiology,* 2nd edition. London: Wright.
4. Osborn, J.W. and Ten Cate, A.R. (1983) *Applied Dental Histology,* 4th edition. Bristol: Wright PSG.
5. Ten Cate, A.R. (1989) *Oral Histology: Development, Structure, and Function,* 3rd edition. St. Louis: C.V. Mosby Company.
6. West, J.B. (1991) editor. *Best and Taylor's Physiological Basis of Medical Practice,* 12th edition. Baltimore: Williams and Wilkins.

Review questions

1. Discuss the factors which influence calcium absorption from the intestinal tract.
2. Describe the synthesis and function of vitamin D_3.
3. Discuss the functions of parathyroid hormone and calcitonin in calcium metabolism.
4. Discuss bone mineralization in general.

5. In what forms is calcium found in the body?
6. What are man's daily requirements of calcium?
7. Enumerate the functions of calcium in the human body.
8. Describe intramembranous osteogenesis.
9. Describe endochondral osteogenesis.

(It is advisable to refer to the chapter on connective tissue, including bone and cartilage (Part I, Chapter 5) for a revision of bone morphology.)

16. The jaws

General remarks

The skull forms the skeleton of the head (Fig 144). The upper part of the skull is the cranium, which houses the brain, while the lower facial part of the skull consists of the bones surrounding the orbits, nose, and mouth. The frontal bone, two parietal bones, occipital bone, two temporal bones, sphenoid bone, and ethmoid bone constitute the cranium, while the facial skeleton is composed of two nasal bones, two maxillary bones, two lacrimal bones, two zygomatic bones, two palatine bones, two inferior turbinates, the vomer, and the mandible. The maxilla and the mandible will be described.

The maxilla

The maxillae form the major part of the upper jaw (Fig 144). For the sake of convenience, only one maxillary bone will be described but relations of the maxillae to one another and to other bones will be mentioned where applicable.

The maxilla consists of a body and four processes: frontal, zygomatic, alveolar, and palatine. The body of the bone is partly hollowed out and contains the maxillary air sinus, or antrum, which communicates with the middle meatus of the nose. The deepest (inferior) part of the sinus is usually at the level of the root apices of the second premolar and first molar teeth. This can be of clinical significance since it is not unusual to expose the lining of the sinus during extraction of the upper premolar and molar teeth. For this reason, an extraction socket in this region should be inspected routinely and adequate measures taken if exposure of the lining, or the sinus itself, has resulted. Furthermore, the proximity of the sinus to these teeth often makes it difficult to distinguish between true dental pain and pain resulting from inflammation of the mucous membrane of the sinus which very often has an irritating effect on nearby dental nerves.

The anterior, or facial, surface of the maxilla shows vertical ridges caused by underlying roots of teeth. The most prominent of these ridges is caused by the root

Fig 144
The jaws
Anterior view of the skull (norma frontalis)

The relation of the maxilla (oblique lines) to the other bones of the skull is illustrated.

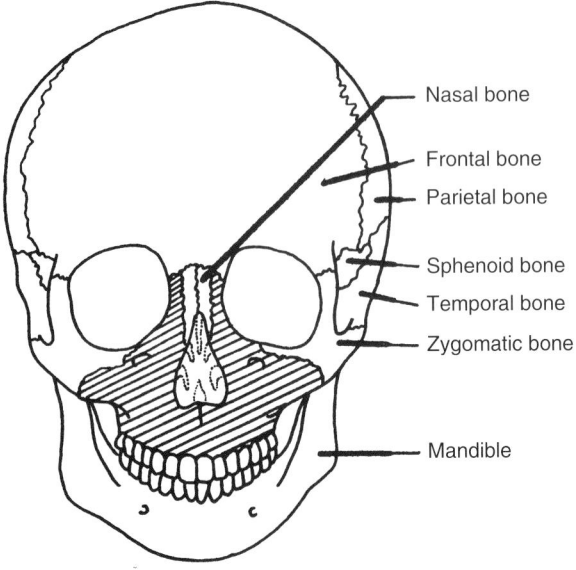

Maxilla (facial surface)

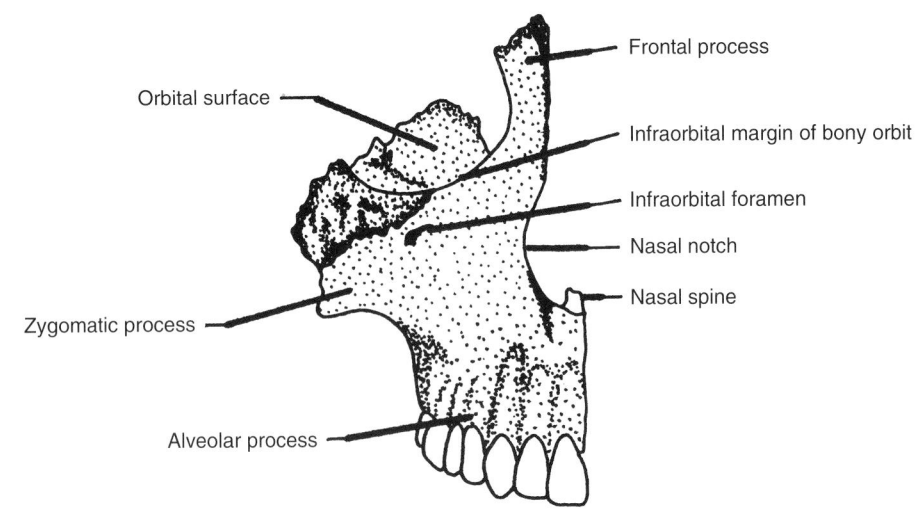

Fig 145 **Orbit**

Fig 146 **Maxilla (posterolateral surface)**

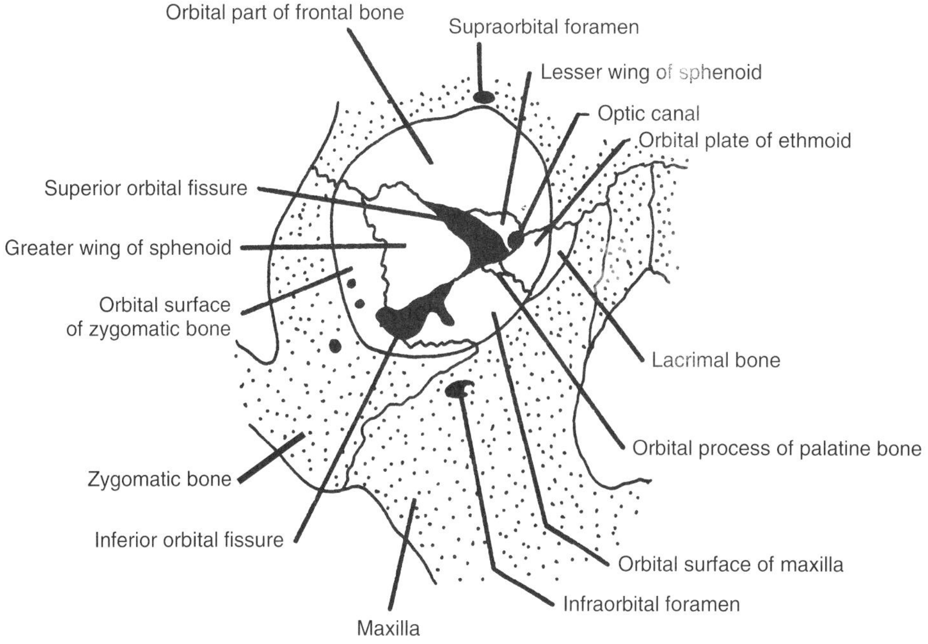

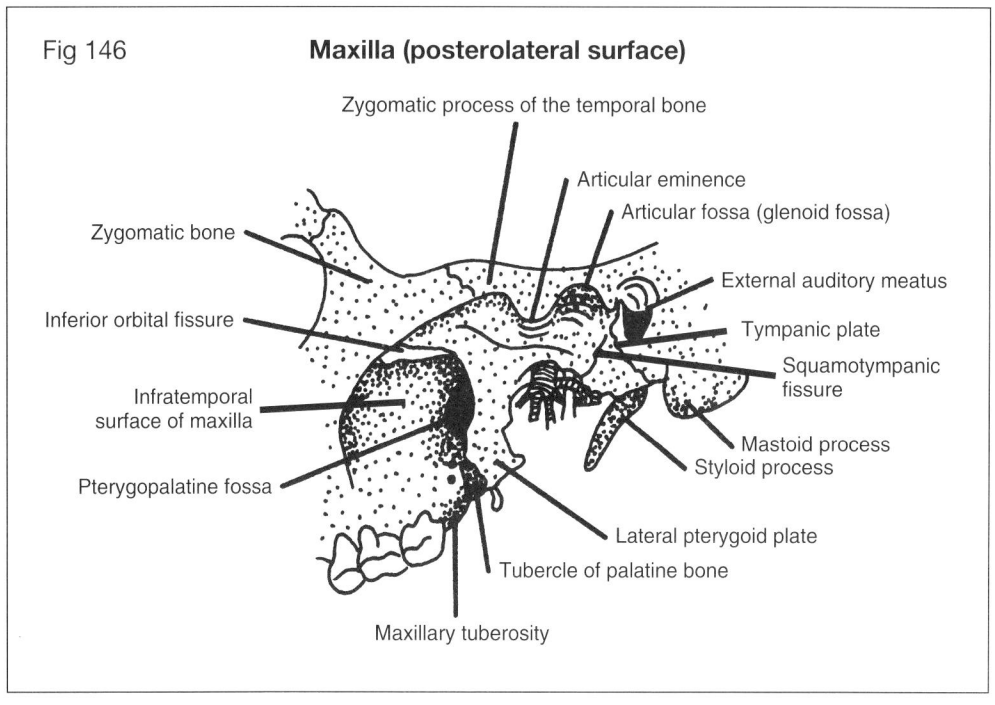

of the canine tooth. The infraorbital foramen, which transmits the infraorbital neurovascular bundle (the terminal branches of the maxillary nerve, the maxillary (retromandibular) artery, and veins), is situated high up on the facial surface. The upper border of the facial surface is formed by the inferior margin of the bony orbit. The inferior border of this surface bears the teeth and constitutes the alveolar process. The medial border of the maxilla is the nasal notch which forms the sides and floor of the external nasal opening and ends in the nasal spine below. The lateral part of the facial surface of the maxilla is formed by its zygomatic process.

The posterior (infratemporal) surface of the maxilla forms one of the walls of the infratemporal fossa laterally (Fig 146), while it faces the pterygopalatine fossa in its medial part. On its infratemporal surface several small foramina transmit the branches of the posterior superior alveolar nerve and blood vessels which are distributed to the molar teeth, while the maxillary tuberosity is situated lower down. This articulates with the tubercle of the palatine bone which is wedged between the tuberosity and the pterygoid process of the sphenoid bone.

The nasal surface of the maxilla is directed towards the nose and contains the opening of the maxillary sinus which is overlapped posteriorly by the vertical plate of the palatine bone. The sinus is closed anterosuperiorly by the lacrimal bone. The nasolacrimal canal, containing the nasolacrimal duct, or tear duct, is found in the articulation between the maxilla and the lacrimal bone. The sinus is closed above by the labyrinth of the ethmoid bone, while below it is closed by the inferior turbinate bone.

The major part of the floor of the bony orbit (Fig 145) is formed by the orbital surface of the maxilla. The infraorbital groove is found in the posterior aspect of this surface and is continued forwards into the infraorbital canal which transmits the corresponding neurovascular tissue.

The frontal process of the maxilla (Fig 145) extends superiorly and articulates with the frontal bone above, the nasal bone anteriorly, and the lacrimal bone posteriorly. The lacrimal fossa for the lacrimal gland is found on the orbital surface of the frontal process, partly on both the maxilla and the lacrimal bone.

The zygomatic process articulates with the zygomatic bone. The two alveolar processes constitute the maxillary dental arch which is occupied by the maxillary teeth. The alveolar processes are discussed fully in the chapter on the form, arrangement and chronology of teeth (Part II, Chapter 11), while the histology of the alveolar process is dealt with in the chapter on the alveolar process (Part II, Chapter 17).

The palatine process (Fig 147) is directed medially to form two-thirds of the hard palate. Its palatal surface is roughened by the attachment of the mucoperiosteum and shows an indistinct shallow groove in the angle formed between the alveolar and palatine processes. This groove leads forward from the greater palatine foramen which is found in the articulation between the maxilla and palatine bone. The greater palatine canal and foramen transmit the corresponding neurovascular bundle. The medial edge of the palatine process articulates with its fellow of the opposite side to form a midpalatal suture. The incisive canal opens in the anterior part of this suture behind the central incisors through the incisive foramen which transmits the greater palatine artery from the hard palate to the nose, and the terminal branches of the nasopalatine nerve and blood vessels from the nasal cavity to the hard palate. The two incisive canals commence in the anterior nasal floor and open through a single foramen. A suture running from the incisive canal towards the space between the lateral incisor and canine teeth can be seen on both sides in a young skull, separating the premaxillary area from the main mass of the maxilla.

The reader is advised to revise the chapter on development of the septomaxillary complex (Part I, Chapter 15).

The mandible

At birth the mandible (Fig 144) consists of right and left halves which are connected in the symphyseal area by cartilage and fibrous tissue. The two halves undergo bony union during the first year of life.

The mandible consists on each side of a horizontal body and a vertical ramus (Figs 148, 149). The body is horseshoe-shaped with a subcutaneous lower border which shows a shallow groove across which the facial artery runs. The upper border of the body is the alveolar process which houses the teeth. On the facial aspect of the symphysis of the adult mandible a low ridge is found. It broadens inferiorly on both sides to form the central triangular mental protuberance and is continued downwards to form the mental tubercles.

On the internal aspect of the symphysis a small protuberance, the mandibular spine, is present but does not have a constant shape. It often consists of four genial tubercles, or they may be fused, and lies above a shallow digastric fossa.

The outer aspect of the body (Fig 148) has an oblique line extending downwards and forwards from the anterior border of the ramus. The mental foramen, which transmits the mental neurovascular bundle, is found approximately midway between the upper and lower borders of the body in the region of the apex of the second premolar tooth.

The inner surface of the body (Fig 149) is divided by the mylohyoid line, which runs obliquely forwards and downwards from just below the last molar tooth to the region of the genial tubercles, into a sublingual fossa for the sublingual salivary gland above, and a submandibular fossa, for the submandibular gland below.

The ramus of the mandible ends above in a pointed coronoid process anteriorly and a condylar, or condyloid, process which bears the mandibular condyle, posteriorly (Fig 148). These two processes are separated by the mandibular notch. The lateral surface of the ramus is relatively smooth, except in the area of the

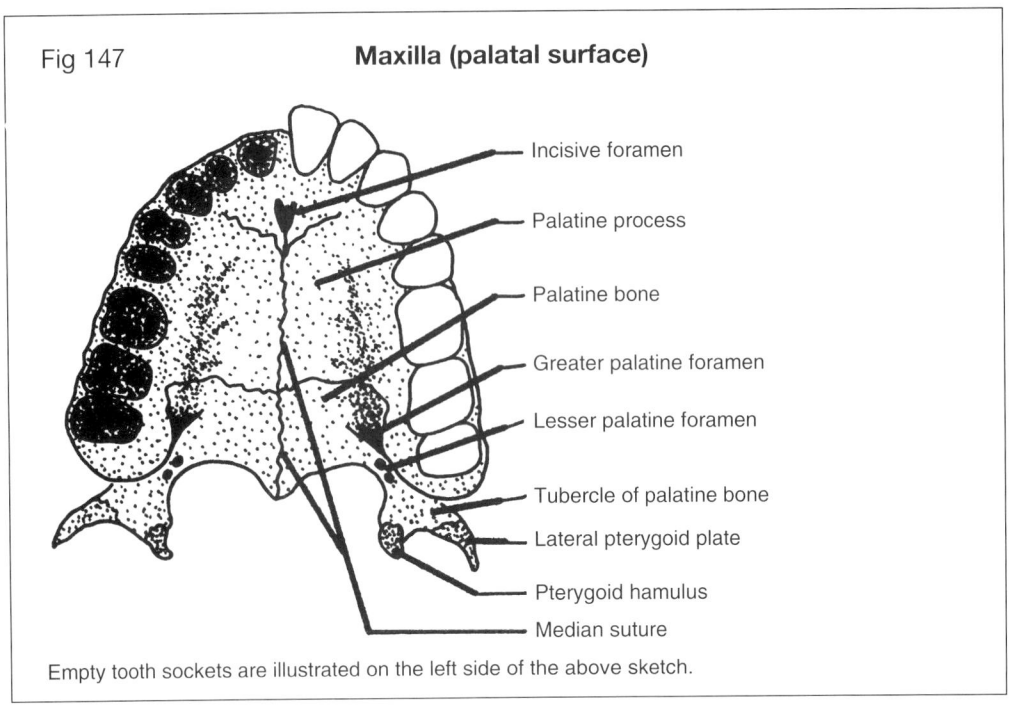

Fig 147 Maxilla (palatal surface)

Empty tooth sockets are illustrated on the left side of the above sketch.

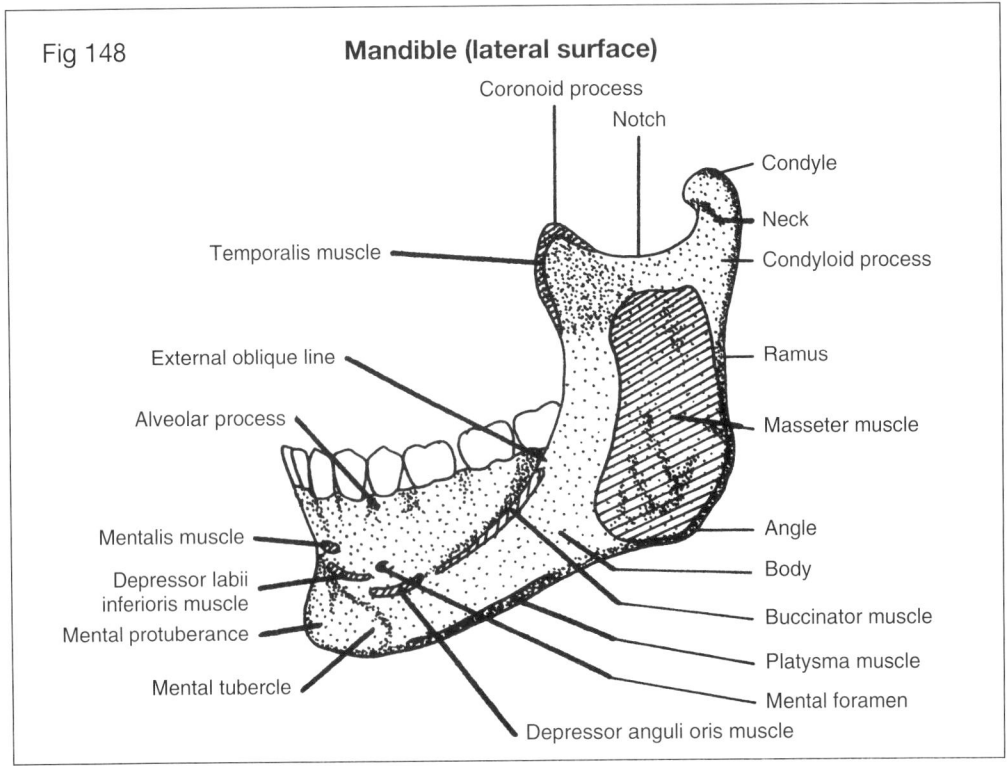

Fig 148 Mandible (lateral surface)

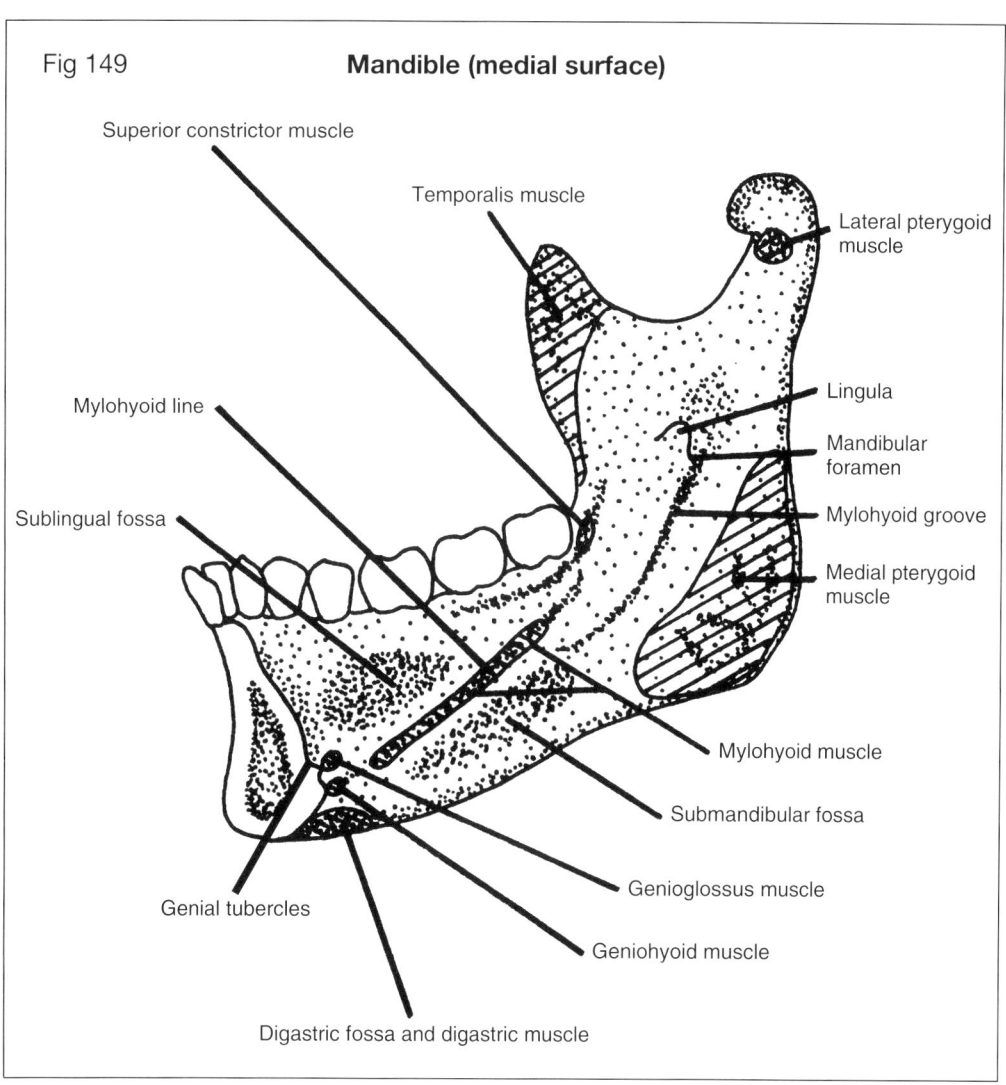
Fig 149 Mandible (medial surface)

angle where it is roughened by the attachment of the masseter muscle. The medial surface is roughened inferiorly by the attachment of the medial pterygoid muscle. The mandibular foramen, which admits the inferior alveolar nerve and vessels to the mandibular canal to supply the bone and the lower teeth, is situated in the middle of the medial surface and is overlapped by a small pointed protuberance, the lingula. This affords attachment to the sphenomandibular ligament. From the mandibular foramen a shallow groove runs downwards and forwards parallel to the mylohyoid line for a short distance. This mylohyoid groove transmits the mylohyoid branches of the inferior alveolar nerves and vessels. The posterior border of the ramus meets the lower border of the body at the angle of the mandible.

The condylar process is narrowed to form the neck which ends in the expanded condyle. This articulates with the mandibular, or glenoid, fossa and the posterior slope of the articular eminence of the temporal bone by means of the articular disc, or meniscus, to form the temporomandibular joint. This joint is fully discussed in the chapter on the temporomandibular joint (Part II, Chapter 21).

The position of muscle attachments are illustrated in the accompanying sketches.

Selected bibliography

1. Dixon, A.D. (1986) *Anatomy for Students of Dentistry*, 5th edition. Edinburgh: Churchill Livingstone.
2. McMinn, R.M.H. (1990) editor. *Last's Anatomy*, 8th edition. Edinburgh: Churchill Livingstone.

Any general text on gross anatomy can be consulted for a detailed description of the maxilla and mandible.

Review questions

1. Describe the maxilla.
2. Describe the mandible.
3. Describe the bony architecture of the orbit.

17. The alveolar process

General remarks

The alveolar process (Fig 150) is that part of the jaw in which the teeth are found. Most authors use the term alveolar bone to describe only the compact cortical bone lining the tooth sockets, and reserve the term supporting bone for the remainder of the tooth-bearing area, or alveolar process. The alveolar process is therefore composed of both alveolar bone and supporting bone. The remainder of the jaw-bone supporting the alveolar process is basal bone. In orthodontics a distinction is often made between genetically determined unchangeable basal bone and "functional" bone of the alveolar process which may be altered by means of orthodontic procedures. Some authors use the term alveolar bone in a broader sense to encompass those parts of the jaws bearing the teeth as well as the alveoli and crypts of developing teeth.

After removal of teeth the alveolar process undergoes resorption which, if severe, results in an expansion of the floor of the maxillary sinus into the edentulous alveolar process. Only a thin layer of cortical bone may eventually separate the oral mucosa from the sinus lining in the maxilla. In the edentulous mandible, alveolar resorption may be so marked that the mylohyoid and external oblique lines approach the upper border of the bone. In extreme cases the mental foramen may be found on the upper border. The clinical significance of this is obvious.

Structure of the alveolar process and the bone of the jaws

Bone formed in the fetal jaws is irregular and a cortical layer of compact bone is absent. A thick, very active, periosteum surrounds fetal bone which is characterized by coarse fibers and is sometimes known as woven bone.

After eruption of the teeth the alveolar bone assumes an adult pattern and consists of a definite surface layer of cortical bone covering a mass of cancellous, or spongy, bone. Both are largely made up of orderly arranged bone lamellae with

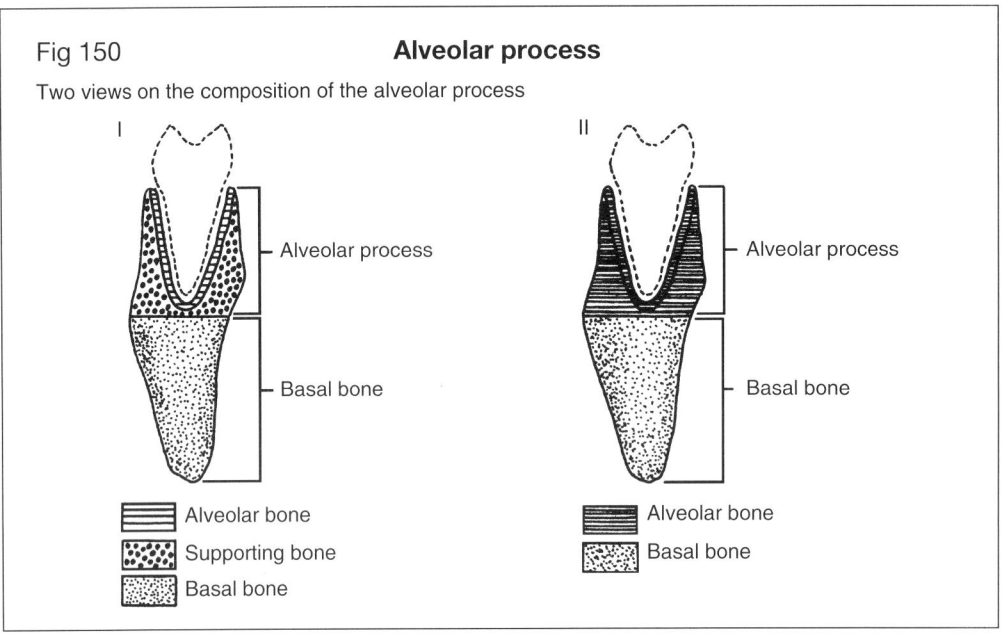

Fig 150 **Alveolar process**
Two views on the composition of the alveolar process

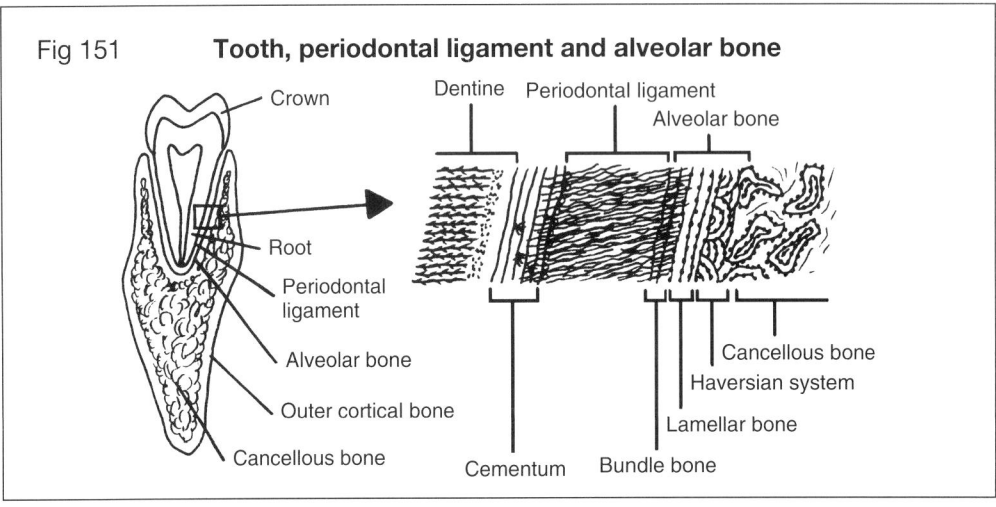

Fig 151 **Tooth, periodontal ligament and alveolar bone**

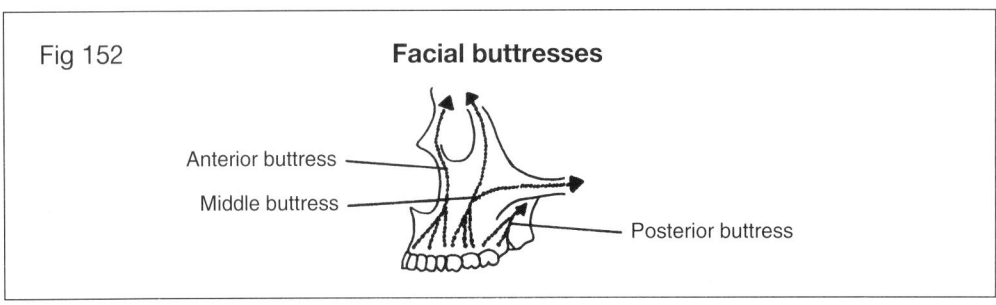

Fig 152 **Facial buttresses**

osteocytes, but the arrangement of the lamellae is different in the two types of bone. Lamellar bone contains fine intrinsic fibers.

The surface cortical bone on the buccal and lingual surfaces of the jaws is continuous with the cortical bone lining the sockets (alveolar bone) at the openings of the tooth sockets. Alveolar bone is radiologically known as the lamina dura because of its radiopacity. Alveolar bone is slightly thinner than outer cortical bone and contains numerous small openings (Volkmann's canals) through which nerves and blood vessels reach the periodontal space from the interior of the bone. For this reason it is also known as the cribriform plate.

The superficial layers of both the alveolar bone and the outer cortical bone under the periosteum consist of lamellar bone. In the sockets, however, the rough-fibered surface lamellar bone contains the Sharpey's fibers of the periodontal ligament (Fig 151), and is known as bundle bone. Sharpey's fibers are also found where periodontal ligament fibers are embedded in cementum. The deeper layers of alveolar bone are made up of Haversian systems, or osteons, in which concentric lamellae surround the blood vessels of a Haversian canal.

Cancellous bone is found deep to the Haversian system between the alveolar bone and the surface cortical layer, and between the alveolar bone of adjacent tooth sockets. It is composed of irregular interlacing bone trabeculae, each consisting of one or more lamellae with osteocytes enclosed in lacunae with radiating canaliculi. Bone marrow separates the trabeculae. The relative amounts of cancellous bone present are largely determined by functional forces applied to the bone.

The bone forming the outer wall of the sockets in the incisor and canine areas of both jaws is thin and composed of very little more than outer cortical bone and inner alveolar bone without intervening cancellous bone. The quantity of cancellous bone increases progressively in a posterior direction.

Trabeculae are constantly remodeled to adapt to changing circumstances and forces applied to the bone. For example, the quantity of cancellous bone is reduced around unopposed teeth although the laminae dura remain unchanged. On the other hand, the cancellous bone surrounding isolated teeth upon which excessive forces are exerted during mastication, is dense with numerous and thicker trabeculae.

The facial buttresses (Fig 152)

Although the maxillary alveolar process is wholly supported by the upper jaw, the forces exerted by mastication are transmitted to the cranium by way of three buttress systems on either side. In these buttresses the cancellous bone is well developed and the cortical bone is thicker than in the rest of the bone. Maxillary cortical bone never reaches the thickness of cortical bone in the mandible, except in the buttresses, while the amount of cancellous bone in the maxilla is largely determined by the size of the maxillary air sinus.

The frontal process of the maxilla, which lies anterior to the maxillary sinus and lateral to the nasal cavities, forms the anterior buttress of the upper facial skeleton. Forces exerted on the incisor and canine teeth during mastication are transmitted to the frontal bone through this buttress.

The zygomatic process of the maxilla, and the zygomatic bone, form the middle buttress and transmit forces to the frontal bone (via the frontal process) around the outer margin of the orbital cavity, and to the base of the skull via the zygomatic arch, from the molar teeth.

Posteriorly, the maxilla is supported mainly by the pterygoid plates of the sphenoid bone and these, together with the palatine bone, constitute the posterior buttress. The upper alveolar process is also greatly reinforced by the arch of the hard palate.

The buttresses described above ensure an efficient spread of masticatory forces in a bone which is closely associated with several large cavities such as the orbits, nasal cavities, and maxillary air sinuses, and at the same time provide a firm base against which the mandible can act during mastication.

Selected bibliography

1. Bhaskar, S.N. (1991) editor. *Orban's Oral Histology and Embryology,* 11th edition. St. Louis: Mosby-Year Book, Inc.
2. Dixon, A.D.(1986) *Anatomy for Students of Dentistry,* 5th edition. St. Louis: C.V. Mosby Company.
3. Mjör, I.A. and Fejerskov, O. (1986) editors, *Human Oral Embryology and Histology.* Copenhagen: Munksgaard.
4. Osborn, J.W. (1981) editor. *Dental Anatomy and Embryology.* Oxford: Blackwell Scientific Publications.
5. Osborn, J.W. and Ten Cate, A.R. (1983) *Advanced Dental Histology,* 4th edition. Bristol: Wright PSG.
6. Scott, J.H. and Symons, N.B.B. (1982) *Introduction to Dental Anatomy,* 9th edition. Edinburgh: Churchill Livingstone.
7. Ten Cate, A.R. (1989) *Oral Histology: Development, Structure, and Function,* 3rd edition. St. Louis: C.V. Mosby Company.

Review questions

1. Explain the meaning of the following terms:
   (a) alveolar bone;
   (b) supporting bone;
   (c) alveolar process; and
   (d) basal bone.
2. Describe the histological features of alveolar bone.
3. Describe the buttresses of the maxilla.

18. The periodontium

General remarks

The periodontium is composed of those tissues that surround and support the tooth in its natural and healthy functional state, and consists of the gingiva and gingival fiber groups (often referred to as the gingival ligament), cementum with embedded Sharpey's fibers, periodontal ligament fiber groups (the dento-alveolar fibers, or principal fibers) and alveolar bone with embedded Sharpey's fibers. The tooth is attached to the alveolar bone by a resilient suspensory fibrous attachment mechanism, the periodontal ligament fibers, which resist normal functional forces of mastication and enable the tooth to adapt to stress. The gingiva provides an epithelial collar around the neck of the tooth and separates those parts of the tooth exposed to the oral environment from the underlying delicate connective tissues. The relationship between the tooth and the gingiva is of major importance due to the essential role played by the dentogingival junction in the maintenance of healthy tooth support. It is of the utmost importance to regard the periodontium as a structural and functional entity and to describe it as such.

All the components of the periodontium, except the gingiva which is of ectodermal origin, arise in the dental follicle surrounding the developing tooth. The undifferentiated ectomesenchymal cells of the follicle differentiate into cementoblasts, fibroblasts, and osteoblasts, and in this way give rise to cementum, gingival, and periodontal fibers, and alveolar bone. This has been substantiated experimentally. In one such experiment tooth germs, including the follicles, of newborn mice were grafted subcutaneously in adult mice. The tooth germs continued forming the hard dental tissues as well as surrounding periodontal ligament and alveolar bone. After these findings, further experiments showed that tooth germs transplanted into a prepared defect in the parietal bone of adult mice, formed alveolar bone which united with the bony edges of the defect, thus simulating tooth socket development. The dental follicle has considerable potential in the above regard.

The gingiva - a general description

The gingiva (Fig 153) is that part of the oral mucosa surrounding the teeth. It is relatively firmly bound to the buccal and lingual plates of the alveolar processes. On the facial aspect of both jaws, the gingiva is separated from the more mobile lining mucosa by a scalloped line, the mucogingival junction. A similar but less distinct demarcation line separates the mandibular gingiva from the lining mucosa of the floor of the mouth, but in the maxilla the gingiva imperceptibly merges with palatal masticatory mucosa. The gingiva normally has a pale pink color but may show areas of endogenous pigmentation.

The most coronal part of the gingiva is not directly attached to the tooth or the alveolar process, and is known as the free, or marginal, gingiva. The free gingiva is separated from the attached gingiva in about 50% of subjects by a shallow free gingival groove which is, at best, not very noticeable. The free gingiva is separated from the tooth surface by a shallow slit-like space, the gingival sulcus (Fig 154). The deepest part of the sulcus is at the level of, or slightly coronal to, the free gingival groove (when present). An interdental papilla is present on the lingual and buccal sides of each interproximal space and the two papillae are connected by a thin mucosal septum with a central depression below the contact area. This depression is known as a col.

At the gingival crest the histological architecture of the gingiva undergoes a fairly abrupt change from a keratinized epithelium on the buccal and lingual aspects, to a non-keratinized crevicular, or sulcular, epithelium lining the sulcus. The crevicular epithelium is continued towards the enamel-cementum junction as the junctional, or attached, epithelium which is in intimate contact with the enamel surface by means of hemidesmosomes. The free (coronal) surface of the junctional epithelium forms the floor of the sulcus.

In conditions of health, the clinical depth of the gingival sulcus is 0.5-2.0 mm, although many clinical researchers maintain that a sulcus is the result of irritation of the dentogingival junction and can be virtually eliminated under conditions of perfect oral heath. This is slightly more than the histological depth (0.5 mm) and the difference can be attributed to a slight penetration of the dentogingival attachment by measuring instruments.

In approximately 40% of subjects, the attached gingiva on the facial aspect of the maxilla (Fig 153) shows fine pin-point hollows which give the appearance of stippling. These stipples may be caused by traction on the mucosa by underlying fibrous attachments to the bone. Disappearance of these stipples may be caused by inflammation of the gingiva (gingivitis) and edema.

Masticatory gingival epithelium is stratified squamous with an underlying connective tissue lamina propria. These two tissues are separated by a basal lamina which follows the undulating contours of the many long and narrow rete ridges which are characteristic of attached gingiva. These ridges become shorter in the transition from attached gingiva to lining mucosa.

Fig 153   **Gingiva**

Anterior view of the maxillary incisor region

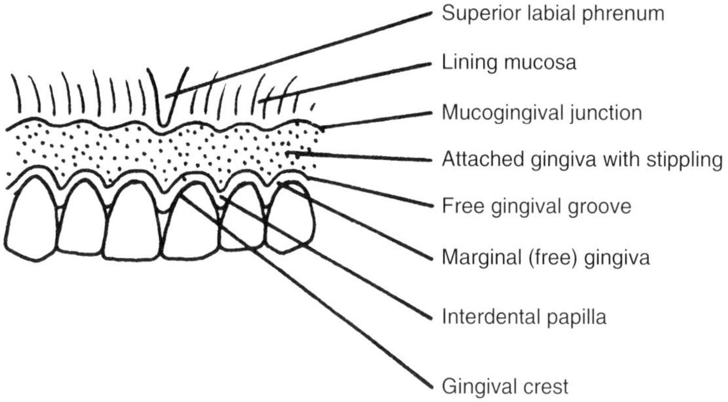

- Superior labial phrenum
- Lining mucosa
- Mucogingival junction
- Attached gingiva with stippling
- Free gingival groove
- Marginal (free) gingiva
- Interdental papilla
- Gingival crest

Fig 154   **Longitudinal section through a tooth, alveolar bone, and gingiva**

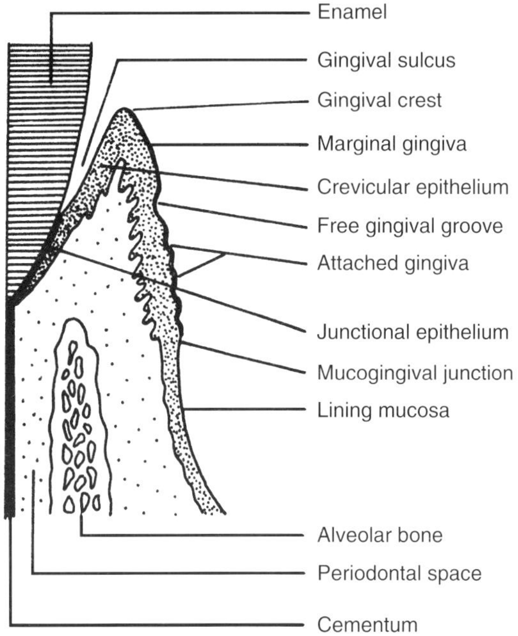

- Enamel
- Gingival sulcus
- Gingival crest
- Marginal gingiva
- Crevicular epithelium
- Free gingival groove
- Attached gingiva
- Junctional epithelium
- Mucogingival junction
- Lining mucosa
- Alveolar bone
- Periodontal space
- Cementum

(Details of general connective tissue have been omitted in this sketch)

The cell population of gingival epithelium undergoes constant renewal at a slower rate than intestinal epithelium, but more rapidly than skin. Mitotic division occurs in the basal cell layer and probably also in the deeper cells of the stratum spinosum. New cells continually move to the surface to replace the cells which desquamate.

The basal lamina which separates the epithelium from the lamina propria appears as a thin amorphous line, the basement membrane, by means of a light microscope, but several zones are seen under an electron microscope. It is composed of an extracellular electron dense finely granular or filamentous layer, the lamina densa, which is separated from the basal cell membrane by a clear zone, the lamina lucida. The lamina densa and lamina lucida constitute the basal lamina, while the term basement membrane includes a third layer, a deeper reticular lamina, containing finely banded reticulin fibrils (anchoring filaments). These filaments are inserted into the lamina densa. Collagen fibers from the lamina propria interlace with the reticulin fibers to attach the epithelium firmly to the lamina propria.

The basal cell layer (stratum germinativum) of masticatory epithelium consists of a row of low columnar or cuboidal cells which rest on a basal lamina. They contain a fairly large oval to round nucleus and many tonofibrils (bundles of tonofilaments) which are inserted into an attachment plaque, an electron dense intracellular thickening associated with the inner aspect of the cell membrane. The tonofibrils form loops within the attachment plaque and return to the interior of the cell. The plaques are associated with desmosomal attachments between cells (Fig 155), and hemidesmosomal junctions between the basal cells and the lamina propria (Fig 156).

The intercellular desmosomal attachments are more obvious in the stratum spinosum, since the intercellular spaces are wider, giving prominence to the desmosomes. This results in a spiny or prickle-like profile. The stratum granulosum is found superficial to the stratum spinosum in orthokeratinized epithelium and consists of cells which are flattened and contain dark keratohyalin granules. As the cells approach the surface, the nuclei become smaller and appear pyknotic, while many cytoplasmic organelles are lost. In the surface layers of orthokeratinized epithelium, the cells have flattened and lost their nuclei and organelles, and are packed with keratin. This represents the final stage in epithelial cell differentiation and gives rise to the stratum corneum.

In parakeratinized epithelium, the stratum granulosum is usually absent and the squames of the stratum corneum retain flattened and pyknotic nuclei.

When present, pigmentation of the gingiva is usually present in the attached gingiva in line with the interdental papillae. The color of the pigment (melanin) varies from light brown to almost black. Melanin is the product of melanocytes found in the basal layer and is transmitted to epithelial cells as small granules (melanosomes) through dendritic cell processes.

The reader is referred to the chapter on oral mucosa (Part II, Chapter 14) for a detailed description of the various cell types and pigmentation.

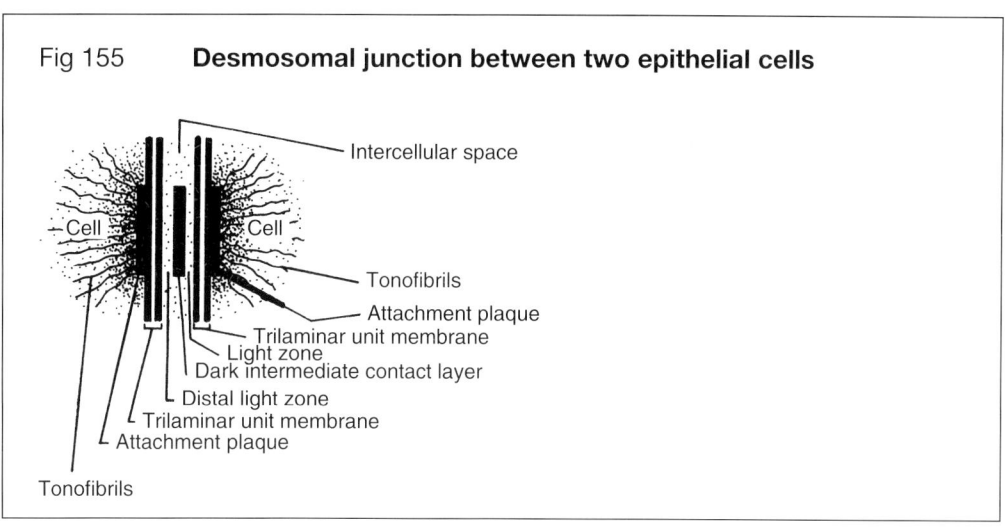

Fig 155 **Desmosomal junction between two epithelial cells**

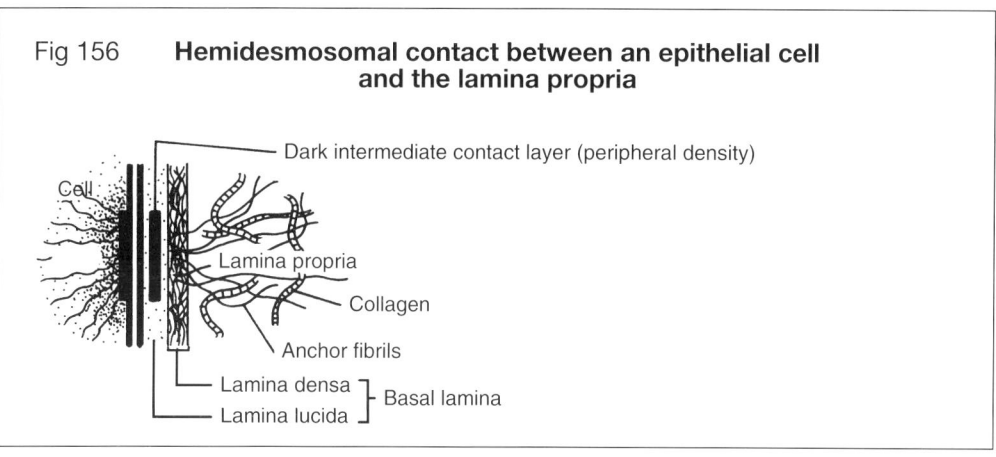

Fig 156 **Hemidesmosomal contact between an epithelial cell and the lamina propria**

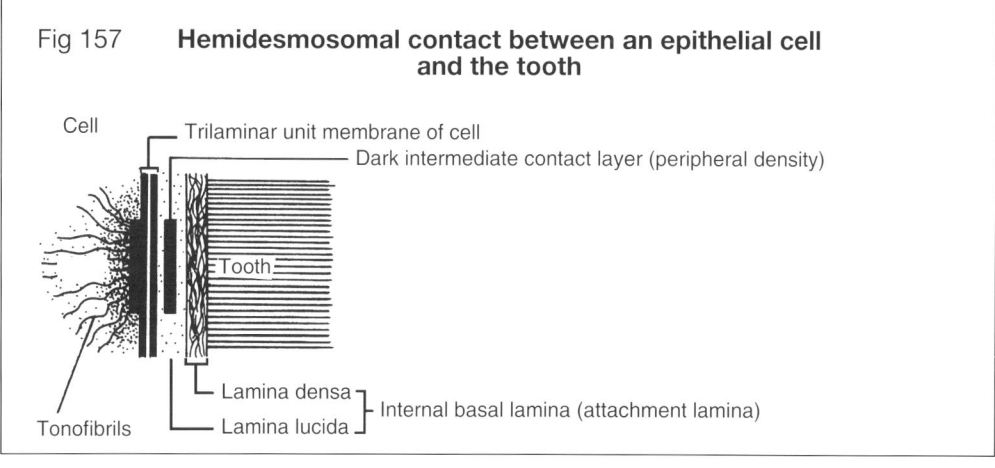

Fig 157 **Hemidesmosomal contact between an epithelial cell and the tooth**

The gingival collar

The junctional epithelium and crevicular epithelium, together with the related gingival fiber groups and lamina propria, constitute the gingival collar. The tightness of the collar is dependent upon factors such as hemidesmosomal adherence of junctional epithelium to enamel (Fig 157), and later possibly to cementum, tissue tension resulting from blood pressure and tension exerted by the gingival fiber groups.

The epithelium of the gingival collar consists of two structurally different regions (Fig 154). Coronally, the crevicular epithelium forms the outer lining of the gingival sulcus. This non-keratinized epithelium merges with keratinized masticatory epithelium of the gingiva at the gingival margin (crest). Rete ridges are absent in crevicular and junctional epithelium. Junctional epithelium is the apical extension of the crevicular epithelium and is attached to the tooth by hemidesmosomes.

Crevicular (sulcular) gingival epithelium

Crevicular epithelium is a non-keratinized stratified squamous epithelium. The cells are relatively small with little cytoplasm and endoplasmic reticulum, and are separated by very narrow intercellular spaces. The long axes of the more superficial cells are arranged parallel to the free surface.

Junctional epithelium

Junctional epithelium represents the apical extension of the crevicular epithelium and is attached to the tooth surface by hemidesmosomes. It is embryologically derived from the reduced enamel epithelium. It varies considerably in thickness (2-15 cells), being thinnest at its cervical end, and has a straight junction with the lamina propria from which it is separated by a basal lamina (sometimes referred to as an external basal lamina or external attachment lamina). Typical hemidesmosomal attachments are present. Junctional epithelium is regarded as an immature epithelium since it does not appear to undergo terminal differentiation. The cells of junctional epithelium are flattened cells which lie with their long axes parallel to the tooth surface to which they are attached by means of an attachment apparatus consisting of a basal lamina and hemidesmosomes. This basal lamina is sometimes referred to as the internal basal lamina or internal attachment lamina and was in earlier years known as the secondary enamel cuticle.

Junctional epithelium is characterized by widened intercellular spaces and the relative absence of desmosomes. Monocytes and polymorphonuclear cells are commonly found in the intercellular spaces as they migrate from the lamina propria to the gingival sulcus. On the other hand, the wide spaces allow antigens to enter the epithelium and reach the lamina propria. There is a continuous coronal migration of cells of the junctional epithelium which are ultimately shed into the sulcus.

In many subjects a gradual apical displacement of the dentogingival attachment occurs with time. Previously this apical shift was regarded as a physiologic process and called passive eruption, but it is currently regarded as resulting from mechanical and bacterial irritation of the supporting tissues of the tooth.

Gingival fibers (gingival ligament)

The lamina propria supporting the junctional epithelium and deeper parts of the crevicular epithelium is different from that underlying oral gingival epithelium due to the presence, even in clinically normal gingiva, of a population of inflammatory cells. It is thought that this inflammatory process arises at the time of tooth eruption.

Collagen fibers, which play a very important role in maintaining the integrity of the supporting apparatus of the tooth, are found in the lamina propria underlying the crevicular and junctional epithelia. Although these gingival fibers intermingle to a large extent and act as a functional unit, they are nevertheless arranged in the following five groups (Fig 158):

1. The *dentogingival group*. These extend from cervical cementum into the lamina propria underlying both the junctional epithelium and the marginal gingiva.
2. The *dentoperiosteal group*. These fibers arise in cementum and run over the bony alveolar crest to insert into either the periosteum or muscle slips in the floor of the mouth or vestibule.
3. The *trans-septal group*. These fibers arise in cervical cementum and connect adjacent teeth over the interdental bony alveolar crest. The trans-septal fibers connecting adjacent teeth together constitute an interdental ligament system connecting all the teeth of the dental arch. Delayed reorganization of these fibers has been implicated in post-retention relapse of orthodontically repositioned teeth in cases of insufficient retention periods.
4. The *alveologingival group*. These fibers arise in the bony alveolar crest and extend into the lamina propria underlying the attached and free (marginal) gingiva.
5. The *circular group*. These fibers run around the neck of the tooth in the lamina propria. They intermingle with the other fiber groups and, together with the dentogingival fibers, play an important role in maintaining a tightly-fitting gingival collar. These fibers are also known as the marginal ligament.

Fig 158 **Gingival fibers**

The gingival collar area (longitudinal section)

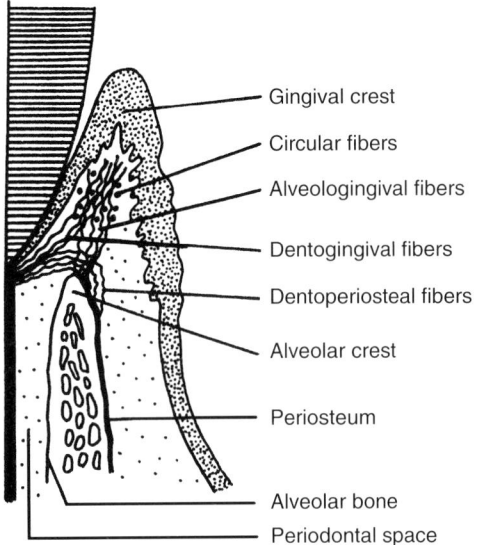

- Gingival crest
- Circular fibers
- Alveologingival fibers
- Dentogingival fibers
- Dentoperiosteal fibers
- Alveolar crest
- Periosteum
- Alveolar bone
- Periodontal space

The interdental gingiva (longitudinal section)

The interdental gingiva contains all the fiber groups illustrated above, plus the trans-septal fibers.

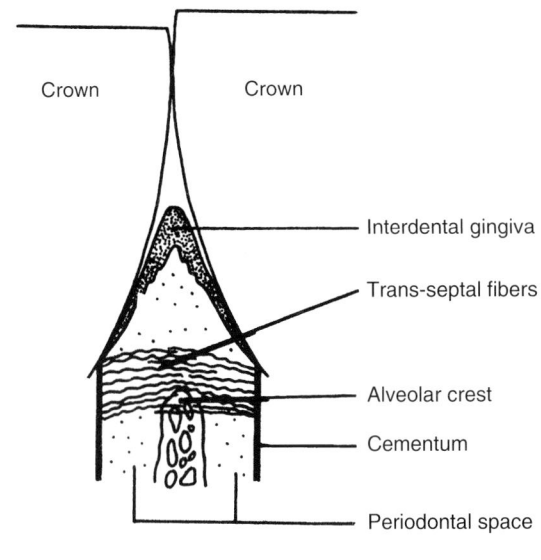

Crown   Crown

- Interdental gingiva
- Trans-septal fibers
- Alveolar crest
- Cementum
- Periodontal space

Fibroblasts are the most numerous connective tissue cells in the region and are found in large numbers between the collagen fiber bundles. Few elastic fibers are present and are found in association with blood vessels. In most subjects a leukocyte infiltration of varying intensity is present in the junctional area between the attached and free epithelia. These cells, especially polymorphonuclear leukocytes, are frequently found in the intercellular spaces between the epithelial cells and migrate from the lamina propria to the gingival sulcus, probably in response to chemotactic infuences exerted by plaque and bacteria. Some authors, however, maintain that such a subclinical inflammation is not present in the gingiva of subjects maintaining a strict state of oral hygiene, and that it should not be regarded as part of the normal histology of the area. This matter is more fully discussed later in this chapter.

The periodontal ligament - a general description

The periodontal ligament is the connective tissue which fills the periodontal space and binds the tooth to the alveolar bone. It is continuous with the connective tissue of the gingival collar and communicates with the bone marrow of the alveolar process through vascular channels (Volkmann's canals) in the alveolar bone. It communicates with the dental pulp at the apical foramen. Similar to connective tissue in general, the periodontal ligament consists of cells and an extracellular compartment of ground substance and fibers. The cell population is made up of fibroblasts, epithelial cell rests of Malassez (the remains of the epithelial root sheath of Hertwig), osteoblasts and osteoclasts associated with alveolar bone, cementoblasts associated with the cementum, macrophages, and undifferentiated mesenchymal cells. Glycosaminoglycans, glycoproteins, and glycolipids are the main constituents of the ground substance.

The collagen fibers crossing the periodontal space are embedded in alveolar bone on the one side and in cementum on the other and the embedded portions of the fibers are known as Sharpey's fibers. Through the collagen spanning the periodontal space, the ligament has a suspensory function and possesses proprioceptive nerve endings which respond to occlusal pressure changes. This property enables a person to apply masticatory forces in such a way that the threshold of tolerance of tooth supporting tissue is not overtaxed. The blood supply of the periodontium not only supplies the ligament itself, but is also important for the maintenance of cementum, alveolar bone, and the gingiva.

The width of the periodontal space varies from 0.1 to 0.4 mm in different subjects, in individual teeth of the same subject, and in different parts of the same tooth. Teeth subjected to considerable forces of mastication have a wider periodontal ligament than those with light occlusal forces, such as unopposed or impacted teeth. It is wider in primary teeth than in secondary teeth and progressively becomes narrower with increasing age.

The periodontal fibers

Collagen constitutes the most important structural component of the periodontal ligament. It is customary to describe the fiber bundles, often referred to as the principal fibers of the periodontium, in five functional and morphologic groups, as follows (Fig 159):

1. The *alveolar crest group.* These fibers arise in the cementum close to the amelocemental junction and run outwards to attach into the bony alveolar crest surrounding the tooth. They often pierce the alveolar crest and attach to the cementum of an adjacent tooth. In this respect they serve a function similar to that of the gingival trans-septal fibers.
2. The *horizontal group.* These fibers are found just apical to the alveolar crest fibers and run a horizontal course from the cementum to the alveolar bone below the alveolar crest. These fibers resist lateral pressure on a tooth.
3. The *oblique group.* These are the most numerous fibers in the periodontal ligament and run an oblique course from the cementum to the alveolar bone at a more coronal level. These fibers resist axial pressure on a tooth.
4. The *apical group.* This group consists of fibers radiating in all directions from the cementum surrounding the root apex to the apical alveolar bone.
5. The *inter-radicular group.* These fibers are found running from the cementum in the area of division of the roots of multirooted teeth to the crest of the inter-radicular bony septum.

In addition to the principal fibers described above, oxytalan fibers are found in the periodontal ligament, as well as in the gingival lamina propria. They are thought to be an immature type of elastic fiber with a distribution related to blood vessels. They run oblique from cementum, or bone, to the walls of blood vessels, generally perpendicular to the occlusal plane and the general principal fiber direction. The function of these fibers is unknown but it is thought that they may anchor and support blood vessels when the ligament is distorted during function.

The collagen fibers of the periodontal ligament follow a wavy course from cementum to bone. The fibers elongate when functional forces are applied to a tooth, thus allowing slight movements of the teeth despite the fact that collagen is

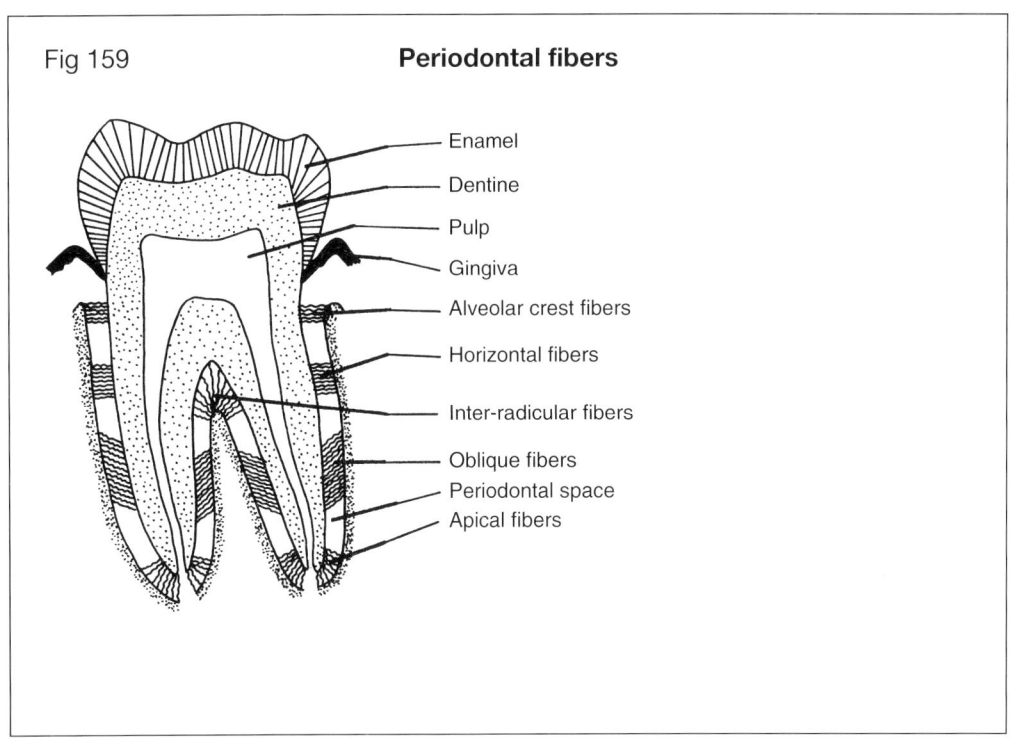

Fig 159 **Periodontal fibers**

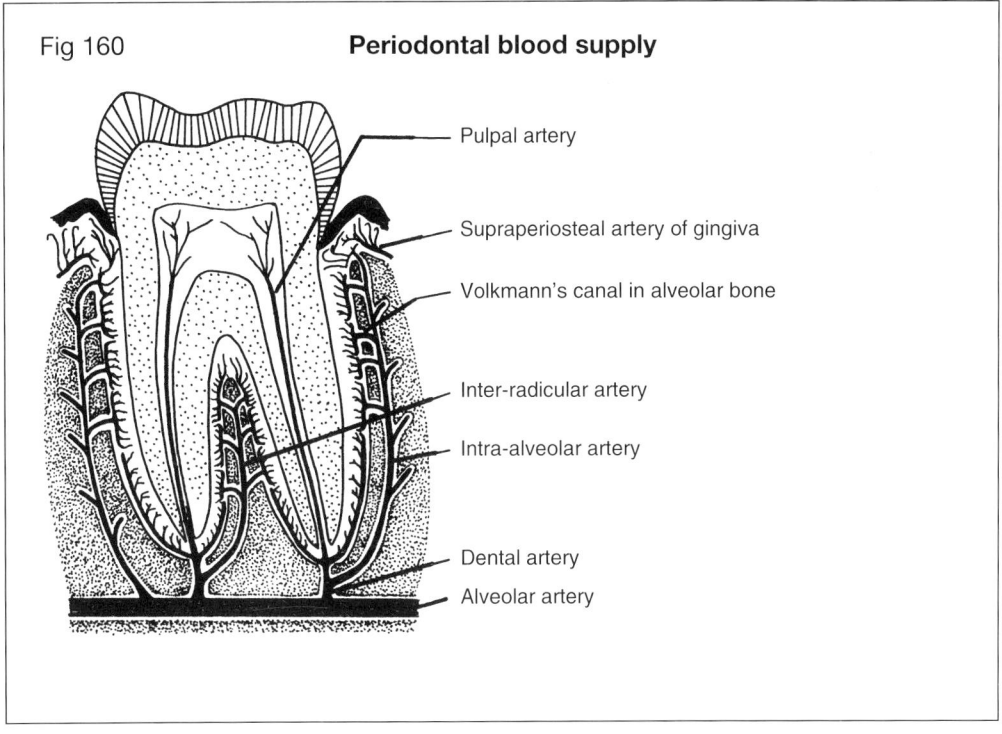

Fig 160 **Periodontal blood supply**

inelastic. Blood and ground substance in the periodontal ligament probably also resist and absorb light occlusal impacts.

All the fibers do not always follow the shortest course from the cementum to the alveolar bone. Some fiber bundles run obliquely to the sides on a horizontal plane and it is speculated that these fibers resist rotational forces applied to a tooth.

The electron microscope reveals that each collagen fiber is made up of numerous fibrils of uncertain lengths. The tropocollagen molecules within each fibril are longitudinally arranged in a staggered fashion, being held together by intermolecular bonds. The number of these bonds increases with age of the fiber, resulting in a progressive resistance by the fibers to environmental changes. Fibroblasts are the predominant cell type in the periodontal ligament and are seen as spindle-shaped cells between collagen fibers.

Blood and nerve supply of the periodontium

Blood supply to the gingiva (Fig 161)

Arteries which course superficial to the periosteum on the facial and lingual (palatal) aspects of the gingiva reach the gingival collar from branches of regional arteries found in the respective areas. In this way, the gingiva is supplied by branches of the infraorbital, nasopalatine, buccal, mental, lingual, and palatine arteries. Branches of these arteries form capillary loops in the connective tissue papillae between the rete ridges of the gingival epithelium before draining into veins. Vessels to the crevicular and junctional areas are mainly derived from intrabony arteries which perforate the alveolar bone close to the alveolar crest through very numerous Volkmann's canals. They form a rich capillary anastomosis with the gingival vessels previously described. The gingival collar can thus be regarded as having a dual blood supply.

Blood supply to the periodontal ligament (Fig 160)

Larger blood vessels of the ligament are found in the interstitial spaces between collagen fiber bundles and are located mainly in the peripheral parts of the periodontal ligament where a rich network of capillary arcades is formed. The arterial blood supply is derived from the superior and inferior alveolar arteries.

The apical region of the periodontal ligament is supplied by branches of the pulpal arteries of individual teeth, and arise immediately before the latter enter the apical foramen.

The inter-radicular interdental alveolar bone is supplied by branches of the

Fig 161 — Blood supply to the gingiva

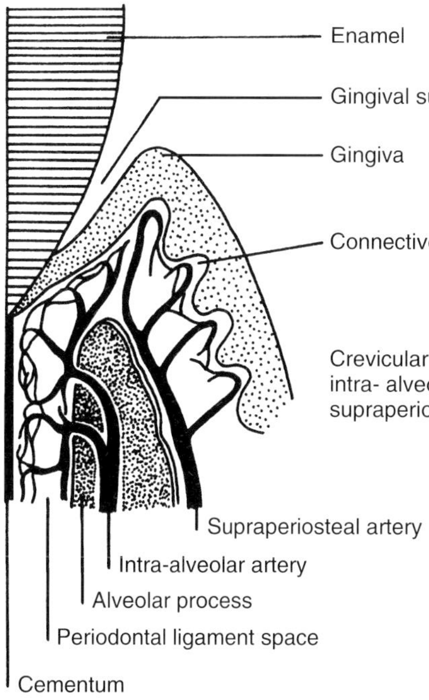

- Enamel
- Gingival sulcus
- Gingiva
- Connective tissue papillae (lamina propria)
- Supraperiosteal artery
- Intra-alveolar artery
- Alveolar process
- Periodontal ligament space
- Cementum

Crevicular and junctional epithelia are supplied mainly by intra-alveolar blood vessels, while the gingiva is supplied by supraperiosteal vessels. These vessels anastomose freely.

Fig 162 — Development of the gingival sulcus and epithelial attachment (initial changes)

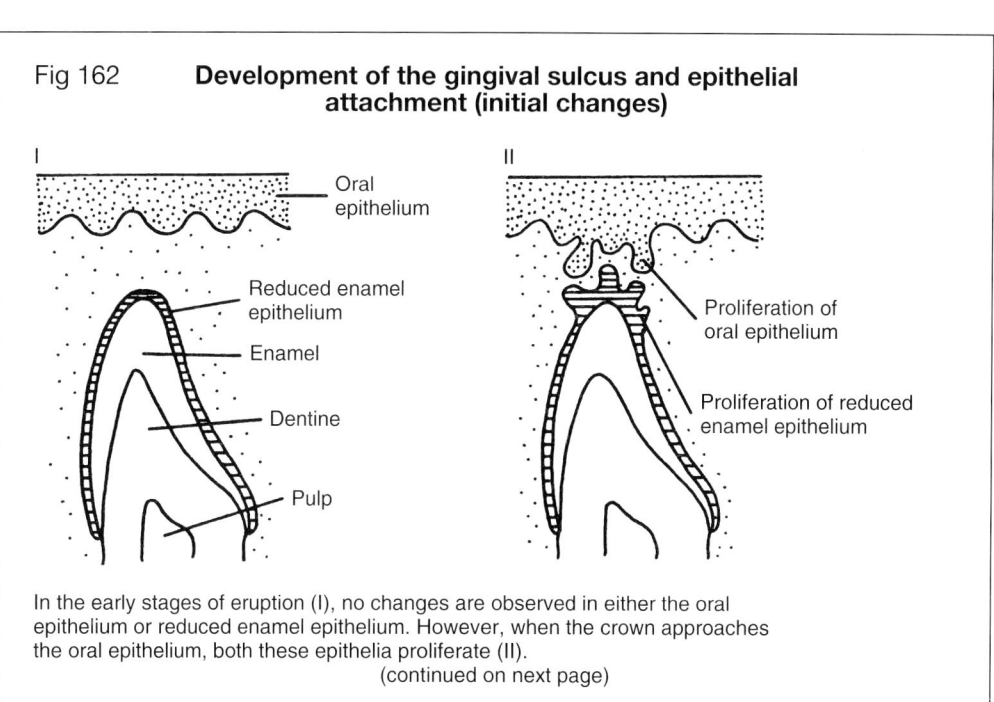

I
- Oral epithelium
- Reduced enamel epithelium
- Enamel
- Dentine
- Pulp

II
- Proliferation of oral epithelium
- Proliferation of reduced enamel epithelium

In the early stages of eruption (I), no changes are observed in either the oral epithelium or reduced enamel epithelium. However, when the crown approaches the oral epithelium, both these epithelia proliferate (II).

(continued on next page)

Fig 163

Development of the gingival sulcus and epithelial attachment (later changes)

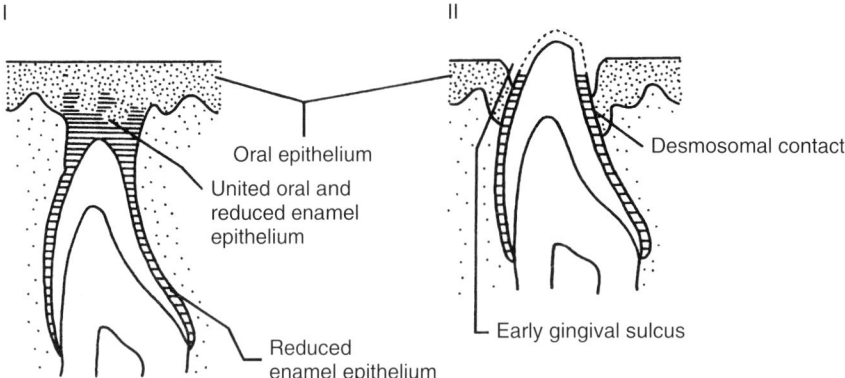

Desmosomal junctions between oral and reduced enamel epithelial cells are established to form a united epithelium (I). The tooth crown erupts through this united epithelium (II). The outer gingival epithelium maintains desmosomal contact with the reduced enamel epithelium which constitutes the early junctional epithelium. With further tooth eruption, cells of the reduced enamel epithelium are shed into the mouth.

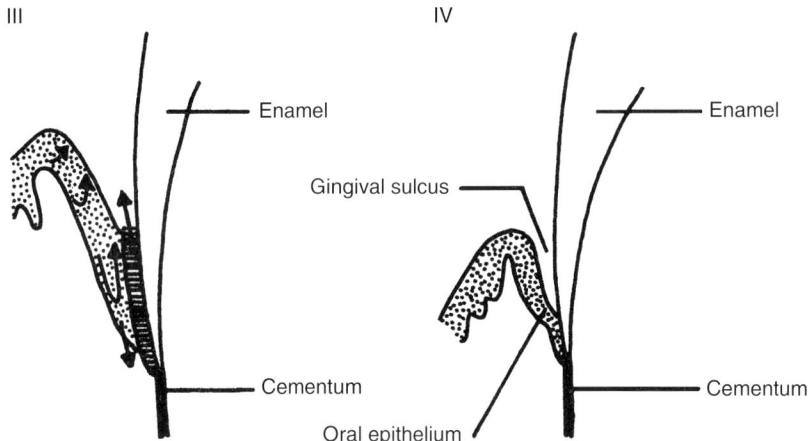

More advanced stages are depicted in III and IV. According to one view, the oral epithelium migrates cervically and replaces the reduced enamel epithelium in due course (arrows in III). In such an event, the junctional epithelium is eventually formed by oral epithelium (IV), which also forms the lining of the gingival sulcus (the crevicular epithelium). Other views are discussed in the text.

dental artery which are given off inside the bone before this artery reaches the apical periodontal tissues. These arteries, sometimes called perforating canals of Zukerkandl and Hirschfeld, course vertically in the alveolar process and provide the main arterial supply to the middle and cervical parts of the periodontal ligament via small perforations (Volkmann's canals) in the alveolar bone. Extensive anastomoses are formed between all the blood vessels of the periodontium. The veins of the region follow a course similar to that of the arteries. The lymphatics of the periodontium follow the course of blood vessels and drain to regional nodes.

Nerve supply to the periodontal ligament

The periodontal ligament has a rich nerve supply derived from either the superior alveolar nerve (maxillary teeth) or inferior alveolar nerves in mandibular teeth. These are branches of the maxillary and mandibular divisions of the trigeminal nerve.

It contains nerves which course from the apical region of the teeth towards the gingival collar. These are joined by nerves entering the ligament through foramina (Volkmann's canals) in the alveolar bone. Once inside the ligament, these fibers branch to run gingivally and apically. All the fibers branch extensively to end as pain receptors or mechanoreceptors.

The nerves of the periodontal ligament accompany blood vessels and are myelinated (large fibers) or unmyelinated (small fibers). It is believed that the small unmyelinated fibers, probably C fibers, end as free nerve endings and are concerned with pain perception, while the thicker myelinated fibers, probably A-δ fibers, end in specialized endings in the form of spiral coils around ligament fibers, or as encapsulated knob-like endings, or as combinations of the above. These endings are believed to be mechanoreceptors concerned with appreciation of touch and pressure, and are stimulated by movement of the tooth. These proprioceptors play an important role in masticatory physiology.

Metabolism of the periodontal ligament

The periodontal ligament is in a constant state of turnover in order to maintain a healthy and functional supporting apparatus for the tooth. New collagen fibrils form and are spliced into older fibers, or completely new fibers are formed by bundles of fibrils in the place of older fibers which are broken down and phagocytosed. Fibroblasts are responsible for both the synthesis and breakdown of collagen in the gingiva and the periodontal ligament. This activity takes place throughout the ligament but is reported to occur at a faster rate close to the alveolar bone. This is perhaps not surprising if it is borne in mind that the alveolar bone constantly undergoes alternating periods of resorption and apposition to minor changes in tooth position, while the cementum is probably not affected to the same extent.

Development of the periodontium

Development of the gingival sulcus and epithelial attachment

Before the tooth erupts, the crown is closely covered by the reduced enamel epithelium consisting of the remains of the dediferentiated ameloblasts and the other cells of the enamel organ. The reduced enamel epithelium is separated from the oral cavity by connective tissue and bone (Fig 162, I). These tissues have to be removed for normal tooth eruption to take place, but the precise mechanism of this process is not fully understood. It is believed that fibroblasts and osteoclasts are ultimately responsible for the removal of the intervening connective tissue and bone and the following discussion is a synopsis of current views.

Pressure created by the erupting tooth may cause cytoskeletal changes in fibroblasts with resulting changes in the function of these cells to facilitate collagenase activity. Lysosomal hydrolytic enzymes are secreted. A zone of lowered pH forms around these cells and the action of hyaluronidase and collagenase results in breakdown of ground substance and degradation of connective tissue fibers which are phagocytosed by the fibroblasts.

In a similar manner an osteoblast may be influenced by factors such as pressure to produce collagenase. By degrading the osteoid and creating an acid environment, bone minerals are partly exposed and exert a stimulating and chemotactic affect on osteoclasts. For the above to occur, the osteoblast must be induced to release osteoclast resorption stimulation activity (ORSA) and this could be brought about by pressure, and certainly is by the action of parathyroid hormone (PTH). It is thought that osteoclasts release proteolytic enzymes which reduce the matrix to smaller molecules which are taken up by the cells, together with mineral ions. Intracellular vesicles containing acid phosphatase further degrade the matrix. The osteoclasts contain numerous mitochondria and it is thought that they are instrumental in (a) providing the energy requirements (ATP) of the cell, (b) producing citric acid for resorption of mineral salts and (c) providing temporary storage for the released apatite ions. The reader is referred to the bibliography for detailed discussions on fibroblastic and osteoclastic activity.

As the tooth crown approaches the oral epithelium, marked localized proliferative activity occurs in both the oral epithelium and the outer layers of the reduced enamel epithelium (Fig 162,II). The two epithelia establish desmosomal contacts and fuse (the united enamel epithelium) (Fig 163,I). The cusps or incisal edge of the erupting tooth are at this stage separated from the oral cavity by a solid plug of epithelium. The central cells of the epithelial plug die, possibly as a sequel to pressure atrophy, creating an epithelium-lined channel through which the crown erupts further (Fig 163, II). The oral epithelium is in continuity with the reduced enamel epithelium around the periphery of this channel, thus protecting the underlying delicate lamina propria. A measure of uncertainty surrounds further developments.

A generally held view is that the reduced enamel epithelium is involved in the establishment of the initial junctional epithelium in a newly erupted tooth, and that the crevicular (sulcular) epithelium is derived from the oral mucosa.

Some authors believe that the reduced enamel epithelium component of the junctional epithelium is replaced by oral epithelium in due course. This view holds that the oral epithelium migrates apically around the outside of the reduced enamel epithelium (primary, or temporary, junctional epithelium) and unites with the reduced enamel epithelial cells (Fig 163, III). As the cells of the reduced enamel epithelium mature they migrate in the direction of their long axes along the tooth surface until they are shed into the gingival sulcus. According to this view these cells do not possess any mitotic ability and are ultimately replaced by cell division in the surrounding oral epithelium which is then responsible for a secondary, or permanent, junctional epithelium (Fig 163, IV). Other investigators maintain that the reduced enamel epithelium in this situation is a reproducing population of unique cells which form a permanent collar around the neck of the tooth, and that oral epithelium is responsible only for the crevicular epithelium lining the normally very shallow gingival sulcus. This view is to some extent based on differences in the orientation of the cells in these two regions. Cells of the crevicular epithelium are orientated with their long axes parallel to the surface of the shallow gingival sulcus, and mature in the same way as gingival epithelium in general, while the cells of the junctional epithelium have their long axes parallel to the tooth surface and migrate coronally along the tooth surface to be shed into the gingival sulcus. There is histologically a fairly distinct demarcating line between these two regions in the floor of the sulcus. The latter view is currently favored.

Both the junctional epithelium and the crevicular epithelium, except close to the gingival margin (gingival crest), are non-keratinized and it is believed that the underlying lamina propria plays a key role in the maturation process of these non-keratinized epithelia, being a further manifestation of epithelium-mesenchyme interaction. It has, for example, been shown experimentally that when a keratinized epithelium, together with its underlying connective tissue, is grafted onto a lamina propria supporting a normally non-keratinized epithelium such as the floor of the mouth, the graft remains keratinized. On the other hand, seeding of only basal epithelial cells from a keratinized epithelium onto a connective tissue site which previously supported a non-keratinized epithelium, results in a non-keratinized epithelium. The lamina propria supporting junctional and deep crevicular epithelium is different from that supporting the gingiva and contains less collagen and varying numbers of inflammatory cells.

It has further been shown experimentally that the lamina propria (superficial connective tissue) permits the normal maturation of stratified squamous epithelial cells with surface loss of desmosomes and keratinization, while deep connective tissue did not have this permissive influence, resulting in a relatively immature epithelium similar to junctional epithelium. Such an epithelium is characterized by prominent desmosomes. It is therefore maintained that junctional epithelium is supported by deep connective tissue (the periodontal ligament). The inflammatory cells in the lamina propria below the junctional epithelium and deeper parts of the crevicular epithelium may be a further reason for the non-keratinized nature of these epithelia, since it has been experimentally shown that strict oral hygiene measures, combined with antibiotic therapy, can eliminate the inflammatory response and cause this region to keratinize. Some authors maintain that the effect of the low-grade inflammation is also manifested in the slow apical migration of the junctional epithelium since it has been experimentally shown that inflammation can induce proliferation in epithelia maintained on deep connective tissue.

The reason for the presence of the inflammatory cells is not clear. The inflammation may be initiated at the time of eruption. It is known that an acute inflammatory response occurs in the connective tissue around an erupting tooth crown when the reduced enamel epithelium fuses with the oral epithelium. This response may be caused by antigens passing through widened spaces between the epithelial cells into the lamina propria. These spaces are perhaps caused much earlier by the biochemical activity necessary to remove the bone and connective tissue overlying the erupting tooth crown. The continued ingress of antigens maintains the epithelium in a reactive state with wide intercellular spaces which also permit the outward movement of leukocytes and crevicular fluid.

Development of the periodontal ligament

The tooth bud develops in a bony crypt where it is enclosed by the mesenchymal follicle. The follicle is made up of an inner cell-rich fibrous layer which encapsulates the tooth, and an outer more loosely constructed layer. The inner layer gives origin to the cementum while the periodontal ligament and alveolar bone arise in the outer layer.

During active tooth eruption the periodontal fibers are in a constant state of readjustment in the middle region of the periodontal ligament. This zone is best seen in a longitudinal section and has been called an intermediate plexus. It has been shown experimentally that a vitamin C deficiency, which causes a disturbed collagen synthesis, shows the greatest effects in this central area, an indication that is an area of active collagen synthesis. Other studies with radioactive proline have shown that collagen remodeling takes place throughout the entire ligament, but at a slightly reduced rate adjacent to cementum. Electron microscopic studies have shown that collagen synthesis and degradation by fibroblasts takes place throughout the ligament and have failed to prove the existence of an intermediate plexus. The appearance of an intermediate plexus is probably due to extensive branching and splicing of collagen fibers in this area.

The remaining components of the periodontium, namely the cementum and alveolar bone, have been dealt with, respectively, in Part II, Chapters 8 and 9, and in Part II, Chapter 17.

Selected bibliography

1. Berkovitz, B.K.B., Moxham, B.J. and Newman, H.N. (1982) editors. *The Periodontal Ligament in Health and Disease.* Oxford: Pergamon Press.
2. Bhaskar, S.N. (1991) editor. *Orban's Oral Histology and Embryology*, 11th edition. St. Louis: Mosby-Year Book, Inc.
3. Davidovitch, Z. (1988) editor. *The Biological Mechanisms of Tooth Eruption and Root Resorption.* Proceedings of the International Conference held at the Great Southern Hotel, Columbus, Ohio, April 28-30, 1988. Sponsored by the Ohio State University.
4. Freeman, E., Ten Cate, A.R. and Dickinson, J. (1975) Development of gomphosis by tooth germ implants in the parietal bone of the mouse. *Archives of Oral Biology,* 20, 139-140.
5. Fullmer, H.M., Sheetz, J.H. and Narkates, A.J. (1974) Oxytalan Connective tissue fibers: a Review. *Journal of Oral Pathology,* 3, 291-316.
6. Grant, D. and Bernick, S. (1972) Formation of the periodontal ligament. *Journal of Periodontology,* 43, 17-25.
7. Genco, R.J., Goldman, H.M. and Cohen, D.W. (1990) editors. *Contemporary Periodontics.* St. Louis: C.V. Mosby Company.
8. Lindhe, J. (1989) *Textbook of Clinical Periodontology,* 2nd edition. Copenhagen: Munksgaard.
9. Mjör, I.A. and Fejerskov, O. (1986) editors, *Human Oral Embryology and Histology.* Copenhagen: Munksgaard.
10. Osborn, J.W. (1981) editor. *Dental Anatomy and Embryology.* Oxford: Blackwell Scientific Publications.
11. Osborn, J.W. and Ten Cate, A.R. (1983) *Advanced Dental Histology.* Bristol: Wright PSG.
12. Sims, M.R. (1975) Oxytalan-vascular relationships observed in histological examination of the periodontal ligaments of man and mouse. *Archives of Oral Biology,* 20, 713-716.
13. Ten Cate, A.R. (1989) *Oral Histology: Development, Structure, and Function,* 3rd edition. St. Louis: C.V. Mosby Company.

Review questions

1. What is the periodontium?
2. Discuss the role of ectomesenchyme in the development of the periodontium.
3. Describe the gingiva with reference to the following:
    (a) macroscopic appearance at clinical examination; and
    (b) microscopic appearance of the attached and free (marginal) gingiva.
4. Describe the dentogingival junction with reference to the following:
    (a) depth of the gingival sulcus;
    (b) crevicular epithelium; and
    (c) junctional epithelium.
5. Describe the gingival fiber groups.
6. Discuss the periodontal ligament with reference to the following:
    (a) the width of the periodontal ligament and factors influencing its width;
    (b) periodontal fiber groups; and
    (c) contents of the periodontal space.
7. Describe the blood and nerve supply of the periodontium.
8. Discuss the development of the dentogingival junction.
9. Discuss the development of the periodontal ligament.
10. Why is the presence of inflammatory cells in the lamina propria deep to the junctional epithelium regarded by many as a fairly normal feature?
11. Write brief notes on the functions of the periodontal ligament.

19. Nerve supply and sensitivity of teeth

General remarks

Healthy exposed dentine is normally exceedingly sensitive to touch, cold, heat, a blast of air, osmotic changes, some chemicals, and electricity, and there is little doubt that pain is the only sensation elicited by these stimuli. It has long been the experience of clinicians that the peripheral parts of the dentine close to the amelodentinal junction are more sensitive to operative procedures than dentine at a slightly deeper level, but that pain perception then increases as the dental pulp is approached. It is further commonly experienced that, while teeth with a carious lesion involving the dentine are often spontaneously painful, a deep cavity may develop so painlessly that the patient is unaware of its existence. The presence, or absence, of pain in these cases is dependent on the rate at which the carious lesion develops and the extent to which a dentinal barrier, in the form of secondary dentine, transparent (sclerotic) dentine or dead tracts, develops. For the same reason, dentine deep to areas of attrition is normally painless.

Nerve supply to the dentine and pulp

Three types of nerves are found in the dental pulp:

1. Unmyelinated efferent fibers of the autonomic (sympathetic) nervous system accompany the blood vessels and are concerned with vasoconstriction. These arise in the superior cervical sympathetic ganglion.
2. Afferent myelinated A fibers of the somatic sensory nervous system.
3. Afferent unmyelinated C fibers of the somatic sensory nervous system. All the sensory afferent fibers are branches of the trigeminal (fifth cranial) nerve.

The dental pulp has a rich somatic sensory nerve supply (Fig 164). The nerves enter the pulp through the apical foramina and course towards the plup chamber in the form of two or three trunks. Small branches are given off in the root canals

Fig 164 — Nerve supply to dentine and pulp

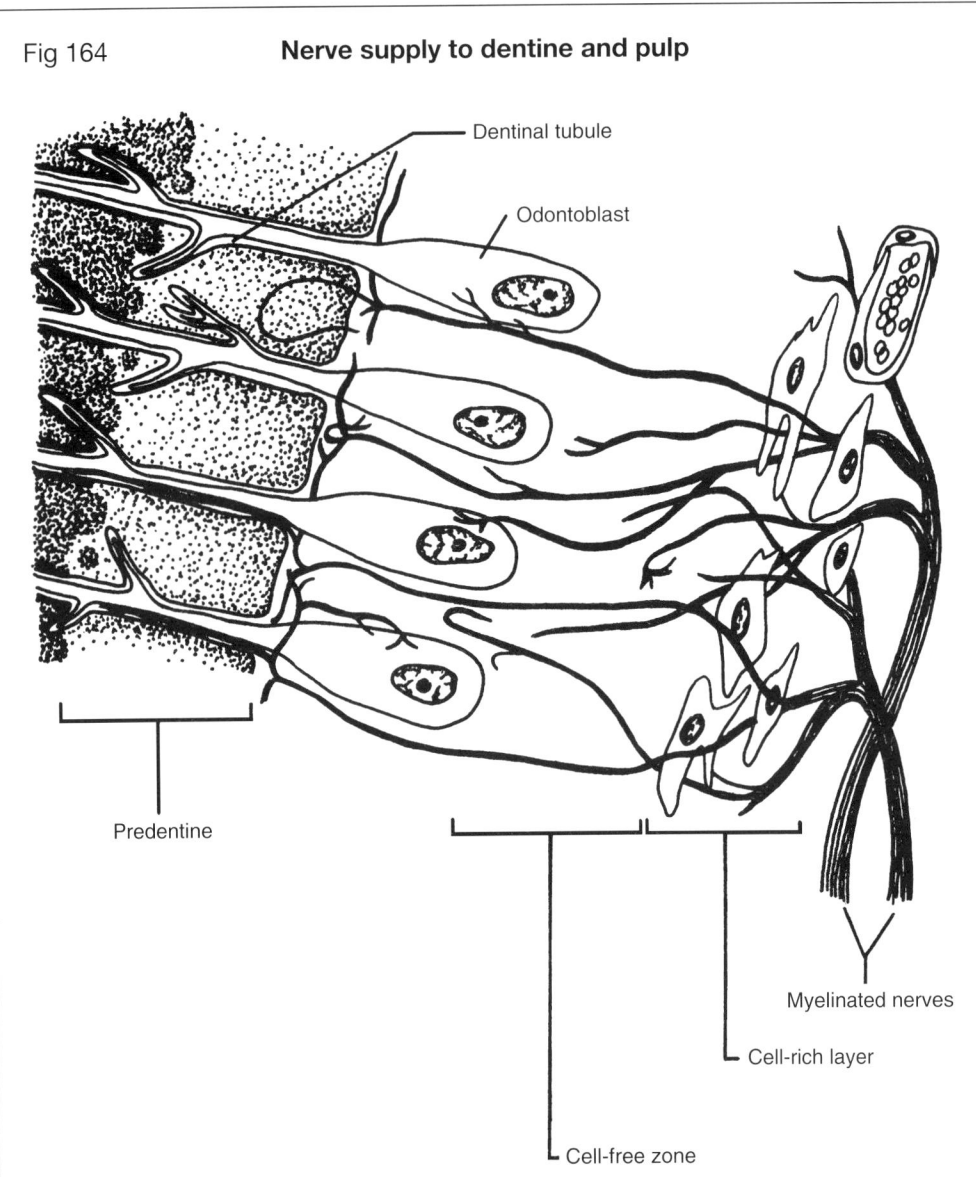

Myelinated as well as unmyelinated somatic sensory nerves enter the dental pulp. Myelinated fibers loose their myelin sheaths in the region of the subodontoblastic cell-rich layer, where a plexus (of Raschkow) is formed by both myelinated and unmyelinated fibers. This plexus is not established before completion of the root. All nerves cross the relatively cell-free zone and terminate as free nerve endings in various ways. Some terminate in the cell-free zone, while others may form gap junctions with odontoblasts. According to some authors, a second (marginal) plexus is formed on the surface of the predentine. Branches may form loops within the predentine, or enter dentinal tubules for varying distances to lie in close proximity to odontoblast processes.

but the nerves branch extensively in the pulp chamber, especially in the pulp horns. An interlacing network of nerve fibers, the subodontoblastic plexus (of Raschkow), is found in the relatively cell-rich layer below the roof and walls of the pulp chamber. Myelinated nerves lose their myelin sheaths in this plexus from where unmyelinated sensory nerves course through the relatively cell-free zone towards the odontoblast layer. These fine branches are covered only by Schwann cells and cannot be distinguished from fibers of the autonomic nervous system. These nerves pass between the odontoblasts and form a second, or marginal, plexus on the pulpal surfaces of the predentine, according to some authors. It is generally accepted, however, that terminal nerve fibers with free nerve endings are found in the odontoblast region. These fibers may end free, may form gap junctions with odontoblasts, may loop into predentine where they are trapped by continued dentine formation, or may enter the tubules in limited numbers and for a limited distance before terminating as free endings. An electronmicroscopic investigation of the nerve supply of human premolars revealed that less than 10% of all dentinal tubules contain nerve fibers up to 100 μm from the pulpal border, that no nerve fibers are present beyond this distance, and that nerve fibers accompany odontoblast processes in only 27% of tubules in the pulp horns. The extent to which dentine is innervated remains controversial.

Dental pain

Pain is perceived when pain receptors (nociceptors) in the form of free nerve endings are stimulated. Excitation of a nerve ending generates a receptor potential, giving rise to a transmembrane action potential which is propagated along the nerve fibers as an impulse. Nerve fibers, in common with all cells, possess a resting membrane potential, characterized by the presence intracellularly of large numbers of anions, including anions to which the cell membrane is impermeable, as well as potassium cations, while the opposite applies to extracallular fluids (more cations, especially sodium). There is thus a negative resting potential inside the cell compared to the outer surface of the cell membrane. An impulse is conducted as follows: The cell membrane is relatively resistant to the passage of ions in the resting state. On excitation, an increased membrane permeability to the passage of sodium ions into the cell occurs. As the concentration of sodium ions inside the cell increases, the interior becomes progressively less negatively charged. When a critical level in potential difference is reached (the firing level, or threshold potential), the potential is reversed and the interior of the cell becomes positively charged by comparison with the outer surface of the cell membrane. The influx of further sodium ions is arrested and potassium ions pass out. That very small segment of the nerve is said to be depolarized, in contrast to the resting segment proximal to it, and the localized difference in electrical potential is continued as a chain reaction producing sequential depolarizations, followed by repolarizations, along the length of the nerve fiber. That part of the membrane initially depolarized, starts to actively repolarize as sodium ions pass out of the cell and potassium ions return to the interior of the cell. In this way the segment returns to its normal resting membrane potential and is once again polarized. Conduction of a nerve impulse is faster along a myelinated nerve because the changes in electrical potential occur at the nodes of Ranvier and the impulse therefore "jumps" from one node to the next.

Pain impulses are conducted from the dental pulp to the central nervous system mainly by means of two types of fibers, designated A-(delta) fibers and C-fibers. A-δ fibers are myelinated, have a thickness of 2-5 μm and conduct impulses at a velocity of 12-30 m/s, while C-fibers are unmyelinated fibers with a thickness of 0.4 - 1.2 μm and conduct at 0.5 - 2.0 m/s.

The thicker A-fibers are concerned with conduction of acute localized pain, so-called "fast" pain, while C-fibers conduct poorly localized dull pain. Type C fibers are considered more susceptible to nerve block anesthesia. Efferent pain fibers from the trigeminal ganglion enter the pons and course caudally in the brain stem in the spinal tract of the trigeminal nerve, situated in the medulla oblongate, to synapse mainly in the spinal nucleus of the nerve (subnucleus caudalis). From this sensory nucleus the fibers cross the midline in the brain stem and form the ascending anterior trigeminothalamic tract (in close anatomical association with the lateral spinothalamic tract) which passes through the midbrain to end in the ventroposterior nucleus of the thalamus, from where fibers are projected to the postcentral gyrus of the somatosensory cerebral cortex. Pain perception probably occurs at the level of the thalamus but fine localization and discrimination is a function of the cortex.

Three categories of neurons are described in the projection from the subnucleus caudalis to the thalamus. Category I neurons respond only to noxious stimuli. category II neurons respond to heat and other noxious stimuli, but also to mechanical stimuli, while category III neurons respond only to low threshold mechanical stimuli. Activation of category III neurons may inhibit or modify activity in the other categories. When, for example, the lip is pinched (activation of category III neurons) during administration of a local anesthetic by injection, this mechanical stimulation may reduce or eliminate any pain associated with penetration of the needle (category I neuron activity). Activity at higher cortical levels may, in addition, modify the capacity of peripheral nerves to signal noxious stimuli to the thalamus and cortex. In this way, descending impulses from higher levels may, in states of panic, impose an almost total block on peripheral pain perception, while states of anxiety may enhance a painful sensation.

Theories of pain perception in dentine and tooth pulp

The mechanism whereby a tooth perceives pain has daunted dental histologists and physiologists for many years. Four possibilities have been widely investigated. Firstly, dentine may have its own (intratubular) nerve supply; secondly, the odontoblast cell bodies and processes may conduct an impulse and be connected by, for example, synapse-like junctions, to a more normal neuro-anatomical nerve pathway starting in the dental pulp; thirdly, movement of the odontoblasts themselves may stimulate modified mechanoreceptors in the pulp, and fourthly, nociceptors may be largely confined to the peripheral pulp but possess the ability to react to local changes brought about by mechanical factors, such as fluid movement, in the dentine.

Nerves inside the dentine

The presence of an intradentinal nerve supply remains controversial. Although no doubt exists regarding the presence of nerves and nerve endings in the peripheral pulp and within the dentinal tubules, it is fairly generally accepted that dentine does not have a rich nerve supply and that any nerves present extend only for a limited distance into the tubules. The relative paucity of intradentinal nerves extending into the tubules for significant distance makes the extreme sensitivity of exposed dentine inexplicable. Furthermore, newly erupted teeth do not contain intratubular nerves but nevertheless are as sensitive as older teeth. There is, in addition, no satisfactory explanation for the marked sensitivity of peripheral dentine under the amelodentinal junction if the absence of nerves in this region is taken into account. Indications are that when pain is experienced in a tooth no intradentinal nerve endings are necessarily stimulated, since treatment of exposed dentine with topical anesthetic ointments or a protein precipitant such as silver nitrate does not eliminate its sensitivity.

Conduction by the odontoblast cell body and process

An alternative theory suggests that the odontoblast process conducts an impulse in the same way as a nerve process, and that the cell body functions as a receptor cell with a functional connection to pulpal nerves. This theory is to some extent based on the neural crest origin of an odontoblast.

Previous studies have failed to show a synaptic relationship between pulpal nerves and odontoblasts and, furthermore, it was found *in vitro* that the membrane potential of the odontoblasts and their processes was too low to permit conduction of nerve impulses. Further factors which negatively influence this theory are the limited extend to which the processes extend into the dentine and the finding that topical anesthetics and protein precipitants do not abolish sensitivity.

This theory has to some extend been revived by some recent studies showing that the processes extend further into dentine than previously believed, and that gap junctions may be present between pulpal nerves and odontoblasts.

Substances causing pain when applied to exposed nerve endings do not cause pain when applied to exposed dentine. On the other hand, substances causing pain when applied to exposed dentine do not have any effect on exposed nerve endings.

Movement of the odontoblasts

Some painful stimuli applied to exposed dentine, such as reduced pressure or an air blast, cause a physical displacement of odontoblasts into the tubules. Surface desiccation of the dentine is responsible for this movement of the nuclei and has led investigators to surmise that this movement stimulates mechanoreceptors in close proximity to the odontoblasts. Positive pressure on dentine, causing a pulpal displacement of tubular contents, has the same painful effect. It has, however, been experimentally established that the sensitivity of dentine is not dependent on the presence of odontoblasts or their processes.

Fluid movement inside dentine (hydrodynamic theory)

From the foregoing it is apparent that it is unlikely (a) that intratubular nerve endings perceive a painful stimulus, (b) that the odontoblast and its process functions as a nerve fiber and (c) that displacement of odontoblasts is *per se* responsible for initiating nerve impulses.

The hydrodynamic mechanism is based on the assumption that fluid movement in dentinal tubules disturbs the peripheral pulpal environment and that this is sensed by free nerve endings in the region. Rapid fluid movement in dentine is caused by application of hypertonic solutions, touch, dehydration by means of cotton wool or an airblast, or operative procedures such as probing or cavity preparation. All the above procedures cause pain *in vivo*, the intensity of which is matched by a directly proportional fluid flow as measured experimentally *in vitro*.

It is provisionally accepted that the movement of fluid through the dentinal tubules disturbs the hydrostatic pressure equilibrium in the peripheral extracellular compartment of the pulp. Pressure changes in this area stimulate the nerve endings (modified mechanoreceptors) in the vicinity of the odontoblasts and initiate a pain impulse. This theory may offer an explanation for the marked sensitivity of dentine at the amelodentinal junction since it is known that dentinal tubules branch extensively in this region and any irritation of this area may result in the sudden displacement of a large volume of intratubular fluid.

In summary, the precise location of nociceptors responsible for pulpal (dentinal) pain has not been established beyond any doubt, nor has the exact mechanism whereby these receptors are stimulated. Although the hydrodynamic theory currently enjoys the most support, the other mechanisms cannot be disregarded.

Selected bibliography

1. Bhaskar, S.N. (1991) editor. *Orban's Oral Histology and Embryology*, 11th edition. St. Louis: Mosby-Year Book, Inc.
2. Brännström, M. (1986) The hydrodynamic theory of dentinal pain: sensation in preparations, caries, and the dentinal crack syndrome. *Journal of Endodontics*, 12, 453-457.
3. Evers, H. and Haegerstam, G. (1981) *Handbook of Dental Local Anaesthesia*. Copenhagen: Schultz Medical Information.
4. Lavelle, C.L.B. (1988) *Applied Oral Physiology*, 2nd edition. London: Wright.
5. Lilja, J. (1979) Innervation of different parts of the predentine and dentine in young human premolars. *Acta Odontologica Scandinavica*, 37, 339-346.
6. Mjör, L.A. and Fejerskov, O. (1986) editors. *Human Oral Embryology and Histology*. Copenhagen: Munksgaard.
7. Osborn, J.W. and Ten Cate, A.R. (1983) *Advanced Dental Histology*, 4th edition. Bristol: Wright PSG.
8. Scott, J.H. and Symons, N.B.B. (1982) *Introduction to Dental Anatomy*, 9th edition. Edinburgh: Churchill Livingstone.
9. Sessle, B.J. (1986) Recent developments in pain research: central mechanisms of orofacial pain and its control. *Journal of Endodontics*, 12, 435-444.
10. Ten Cate, A.R. (1989) *Oral Histology: Development, Structure, and Function*, 3rd edition. St. Louis: C.V. Mosby Company.
11. Trowbridge, H.O. (1986) Review of dental pain–histology and physiology. *Journal of Endodontics*, 12, 445-452.
12. West, J.B. (1991) editor. *Best and Taylor's Physiological Basis of Medical Practice*, 12th edition. Baltimore: Williams and Wilkins.

Review questions

1. Describe the nerve supply to the pulp and dentine.
2. Briefly discuss the propagation of a nerve impulse.
3. Briefly describe the neural pathway followed by a pain impulse from its origin in a pulpal nerve ending to the cerebral cortex where it is consciously perceived.
4. Give a brief outline of the different theories of pain perception by the pulp-dentine complex.

20. Tooth eruption

General remarks

The process of tooth eruption not only entails the axial movement of a tooth which enables it to appear in the oral cavity, but also further movements towards opposing teeth, creation, and maintenance of a functional occlusion, completion of root development and establishment of the periodontium. Eruption, perhaps better described as physiologic tooth movement, is often described as occurring in a pre-eruptive phase, a prefunctional phase, and a functional phase.

The pre-eruptive phase is characterized by differentiation of the tooth germ in its bony crypt, followed by completion of mineralization of the crown. Secondary tooth germs are enclosed by bone except at small openings, the gubernacular canals, where the crypts communicate with the submucosa of the lingual (palatal) gingiva close to the alveolar crest in the case of anterior teeth. The gubernacular canals of developing premolars are situated between the roots of their deciduous predecessors, while secondary molars have their own canals on the crest of the alveolar ridge. This canal usually contains a few strands of connective tissue fibers and remnants of the dental lamina and is known as the gubernacular cord. Previously some importance was attached to these fibers as a mechanical aid during tooth eruption. As the secondary tooth erupts, this canal is widened by osteoclastic resorption.

The prefunctional phase involves the eruptive movement of a tooth till the occlusal plane is reached and opposing teeth are met. This phase includes completion of the root and partial establishment of the occlusion and periodontium. The prefunctional phase is continued as the functional phase in which the occlusion and periodontium become stabilized.

It is a well recognized fact that a tooth does not remain static and immobile during the entire life of an individual, but changes its position to compensate for growth of the jaws, attrition and proximal wear, and loss of neighboring and opposing teeth. These movements of the tooth and its periodontium may involve axial and mesial migration (drift). These posteruptive movements are seen by some authors as a continuation of the functional phase of eruption, while others regard them as reactivated eruption.

The forces acting on a tooth during its eruption are not fully understood and have in the past been the subject of much research. Several theories to explain the phenomenon of eruption have been put forward and it may well be that more than one mechanism operates at any one time.

Three theories have enjoyed popular support. These are (a) proliferation of cells and tissues responsible for root growth, (b) pressure exerted by tissue fluid or blood pressure, and (c) contraction in the developing periodontal ligament fibers.

The role of cell proliferation

Proliferation of cells in the region of the tooth apex is responsible for all the processes involved in elongation of the root, namely growth of the dental pulp, growth of the root sheath, formation of dentine, cementum, and the periodontal ligament. The majority of experiments to investigate the mechanisms responsible for generating the forces necessary for tooth eruption have involved the continuously erupting incisors of rodents. These investigations have revealed that administration of antimitotic drugs, such as demecolcine, and cytotoxic drugs, such as triethanomelamine, significantly retard eruption. Although these findings imply that cell proliferation is an important factor in eruption, it is impossible to assess which cell population is affected since these drugs interfere with a wide range of metabolic processes. No known drug interferes solely with mitosis of an isolated cell population and no definite conclusion can therefore be drawn from the above experiments.

Regarding the above theory of cell proliferation, it is significant that fully formed but impacted human incisors and canines often erupt spontaneously once the obstruction is removed. Apical cell proliferation and root growth are obviously not involved in eruption of such teeth.

The role of tissue fluid pressure

When the vascular system is considered as generating axial eruptional forces, it is assumed that pressure is exerted in the pulp and around the tooth apex, and that this pressure must be greater than that in the tissues between the crown and the oral cavity. Such a pressure gradient has been shown to exist. In the erupting canines of dogs, it was found that the tissue fluid pressure above the crown was 10 ± 5 mmHg, while the pressure in the pulp chamber was 23 ± 6 mmHg. If it is assumed that this intrapulpal pressure applies also to the apical region of the pulp, it has been calculated that an axial force of 15 g could be generated by such a system. It has also been reported that vasoactive drugs have been successfully used experimentally to increase or decrease rates of eruption. Other factors, such as fluid pressure in the periodontal ligament, should be considered before giving the above theory unqualified support.

The role of the periodontal ligament

The role of the periodontal ligament in tooth eruption has been investigated in the continuously erupting incisors of rats and rabbits. These teeth have a unique morphology. The dentine is covered buccally by enamel along the full length of the tooth from the incisal edge to the continuously developing root apex. No periodontal fibers attach to the buccal aspect of these teeth. The lingual dentine is covered by cementum along the full length of the tooth. Periodontal ligament fi bers attach into the cementum. The chisel-shape of these teeth is the result of differential wear of enamel, dentine, and cementum.

In well-documented experiments on these teeth, one incisor is removed from occlusion by cutting away the incisal edge, thus allowing unimpeded eruption. The effect of various procedures on the rate of eruption of this tooth is subsequently assessed.

In an experiment to assess the eruptive force generated by cellular proliferation in the forming root, the apical enamel organ and dental papilla were surgically removed. Any eruptive force which might have been generated by cellular proliferation was thus eliminated. Following this procedure, no significant effect on the normal rate of eruption was recorded. Since no new dental tissues were formed, the tooth was exfoliated. The only tissue remaining after this surgical intervention was the periodontal ligament, suggesting that it exerted some form of traction on the tooth. Very similar results were obtained when the tooth was transected and the proximal (basal) portion immobilized. The distal portion erupted until it was exfoliated.

A further experiment to investigate the effect of pulpal pressure involved surgical removal of the enamel organ and adjacent papilla on the buccal side of the tooth. No further enamel or dentine could therefore form buccally. With continued eruption, only the newly developed lingual cementum and dentine remained and the pulp was exposed buccally. The effect of pressure in the dental pulp on eruption was thus eliminated, a further indication of the positive role of the periodontal ligament in tooth eruption. It has also been shown that if the apical tissues are destroyed by radiation, no new hard tissue forms but the distal (coronal) portion of the tooth continues to erupt.

Fig 165   **Collagen synthesis**

Polypeptide chain

Individual amino acids form polypeptide chains on the ribosomal complexes of the cells responsible for collagen synthesis. These so-called protocollagen chains contain unhydroxylated proline and lysine and other amino acids.

Procollagen macromolecule

Distinctive amino acids

Hydroxylation of amino acid takes place and three polypeptide chains (protocollagen) are bound together by hydrogen cross-links. They form a triple helix, resembling a rope with three strands. The macromolecule is known as procollagen. Each chain in the chain has a group of distinctive amino acids at its end. Before procollagen can be secreted by the cell, these terminal amino acids have to be removed by a hydrolytic enzyme called procollagen peptidase. These procollagen molecules are extracellularly arranged end to end to form a protofibril.

End to end arrangement of procollagen macromolecules to form a protofibril.

Protofibrils group together with adjacent protofibrils to form a collagen fibril. This grouping is initially fairly loose but is strengthened by strong covalent cross-linkages as collagen matures. The grouping takes place in a quarter-stagger arrangement and results in the typical 64 nm collagen fibril periodicity. Fibrils group together to form fibers of varying size.

Collagen fibril

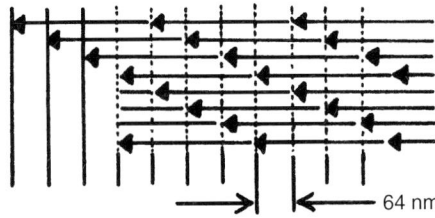

64 nm

Most studies to date have indicated that the periodontal ligament plays an important role in tooth eruption, either by means of the cells of the ligament or the extracellular collagen.

The name collagen is derived from the Greek, meaning a glue producer. When insoluble collagen is heated with water the soluble derived protein, gelatin, is produced. It forms a gel in water.

Collagen is the main constituent of the periodontal ligament and is a fibrous protein composed of 18 amino acids, including glycine, proline, hydroxyproline, lysine, and hydroxylysine. The procollagen macromolecule consists of three long polypeptide chains which are twisted around one another in a triple helix. Each individual chain is wound around its own axis in a left-handed helix and is cross-linked to the other chains by hydrogen bonds. Collagen is synthesized in the ligament by fibroblasts (Fig 165), and the macromolecules are initially present extracellularly in a random disordered fashion. Organization of procollagen macromolecules to form fibrils and later fibers, is accompanied by development of axial forces along the length of the fibril. These forces prevent the macromolecules from reverting to their previously distorted state. It has been postulated that a collagen fiber contracts by as much as 10% during its formation and maturation, due to cross-linking and aggregation, and that the fibers of the ligament exert traction on a tooth in an axial direction.

For a tooth to erupt by the above mechanism, a constant rapid turnover of collagen in the ligament is required. This has been shown to be the case. If contraction of collagen is a factor in eruption, interference with intermolecular links and maturation should have a retarding effect on eruption. Cross-linkings are inhibited by the lathyritic agent aminoacetonitrile. Administration of this drug had an impeding effect on eruption in one study, while other studies report no effect on unimpeded eruption, though impeded rates were retarded. The latter finding could be due to the effect of occlusal stress on a weakened periodontal ligament. The propulsive force of such a ligament is much reduced. These findings do, however, cast doubts on the validity of the theory that shrinkage of collagen fibers during their maturation provides an eruptive force. Another important factor in collagen synthesis is ascorbic acid (vitamin C) which plays an important role in the hydroxylation of protocollagen proline and lysine. Synthesis of collagen is depressed in states of vitamin C deficiency and one study showed that in scorbutic guinea pigs, both unimpeded and impeded eruption rates of lower incisors were significantly retarded.

It has also been suggested that periodontal ligament fibroblasts generate an eruptive force, either by their contractility or recognised locomotor ability. This process may be regarded as analogous to the contraction observed in scar tissue in healing wounds. If fibroblasts were to exert traction on a tooth, this would have to be indirectly through collagen fibers, or through a direct connection with the tooth. *In vitro* studies have lent some support to fibroblast activity in tooth eruption. There is, however, no convincing evidence to date that periodontal fibroblasts are directly involved in generating a force in the eruption process of human teeth. Their vital role in collagen metabolism is, however, above dispute.

In summary, although the force generated by contraction of collagen fibers currently enjoys the most support, no single biological system within the periodontal ligament can be clearly identified as providing the only force causing a tooth to erupt. Tooth eruption must in the light of present knowledge be regarded as a multifactorial phenomenon.

The reader is referred to the bibliography for detailed discussions of collagen biosynthesis and structure, and of other views on tooth eruption.

Selected bibliography

1. Bhaskar, S.N. (1991) editor. *Orban's Oral Histology and Embryology*, 11th edition. St. Louis: Mosby-Year Book.
2. Davidovitch, Z. (1988) editor. *The Biological Mechanisms of Tooth Eruption and Root Resorption.* Proceedings of The International Conference, held at the Great Southern Hotel, Columbus, Ohio, April 28-30, 1988. Sponsored by The Ohio State University.
3. Cole, A.S. and Eastoe, J.E. (1988) *Biochemistry and Oral Biology*, 2nd edition. London: Wright.
4. Lavelle, C.L.B. (1988) *Applied Oral Physiology*, 2nd edition. London: Wright.
5. Mjör, I.A. and Fejerskov, O. (1986) editors. *Human Oral Embryology and Histology.* Copenhagen: Munksgaard.
6. Osborn, J.W. (1981) editor. *Dental Anatomy and Embryology.* Oxford: Blackwell Scientific Publications.
7. Osborn, J.W. and Ten Cate, A.R. (1983) *Advanced Dental Histology*, 4th edition. Bristol: Wright PSG.
8. Poole, D.F.G. and Stack, M.V. (1976) editors. *The Eruption and Occlusion of Teeth.* Proceedings of the 27th Symposium of the Colston Research Society, held in the University of Bristol, April 3-7, 1975.
9. Schroeder, H.E. (1991) *Oral Structural Biology.* New York: Thieme Medical Publishers.
10. Ten Cate, A.R. (1989) *Oral Histology: Development, Structure, and Function*, 3rd edition. St. Louis: C.V. Mosby Company.

Review questions

1. What is tooth eruption?
2. Briefly describe the different phases in which tooth eruption occurs.
3. Write brief notes on the theories of tooth eruption, with special emphasis on the role of the periodontal ligament in this process.

21. The temporomandibular joint

The temporomandibular joint (TMJ) will be discussed under the following main headings:

- A. The articular surface of the temporal bone
- B. The condyle of the mandible
- C. The capsule of the joint
- D. The articular disc (meniscus)
- E. The joint cavities
- F. Mandibular movements

The articular surface of the temporal bone

The articular surface of the temporal bone (Figs 166, 167) consists of a concave posterior part (the glenoid, or mandibular fossa), and a convex anterior part (the articular eminence). The concave posterior part is bounded posteriorly by the squamotympanic fissure separating the squamous part of the temporal bone from the tympanic part. The thin anterior edge of the tegmen tympani projects into the medial part of the squamotympanic fissure, dividing it into an anterior petrosquamous fissure and a posterior petrotympanic fissure. Behind the fissure, the tympanic plate of the temporal bone forms the anterior wall and floor of the auditory canal and of the external auditory meatus.

The articular eminence (the anterior origin, or root, of the zygomatic process) is an elongated transverse bony ridge. The articular tubercle is situated at the lateral end of the eminence and provides attachment to the temporomandibular ligament on its lateral surface. The zygomatic process is elongated posteriorly to form the lateral border of the glenoid fossa anteriorly, while its posterior part forms the suprameatal bony ridge above the external auditory meatus. A postglenoid tubercle, sometimes called the intermediate origin of the zygomatic process, forms the posterior border of the glenoid fossa and is situated in front of the squamotympanic fissure. The medial border of the articular surface is formed by a suture between the temporal bone and the greater wing of the sphenoid bone. The spine of the sphenoid bone is situated on the medial side of the glenoid fossa. The glenoid fossa represents the sole articular surface at birth and is represented by a flat, round, area, about 10 mm in diameter, without any sign of the articular eminence.

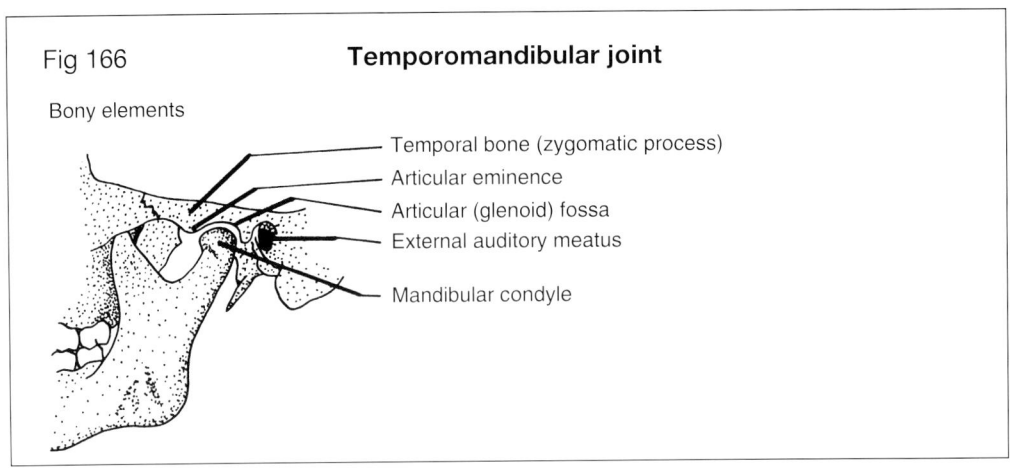

Fig 166 — Temporomandibular joint

Bony elements
- Temporal bone (zygomatic process)
- Articular eminence
- Articular (glenoid) fossa
- External auditory meatus
- Mandibular condyle

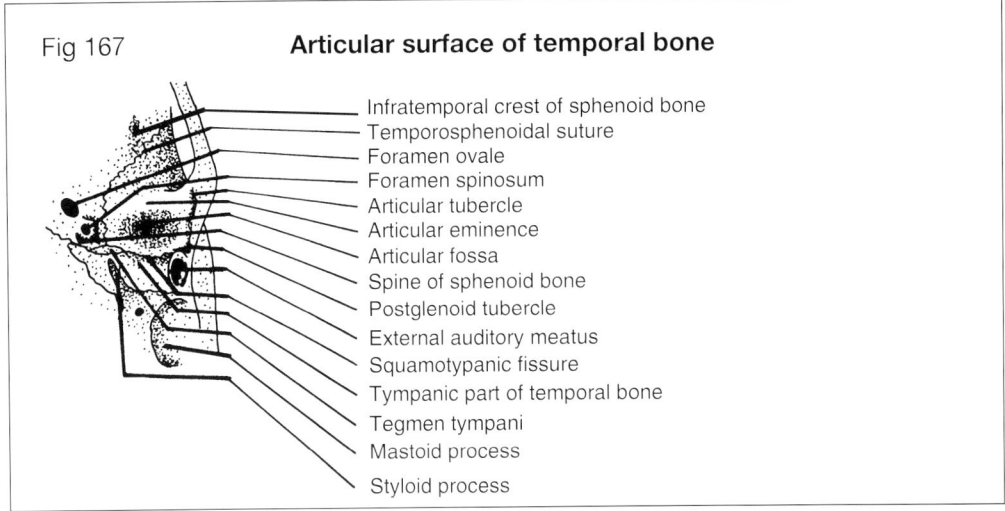

Fig 167 — Articular surface of temporal bone

- Infratemporal crest of sphenoid bone
- Temporosphenoidal suture
- Foramen ovale
- Foramen spinosum
- Articular tubercle
- Articular eminence
- Articular fossa
- Spine of sphenoid bone
- Postglenoid tubercle
- External auditory meatus
- Squamotypanic fissure
- Tympanic part of temporal bone
- Tegmen tympani
- Mastoid process
- Styloid process

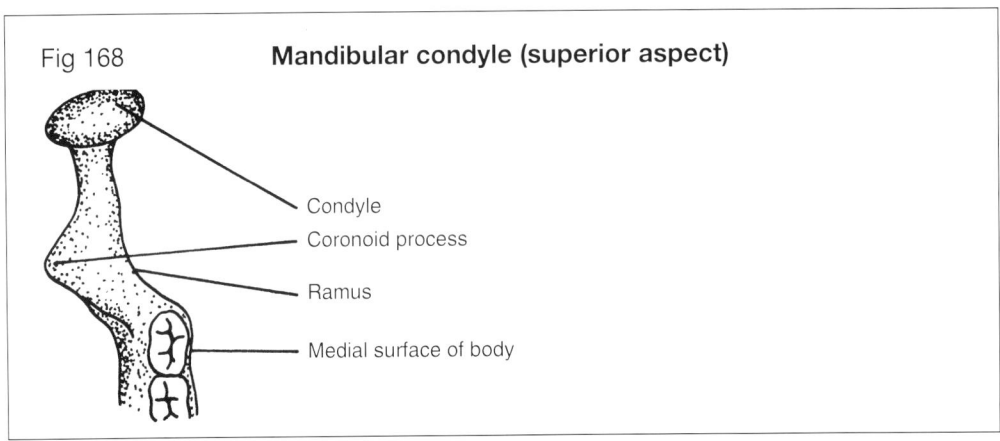

Fig 168 — Mandibular condyle (superior aspect)

- Condyle
- Coronoid process
- Ramus
- Medial surface of body

The articular eminence only becomes well-developed at the age of about 12 years.

In the skulls of most adults, the central part of the glenoid fossa consists of very thin cortical bone that separates the fossa from the middle cranial fossa containing the temporal lobe of the cerebral hemisphere.

The articular surface of the temporal bone is normally completely covered by a layer of collagenous fibrous tissue. The thickest part of this fibrous tissue is on the posterior incline and crest of the articular eminence.

The condyle of the mandible

The articular surface of the condyle (Fig 168) is about 2-2.5 times longer in a medio-lateral direction than in an antero-posterior direction. It is noticeably convex when viewed from the side, but only slightly convex when viewed from the front. The long axis of each condyle inclines slightly backwards and medially, so that an imaginary elongation of these axes meet close to the anterior margin of the foramen magnum at an angle of less than 180° (140°-160°) facing forwards. The size of this angle varies from person to person. Immediately below a faint line demarcating the posterior border of the articular surface of the condyle, a small roughened, triangular area indicates the attachment of the posterior part of the capsule and articular disc.

The anterior border of the articulating surface is distinctly marked. Beneath this border, which often takes the form of a bony ledge, a triangular depression marks the insertion of the lower fibers of the lateral pterygoid muscle. The lateral and medial poles of the condyle are usually distinct bony tubercles for attachment of the articular capsule and disc. The condyle is connected to the coronoid process by a sharp, sagittally flattened, crescent-shaped bony margin which forms the mandibular notch.

The inter-condylar distance has almost reached adult dimensions by the age of about 7 years, but the condyle still remains at a fairly small distance above the level of the teeth. The later growth and remodeling of the ramus increases this vertical dimension to adult proportions, but the ultimate height will vary considerably from person to person.

The articular surface of the condyle is covered by a thin layer of hyaline cartilage with a fibrous perichondrium until the age of about 20 years. After this age, the cartilage ossifies and the cancellous bone of the condyle is subsequently covered by compact bone with overlying layers of periosteum and collagenous fibrous tissue.

During the period of endochondral growth in the condylar cartilage, the condyle consists microscopically of the following layers from the articulating surface inwards:

1. a superficial layer of fibrous tissue (the fibrous perichondrium);
2. a layer in which active mitosis of cells (chondrogenesis) occurs;
3. hyaline cartilage;
4. a zone consisting of degenerating cartilage (after mineralization of the cartilage matrix) in the process of replacement by bone; and
5. cancellous bone which has replaced the cartilage.

The capsule of the joint

The capsule (Figs 169, 170) is firmly attached to the articular area of the temporal bone posteriorly and laterally, but rather weakly anteriorly. On the lateral side, the capsule is attached to the articular margin from the post-glenoid tubercle posteriorly to the articular tubercle anteriorly. Posteriorly it is attached to the anterior margin of the squamotympanic fissure. Structures that pass through this fissure, namely the chorda tympani nerve and the tympanic branch of the first part of the maxillary artery, are outside the joint. On the medial side, the capsule is attached to the temporal bone, close to the suture between the temporal bone and the sphenoid bone, and in front it is attached to the anterior margin of the articular eminence.

Below, the capsule is attached to the margin of the articular surface of the condyle, firmly at the sides to the poles, and at the back to the triangular attachment area.

The capsule is strengthened by means of medial and lateral collateral ligaments. The temporomandibular ligament is a strong, fan-shaped ligament found laterally, extending from the articular tubercle to the lateral pole and neck of the condyle. When the condyle is in a resting position in the glenoid fossa, fibers of this ligament run in a downward and backward direction, thus preventing further backward displacement of the condyle. The ligament becomes taut when the condyle moves forward on the articular eminence, for example, when the mouth is opened wide. The deeper fibers of this ligament are entwined with the fibers of the capsule.

The medial collateral ligament is weaker than the temporomandibular ligament. It is attached superiorly to the medial end of the articular eminence, and below to the medial pole of the condyle, just inferior to the attachment of the capsule.

It remains doubtful whether the sphenomandibular ligament (from the spine of the sphenoid bone to the lingula of the mandible) and stylomandibular ligament (from the styloid process to the posterior margin of the ramus near the angle of the jaw), sometimes also referred to as accessory joint ligaments, play any significant role in the mechanics of jaw movement, except in those cases where ossification of these ligaments occurs.

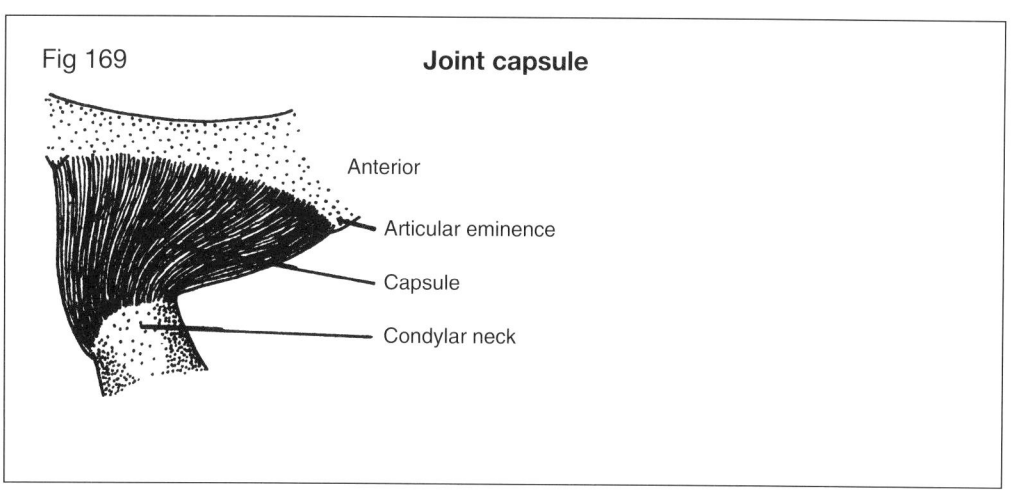

Fig 169 **Joint capsule**

- Anterior
- Articular eminence
- Capsule
- Condylar neck

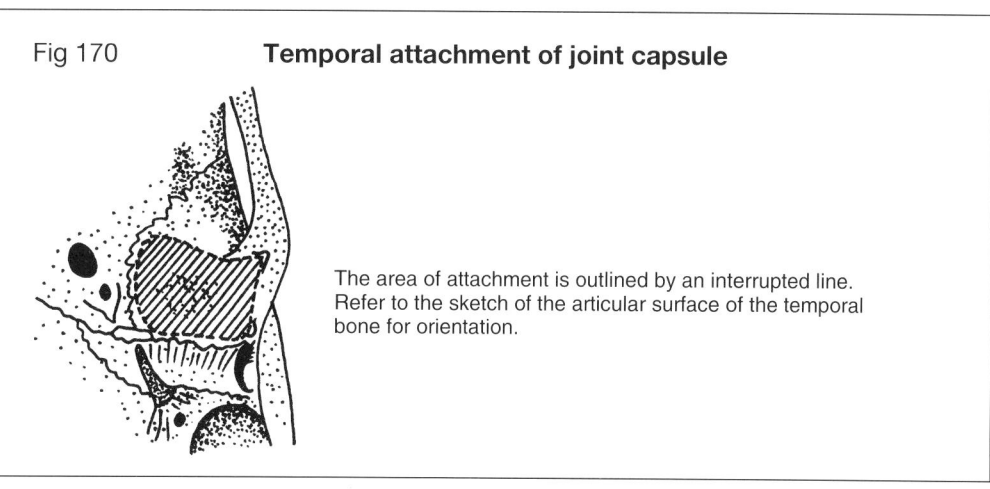

Fig 170 **Temporal attachment of joint capsule**

The area of attachment is outlined by an interrupted line. Refer to the sketch of the articular surface of the temporal bone for orientation.

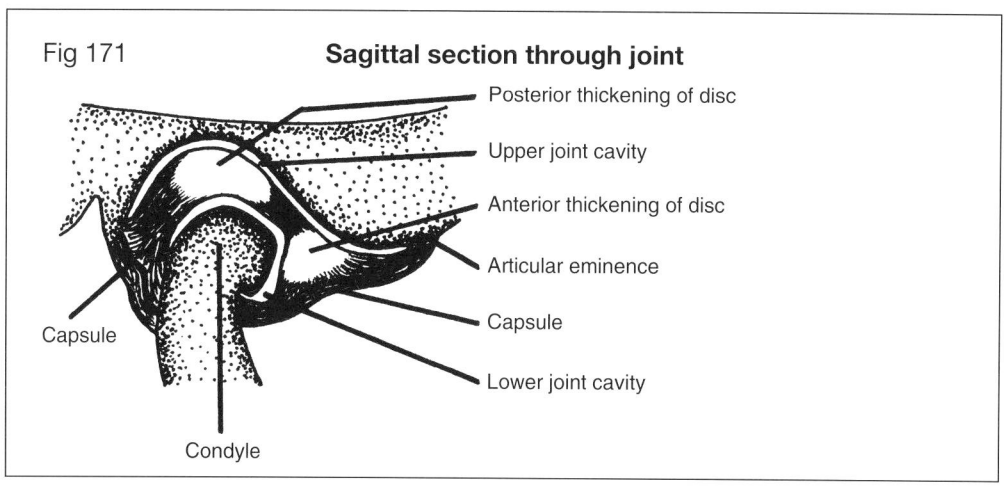

Fig 171 **Sagittal section through joint**

- Posterior thickening of disc
- Upper joint cavity
- Anterior thickening of disc
- Articular eminence
- Capsule
- Lower joint cavity
- Capsule
- Condyle

The articular disc (meniscus)

The articular disc (Fig 171) consists mainly of dense collagenous connective tissue fibers and fibroblasts. Although it sometimes contains isolated groups of cells surrounded by cartilagenous matrix, it is incorrect to describe it as fibrous cartilage. The amount of cartilage-like tissue increases with age.

According to another view, some of the fibroblasts change in course of time to form chondroid cells. These cells may, at a later stage, differentiate into true chondrocytes. In isolated cases, hyaline cartilage may be present in the discs of elderly persons.

The central part of the disc is very dense and usually shows thickened anterior and posterior parts, separated by a thinner intermediate part. On its sides the disc is firmly attached to the medial and lateral poles of the condyle, together with the capsule and collateral ligaments. Occasionally the central part of the disc shows a perforation which does not appear to be of any functional significance.

Posteriorly, the composition of the disc is of a looser nature, and it is divided into an ascending temporal lamina, attached to the anterior margin of the squamotympanic fissure together with the joint capsule, and a descending mandibular lamina, which is attached to the triangular area below the posterior margin of the articular surface of the condyle. This bilaminar part of the disc contains a moderate amount of elastic connective tissue, blood vessels, and nerves (branches of the auriculotemporal nerve) with free nerve endings. These endings are possibly a source of pain in joint abnormalities. The central part of the disc is usually bloodless and does not contain nerves.

The front part of the disc is also of a looser composition and is divided into an ascending lamina, which is attached to the anterior part of the articular eminence, and a descending lamina, which is attached to the margin of the articular surface of the condyle. The upper fibers of the lateral pterygoid muscle are attached to the disc between these two laminae. The masseter and temporalis muscles have a slight attachment to the lateral part of the disc.

The joint cavities (joint compartments)

The joint cavities of the temporomandibular joint consist of two individual compartments which are normally completely separated from each other by the articular disc. The upper compartment is larger than the lower compartment since the temporal attachment of the joint capsule is much more extensive than its condylar attachment.

When the mouth is opened, the disc and the condyle move anteriorly, the latter against the slope of the articular eminence. The posterior part of the capsule, together with the posterior superior lamina, stretches and some of the retrocapsular adipose tissue is drawn into the back part of the articular cavity.

Further attention must be paid to the coverings of the articular surfaces. The condyle is covered by a fairly thick layer of fibrous tissue, containing fibroblasts. The fibrous tissue, which covers the articular surface of the temporal bone, is thin in the glenoid fossa, but thickens on the posterior incline of the eminence. Fibrous tissue covering the articular surfaces is devoid of a blood or nerve supply, and only has a limited regenerative capacity.

The interior of the joint cavities is lined by interrupted endothelium. This is reflected from the capsule, of which it forms a part, to the upper and lower surfaces of the disc and covers the fibrous tissue on the articular surfaces of the condyle and temporal bone. This endothelium forms a synovial lining, or joint membrane, in the joint cavities. A unanimous opinion has, as yet, not been reached regarding the extent of the synovial membrane.

Mandibular movements

In the lower joint compartment, the condyle rotates (a hinge-like movement) on the lower surface of the articular disc which is connected to the medial and the lateral poles of the condyle. In the upper compartment, the disc moves forward (a translational movement), together with the condyle, against the incline of the articular eminence, a movement which is possibly limited by the elasticity of the temporal attachment of the joint capsule, the collateral ligaments of the joint, and the posterior fibers of the temporalis muscle.

When the mouth is opened, the condyle initially lies in contact with the posterior thickened part of the disc, then with the intermediate thinned part, and lastly with the anterior thickened part of the disc. When the disc and condyle move together against the incline of the eminence, the back part of the upper compartment is evacuated and filled with soft capsular tissue as previously described.

When the mouth is opened wide, the condyle and the anterior thickening of the disc lie in relation to the articular eminence. When the mouth is closed, and the teeth are in normal resting articulation, the condyle and the posterior thickening of the disc return to the posterior part of the glenoid fossa.

The movements of the mandible may be briefly classified and described as follows:

Protrusion

Both condyles move forward against the incline of the articular eminence, and the teeth maintain sliding contact. This movement is made possible by the bilateral contraction of the medial and lateral pterygoid muscles. The posterior fibers of the temporalis muscle relax. The masseter and anterior fibers of the temporalis muscle maintain a condition of tonic contraction.

Retrusion

Both condyles move backward into the posterior part of the glenoid fossa, while the teeth maintain sliding contact. The posterior fibers of the temporalis muscle are responsible for this movement, while the pterygoid muscles relax. Extreme retrusion is resisted by the deeper, more horizontal, fibers of the temporomandibular ligament, and also by the bony posterior wall of the glenoid fossa.

Opening

In this movement, the condyles are pulled forward by the lateral pterygoid muscles against the incline of the articular eminence. The posterior fibers of the temporalis muscle relax, followed by relaxation of the masseter and medial pterygoid muscles and the anterior fibers of the temporalis muscle.

The mandible is permitted to rotate around a horizontal (coronal) axis, resulting in the angle of the mandible moving backwards (towards the neck), while the condyle glides forward. Opening of the mouth is assisted by the mass of the mandible (gravity), and the digastric muscles. A stable hyoid bone is assured by simultaneous contraction of the infra-hyoid strap muscles of the neck.

Closing

This movement is made possible by contraction of the medial pterygoid, masseter and temporalis muscles. Closure can be effected with the jaws in a variety of positions. Closure in protrusion is made possible by contraction of chiefly the lateral pterygoid muscles. In a retrusive closing, the masseter muscles and posterior fibers of the temporalis muscles ensure that the condyles return to the glenoid fossae to enable the teeth to meet in normal contact.

Side to side movements

When moving the chin to one side, away from the midline, to bring about a grinding movement between the posterior teeth, the condyle on the side to where the movement is directed, is maintained in position in the glenoid fossa by the tonic contraction of the muscles on that side.

The "resting" condyle rotates around a shifting vertical axis. This condyle slides slightly laterally, forwards and downwards on the incline of the articular eminence, while rotating. This component of lateral movement is called the movement of Bennett.

On the other side, the condyle is drawn forward by contraction of the lateral pterygoid muscle. The posterior fibers of the temporalis muscle relax. When the chin is again returned to the midline, the posterior fibers of the temporalis muscle contract, while the lateral pterygoid muscle relaxes to enable the condyle to move back into the glenoid fossa. An identical mechanism is followed when moving the chin to the other side.

Functional movements of the mandible are fully discussed in the chapter on mastication (Part II, Chapter 25).

Selected bibliography

1. Bhaskar, S.N. (1991) editor. *Orban's Oral Histology and Embryology*, 11th edition. St. Louis: Mosby-Year Book.
2. Dixon, A.D. (1986) *Anatomy for Students of Dentistry*, 5th edition. Edinburgh: Churchill Livingstone.
3. Mjör, I.A. and Fejerskov, O. (1986) editors, *Human Oral Embryology and Histology*. Copenhagen: Munksgaard.
4. Scott, J.H. and Symons, N.B.B. (1982) *Anatomy for Students of Dentistry*, 9th edition. Edinburgh: Churchill Livingstone.
5. Ten Cate, A.R. (1989) *Oral Histology: Development, Structure, and Function*, 3rd edition. St. Louis: C.V. Mosby Company.

Review questions

1. Describe the osteology of the articular surface of the temporal bone.
2. Describe the osteology of the mandibular condyle.
3. Describe the histological appearance of the mandibular condyle in a child of 12 years.
4. Describe the capsule of the temporomandibular joint.
5. Describe the macroscopic and microscopic appearance of the articular disc (meniscus).
6. Describe the cavities of the temporomandibular joint.
7. Describe the movements executed by the mandible and the role of muscles in this process.

22. Salivary glands

Embryological development

True salivary glands are a distinctive feature of mammals, the only animals that perform purposeful masticatory movements. They are exocrine glands and discharge their secretions directly into the oral cavity. The parotid salivary gland is the largest and is situated anterior to the ear where it is adapted to the posterior border of the ramus of the mandible. Its excretory duct (Stensen's duct) passes forward across the masseter muscle. At the anterior border of the muscle it turns inwards to pierce the buccinator muscle and opens into the oral vestibule opposite the maxillary second permanent molar. The submandibular gland is the second largest salivary gland. It lies in relation to the free posterior edge of the mylohyoid muscle in the floor of the mouth against the medial aspect of the body of the mandible. Its duct runs forwards and opens into the floor of the mouth lateral to the lingual frenum. The sublingual gland is the smallest of the three major salivary glands and lies beneath the mucous membrane in the anterior part of the floor of the mouth. Each gland has from 10 to 20 small ducts which pierce the mucous membrane and open into the floor of the mouth or into the duct of the submandibular gland. Apart from the major salivary glands, very numerous minor glands are scattered in the submucosa throughout the oral mucous membrane, except in the anterior part of the hard palate and on the gingiva.

Salivary glands are usually regarded as arising in ectoderm (Fig 172), although a measure of uncertainty exists regarding the tissue of origin of the submandibular and sublingual glands. Nevertheless, they all develop in a similar manner. The first sign of a gland is the appearance of an epithelial bud which proliferates as a solid cord of cells into the underlying ectomesenchyme. This cord branches extensively but is initially not canalized. The end-twigs of the solid cords show the development of berry-like swellings in some glands, the future secretory acini.

The cords canalize shortly before the secretory end pieces by degeneration of their central cells to give rise to the ductal system.

Oral ectomesenchyme plays and essential role in the differentiation of salivary glands, a further example of epithelium-ectomesenchyme interaction. While the epithelial ingrowth ultimately forms the parenchyma of the salivary gland, the ectomesenchyme differentiates to form the supporting connective tissue, namely the fibrous capsule and septae which divide the gland into lobes and lobules and carry ducts, blood vessels, lymphatics, and nerves.

Slight differences of opinion exist regarding the time of onset of development of the salivary glands. The bud of the parotid gland probably arises at about 5 weeks of embryonic life, closely followed by the submandibular gland. The sublingual and minor glands arise at about 10 weeks.

Although differentiation of acini is not completed before birth, the fetus secretes salivary amylase before this event.

Classification of salivary glands

Salivary glands are classified in two main ways:

1. *Size.* A distinction is drawn between paired major glands such as the parotid gland, submandibular gland and the sublingual gland, and the minor glands scattered throughout most of the oral mucosa.
2. *Nature of the secretion.* A distinction is drawn between those glands producing a serous secretion (watery and thin, rich in both non-enzymatic and enzymatic proteins, and containing some polysaccharides), a mucous secretion (ropey and thick, rich in polysaccharides and containing some non-enzymatic proteins), and glands producing a mixed secretion.

The latter classification is more commonly used in histological descriptions. A third classification used in the past was based on the location of the openings of the secretory ducts. In this way a distinction is drawn between those glands discharging into the oral vestibule, such as the parotid gland and minor glands of the vestibule, and those draining into the oral cavity proper such as the submandibular, sublingual and minor glands.

Components of salivary glands (Fig 173)

1. *Connective tissue.* An important distinguishing feature between major and minor salivary glands is the presence of a distinct connective tissue capsule in the former. Connective tissue septae from the capsule extend into the gland and divide it into lobes, macrolobules and microlobules, and carry the excretory ducts, nerves, blood vessels, and lymphatics.

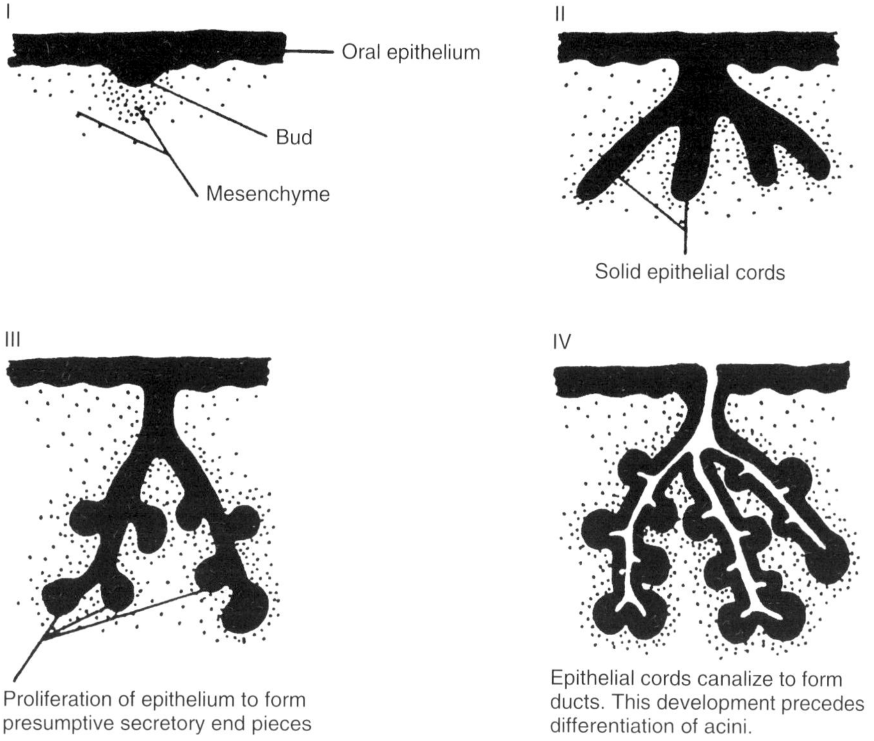

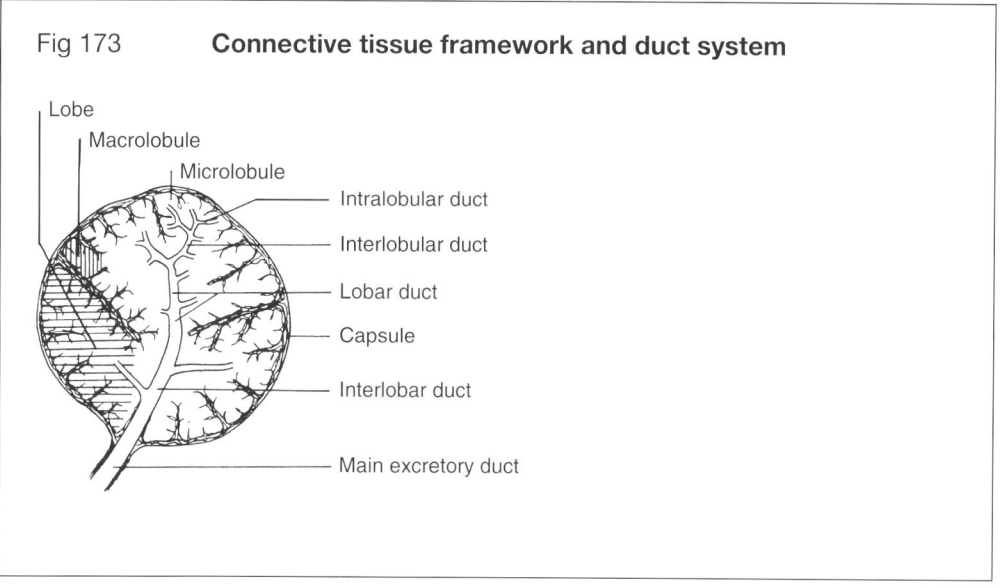

2. *Secretory ducts.* The larger excretory ducts divide into ducts of decreasing size to form a complex system. The smallest branches of the ductal system, the intralobular ducts, are continuous with the terminal secretory units (mucous tubules or serous acini) and are composed of both striated and intercalated ducts (Fig 175). More than one acinus or mucous secretory tubule may be connected to a single intercalated duct. The ductal system in minor salivary glands is much reduced and some elements may be absent.
3. *Terminal secretory cells.* These cells are found in the microlobules where they are arranged around a narrow lumen. They are separated from the adjacent connective tissue by a basal lamina.

Pure mucous glands are compound tubular glands in which the secretory cells are radially arranged around a central lumen. These mucous tubules branch extensively. Pure serous glands, as well as glands with a mixed secretion, are compound tubulo-acinar glands, composed of branching tubules (some being composed of mucous cells in mixed glands) with globular outpouchings of serous cells. The term acinus, or alveolus, is commonly applied to both mucous and serous secretory units, while some authors prefer to use the term mucous tubule for a mucous secretory unit.

General remarks on mucous cells, serous cells and the arrangement of cells in a mixed gland

Mucous cells

The histologic appearance of mucous and serous cells (Fig 174) depends on the state of functional activity of the cells. In the resting state, routinely stained by means of hematoxylin and eosin, a mucous cell is a wedge-shaped cell containing a densely stained basophilic oval or flattened nucleus located adjacent to the basal cell membrane. The cytoplasm is faintly eosinophilic and packed with droplets of mucinogen, the precursor of the glycoprotein mucin. The presence of these droplets give the cell a foamy appearance. The cytoplasm is usually so full of these droplets that the morphology of the cell is distorted and cell outlines indistinct. The droplets referred to above are in fact membrane-bound secretory vesicles and are disrupted by ordinary fixation. Fusion of these droplets commonly occurs. The electron microscope reveals the presence of rough endoplasmic reticulum and mitochondria in the basal and lateral parts of the cell, while it differs from a serous cell in that it contains very prominent golgi complexes on the apical side of the nucleus. The major part of the cytoplasm is filled with secretory vesicles. Secre-

Fig 174 — Mucous and serous cells

Both types are illustrated in a resting state, stained with hematoxylin and eosin.

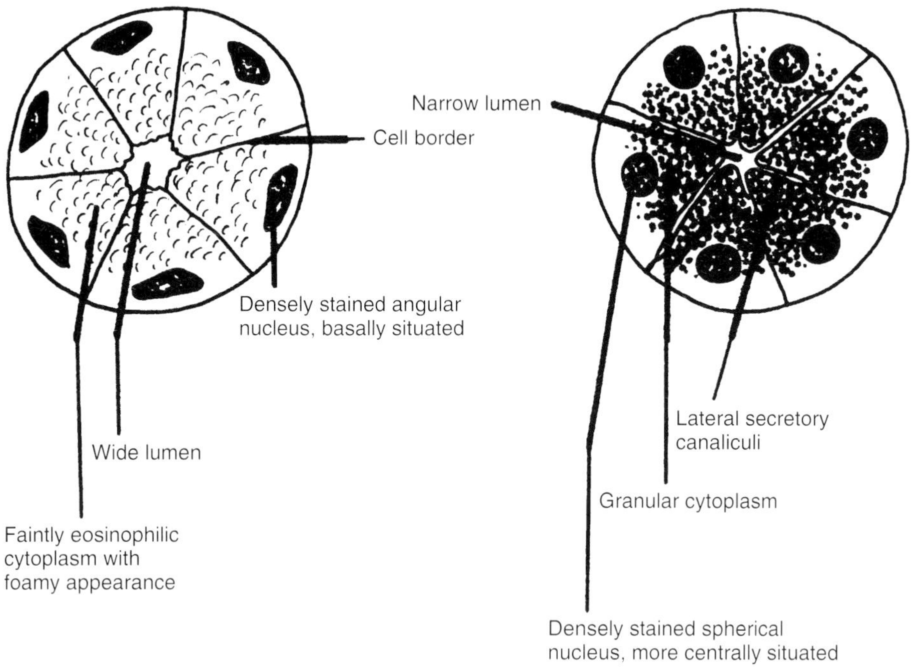

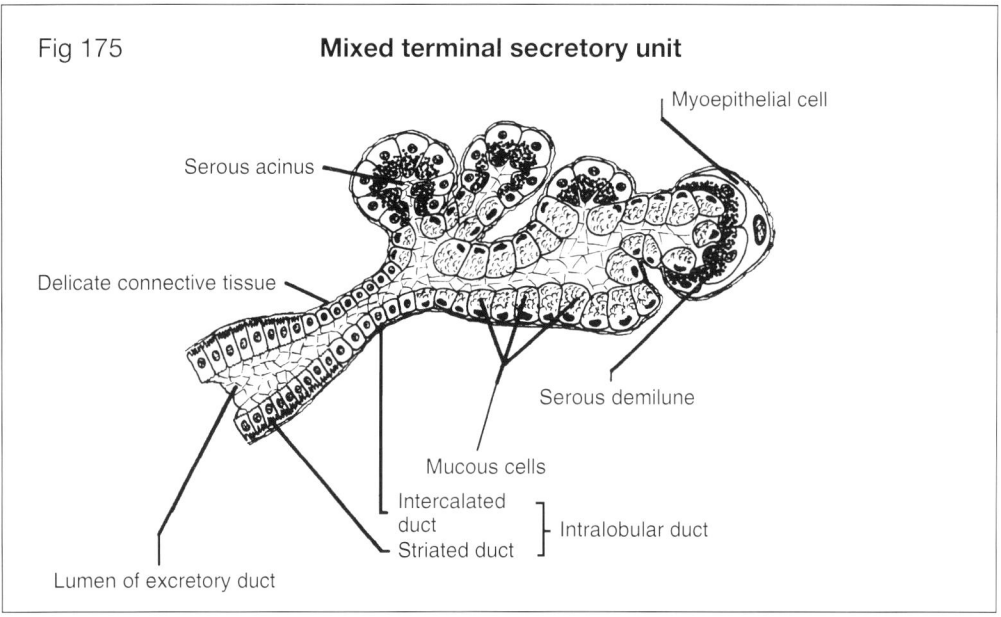

- Narrow lumen
- Cell border
- Densely stained angular nucleus, basally situated
- Wide lumen
- Faintly eosinophilic cytoplasm with foamy appearance
- Lateral secretory canaliculi
- Granular cytoplasm
- Densely stained spherical nucleus, more centrally situated

Fig 175 — Mixed terminal secretory unit

- Myoepithelial cell
- Serous acinus
- Delicate connective tissue
- Serous demilune
- Mucous cells
- Intercalated duct
- Striated duct
- Intralobular duct
- Lumen of excretory duct

tion into the relatively wide lumen occurs when the membrane of the secretory vesicle fuses with the apical membrane of the mucous cell, followed by rupture of the membrane and discharge (exocytosis) of the contents of the vesicle. The cytoplasm of these cells, because of the high carbohydrate content of the vesicles, stains brilliant red by means of the periodic acid-Schiff technique. Some intercellular secretory canaliculi are found, while both the apical and basal cell membrane shows basal folds.

When the cell is actively secreting, the secretory vesicles decrease in number, while the nucleus becomes rounded and assumes a more central position in the cell.

The secretory product of mucous cells have little or no enzymatic activity and mainly serve a lubricating function.

Serous cells

Stained by means of hematoxylin and eosin, a serous acinus (Fig 174) is composed of wedge-shaped cells arranged around a very small lumen. The nuclei are spherical, densely stained (basophilic) and situated in the basal third of the cell. The infranuclear (basal) cytoplasm is intensely basophilic (large quantities of rough endoplasmic reticulum), while the apical cytoplasm is granular and faintly eosinophilic. The presence of abundant rough endoplasmic reticulum, as seen by means of the electron microscope, is a reflection of their protein-secreting function. Mitochondria are found in the basal part of the cell and the Golgi apparatus is in a supranuclear position. The granules are membrane-bound vesicles containing α-amylase and other substances. Since these cells also secrete demonstrable quantities of polysaccharide, some authors prefer to call them seromucous cells. The lumen of a serous acinus is in communication with many intercellular secretory canaliculi, and both are lined by many short microvilli. The basal cell membrane of a serous cell shows numerous folds and rests on a basal lamina.

Arrangement of cells in a mixed gland

Mixed glands (Fig 175) contain both mucous and serous cells, and the proportion of each varies. The submandibular gland is predominantly serous, while the sublingual gland is predominantly mucous. These glands are composed of a mixture of pure mucous, pure serous, and mixed terminal secretory units. In a typical mixed secretory unit in glands in which mucous cells predominate, such as the sublingual gland, the mucous terminal portion is capped at its blind end by a crescent of serous cells, known as a serous demilune (of Gianuzzi or von Ebner). These serous cells discharge their secretions into the tubule lumen via minute intercellular canaliculi which communicate with small channels between adjacent mucous cells. These demilunes were known as basket cells in older literature.

A cell type not easily demonstrated by ordinary histological procedures is the myoepithelial cell. These cells, which have been likened to an octopus, are mainly associated with terminal secretory units but may be found around the intercalated ducts, even extending to striated ducts. Although one myoepithelial cell is usually associated with a secretory endpiece, it is not uncommon to find two or three such cells per unit. The myoepithelial cells lie between the basal cell lamina and the basal cell membranes of the secretory cells to which they are attached by desmosomes. The cell body is relatively small with a flattened nucleus and four to eight branching cytoplasmic processes. These processes, which bear some resemblance to smooth muscle cells, embrace the secretory acini and tubules and contract when stimulated to squeeze the products of the cells into the ducts, causing saliva to reach the mouth.

The duct system

The duct system of the salivary glands (Figs 173, 175) consists of the intralobular ducts (intercalated and striated ducts) which join larger interlobular ducts. The interlobular ducts join to form a lobar duct, which drains a lobe of the gland. The lobar ducts join to form the interlobar duct which runs in the connective tissue between lobes and is continued as the main excretory, or terminal, duct.

The terminal secretory units drain into intercalated ducts lined by a layer of faintly eosinophilic simple cuboidal epithelium. The cells have a fairly centrally placed nucleus with a light-colored cytoplasm containing a small amount of rough endoplasmic reticulum basally, and a Golgi apparatus apical to the nucleus. A few small secretory granules may be present in cells close to the secretory units. These inconspicuous ducts vary in length among the major glands, being difficult to identify in the sublingual gland, slightly longer in the submandibular gland, and longest in the parotid gland. Their contribution to the composition of saliva is insignificant.

Intercalated ducts are continuous with striated ducts. These have a comparatively large lumen lined by either simple columnar or pseudostratified columnar epithelial cells with centrally placed spherical nuclei and abundant eosinophilic cytoplasm. The name of these ducts is derived from the numerous infoldings of the basal plasma membrane of the cells, giving the appearance of striations. Numerous large rod-shaped mitochondria are vertically orientated between the infoldings. Many short microvilli are present on the luminal surface of these cells. The cells of the striated duct play an important role in modifying the composition of the saliva. This will be discussed in the following chapter.

The striated ducts are followed by larger excretory ducts, as previously described. The epithelium of these ducts gradually becomes pseudostratified columnar with occasional mucous goblet cells and surface cilia, changing to a stratified epithelium as the orifice of the duct in the oral cavity is approached.

Description of the salivary glands with reference to the nature of the saliva

The major salivary glands

The parotid gland, submandibular gland, and sublingual gland are the major salivary glands.

The parotid is a pure serous gland in the adult human, although mucous cells are occasionally present in the glands of young children.

The submandibular gland is a mixed, but predominantly serous, gland. Serous elements outnumber mucous endpieces in a ratio of approximately 12 to 1.

The sublingual gland is also a mixed gland but predominantly mucous. There are few pure serous acini and the serous cells present are usually arranged in demilunes.

The minor salivary glands

Depending on their location, the minor salivary glands are classified as lingual, buccal, labial, palatine, and glossopalatine glands.

The *lingual glands* are found bilaterally and are divided into several groups. The anterior lingual glands are located on the inferior surface of the tongue near its tip and are divided into anterior mucous glands and a mixed posterior portion. The posterior lingual glands are found in association with the lingual tonsil and the lateral margins of the tongue. They are pure mucous glands. The serous glands (of von Ebner) draining into the moat surrounding the circumvallate papillae, form part of this latter group.

The *buccal and labial glands* are found in the cheeks and lips. Their terminal secretory units may contain both mucous and serous elements, the latter being predominantly present as demilunes.

The *palatine glands* are pure mucous glands and found in the soft palate and uvula, and in the posterolateral region of the hard palate.

The *glossopalatine glands* are pure mucous glands located on the glossopalatine fold.

Selected bibliography

1. Bhaskar, S.N. (1991) editor. *Orban's Oral Histology and Embryology*, 11th edition. St. Louis: Mosby-Year Book
2. Dixon, A.D. (1986) *Anatomy for Students of Dentistry*, 5th edition. Edinburgh: Churchill Livingstone.

3. Kelly, D., Wood, R.L. and Enders, A.C. (1984) *Bailey's Textbook of Microscopic Anatomy,* 8th edition. Baltimore: Williams and Wilkins.
4. Krause, W.J. and Cutts, J.H. (1986) *Concise Text of Histology,* 2nd edition. Baltimore: : Williams and Wilkins.
5. Lavelle, C.L.B. (1988) *Applied Oral Physiology,* 2nd edition. London: Wright.
6. Mjör, I.A. and Fejerskov, O. (1986) editors, *Human Oral Embryology and Histology.* Copenhagen: Munksgaard.
7. Scott, J.H. and Symons, N.B.B. (1982) *Anatomy for Students of Dentistry,* 9th edition. Edinburgh: Churchill Livingstone.
8. Ten Cate, A.R. (1989) *Oral Histology: Development, Structure, and Function,* 3rd edition. St. Louis: C.V. Mosby Company.

Review questions

1. What is the location of the salivary glands?
2. Describe the embryologic development of the salivary glands.
3. Give your opinion on the classification of salivary glands.
4. Give a general description of the components of salivary glands (no histology required).
5. Contrast the histologic features of mucous and serous secretory units.
6. Describe a mixed terminal secretory unit consisting of mucous cells and a serous demilune.
7. Describe the duct system of salivary glands.
8. Classify the minor salivary glands.

23. Saliva

General remarks

Saliva is secreted by the three major salivary glands, the parotid gland, submandibular gland, and the sublingual gland, and by the very numerous minor salivary glands classified as lingual glands (on the tongue), buccal and labial glands (in the cheeks and lips), palatine glands (in the palate), and glossopalatine glands (on the glossopalatine folds).

Saliva contains a large number of constituents, such as both enzymatic and non-enzymatic proteins, calcium, phosphorus, sodium, and other salts, dissolved gases such as nitrogen, oxygen, and carbon dioxide, and cells. Microscopic examination of whole saliva found in the mouth always shows the presence of desquamated epithelial cells of the mouth, as well as leukocytes (mostly polymorphonuclear leukocytes) which enter the saliva from the gingival sulcus. They are not present in whole saliva obtained from edentulous subjects.

The pH of saliva is in large measure dependent on its rate of secretion. The faster the rate of secretion, the more alkaline is the saliva, whatever the nature of the stimulus. Most authors agree that the pH of saliva varies during the day, probably as a direct result of differences in the rate of flow. During sleep, when little saliva is produced, the pH is low, while the pH of saliva produced during meals, when the rate of flow is increased, is high. The pH of resting unstimulated saliva produced by the parotid gland is reported to be 5.81 (range 5.45 - 6.06), and of saliva produced by the submandibular gland 6.39 (range 6.02 - 7.14). Figures in respect of the pH of whole saliva given in the literature vary. One recent source states that the average pH is 6.7 with a range of 6.2 - 7.6, while an average of 5.97 (range 5.73 - 6.15) is quoted in older literature. There is fairly general agreement that the pH of fast-flowing saliva can rise to 8.0, probably as a result of an increased bicarbonate content.

Diurnal rhythms, regular fluctuations in the functions of the body over a 24 hour period, are also reflected in the rate of flow of saliva. The total volume of saliva produced in this period is calculated to be between 600 and 700 ml, although figures up to 1.5 litres per day have been cited in the past. There is an average flow of 20 ml/h in a resting (unstimulated) state for 15 hours, totalling 300 ml, 150 ml/h for 2 hours during meals (300 ml) and 20-50 ml during sleep. It has been estimated that the surfaces of the mouth, including the teeth, are covered by a 100 μm thick film of saliva, providing for a dynamic relationship between saliva and tooth enamel.

The contribution of the different glands to the volume of saliva is reported to be 60 - 65% by the parotid glands, 20 - 30% by the submandibular glands and only 2 - 5% by the sublingual glands. The minor glands provide 6 - 7%. Crevicular (sulcular) fluid is reported to contribute 10 - 100 μl per hour to the volume of whole saliva. Figures provided in the literature vary regarding the contribution of the various glands.

General factors influencing the secretion of saliva

Secretion of saliva is under nervous control and to date no hormones directly affecting the rate of salivary secretion have been identified, although it is reported that testosterone and thyroxine stimulate its rate of flow. The composition of saliva is, however, altered by antidiuretic hormone (water reabsorption is facilitated), while aldosterone results in increased sodium reabsorption in ducts, and others.

An increased secretion may result from a conditioned reflex. This psychic flow may be initiated by the noise of food being prepared, talking about food or the sight of food. On the other hand, thought about disliked food may reduce salivary secretion.

The flow of saliva is largely controlled by unconditioned reflexes and an increased secretion may be caused by the following:

1. *Taste.* Different tastes vary in their stimulating effect on the flow of saliva.
2. *Smell.* The influence of the smell of food on salivary flow is not disputed but is probably less than previously thought.
3. *Mechanical stimulation of the oral mucosa* by especially coarse food.
4. *Mechanical irritation of the gingiva* by, for example, tooth scaling and polishing procedures.
5. *Mastication of food.* The chewing of food is responsible for a variety of sensory impulses arising, for example, from mechanical stimulation of the oral mucosa, pressure on the teeth involving periodontal receptors, and impulses from the temporomandibular joint and masticatory muscles.
6. *Chemical irritation of the oral mucosa.* Acids, especially citric acid, markedly stimulate salivary flow, followed in order of effectiveness by table salt, sucrose, and bitter tastes.
7. *Distention or irritation of the esophagus* by, for example, a foreign body.
8. *Chronic irritation of the esophagus* by, for example, esophageal carcinoma.
9. *Chemical irritation of the stomach* wall leading to nausea.
10. *Pregnancy* is usually accompanied by an increased salivary flow.

Nervous control of secretion of saliva

The salivary glands have both a sympathetic and a parasympathetic secretomotor innervation (Fig 176).

The otic ganglion is a parasympathetic ganglion located just below the foramen ovale and medial to the mandibular nerve to which it is connected. The lesser superficial petrosal nerve, a branch of the glossopharyngeal (ninth cranial) nerve, carries preganglionic parasympathetic fibers from the inferior salivatory nucleus in the brain stem to synapse in the otic ganglion. Postganglionic fibers reach the parotid gland via the auriculotemporal branch of the mandibular nerve.

The sympathetic innervation of the parotid gland arises in the first two thoracic segments (T_1 and T_2) and synapse in the sympathetic superior cervical ganglion, from where postganglionic fibers reach the otic ganglion via a plexus on the middle meningeal artery. The sympathetic fibers pass through the otic ganglion without synapsing and accompany parasympathetic fibers to the gland.

The submandibular ganglion is a small parasympathetic ganglion located in the floor of the mouth and is connected to the lingual nerve. Preganglionic fibers from the superior salivatory nucleus in the brain stem reach the ganglion via the chorda tympani branch of the facial (seventh cranial) nerve which joins the lingual nerve. Postganglionic fibers from this ganglion are secretomotor to both the submandibular and sublingual salivary glands.

Sympathetic nerves to the submandibular and sublingual salivary glands initially follow a similar route to those supplying the parotid gland. Postganglionic fibers, however, reach the submandibular ganglion via plexuses on the facial and lingual arteries and pass through the ganglion without synapsing to supply the submandibular and sublingual salivary glands, as well as the minor salivary glands of a large part of the oral cavity.

The minor salivary glands of most of the palate are supplied by parasympathetic fibers arising in the superior salivatory nucleus. Preganglionic fibers run to the parasympathetic sphenopalatine ganglion, situated in the pterygopalatine fossa and connected to the maxillary nerve, via the greater superficial petrosal branch of the facial nerve and later the lesser superficial petrosal branch. Postganglionic fibers from the sphenopalatine ganglion reach the glands of the palate via palatine branches of the maxillary nerve.

Sympathetic fibers pass to the glands of the palate from the first two thoracic segments (T_1 and T_2). The preganglionic fibers synapse in the superior cervical ganglion, from where postganglionic fibers reach the parasympathetic sphenopalatine ganglion via a plexus on the maxillary artery. They pass through this ganglion without synapsing to reach the palate together with parasympathetic fibers.

Both the superior and inferior salivatory nuclei are found in the medulla oblongata. The former is associated with the brain stem nucleus of the facial nerve, while the latter is found in association with the nucleus of the glossopharyngeal nerve.

Fig 176 **Autonomic nerve supply to salivary glands**

Parotid gland

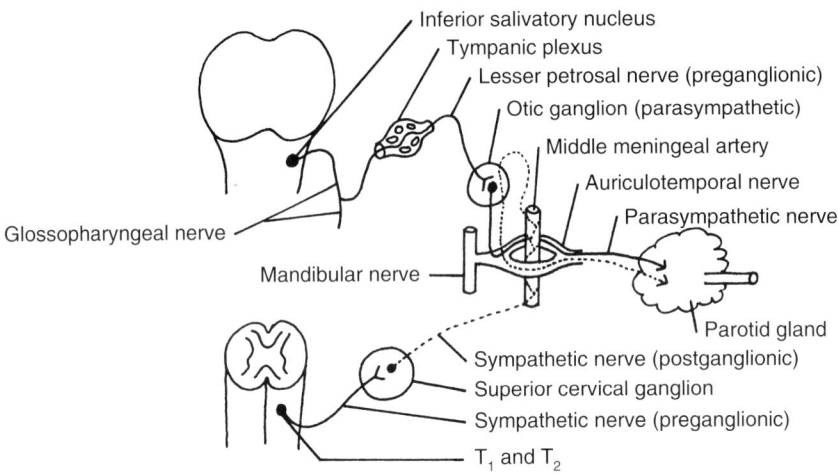

Submandibular and sublingual glands

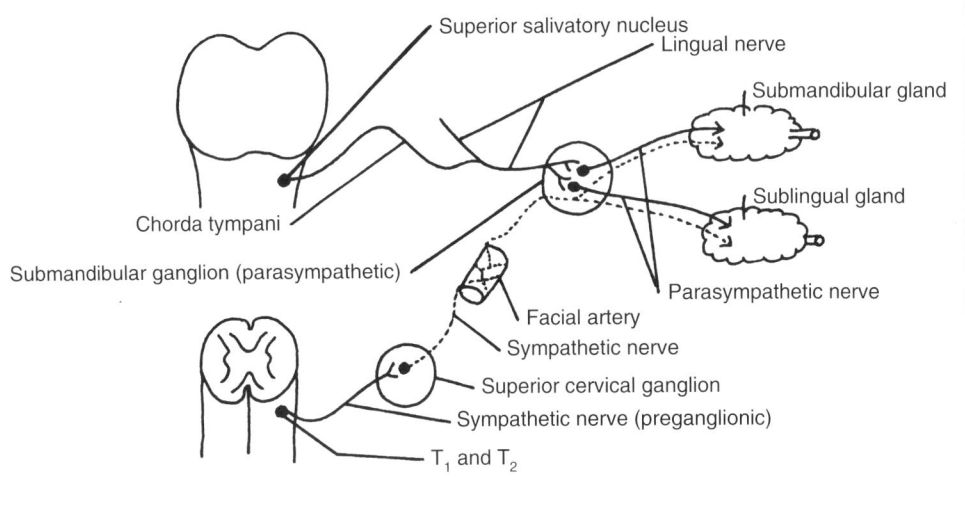

Parasympathetic innervation is secretory and vasodilatory, while sympathetic innervation is vasoconstrictive, although stimulation of the latter also promotes secretion in some cases. Secretory activity of gland cells is mediated by cholinergic (parasympathetic system) and adrenergic (sympathetic system) agents. Secretomotor nerves terminate in association with the cells of the striated and intercalated ducts, the myoepithelial cells, the smooth muscle of the arterioles, and the terminal secretory cells.

The following statements can be made with regard to the autonomic secretomotor innervation of the salivary glands in the light of present knowledge.

1. Secretory cells are supplied by both parasympathetic and sympathetic nerves.
2. Impulses conducted via the parasympathetic system are more regular than impulses along sympathetic nerves. The latter tend to be more intermittent.
3. The effect of stimulation by nerves of these two systems is not necessarily antagonistic.
4. A regular flow of impulses is necessary to maintain the normal metabolism of secretory cells.
5. Both parasympathetic and sympathetic stimulation causes a contraction of myoepithelial cells to promote salivary flow, but the effect of the former is greater in this respect.
6. Blood capillaries receive stimuli from both systems, but parasympathetic stimulation results in vasodilatation, while the vasoconstriction resulting from sympathetic stimulation forms part of a more generalised vascular control system and does not have a direct influence on the reflex secretory activity of the sympathetic system.
7. Parasympathetic stimulation is mainly responsible for secretion of a larger volume of saliva by the secretory cells. Sympathetic stimulation has a greater influence on the composition of saliva, and results in a higher concentration of organic substances due to increased exocytosis in the cells with a concomitant decreased movement of water.
8. There is no direct inhibition of the salivary glands by nerves. The "dry mouth" syndrome in states of nervous tension which was for a long time thought to result from sympathetic inhibition (overaction), is due to a direct inhibitory influence of higher centers on the brain stem salivatory nuclei.

Functions of saliva

Although not essential for life, saliva has a diversity of important functions which contribute to efficient working of the body and general well-being.

Digestive function

α-Amylase (α-1:4 glucan - 4 - glucanohydrolase), previously known as ptyalin, is present in relatively high concentrations in parotid saliva, where it forms 30% of the protein fraction, and in the secretion of submandibular gland cells, but in insignificant quantities in the saliva from the sublingual and minor salivary glands. This enzyme disrupts the α-1:4 linkages between glucose molecules in amylose and amylopectin, the constituents of starch (the latter being the main component), provided the starch has been previously cooked to break up the insoluble outer envelope of the starch granules. The end product of the process is maltose with little or no free glucose being formed.

The ability of salivary amylase to act upon starch is limited since optimal activity takes place at pH 6.8 and it is readily inactivated by gastric hydrochloric acid at pH 4.0. After a large meal, however, the contents of the stomach remain relatively neutral for up to 30 min, allowing the enzyme to hydrolyse starch. The major activity of amylase in the mouth is probably to digest remaining fragments of starch after a meal.

The salivary amylase activity of different individuals varies but a weak action is compensated for by the strong action of pancreatic amylase in the duodenum. The concentration of α-amylase is low before breakfast and rises during the morning to reach a peak in the middle of the day, after which it gradually decreases. The concentration rises with an increased rate of flow of saliva.

Antibacterial function

Several mechanisms are responsible for the antibacterial action of saliva. The following are perhaps the most important.

Secretory IgA

Immunoglobulins, or antibodies, principally IgA, are present in saliva. IgA and IgM are present in saliva in small amounts. Most of the secretory IgA and IgM are formed in the salivary glands, while IgG and a small proportion of IgM are found in the crevicular fluid. Salivary IgA is chemically distinguished from plasma IgA by the presence of a secretory component in salivary IgA which renders it more resistant to proteolysis by bacterial enzymes. IgA-specific proteases are produced by, for example, *Bacteroides*, *Streptococcus sanguis* and *Streptococcus mitis*. IgA possessing a secretory component is designated sIgA to distinguish it from plasma

IgA. Approximately 90% of the IgA of parotid saliva, and 85% of whole saliva in the mouth, is sIgA. The concentration of sIgA in saliva from the minor glands is exceptionally high, but their contribution to the volume of whole saliva is small. IgA is formed mainly by plasma cells present interstitially in the glands and taken up by the glandular cells, while the secretory component is synthesized by the glandular cells. During the intravesicular transport of IgA through a glandular cell, IgA becomes bound to the secretory component and is secreted in the saliva as sIgA.

The antibacterial activity of sIgA is chiefly the prevention of bacterial colonization by binding to specific antigens responsible for adhesion. On the other hand, colonization may be prevented by agglutination, or clumping, of bacteria which are then washed into the esophagus, or by affecting specific enzymes necessary for bacterial metabolism. Some oral bacteria are more easily phagocytosed after coating with IgA. This does not happen to the same extent when the bacteria are coated with IgG or IgM.

Peroxidase

The antibacterial peroxidase system, mainly present in parotid saliva, consists of hydrogen peroxide, formed by a variety of micro-organisms, thiocyanate, and lactoperoxidase. This system inhibits acid production and growth of many micro-organisms including lactobacilli, streptococci, and fungi. In the presence of hydrogen peroxide, thiocyanate oxidation by lactoperoxidase is catalysed, resulting in the formation of hypothiocyanate anions which oxidize certain bacterial enzymes.

Lysozyme

Salivary lysozyme is active against the cell walls of Gram positive organisms. It has been stated that the concentration of lysozyme in oral fluids is too small to have an independent effect on bacteria. Its effect results from a concerted action with other antibacterial systems such as thiocyanate. Hypothiocyanate has been shown to be very effective in the lysis of bacteria treated with lysozyme.

Lubrication

The glycoprotein content of saliva is responsible for its slimy character. It facilitates the chewing process, food bolus formation, swallowing, and speech. In addition, it protects the soft mucosal surface from damage by coarse foods.

Taste

A substance can only be tasted when dissolved and the solvent action of saliva is most important. It is almost impossible to taste absolutely dry food in the absence of saliva. Fruit can usually be tasted immediately because the taste elements are already in a virtually dissolved state.

Buffering action

The pH and buffering properties of saliva are largely dependent on its bicarbonate content, but inorganic phosphates also contribute to its buffering capacity. In the presence of a high rate of salivary flow, bicarbonate is a very effective buffer against acids and its action can be simply summarized as follows. When bicarbonate ions (HCO_3^-) come into contact with acid ions (H^+), weak carbonic acid (H_2CO_3) is formed. This rapidly dissociates to form water and carbon dioxide.

Hygienic action

Desquamated oral epithelial cells, clumps of bacteria, and food debris are loosened by the cleansing action of saliva, and subsequently swallowed.

Blood coagulation and tissue repair

Clotting time is reduced by the presence in saliva of proteins similar to the blood clotting factors VII, IX and platelet factor. The clot formed when blood is experimentally mixed with saliva is, however, less solid than a normal clot. It has also been experimentally shown that saliva, especially from the submandibular gland, accelerates the rate of wound contraction in mice, probably due to the presence in saliva of epidermal growth factor.

Inhibition of dental caries

In summary, saliva has a mechanical action by cleansing the tooth surfaces; an immunologic action by means of secretory IgA, an enzymatic action through its peroxidase and lysozyme systems, and, by virtue of the fact that it contains fluoride, calcium, and phosphate ions, promotes remineralization of carious lesions.

Water balance

In states of dehydration, salivation decreases to conserve water, causing thirst. Fluid is then normally replaced by drinking and water balance is restored.

Composition of saliva

Parotid saliva (serous) is watery while submandibular and sublingual saliva is mixed. Saliva is composed of water (94.0 - 99.5%) and solids (6.0% in unstimulated saliva, 0.5% in stimulated saliva).

Organic constituents

The following are the main organic constituents of whole saliva: urea, uric acid, free glucose, free amino acid, lactate, and fatty acids. The following macromolecules are found in saliva: proteins, amylase, peroxidase, thiocyanate, lysozyme, lipid, IgA, IgM, and IgG.

Inorganic constituents

The most important inorganic substances found in whole saliva are the following ions: Ca, Mg, F, HCO_3, K, Na, Cl, NH_4.

Gases

CO_2, N_2 and O_2.

Water

Constituents derived from the oral cavity

These include desquamated epithelial cells, polymorphonuclear leukocytes from crevicular fluid, and bacteria.

Modification of salivary composition by the ductal system

In passing through the ducts, acinar fluid is modified from an isotonic, or slightly hypertonic, fluid to a hypotonic fluid containing low concentrations of sodium and chloride ions. This electrolyte exchange activity takes place mainly in the striated duct and is most marked in stimulated saliva. The osmotic character of saliva is altered by active transport of sodium from the saliva through the striated duct cells to the extracellular compartment. This reabsorption of sodium occurs against a concentration gradient and energy for the process is provided by mitochondria in these cells. Chloride ions are simultaneously passively resorbed, while bicarbonate and patassium ions actively move from these cells into the saliva. The duct cells appear largely impermeable to water which remains in the duct lumen. The above process result in a hypotonic solution.

Halitosis (Fetor oris)

Halitosis is a fairly common condition and is experienced to some extent by virtually everyone after a night's sleep. It is reduced by eating and tends to increase as a person ages. The common form of halitosis after a night's sleep is produced in healthy mouths by normal bacterial activity which is not counteracted by the cleansing action of a sufficient flow of saliva during sleep. In cases of gingivitis, periodontal disease, and carious teeth, the odor is caused by products of tissue degradation, while halitosis may also originate in the paranasal sinuses, the lungs, or in the digestive tract.

Methyl mercaptan and H_2S are reported to be the constituents most often responsible for offensive breath. The odor-producing substances are the result of the proteolytic action of oral bacteria on desquamated epithelial cells and necrotic tissue debris. Odoriferous constituents of certain foods, such as garlic, may become adsorbed on to the oral mucosa under quite normal circumstances and lead to bad breath, while the condition is fairly common in mouthbreathers, being caused by bacterial action on organic constituents of saliva which become concentrated due to rapid evaporation of water.

When systemic conditions such as uremia and diabetes mellitus cause halitosis, the nature of the odor may have diagnostic significance.

Selected bibliography

1. Cole, A.S. and Eastoe, J.E. (1988) *Biochemistry and Oral Biology*, 2nd edition. London: Wright.
2. Garrett, J.R. (1987) The proper role of nerves in salivary secretion: a review. *Journal of Dental Research,* 66, 387-397.
3. Lavelle, C.L.B. (1988) *Applied Oral Physiology,* 2nd edition. London: Wright.
4. Roitt, I.M. and Lehner, T. (1983) *Immunology of Oral Diseases,* 2nd edition. Oxford: Blackwell Scientific Publications.
5. Roth, G.I. and Calmes, R. (1981) *Oral Biology.* St. Louis: C.V. Mosby Company.
6. Ten Cate, A.R. (1989) *Oral Histology: Development, Structure, and Function.* St. Louis: C.V. Mosby Company.

Review questions

1. Write brief notes on saliva with reference to the following:
   (a) pH;
   (b) volume produced in 24 hours; and
   (c) factors influencing the secretion of saliva.
2. Describe the nervous control of saliva secretion.
3. Discuss the functions of saliva.
4. What are the constituents of saliva.
5. Write brief notes on halitosis.

24. Fluoride

A historical review

Early this century attention in the United States of America was focussed on the incidence of mottled teeth, an abnormality in which unsightly brown pigment was present in tooth enamel. Careful observation lead to the conclusion that mottling occurred only in residents of certain districts and, furthermore, only in those teeth which mineralized during the period that affected persons resided in those districts. Teeth which mineralized when subjects resided in other areas were not affected even in the case of persons who later lived in affected areas where endemic pigmentation occurred.

It later became clear that the boundaries of affected communities corresponded with the water supply and it became apparent that a factor in the drinking water was responsible for the pigmentation. Routine investigations of the water did not reveal any causative factors until 1931, when excessive fluoride was detected in the water supply of affected communities. In 1932 it was experimentally found that when fluoride was added to the laboratory diet of rats, they developed enamel lesions in their incisors of a similar nature to those found in human subjects, an effect also obtained when fluoride was added to their drinking water. It became clear that fluoride was the cause of the mottling.

It was previously observed that mottled teeth were not more susceptible to dental caries than perfectly formed teeth, and it was determined that teeth which mineralized in areas with a water supply containing 0.5 - 1.0 part per million (0.5 - 1.0 mg per liter) fluoride were less susceptible to dental caries than those mineralizing in fluoride-free areas.

Metabolism of fluoride

General remarks

It is a well-known fact that ingestion of optimum amounts of fluoride in the food or drinking water has a beneficial effect on teeth with regard to caries.

Most of the fluoride ingested is absorbed in the digestive tract, while a small amount remains unabsorbed and is excreted in the feces. After absorption, fluoride enters the blood.

Fluoride is cleared from the plasma along different routes. It is mainly deposited in the bones and teeth, but also in small quantities in soft tissues. It is excreted in urine, an important excretory pathway, and in sweat, while small quantities are found in digestive juices, saliva and milk. It crosses the placenta in small amounts and can in this way reach the fetus.

If the rate of ingestion is far greater than the rate of elimination, fluoride will accumulate in the body, but this is normally prevented by an increased rate of excretion. Excretion of fluoride is extremely effective and is responsible for maintaining plasma fluoride levels within narrow limits.

The processes involved in assimilation, excretion, and storage of fluoride can be illustrated by the following diagram:

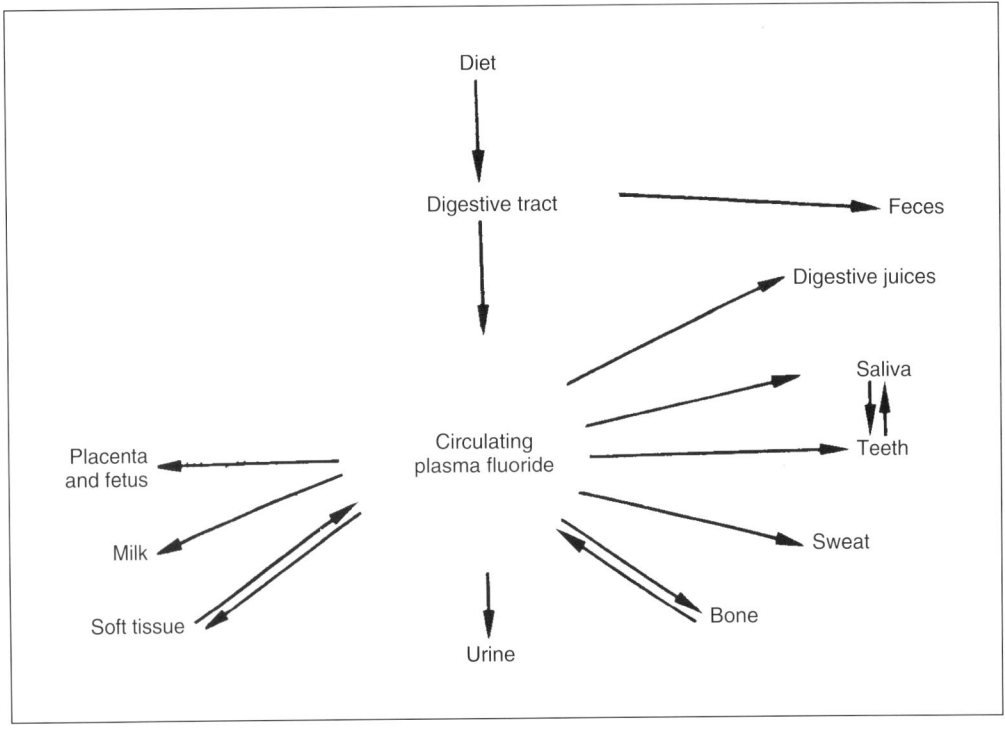

Chemical nature and natural sources of fluoride

Fluoride, together with elements such as chlorine, bromine, and iodine, belong to a group of elements known as the halogens. Fluoride is the most electronegative element and readily forms ionic compounds with metallic elements. Mutual electrostatic forces hold such a molecule together and, while most fluoride-containing minerals are not readily soluble in water, they may dissociate to form negatively charged fluoride anions and positively charged metallic cations. In this way calcium fluoride, or fluorspar (CaF_2), which is widely found in soils and rocks, can dissociate to form Ca^{++} (cation) and two F^- anions. It is important to realize that all fluoride ions are identical, notwithstanding the chemical nature of the original compound.

Fluoride occurs widespread in nature. It is present in most soils, with an average concentration of 250 ppm, and in virtually all plants and animal tissues. Surface waters do not have high concentrations of fluoride (usually less than 1.0 mg per liter), but levels up to 2800 mg per liter in underground waters have been reported. Fluoride is present in virtually all foods and it is regarded as virtually impossible to prepare a fluoride-free diet which is nutritionally adequate. Fish is a rich source of fluoride and types mentioned in this regard include salmon, sardines, and mackeral. This is probably due to a relatively high fluoride concentration in sea water, estimated to average 1.4 ppm. Tea, some mineral waters, and wine are mentioned as fluids containing appreciable amounts of fluoride. Tea leaves are very rich in fluoride of which up to 90% may be extracted during infusion. A concentration of 0.5 - 1.5 ppm per cup of tea, providing 0.1 - 0.3 mg fluoride, is cited in the literature. The total daily intake of fluoride in food and drinking water is approximately 3.2 mg for males and 2.2 mg for females, and it is estimated that water with a fluoride concentration of 1.0 ppm contributes 1.0 - 1.5 mg to a person's daily intake.

A question commonly asked is whether fluoride is an essentail nutrient. Many opinions regarding this question have been expressed. It is, however, significant that in 1973 the Federal Register of the US Food and Drug Administration listed fluoride as an essential trace element, a view endorsed by the WHO Expert Committee on Trace Elements in Human Nutrition. The beneficial effects of fluoride cannot be disputed.

Absorption of fluoride

The absorption of fluoride has been extensively studied, both in humans and in experimental animals. The general conclusion has been that small to moderate amounts of soluble fluorides are almost completely and rapidly absorbed from the digestive tract. The relatively insoluble fluorides are more incompletely absorbed and at a slower rate. Sodium fluoride, which is about 4% soluble at room temperature, together with stannous fluoride (10% solubility) and sodium monofluorophosphate (25% solubility) are listed as highly soluble, while calcium fluoride, with a 0.0016% solubility, is regarded as virtually insoluble.

Absorption of fluoride occurs by passive diffusion in the stomach and duodenum. The simultaneous presence of substances such as calcium, aluminum, and magnesium in the gut reduces absorption, probably by the formation of insoluble complexes with fluoride. On the other hand, molybdenum has experimentally been found to increase absorption and retention of fluoride, especially in older rats.

Fluoride in the blood

Absorbed fluoride rapidly enters the blood. Experiments in dogs with F^{18}, the radioisotope of fluoride, showed that approximately 75% of blood fluoride is in the plasma, while 25% is bound to red blood corpuscles. The total human plasma fluoride concentration is about 0.1 - 0.2 ppm (mg/liter) but only 20% of this fluoride is present in ionic form (0.01 - 0.02 ppm). The plasma level is maintained within narrow limits and shows only a transient rise after ingestion of fluoride, due to the very efficient clearing mechanisms.

Fluoride in saliva and digestive juices

The concentration of fluoride in saliva is slightly lower than in plasma, or of the same magnitude, and little is known of the efficiency of saliva as a route for fluoride excretion. The exchange of fluoride between tooth enamel and saliva will be discussed later in this chapter. Little work has been done on excretion of fluoride by way of digestive juices but indications are that this is an unimportant route for fluoride excretion.

Fluoride in sweat

Sweat is an important excretory route for fluoride which is reported to be present in a concentration of 0.3 - 0.4 ppm. In an air temperature of 29.5 °C, 15% of total excreted fluoride is found in the sweat, while this figure may rise to 50% in warmer climates. It is clear that an increased excretion of fluoride in warm conditions may largely compensate for an increased intake of fluids containing fluoride.

Fluoride in urine

The kidneys are the second most important organs after the bony skeleton for maintaining plasma levels of fluoride. It has been found that 20 - 30% of an oral dose of F^{18} can be recovered from the urine within 4 hours.

Fluoride in milk

The fluoride concentration in milk is probably slightly lower than in plasma and the amount of fluoride provided to an infant in this way is negligible.

Fluoride in soft tissues

The soft tissues and general organs of the body contain minute quantities of fluoride, except where secondary foci of calcification, resulting from disease processes, occur.

Transfer of fluoride to the fetus

It is generally accepted that the placenta provides a barrier to the free and unimpeded transfer of fluoride from the expectant mother to the fetus. It has been found experimentally that the concentration of F^{18} in fetal blood never exceeds 25% of that present in the blood of the mother after intravenous administration to the mother. In a further experiment in which the fluoride concentrations of fetal blood, the placenta shortly before birth, and the blood of the mother were compared, it was shown that the placental concentration was always higher than in the mother's blood, and significantly higher than the concentration in fetal blood. After the above findings it was concluded that the placenta is a semipermeable barrier to fluoride and allows limited quantities of fluoride from the mother's circulation to be incorporated in the fetal teeth and skeleton. More recent investigations have shown that slight dental fluorosis is not uncommon in primary teeth mineralized in utero while the mother was living in a high-fluoride area during pregnancy.

Toxicity of fluoride

Fluoride is, like many substances, toxic if ingested in sufficient amounts. It is estimated that a single dose of 2.5 - 5.0 g of sodium fluoride would be fatal, death probably occurring in 2 - 3 days. It inhibits many enzymes and is an irritant poison which affects all the systems of the body.

Acute poisoning as described above is extremely rare, but continued ingestion of high doses of naturally occurring fluoride will result in signs of chronic intoxication reflected in changes in the teeth and bones. Fluoride has been implicated in the etiology of a wide variety of diseases and these will be briefly considered.

Stunted growth

Many allegations have been made in this regard. Studies purporting to show a relationship between a high fluoride intake and stunted growth in animals have been carried out. It appears that a species-related susceptibility to the effects of fluoride may exist since some animals on a diet containing 100ppm fluoride showed no ill-effects, while cattle are adversely affected by an intake of 40 ppm. There are, however, no indications that humans living in high fluoride areas show any adverse affects in this regard.

Prenatal death of babies

Intake of water which is naturally or artificially fluoridated has never been proven to play any role in prenatal deaths.

Periodontal disease processes

Available evidence subjected to careful scientific scrutiny shows no correlation between fluoride intake and periodontal disease.

Effect on the joints

Fluoride does not cause arthritis. One study did show the unexpected result that a tendency towards hypertrophic changes in the vertebral column was more prevalent in subjects living in fluoride-deprived areas, compared to persons living in high fluoride areas.

Effect on the kidneys

Since especially the glomerular cells of the kidney are directly involved in a main excretory route for fluoride, it could perhaps be expected that they would be adversely affected.

It has experimentally been found that changes in animals subjected to high concentrations of fluoride are limited to a reduced tubular concentration of urine. No similar changes have been observed in man under natural conditions.

Effect on the thyroid gland

Due to the chemical relationship between fluorine and iodine, the opinion has been expressed that fluoride may be taken up by the thyroid gland and incorporated into the thyroxine molecule at the cost of iodine. This does not happen, however, under natural conditions since the gland is strongly selective in favor of iodine.

Effect on the digestive system

Although fluoride is absorbed in the digestive tract, it does not accumulate there, nor does it accumulate in the liver. No adverse effects of fluoride on these organs have been reported.

Effect on the skeleton

Fluoride accumulates in the bones and a daily intake of large amounts of fluoride can result in crippling skeletal fluorosis. Skeletal fluorosis is characterized by an increased density of the bones and can be diagnosed radiologically at an early stage. Crippling skeletal fluorosis as a result of industrial pollution was diagnosed in workers who handled powdered cryolite (Na_3AlF_2) in the aluminum industry and inhaled 20 - 80 mg of fluoride dust per day for periods of up to 20 years. The effect of fluoride in excessive quantities is to stimulate osteoblastic activity resulting in benign exostoses (non-malignant bony swellings or growths). It may also lead to calcification of ligaments and tendons, and occasionally muscle. In addition to these hard tissue changes, excessive fluoride may affect the nervous system, the thyroid gland, and the kidney.

Under natural conditions, effects on the skeleton have been limited to an increased radiopacity with no impairment of function and no clinical signs or symptoms.

Fluoride in teeth

More than 20 elements, including fluoride, have been identified in dental hard tissues. The reaction between the fluoride ion and mineralized tissue is based on an exchange between fluoride present in tissue fluid (derived from the blood plasma) and the apatite crystals of bone and teeth. The uptake of fluoride in the microstructure of these tissues is identical.

Simply stated, biological apatite is a double salt consisting of calcium phosphate and a second calcium salt, such as calcium hydroxide, calcium fluoride, calcium chloride, calcium sulphate, and calcium carbonate. The primary unit of bone and teeth (hydroxyapatite) has the following formula: $3Ca_3(PO_4)_2.Ca(OH)_2$. The formula for a pure hydroxyapatite crystal may be empirically regarded as $Ca_{10}(PO_4)_6.(OH)_2$, since it contains repeating $3Ca_3(PO_4)_2.Ca(OH)_2$ groups. An important feature of apatites is their ability to undergo ionic exchanges.

The basic structure of an apatite crystal can be likened to a honeycomb viewed from above along its long axis. The nucleus, consisting of $Ca_{10}(PO_4)_6.(OH)_2$ is surrounded by a layer, the adsorption layer, or bound ion layer, in which various ions have become adsorbed to the surface of the crystal. External to the ionic adsorption layer, the hydration shell, or layer, is composed of water taken up by the crystal when it is exposed to a fluid environment, or is occupied by enamel matrix. During enamel formation this layer is fairly wide and separates the growing crystals, but becomes narrower as enamel matures and dehydration occurs. The above arrangement may be represented by the following diagram of an apatite crystal:

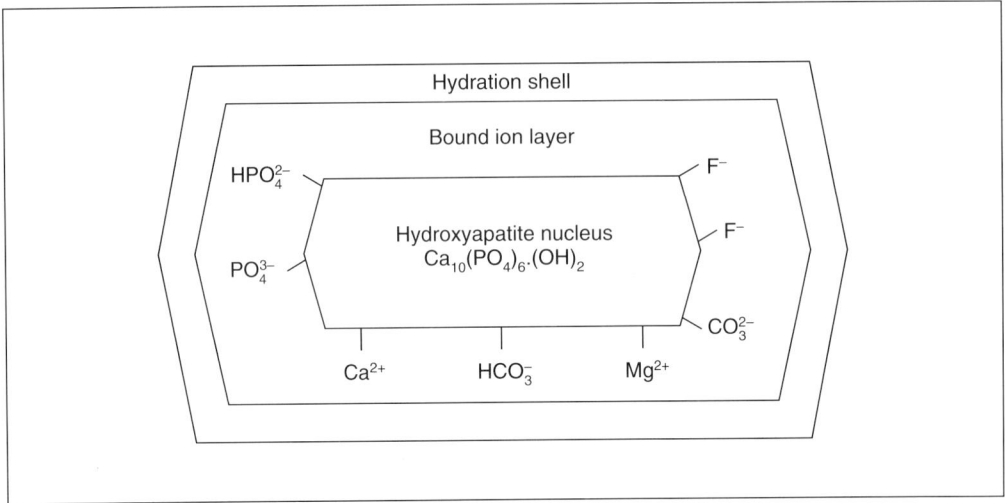

During mineralization of enamel, a continuous fairly rapid ionic exchange takes place among tissue fluid, the hydration layer, and the adsorbed ionic layer, and a slower exchange between the latter and the crystal nucleus.

Fluoride is a "calcium-seeking" ion and becomes systemically incorporated into the crystal lattice of enamel by exchanging positions with the hydroxyl ion to result in the formation of fluorapatite - $Ca_{10}(PO_4)_6 \cdot F_2$.

Similar ionic exchange is possible after completion of enamel maturation but takes place at a slower rate. Fluoride ions incorporated by the above means are semipermanently bound, being subsequently released only by resorptive processes. It has been estimated that substitution of only 10% of hydroxyl groups in enamel apatite by fluoride renders the enamel maximally resistant to caries.

Fluoride can be taken up by enamel during three stages in the life history of a tooth, namely during mineralization, as described above; after mineralization but before eruption by surface incorporation from follicular tissue fluid, and after eruption from oral fluid, again a topical effect.

The concentration of fluoride in erupted teeth is very high in surface enamel, and figures as high as 2100 ppm are given under conditions where the drinking water contains 0.05 ppm. This figure rises to 3000 ppm with 1.0 ppm F in the drinking water. The concentration falls sharply by about 1000 ppm when measured 1.5 µm below the surface. The concentration is higher in erupted than unerupted teeth, and higher in secondary than in primary teeth formed under the same circumstances.

The protective effect of fluoride on tooth enamel

General remarks

In the past, when fluoride was not used prophylactically in the form of toothpaste, tablets, oral rinses, and topical applications on such a large scale as at present, the fluoride content of the drinking water largely determined the natural resistance of tooth enamel to dental caries. The value of fluoride in drinking water has gradually decreased in the above regard in developed communities in modern times due to widespread use of the above methods of supplementing fluoride intake. A real danger of over-supplementing natural fluoride intake has arisen, and an increased incidence of mild fluorosis of secondary teeth has been observed in cases where the fluoride content of drinking water available to a community has not been taken into account in treatment planning.

Theories regarding the beneficial effect of fluoride on tooth enamel

1. Experiments on animals have shown that fluoride administered during tooth formation may result in thinner enamel, presumably due to a reduction in the amount of matrix formed. Shallower and wider fissures are formed, thus reducing the number and size of sites where food and plaque could accumulate. Such teeth may therefore morphologically be less susceptible to caries. Smaller teeth are also reported under these circumstances.

2. A statement generally made is that the presence of fluorapatite clinically reduces the solubility of enamel in the presence of plaque acids. The following experimental findings support this view:

(a) Fluorapatite forms a more compact and regular crystal than hydroxyapatite and thus presents less surface area for the action of acids.
(b) When the solubility of hydroxyapatite and fluorapatite in an acid are compared, the rates of solubility initially correspond, but the rate of solubility of the latter gradually decreases. This is thought to be due to the formation of calcium fluoride on the surface of these crystals from released calcium and fluoride ions. This serves as a protective coating which limits the rate of diffusion of hydrogen ions to the crystal interior since calcium fluoride is less soluble in weak acids than hydroxyapatite.
(c) Fluoride may replace carbonate ions in the apatite structure. Apatite crystals with a low carbonate content are more stable and less soluble than those with a high content.
(d) Fluoride facilitates the precipitation of apatite from saturated solutions. Studies involving extracted human teeth have shown that reprecipitation of mineral ions into enamel lesions caused by caries is enhanced in

the presence of even low concentrations of fluoride. Calcium and phosphate ions are precipitated in the form of apatite. The presence of fluoride in saliva in the oral environment can thus be expected to enhance remineralization of enamel affected by acids produced by plaque bacteria.

(e) Fluoride is known to inhibit many enzyme systems. A direct inhibitory action on enolase activity in oral streptococci adversely affects the transport of sugars into the cells and subsequent acid formation. Although the levels of fluoride ions in saliva are regarded as too low to have the above effect, plaque fluoride concentration is high. Most of the fluoride in plaque is, however, bound to inorganic or organic substances and only 2 - 3% is present in ionic form. Lowering of plaque pH by, for example, bacterial activity, makes more free ions available.

The detrimental effect of fluoride on teeth

Fluorosis of enamel is primarily caused by defective matrix formation (hypoplasia), probably due to a direct effect of fluoride on ameloblast metabolism, followed by abnormal mineralization. The lesion is usually confined to the outer third of the enamel and the optical properties are changed, giving the appearance of opaque white flecks in mild fluorosis. In more severe cases, oral pigments are taken up by the lesion, resulting in a brown mottled appearance.

In a study widely quoted in the literature, it was found that with a water fluoride intake of 2 ppm, fewer than 25% of teeth were completely unaffected. With an intake of 4 ppm, only 5% of teeth were normal, 25% showed moderate changes and mottling in 12% was graded as severe. These teeth displayed opacities, brown stains and pitting, involving all enamel surfaces. No teeth were free from signs of fluorosis at an intake of 6 ppm.

Selected bibliography

1. Cole, A.S. and Eastoe, J.E. (1988) *Biochemistry and Oral Biology,* 2nd edition. London: Wright.
2. Ekstrand, J., Fejerskov, O. and Silverstone, L.M. (1988) editors. *Fluorides in Dentistry.* Copenhagen: Munksgaard.
3. Jenkins, G.N. (1978) *The Physiology and Biochemistry of the Mouth,* 4th edition. Oxford: Blackwell Scientific Publications.
4. Lavelle, C.L.B. (1988) *Applied Oral Physiology,* 2nd edition. London: Wright.
5. Newbrun, E. (1986) editor. *Fluorides and Dental Caries,* 3rd edition. Springfield: Charles C. Thomas.

Review questions

1. Describe the fate of fluoride after its absorption in the digestive tract, with reference to the following:
   (a) plasma concentration;
   (b) excretion in saliva, digestive juices, sweat, urine and milk; and
   (c) transplacental transfer.
2. Write brief notes on the chemical nature and behavior of fluorine.
3. Discuss the toxicity of fluoride.
4. Discuss fluoride in teeth with reference to the following:
   (a) uptake by enamel; and
   (b) effects on the susceptibility of the tooth to caries.
5. Write brief notes on dental fluorosis.

25. Mastication

The mechanics of mastication

The following factors will be briefly discussed:

1. The temporomandibular joint
2. The role of the tongue
3. The role of the hard palate
4. The role of the lips and cheeks
5. Occlusal contacts between opposing teeth
6. Movement of individual teeth

The temporomandibular joint (TMJ)

The human temporomandibular joint allows three types of movement:

(a) A simple hinge-like opening and closing movement of the mandible around an imaginary horizontal axis located in the mandibular condyles. This movement is used during the biting of food, and is the only type of movement of which the mandible of carnivores is capable.
(b) A movement in which the mandible is protruded. This movement is used in the early stages of mastication and is made possible by a forward sliding action, or translation, of the mandibular condyle on the posterior slope of the articular eminence of the temporal bone. This movement corresponds with the movements of the mandible in rodents.
    When a slight hinge movement is followed by a protrusive movement, the cutting edges of the incisors are brought nearer to each other. This movement is used to bite soft foods. It is often necessary to follow the above movement with a fast backward movement of the mandible, producing a tearing action suitable for tougher foods. The result is a sliding scissors-like action of the incisors on each other.
(c) A third movement allowed by the temporomandibular joint is a lateral or sideways movement of the mandible. This is used when food is chewed between the posterior teeth and is a grinding action. This movement

is accompanied by rapid hinge movements to simultaneously crush the food. The sideways movements are nearly symmetrical in most individuals but factors such as a painful teeth may cause one side to be favored, resulting in the possible establishment of persistent unilateral chewing habits.

The role of the tongue

The tongue has several functions in the masticatory process.

(a) It may be used to crush soft food against the hard palate and is aided in this function by the roughness of the palate (the rugae) and of the dorsal surface of the tongue (the papillae).

(b) The tongue mixes the food with saliva and transfers the food from one side of the mouth to the other, selecting appropriate occluding surfaces and ensuring that all parts of the food are properly chewed.

(c) The sensory endings on the dorsal surface of the tongue aid it in the above functions and further enable it to discriminate between those parts of the food ready for swallowing and those parts requiring further mastication. The former are then moved to the oropharynx.

(d) The tongue has a hygienic function by removing residues of food from between the teeth and from the oral vestibule by a sweeping action.

The role of the hard palate

Due to its roughness which prevents the uncontrolled sliding and displacement of food, the hard palate is an ideal surface for the crushing of soft food by the tongue. The large number of very sensitive endings in the hard palate play an important role in gauging the roughness of food and aid the tongue in selecting those portions ready for swallowing. Denture wearers, especially older individuals, not uncommonly report a partial loss of taste. This can probably be attributed to a loss of the sense of touch (texture of the food), the nerve endings being covered by the denture. For the same reason, these individuals are unable adequately to judge the temperature of food. It is also reported that sensitivity to bitter and acid tastes may be impaired since these sensations are perceived mainly on the hard palate.

The role of the lips and cheeks

The sensory function of the lips, especially to temperature and touch, ensures that harmful materials are prevented from entering the mouth. By their mechanical function the lips allow the transfer of food, and especially fluids, into the mouth. By sealing the mouth opening, the lips prevent the loss of fluid and food to the outside.

The cheeks probably do not play an important role in mastication. They may aid transfer of food which has entered the vestibule during mastication back into the mouth proper through the action of the buccinator muscle. This is, however, largely the function of the tongue.

Occlusal contacts between opposing teeth

In individuals with a full complement of healthy teeth, the chewing pattern is more or less as follows. The food is mashed between the occluding posterior teeth. Part of the food may then be pushed forward by the tongue against the hard palate and some may be introduced between the anterior teeth for finer processing. The process is then completed between the posterior teeth. The chewing cycle is seldom completed unilaterally, except in individuals with painful teeth on one side, or with such an established habit.

The relationship between the number of movements and the number of contacts between opposing teeth depends largely on the nature of the food. It was found that more than half the individuals made occlusal contact with every chewing movement in the case of raw carrots, apples, cooked meat, or bread, but that contacts were far fewer in the case of biscuits. A view sometimes expressed is that direct occlusal contact between opposing teeth is of limited value during mastication and merely serves as a warning to stop the masticatory movement.

Movement of individual teeth

Independent slight movements of individual teeth in their sockets occur during mastication but one is normally unaware of these movements. Damage to the teeth would be caused if slight movements were impossible, since the teeth would be unable to withstand sudden occlusal stresses brought about the action of the masticatory muscles. The periodontal ligament and vascular system absorb and withstand some of the forces applied to a tooth. The normal mobility of teeth is largely determined by the periodontal vasculature. It was experimentally found that mobility decreased after local administration of norepinephrine (a vasoconstrictor), probably because movement of the periodontal space was inhibited by local stagnation.

In experiments performed on monkeys, it was found that occlusal stresses can be accommodated by the periodontium in two ways, depending on the magnitude of the forces applied. Light forces result in a rapid stretching of periodontal ligament fibers, while stretching of ligament fibers is followed by elastic deformation of the surrounding alveolar bone when heavier forces are applied.

Nervous control of the masticatory process

Masticatory movements are controlled by voluntary muscles and this control is normally consciously exercised entirely by the will. It is, however, a complicated activity and as such is subconsciously controlled through vital subcortical centers. In this way reflex activity occurs.

In studies performed approximately 70 years ago on anesthetized cats to investigate the reflex nature of masticatory movements, it was found that mechanical irritation of the tongue caused a reflex closing of the jaw. Mechanical stimulation of the occlusal surfaces of the teeth, the gums, and anterior part of the hard palate resulted in reflex opening of the jaw.

The dynamics of mastication

Methods to determine the strength of the bite

Borelli (1681) was the first person to record measurements of masticatory forces. His simple method was to loop a cord over the mandibular molar teeth and to attach weights to the cord. In this way he determined maximum weights that could be lifted by the mandible. In more recent times gnathodynamometers have been used in studies of biting force and factors which influence it. This instrument to measure the strength of the bite basically consists of pads of various sizes (depending on the size of the area over which force is to be recorded) which are placed in contact with opposing teeth in such a way that the force generated in closing the jaw is measured against resistance. The pads are attached to a measuring instrument. In the past, calibrated springs with pointers on scales and hydraulic pressure gauges were used to measure biting forces, but sophisticated electronic instruments, making use of strain gauge transducers built into dentures, bridges, and crowns with graphic recordings of bite force, have taken their place in recent times.

General information on biting force

It has been determined that maximum biting force between the molar teeth of young people is about 50 kg, males being able to exert more force than females, except between the anterior teeth where the forces are reported to be about equal in both sexes. Despite better neuromuscular control and muscular development, athletes do not exert more biting force than non-athletes. In ordinary mastication one uses about a third of the force one is capable of.

The maximum biting force (45 - 50 kg) is measured between the first permanent molars and gradually diminishes on either side of these teeth, being approximately 10 kg less in the incisor region. Other sources report that premolars and incisors exert about one third of the force produced through the molars. The factors limiting the biting force are not quite clear but it is thought that protective reflexes may arise in periodontal receptors generally and inhibit contraction of masticatory muscles when loads become unacceptably high; that the greater periodontal support of especially the first molar teeth distribute the tension more widely, thus causing periodontal mechanoreceptors to react to higher loads than teeth with smaller attachment before reflex inhibition of muscles; that pain receptors in the ligament may be activated when the load applied to the tooth becomes to great, thus limiting the masticatory force, and that the anatomical position of teeth in relation to the muscle insertions may be important (the lever principle). The importance of periodontal mechanoreceptor feedback in muscular control is well illustrated by the finding that complete denture wearers are capable of only one quarter to one third the force exerted by a person with an adequate natural dentition. Partial denture wearers are also not capable of biting as hard as a person with natural teeth.

The effect of practice

An individual's maximum biting force can be increased by exercise. In a study performed on males and females in which the masticatory apparatus was exercised over a period of 50 days by chewing a cube of paraffin wax, males increased their biting force by about 11 kg and females by 10 kg. Two weeks after the termination of the experiment, biting forces in both sexes had almost returned to the starting figures. The fact that biting forces of up to 150 kg have been measured between the molar teeth of traditional Eskimos who have for many generations lived on a very tough diet requiring vigorous mastication in their natural state, emphasizes the underdeveloped condition of the masticatory system in westernized man. On the other hand, it may well be that the robust masticatory system of Eskimos is a manifestation of their genetic inheritance. It has further been shown that persons who habitually favor one side during chewing are able to exert double the force on that side, compared to the other side, a further indication of the role of exercise.

Tests for masticatory efficiency

One of the few methods which have been used to investigate masticatory efficiency is to request the individual to be tested to masticate a weighed quantity of food which does not dissolve completely, such as nuts, under controlled conditions of time and number of chews. The contents of the mouth are then spat out and the mouth carefully rinsed. The water, saliva, and chewed food are collected and poured through sieves of decreasing mesh size. The efficiency of mastication is measured by comparing the mass of the dried food residues passing through the sieve with the smallest mesh with the original mass before chewing.

The efficiency of mastication

The efficiency of mastication is related to the state of the dentition. In an experiment in which standardized particles of hardened gelatine were used as a test material under controlled conditions, the ability of persons to increase the surface areas of the particles by chewing was taken as a coefficient of masticatory efficiency. It was found that individuals with good dentitions (eighteen occlusal contact points) could increase the surface area 11 times, while persons with poor dentitions (five occlusal contact points) were only able to increase the surface area by about seven times. It has also been reported that individuals with a poor occlusion partly compensate for this deficiency by managing the food more skilfully in the mouth, but do not chew for a greater number of times, and that older men chew more efficiently than younger men, probably due to larger occlusal contact areas on the teeth caused by attrition.

Number of masticatory movements

It appears that every individual adopts a fixed habit regarding the number of masticatory movements and that the number of movements is not influenced by the state of the dentition. In this respect it was found that the number of chewing movements of persons with fewer teeth, and consequently a smaller occlusal contact area, did not differ significantly from the number of chews of individuals with complete dentitions. One would have expected the former to require more movements to chew more thoroughly to compensate for a smaller occlusal area. Adult males make slightly more chewing movements than females, and young children make fewer movements than adults. The number of movements of an individual is not altered by small changes in consistency and toughness of food, but where big differences in this regard are encountered, for example, between an egg and tough meat, the number of chews are adapted to the nature of the food.

The importance of mastication for digestion

In a widely reported experiment to investigate the effect of mastication on the digestibility of food, the test persons swallowed two small net bags, one containing unmasticated food and the other a similar quantity (1 g) of the same food which had been chewed 20 times on the preferred side before being placed in the bag. The bags were recovered from the feces and the contents weighed to determine the degree of digestion. On the basis of the results obtained, the 29 foods tested could be divided into two main groups, namely those foods tending to leave large undigested residues if not chewed but were almost completely digested when chewed (group 1), and those foods which are digested whether chewed or not (group 2). The foods in each group are listed below.

Group 1:	roast and fried pork
fried bacon
roast, fried, and stewed beef
roast and stewed mutton
roast and fried lamb
stewed lamb
roast chicken breast and leg
fried potatoes
boiled old and new potatoes
boiled garden peas
boiled carrots

Group 2:	fried and stewed beef fat
fried and boiled cod
fried kipper
hard-boiled egg
boiled rice
white and wholemeal bread
cheddar cheese

The results further showed that the degree of mastication required for optimal absorption of the least digestible foods was slight and could be attained by persons with inadequate dentitions. It is stated that a person with 23% of masticatory efficiency can be expected to masticate sufficiently well to ensure digestion of any of the foods in this experiment.

Efficient mastication does, however, have many beneficial effects, such as optimal development of the jaws and the muscles of mastication, inhibition of plaque formation, prevention of gingival inflammation and improved keratinization of the gingiva.

Selected bibliography

1. Brekhus, P.J., Armstrong, W.D. and Simon, W.J. (1941) Stimulation of the muscles of mastication. *Journal of Dental Research*, 20, 87 - 92.
2. Dahlberg, B. (1946) The masticatory habits. *Journal of Dental Research*, 25, 67-72.
3. Farrell, J.H. (1956) The effect of mastication on the digestion of food. *British Dental Journal*, 100, 149-155.

4. Jenkins, G.N. (1978) *The Physiology and Biochemistry of the Mouth,* 4th edition. Oxford: Blackwell Scientific Publications.
5. Kawamura, Y. (1974) editor. *Physiology of Mastication,* Volume 1 of *Frontiers of Oral Physiology.* Basel: S. Karger.
6. Lavelle, C.L.B. (1988) *Applied Oral Physiology,* 2nd edition. London: Wright.

Review questions

1. Briefly describe the mandibular movements allowed by the human temporomandibular joint.
2. Discuss the masticatory process with reference to the following:
   (a) the role of the tongue;
   (b) the role of the hard palate;
   (c) the role of the lips and cheeks;
   (d) occlusal contacts between opposing teeth;
   (e) movement of individual teeth; and
   (f) nervous control.
3. Write brief notes on methods of determining biting force.
4. Discuss biting force with reference to the following:
   (a) the force transmitted through the teeth; and
   (b) the effect of practice.
5. Briefly discuss a method for testing masticatory efficiency.
6. Briefly discuss the relationship between the state of the dentition and masticatory efficiency.
7. Briefly discuss the importance of mastication in the digestion of food.

26. Taste

General remarks

The sense of taste is probably of greater importance to lower animals than to modern man. It has been shown that animals develop a craving for certain nutrients, such as salt, in which their diets may be deficient, and the mechanism for this phenomenon probably involves taste. In an experiment to test the effect of the body's requirements on tasting ability, it was found that after removal of the adrenal glands in rats, their ability to taste table salt increased to such an extent that they could detect much lower concentrations of table salt than normal rats. Adrenalectomized rats require more sodium chloride than normal rats to survive.

Taste is the detection and recognition of substances in solution. Some of the nerve endings responsible for the change of a chemical stimulus into a nerve impulse are located in the taste buds found on certain parts of the tongue, soft palate, fauces, pharynx, and epiglottis. They are mainly found in the epithelium of the circumvallate and fungiform papillae in the adult tongue, being much reduced in number on the central part of the dorsum of the tongue and near its tip. They are, however, not confined to the tongue. Approximately 9000 taste buds are found in the mouth of an average adult, but many more in a child, where they are located on the whole dorsum of the tongue, the hard and soft palates, and the inner surface of the cheeks. The number of taste buds decreases by as much as 60% in old age and this fact is partly the reason why tasting ability declines in older people.

Primary (basic) tastes

Man can distinguish four tastes, although some investigators maintain that it is possible to identify six primary tastes. The confirmed primary tastes are salt, sweet, bitter, and acid or sour. The tastes that are questionable include the ability to taste an alkaline or a metallic substance. All bitter and sweet substances do not taste the same on different parts of the tongue, and the ability to distinguish taste varies in the different regions. The tip of the tongue is more sensitive to sweet taste. The area where the circumvallate papillae are located is more sensitive to bitter taste, where as the edge of the tongue is more sensitive to acid and salt. The taste buds react to more than one primary taste, but some are relatively specific in their reaction to certain tastes. Even individual taste bud cells generate an action potential when they are brought into contact with different tastes. Such results cast doubt on the existence of specific receptors for each primary taste.

In man, the general preference for sweet and salt tastes is probably due to the fact that these tastes are associated with foods having a high nutritional value. Acid and bitter tastes, on the other hand, are quite often identified with deteriorating foods that should best be avoided. The tongue is, however, not the only structure for the detection of taste. It has been shown experimentally that anesthesia of the palate reduces the ability to detect acid and bitter tastes, but that the detection of sweet and salty substances remains normal. The obvious explanation for the latter finding is that the tongue is responsible for these tastes.

A general clinical observation is that tasting ability is adversely affected by the wearing of full dentures. In some cases the ability to taste returns to normal when the upper denture is removed, for reasons mentioned previously, but in other cases there is no improvement. In the latter group, degeneration of the palatal gustatory nerve endings is probably caused by repeated mechanical irritation of the mucosa by the denture base.

Taste and chemical structure

In the past, many investigations have been carried out regarding the relation between chemical structure and taste, but it still remains difficult to isolate the specific chemical groupings that confer either sweet or bitter tastes to substances. There appears to be little in common between the chemical structure of saccharine, lead acetate, glycine, and sucrose, yet all these substances have a sweet taste. Acids share the ability to release hydrogen ions and this is evidently the reason for a similarity in the taste of different acids. Any difference between the taste of, for example, citric acid and vinegar, could be ascribed to the different anions that are superimposed on the hydrogen ions (citrate ions are sweet, whereas acetate ions are bitter). A further hypothesis is that the taste of some substances differ when brought into contact with different areas on the surface of the tongue, and that the nature of the taste is further determined by the concentration of the substance in question. Saccharine in low concentration, for example, tastes sweet, but bitter in high concentration.

Other stimuli associated with taste

A logical question might be the following: If our tasting mechanism detects so few primary tastes, how is it then possible to identify such a large variety of different tastes? Although a complete answer to the question is difficult, it remains true that many factors influence tasting ability. The most important factor in this regard is smell.

The sensation generally known as taste, is in fact a combination of true taste and smell, and is known as flavor. The importance of smell is shown by the influence of the common cold on tasting ability. The olfactory epithelium in this condition is covered by a mucus secretion which greatly decreases the ability to discriminate between the taste (flavor) of many substances. The double nature of flavor can be very clearly illustrated by holding the nose when a piece of raw onion is placed in the mouth. Almost all the flavour is banished by this procedure, only to return when nose-breathing is again allowed. This observation implies that volatile flavors expired from the mouth are transported to the olfactory epithelium through the posterior nasal openings.

Other factors which contribute to flavor are the temperature and texture (hardness, roughness, smoothness, or softness) of food. These sensations are partly felt on the hard palate and their acuity may diminish with the wearing of artificial dentures that cover the palate. Temperature preferences are probably acquired. For example, hot coffee is preferred to cold coffee by most people, whereas water is preferably drunk fairly cold.

The oral mucosa is also sensitive to general chemical stimulation by irritants, accounting for the pungent flavor of substances such as pepper or mustard. Free nerve endings which are widely distributed in the oral mucosa are responsible for these sensations, and taste buds probably play little, if any, role.

Sensitivity of the tongue to a taste is altered by prior stimulation by another taste. A good example is that a bitter substance tastes more bitter when taken after sweet substances.

Adaptation to taste

Adaptation to taste and smell is rapid. It is a fairly common experience that the intensity of a taste sensation rapidly diminishes when the stimulus remains unchanged and in one place. When, for example, a sweet is held in one area of the mouth for a minute or two, consciousness of presence and taste diminishes rapidly, but returns immediately when the sweet is transferred to another area which has not been stimulated. If the tongue adapts to one kind of acid and perception diminishes, it was found that the same adaptation applied to various other acids (cross adaptation), but it might then become more sensitive to other tastes (water tastes sweet after application of an acid). However the same cannot be said about bitter and sweet substances, since adaptation to one of these tastes does not necessarily imply adaptation to another substance with a similar taste.

The nerves involved in taste are discussed in the chapter on the development of the pharyngeal arches and the tongue (Part I, Chapter 17), and the histology of a taste bud in the chapter on oral mucosa (Part II, Chapter 14).

Selected bibliography

1. Ganong, W.F. (1987) *A Review of Medical Physiology,* 13th edition. Los Altos: Lange Medical Publications.
2. Jenkins, G.N. (1978) *The Physiology and Biochemistry of the Mouth,* 4th edition. Oxford: Blackwell Scientific Publications.
3. Lavelle, C.L.B. (1988) *Applied Oral Physiology,* 2nd edition. London: Wright.
4. Meyer, J., Squier, C.A. and Gerson, S.J. (1984) editors. *The Structure and Function of Oral Mucosa.* Oxford: Pergamon Press.
5. West, J.B. (1991) *Best and Taylor's Physiological Basis of Medical Practice,* 12th edition. Baltimore: Williams and Wilkins.

Review questions

1. Discuss the preference of man and animals for certain tastes.
2. On what parts of the tongue are each of the primary tastes sensed the best?
3. Discuss the relation between the chemical composition and taste of a substance.
4. What is flavor?
5. To what would you ascribe the loss of tasting ability experienced by some wearers of full dentures?

27. Deglutition

General remarks

Despite the fact that the swallowing process has been extensively investigated, the precise mechanism of the act remains largely uncertain. Parts of the process are so rapid that it is impossible to follow even a radiological image when a radiopaque substance is swallowed.

There are three basic phases in the swallowing process:

1. A voluntary transfer of material from the mouth to the pharynx. The tongue muscles and the muscles of the floor of the mouth play an important role in this phase. This phase is also called the oral phase.
2. An involuntary, or reflex, mechanism transfers the material from the pharynx into the upper esophagus. The pharyngeal constrictor muscles play the most important role in this phase. This phase can, however, be reversed by a conscious effort. It is also known as the pharyngeal phase.
3. In the third, or esophageal phase, the contents of the upper part of the esophagus are transferred to the stomach by involuntary peristaltic contractions of the esophageal musculature.

The first phase (oral phase)

The initial stage is voluntary and begins when the masticated food is selected by the tongue to form a bolus, ready to be swallowed. The bolus is placed on the dorsum of the tongue and moved upwards against the hard palate. Contraction of the mylohyoid muscle, in conjunction with the intrinsic muscles of the tongue (genioglossus, styloglossus, and palatoglossus), makes the above movement possible. The next action is initiated when the tip of the tongue is pressed against the palate, just behind the incisors, followed by contraction of the buccinator muscles to pass any residual food from the vestibule to the oral cavity proper to be incorporated into the bolus. The teeth come into close approximation and the tongue seals against the lingual surfaces of the maxillary teeth and the adjacent palatal mucosa. The lips are pressed together to seal the oral opening. The bolus is now ready to be transferred to the pharynx.

The next step in the oral phase of swallowing begins when the base of the tongue is lowered and the soft palate is elevated by the action of the levator palati and tensor palati muscles to seal the posterior nasal openings. The anterior part of the tongue is then pressed in quick succession against the adjacent maxillary gingiva and the front of the hard palate. The tongue executes a peristalsis-like movement from before backwards to transfer the bolus posteriorly along the dorsum of the tongue towards the fauces. Some authors describe this action as a midpalatal phase. The oral phase terminates when the bolus contacts the fauces.

A second mechanism which has been suggested, by which the bolus is transferred to the pharynx, is based on radiological observations. It was concluded that rapid relaxation of the tongue and pharynx, with the airways closed, creates a space. This leads to negative pressure in the oropharynx which was thought to suck the bolus from the mouth into the pharynx. This theory has not received much support, since most researchers are of the opinion that the pressure in the pharynx is positive during all stages of the swallowing process. Deglutition is not reported to be impaired in persons in whom air can freely enter the pharynx as a result of accidents.

When solids are swallowed, the process is initiated when teeth make contact, the tongue is pressed against the palate behind the incisors and against the palatal gingiva, and the lips seal the oral opening. This mechanism may be disturbed in some children who swallow with the teeth apart, possibly a continuation of the swallowing process used when suckling. The accompanying tongue thrust may lead to the development of an anterior malocclusion. It has, on the other hand, been found that many individuals with good occlusion do not make tooth contact during swallowing.

Pressure exerted by the tongue during swallowing

By means of transducers placed in different parts of an acrylic plate, it was found that pressure exerted by the tongue on the lateral and anterior parts of the palate were higher than on the central part. The shape of the palate was found to be an important factor, since this largely determines the nature of the tongue seal on its different parts. In deep palates, more pressure was exerted on the periphery than on the central part. The reverse was found in a shallow, flatter palate, since the central part of the tongue would make better contact with a flat palate.

Pressure exerted by the tongue on the palate remains far less than pressure exerted by the tongue on the teeth during swallowing. It is therefore not surprising that abnormal function of the tongue is an important factor in the etiology of malocclusion.

The second phase (pharyngeal phase)

The second phase begins when the bolus makes contact with the posterior parts of the oral mucosa and with the mucosa of the pharynx. These contacts on sensitive areas act as stimuli for a series of reflexes that are responsible for the bolus being transferred into the esophagus and not into the trachea or nasopharynx.

The sensitive parts have been determined in a large number of people by stimulating various parts of the mucosa to determine which is followed by a swallowing reflex. No single area could be found that would lead to a reflex swallowing action in all the subjects. The most sensitive areas were, nevertheless, the anterior pillars of the fauces, but even here 23% of individuals did not respond. The soft palate and uvula were insensitive in, respectively, 81% and 68% of individuals. The swallowing reflex on mechanical stimulation was impaired after the sensitive areas were anesthetized. Anesthesia also impaired voluntary swallowing in some individuals. By the above mechanism the pharyngeal phase is initiated.

Two basic movements are carried out by the pharynx. The whole pharyngeal tube is elevated by the stylopharyngeus and palatopharyngeus muscles, followed by a wave of peristalsis of the pharyngeal constrictor muscles which propels the bolus into the esophagus.

The larynx is elevated and pulled up under the tongue and the epiglottis folds down and closes the laryngeal opening. The upward movement of the larynx is initially brought about by its attachment to the hyoid bone which is lifted by the digastric muscle (especially its anterior belly), stylohyoid, mylohyoid, and geniohyoid muscles. This movement stretches and widens the upper anterior part of the esophagus to make it more suitable for receiving the bolus. A cricopharyngeal, or hypopharyngeal, sphincter has been described at the pharyngeal opening of the esophagus. It is normally closed to prevent inhaled air from entering the esophagus. It relaxes and the vocal cords are approximated.

The second phase of deglutition ends when the bolus is transferred from the pharynx into the upper esophagus and the muscles of the tongue, palate, pharynx, and larynx relax. The first and second phases of swallowing last approximately 1 second.

Breathing is inhibited during the second phase of the swallowing process (apnoea of deglutition) to prevent food being sucked into the respiratory passages.

The third phase (esophageal phase)

The final phase of deglutition begins as the bolus enters the pharyngeal opening of the esophagus. Peristaltic contractions of the esophageal musculature propels the bolus to the stomach. The peristaltic wave normally takes 6 - 7s to force the bolus along the entire length of the esophagus. While in the upper part of the esophagus, the bolus can be returned to the mouth by a strong voluntary effort.

The process is slightly different when fluids are swallowed since fluids flow down the esophagus by gravity. The movement of fluid is faster than the peristaltic wave, and fluids tend to pool at the esophageal opening of the stomach before this area relaxes under the influence of the peristaltic wave to allow the fluid to enter the stomach. Although a definite cardiac sphincter has not been demonstrated in man, pressure changes in the lumen of the esophagus have pointed to the existence of a functional sphincter at this site. At rest this sphincter is in a state of tonic contraction.

It was found in one study that a person swallows about 590 times during a typical 24-hour cycle. This includes 146 swallows while eating, 394 times between meals, and 50 during sleep. The low number of swallows recorded during sleep was attributed to a decreased salivary secretion. In the light of other investigations, the above figures may be regarded as too low. Figures of up to 3000 swallows per day have been reported in normal individuals.

The volume of fluid swallowed per act of deglutition was determined. The volumes were found to average 21 ml for adult males, 14 ml for females, and 4.5 ml for children between the ages of 1.25 and 3.5 years.

Nervous control of deglutition

The afferent (sensory) nerve fibers belong to the trigeminal, glossopharyngeal, and vagus nerves (cranial nerves V, IX, and X) and arise in nerve endings in the mucosa of the mouth, fauces, tonsils, and pharynx.

The main swallowing center is situated in the medulla oblongata, close to the nucleus of the vagus, and from here efferent nerve fibers lead to the muscles responsible for swallowing, namely the facial nerve (cranial nerve VII) to the lip muscles and buccinator muscle, the hypoglossal nerve (cranial nerve XII) to the muscles of the tongue, the trigeminal nerve (cranial nerve V) to the mylohyoid muscle in the floor of the mouth, as well as to palatal muscles, the glossopharyngeal (cranial nerve IX) and vagus (cranial nerve X) nerves to the muscles of the pharynx and esophagus.

Selected bibliography

1. Dixon, A.D. (1986) *Anatomy for Students of Dentistry*, 5th edition. Edinburgh: Churchill Livingstone.
2. Ganong, W.F. (1987) *A Review of Medical Physiology*, 13th edition. Los Altos: Lange Medical Publications.
3. Jenkins, G.N. (1978) *The Physiology and Biochemistry of the Mouth*, 4th edition. Oxford: Blackwell Scientific Publications.
4. Lavelle, C.L.B. (1988) *Applied Oral Physiology*, 2nd edition. London: Wright.
5. West, J.B. (1991) *Best and Taylor's Physiological Basis of Medical Practice*, 12th edition. Baltimore: Williams and Wilkins.

Review questions

1. Describe the role of muscles during the first phase of deglutition.
2. What is the motor nerve supply to these muscles?
3. Describe the second and third phases of swallowing with reference also to those sites on the mucosa from where reflex actions might be initiated.

INDEX

A

Actin	147
Actinopterygii	228
Adolescent growth spurt	87, 88, 91
Adrenal glands	41, 81, 83, 90
Age changes of teeth	373, 375-377
Age determination	
chronologic age	88
dental age	89
skeletal age	89
Agglutinin	181
Agglutinogen	181
Agnatha	228, 230
Aldosterone	84, 90
Alimentary canal	165
Allelomorph	175
Alveolar crest	242, 245
Alveolar crest fibers	245, 424
Alveoli (of lungs)	169, 171
Alveologingival fibers	421
Ameloblasts	240, 254, 281, 282
Amelocemental junction	240, 244, 313
Amelodentinal junction	241, 244, 267, 277, 290
Amelogenins	285
Amniotic cavity	36, 95
Amphibia	220, 230
Amylase	222, 474
Androgen	84, 90
Animalculists	30
Anterior horn	108
Anthropoidea	221
Anthropometry	91
Antibody	69, 181
Antigen	181
Anura	230
Aortic sac	199
Apatite (see hydroxyapatite)	
Apical foramen	305, 307
Appendix (vermiform)	166
Aqueduct (of midbrain)	97
Areolar tissue	56
Aristotle	29
Arterial circulation of the head and neck	199-208
Arteries	
aorta	193
aortic arch	194, 199
basilar	200
brachiocephalic	194, 206
buccal	200
central (of retina)	200
cerebral	200
ciliary	206
common carotid	194, 206
common iliac	194
communicating	200
coronary	193, 194
dorsal nasal	200
ethmoidal	206
external carotid	200
facial	200, 211, 405, 471
hyoid	200
inferior alveolar	200
infraorbital	200
infratrochlear	206
internal carotid	200
lacrimal	206
lingual	200, 471
mandibular	200
maxillary (retromandibular)	200, 404

middle meningeal	200, 471
opthalmic	200
palpebral	206
posterior superior alveolar	200
pulmonary	193, 194
stapedial	200
subclavian	194, 206
superficial temporal	200
superior thyroid	200
supraorbital	200
supratrochlear	200
thoracic aorta	194
umbilical	196
Articular eminence	449
Articular tubercle	449
Artiodactyla	220
Arytenoid cartilage	151
Astrocytes	100
Atrioventricular valve	194, 196
Atrium (of heart)	193
Atrium (of nose)	163
Attrition	223, 235, 376
Auricle (of ear)	153
Autonomic nervous system (see nervous system)	
Autophagy	25
Autosomes	29, 35, 175
Aves	220
Axon	109, 112

B

Barr body	26, 177
Basal layer (of Weil)	320
Basement membrane, basal lamina	45, 46, 49, 50, 418, 380
Bipolar neurons	109, 130
Biting force	496
Blastocyst	35
Blastomeres	35
Bleeding time	188
Blood (general)	65-72
clotting	187
factors	188
groups	177, 181-186
plasma	70, 181, 188
platelets	69, 187
serum	181, 182, 188
transfusion	181
values	71

Bolus	505
Bone (general)	58-61
alveolar	42, 242, 409, 423
basal (supporting)	236, 409
cancellous (spongy)	58, 393, 395, 409, 411
cortical (compact)	58, 393, 395, 409
development (see osteogenesis)	
endosteum	61, 395
ethmoid	135, 401
frontal	401
hyoid	151
lacrimal	137, 401
lamellae	58, 60, 411
mandible (see mandible – general)	
marrow (medullary cavity)	60, 411
maxilla (see maxilla – general)	
nasal	163, 401
occipital	401
palatine	130, 135, 138, 404
parietal	401
periosteum	56, 61, 393, 397, 409
sphenoid	143, 401, 449
temporal	133, 145, 401, 449
trabeculae	60, 395, 411
vomer	135, 138, 401
woven	409
zygomatic	401
Bonnet	30
Brachycardia	197
Brain	97, 105
Brain flexures	100
Brain stem	105
Bronchi	169
Brush border	46
Bucconasal membrane	121, 127
Buccopharyngeal membrane	38, 95, 117, 118, 127
Bulbis cordis	199

C

Cainozoic Era	227, 228
Calcitonin	82, 154, 390
Calcium metabolism	389
absorption	390
daily intake	397
dietary sources	397
hormonal influences	390
Calcospherites	267, 273
Cardiogenic area	38

Cardiovascular system	193-197
Carnivora	220
Cartilage	61-63
Ceboidea	221
Cecum	165, 166
Celacanth	228
Cell (general)	19-28
cytoplasm	22
division	19, 26
growth	22
membrane	19, 20
rests (of Malassez)	240, 302, 423
Cementum (general)	309-315
accessory canals	307
acellular	244, 305, 309
age changes	377
amelocemental junction	240, 244, 313
cellular	244, 305, 309
cementoblasts	244, 302, 310, 415
cementocytes	244, 305, 310
classification	305
composition	241, 309
dentinocemental junction	279, 310
development	301-308
functions	313
general structure	309
hypercementosis	307
lamellae	310
matrix fibers	309
mineralization	302
permeability	374
Centrioles	23
Centromere	175
Centrosome	23
Cercopithecoidea	221
Cerebellum	100, 105
Cerebrum	100, 105
Cetacea	220
Character (trait)	176
Charles Darwin	215-218
Chief sensory nucleus (of trigeminal nerve)	114
Chiroptera	220
Chondrichthyes	229
Chondroblasts	53, 61
Chondrocranium	133
Chondrocytes	61, 62, 395
Chondrogenesis	61
Chordata	219
Chorion	38
Chromosomes	26, 173
Cilia	46, 48
Circle of Willis	200
Circular fibers	421
Circumferential lamellae	60
Circumvallate papillae	155, 385, 387
Classification (hierarchial system)	219
Clotting time	190
Co-dominance (genetic)	177
Coelom	38
Collagen	53-56, 392, 393, 421, 426, 447
Colon	165
Conduction of nerve impulses	437
Condylar (condyloid) process	405
Condylar secondary cartilage	145
Connective tissue (general)	27, 53-64
Copula	155
Cornea	41
Corona radiata	33
Coronoid secondary cartilage	145
Coronoid process	405
Cortex (of cerebrum)	114
Cretinism	81
Cribriform plate (of alveolar bone)	242
Cribriform plate (of skull)	130, 135
Cricoid cartilage	151
Crista galli	135
Crossopterygii	228
Cyclostomata	228
Cytoplasm	22

D

De Graaf	30
Deglutition	505-509
esophageal phase	508
frequency	508
nervous control	508
oral phase	505
pharyngeal phase	507
reflex activity	507
Demes	219
Demilune (of Gianuzzi)	464
Dendrite	109
Dental arches	324
Dental follicle	240, 254, 302, 415
Dental formula	
carnivores	234
herbivores	233
man	235, 324
original mammalian	232

rodents	233
Dental lamina	41, 247
Dental pain	437
theories	439-440
Dental papilla	240, 248, 252, 263
Dental pulp (general)	241, 241, 267, 317-322
accessory canals	307, 318
accessory foramina	318
age changes	377
apical constriction	244, 310
apical foramen	244, 305, 309, 318
blood vessels	318, 320
cells	318, 320
chamber	318
fibers	318, 320
functions	317
general structure	318
horns	318
lateral canals	318
lymphatics	321
nerve supply	321, 435
odontoblasts	244
root canal	244, 318
Dentine (general)	240, 271-280
age changes	375
amelodentinal junction	277
calcospherites	273, 275
circumpulpal	267, 272
composition	241, 271
contour lines (of Owen)	277
curvatures of tubules	244, 273
dead tracts	376
dentinocemental junction	279, 310
depth of processes	267, 272
development (dentinogenesis)	240, 254, 263-269
general structure	271
granular layer (of Tomes)	244, 277
hyaline layer	279, 301
incremental lines (of von Ebner)	275
innervation	272, 435-437
interglobular	244, 367, 275
intertubular	268
intratubular (peritubular)	268, 275
lateral branches of tubules	267, 272
mantle	264, 272
matrix fiber orientation	264, 272
mineralization	267, 272
nerve supply	435
odontoblast processes	244, 264, 267, 272
odontoblasts	244, 263, 272, 320
permeability	374
predentine	267, 273
primary	375
secondary	317, 375
terminal branches of tubules	267, 273
transparent (sclerotic)	376
tubule numbers	272
tubules	244, 267, 273
Dentogingival attachment	313, 415
Dentogingival fibers	421
Dentoperiosteal fibers	421
Deoxyribonucleic acid (DNA)	26, 173
Dermatome	42
Desmosomes	45, 49, 418
Diabetogenic hormone	83
Diaphragm	165, 193
Diaphysis	395
Diastema	224, 326
Diencephalon	97
Differentiation (embryonic)	36
Digastric fossa	405
Digestive system (general)	165-168
Dinosaurs	231
Diploid number (of chromosomes)	26, 29, 33, 173
Diplosome	23
Dipnoi	228
Dominance (genetic)	177
Dorsal aortae	199
Driesch	31
Ductus arteriosus	196
Ductus venosus	196
Dumas	30
Duodenum	165, 166

E

Ectoderm	35, 41, 117
Ectomesenchyme	42, 118, 257
Ectomorph	91
Effector cells (neurons)	106
Elastic cartilage	61, 62
Elastic fibers	53, 54
Elastin	54
Embryonic disc	36, 117
Enamel (general)	289-300
age changes	376
amelocemental junction	240, 244, 313
amelodentinal junction	244, 277, 290, 296
appearance	289

composition	241, 289	simple cuboidal	46, 48
cross-striations	244, 292	simple squamous	46
crystals	290	stratified columnar	46, 50
development (amelogenesis)	281-288	stratified cuboidal	46, 50
enamel cuticle	287	stratified squamous	46, 49
fluoride in	289	transitional	46, 50
general structure	290	Epithelium-ectomesenchyme	
gnarled	290	interaction	257-261
Hunter-Schreger bands	295	Erythrocytes (red blood corpuscles)	65, 181
lamellae	299	Esophagus	165
lines of Retzius	244, 292	Estrogen	84
matrix	285, 289	Ethmoidal sinus	130, 135
maturation	286	Eukariotic cell	19
mineralization	285, 286	*Eutheria*	220
neonatal line	295	Evolution, theory of	217
organ	240, 248, 252	Exocytosis	25
pearls	307	Exopthalmos	82
perikymata	244, 295	External auditory meatus, canal	149, 153
permeability	299, 373	External basal (attachment) lamina	420
reduced enamel epithelium	286	External (outer) enamel epithelium	239, 249
rods	241, 244, 290	Exteroceptive stimuli	106
rod sheath	241, 290		
spindles	277, 296		
Tomes process	285, 290	**F**	
tufts	296		
Enamel organ	41	Face (development)	117-126
Enamelins	285	Facial buttresses	411
Endocardium	42	Fainting	196
Endochondral ossification	133, 135, 145	Fauces	159
Endocrine system (general)	81-85	Fertilization	33
Endocytosis	25	Fetal blood circulation	196
Endoderm	36, 41, 117	Fetus	36, 79
Endomorph	91	Fibrin	188
Endomysium	78	Fibrinogen	188
Endoneurium	112	Fibroblasts	45, 53, 56, 415
Endoplasmic reticulum	22, 109	Fibronectin	45
Ependymal layer	100	Fibrous cartilage (fibrocartilage)	61, 62
Epigenesis	30	Filiform papillae	155, 385
Epiglottis	155	Flagella	46
Epimysium	78	Fluorapatite	488
Epinephrine (adrenaline)	83	Fluoride (general)	481-491
Epineurium	112	absorption	483
Epiphysis	395	chemical nature	483
Epithelial diaphragm	305	detrimental effect on teeth	490
Epithelium (general)	45-51	history	481
non-keratinized	382	in blood	484
orthokeratinized	49, 380	in milk	485
parakeratinized	49, 382, 418	in saliva and digestive juices	484
pseudostratified ciliated	48	in soft tissues	485
pseudostratified columnar	46, 48	in sweat	484
simple columnar	46, 48	in urine	484

in teeth	487
protective effect on enamel	489
sources	483
toxicity	485
transfer to fetus	485
uptake in teeth	488
Fluorosis	490
Foramen cecum (of tongue)	155
Foramen ovale (of heart)	196
Foramen ovale (of skull)	112, 141, 471
Foramen magnum	105, 451
Free gingival groove	416
Free (marginal) gingiva	416
Frenulum	159
Frontal sinus	160, 163
Frontonasal process	118, 137
Fungiform papillae	155, 385

G

Gamete	29, 33, 173
Genes	173, 175
Genetics (general)	173-180
Genial tubercles	405
Genotype	91, 176, 217
Genus	219
Geological time	227
Germ layers	35
Germinal centers	211
Gingiva (general)	41, 416-423
attached	416
blood supply	426
col	416
collar	420
crest	245, 416
description	383, 416
development	240, 430
fibers	245, 421
free (marginal)	245, 416
histology	383, 416
junctional epithelium	240, 245, 416, 420
lamina propria	421
sulcular (crevicular) epithelium	240, 245, 416, 420
sulcus	240, 245, 416
Gland (see individual glands)	
Glucagon	84
Glycoprotein	45, 54
Glycosaminoglycan	45, 61, 249, 264, 285
Goblet cells	48

Golgi apparatus	23
Gonads	29, 81, 84, 90
Gonadotrophic hormone	90
Greater palatine canal	404
Greater palatine foramen	404
Grey matter	100, 103, 108
Ground substance	27, 54, 393
Growth and development (general)	87-94
Growth (somatotrophic) hormone	83, 90
Gubernacular canal	443

H

Halitosis	478
Hamm	30
Haplodontism	231
Haploid number (of chromosomes)	29, 173
Hartsoeker	30
Haversian canals (systems)	58, 397, 411
Head fold	38, 95, 117
Heart (general)	193
Heart valves	193
Hemidesmosomes	45
Hemoglobin	65
Hemophilia	178
Hemostasis	187-191
Herbivores	233
Hertwig	30
Heterodontism	222, 232
Heterozygotes	176, 217
History of embryology	29-31
Hominidae	219, 221
Hominoidea	221
Homodontism	222, 231
Homologous chromosones	173
Homo sapiens	219
Homozygotes	176, 217
Hyaline cartilage	61, 62
Hyaluronic acid	56
Hyaluronidase	56
Hydrocortisone	83, 90
Hydrodynamic theory of dental pain	440
Hydroxyapatite	241, 267, 289, 302, 391, 487
Hydroxylysine	54
Hydroxyproline	54
Hyoid arch (see pharyngeal arches)	
Hyperglycemia	84
Hyperparathyroidism	82, 391
Hyperthyroidism	81
Hypobranchial eminence	155

Hypoglossal cord	147, 155
Hypoglycemia	84
Hypoparathyroidism	82
Hypothalamus	97
Hypothyroidism	81, 90

I

Ileum	165
Immune system	68, 83, 211
Immunoglobulins (IgA, IgD, IgE, IgG, IgM)	69
Implantation	35
Incisive canals	130, 161, 405
Incisive foramen	137
Incisive papilla	161
Incus	143, 151
Inferior conchae (turbinates)	130, 135, 160
Inferior meatus	163
Infraorbital canal	404
Infraorbital foramen	404
Infratemporal fossa	404
Inner cell mass	35
Insectivora	220
Insulin	84
Intercalated discs	78
Intercalated ducts	465
Interdental papillae	416
Intermediate cell layer (tooth germ)	239, 249, 282
Intermediary neuron	103, 108
Intermediate plexus	433
Internal ear	154
Internal basal (attachment) lamina	420
Internal (inner) enamel epithelium	239, 248, 281
Interoceptive impulses	106
Inter-radicular fibers	424
Interstitial lamellae	60
Intervertebral foramina	103
Intramembranous ossification	141

J

Janssen	29
Jaws (general)	401-408
Jejunum	165, 166

K

Karyotype	173
Keratin	49
Keratinocytes	380
Keratohyalin	49, 380, 418
Klinefelter syndrome	88

L

Lacrimal gland	108
Lamina	
alar	100
basal (nervous system)	100
densa	45, 418
dental	239, 247
dura	411
lucida	45, 418
propria	379, 382
reticular	45
vestibular	239, 247
Laminin	45
Landsteiner	181, 184
Langerhans cells	382
Laryngopharynx	163
Larynx	163, 507
Lateral nasal folds	118
Leukocytes	20, 65, 66-69
basophil	65, 68
eosinophil	65, 66
neutrophil	65, 66
non-granular	68
polymorphonuclear (granular)	65
Ligaments	56
Ligamentum arteriosum	196, 206
Lingual tonsillar tissue	155, 385
Lining mucosa	380
Linnaeus	219
Liver	38, 41
Lungs	41, 169
Lymph	65, 70, 209
Lymph nodes (glands)	70, 209, 211
buccal	211
deep cervical	212
jugulodigastric	212
jugulo-omohyoid	212
mandibular	211
mastoid	212
occipital	212
parotid	212

submandibular	211
submental	211
superficial cervical	212
Lymph vessels	70, 209
Lymphatic drainage (general)	209-214
Lymphocytes	68, 211, 317
B-	68, 211
T-	69, 211
Lymphokines	69
Lysosomes	25, 66
Lysozyme	66

M

Macroglossia	157
Macrophage	56, 69, 211, 317
Malleus	143, 151
Mammalia	219, 220, 230
Mandible (general)	405-408
alveolar process	405
angle	408
body	141, 405
condylar process	145, 405
condyle	145, 405
coronoid process	145, 405
development	141-145
lingula	143, 408
mandibular foramen	143, 408
mental foramen	141
mylohyoid line	408
notch	405
oblique line	405
ramus	143, 405
symphysis	143, 405
Mandibular arch (see pharyngeal arches)	
Mandibular canal	143, 408
Man's classification	219, 221
Mantle layer	100
Marginal layer	100
Marsupialia	232
Mast cells	56, 317
Mastication (general)	493-500
and digestion	498
dynamics	496
movement of teeth	495
nervous control	496
occlusal contacts	495
role of hard palate	494
role of lips and cheeks	494
role of tongue	494
Masticatory mucosa	380
Maxilla (general)	401-405
alveolar process	401, 409
body	401
buttresses	411
development	130, 135-137
osteology	401
processes	135, 401, 404
sinus	130, 135, 138, 163, 401, 404, 409
surfaces	401, 404
tuberosity	404
Maxillary process	118, 127, 135
Meckel's cartilage	133, 141, 143, 149
Medial nasal fold	118
Mediastinum	169, 194
Medulla oblongata	105, 114
Meiosis	26, 29, 33
Melanin	382
Melanocytes	382
Menarche	92
Mental foramen	141, 405
Merkel cells	382
Mesaxon	112
Mesencephalic nucleus	112, 114
Mesencephalon	97
Mesenchyme (ectomesenchyme)	42, 53, 118, 121, 135, 149, 199, 252, 257, 258
Mesoderm	38, 41
Mesomorph	88, 91
Mesothelium	38
Mesozoic Era	227
Metaphase	175
Metencephalon	100
Microfilaments	25
Microglossia	157
Microtubules	25
Microvilli	22, 46, 48
Midbrain	97, 105
Middle concha (turbinate)	130, 135
Middle ear	41, 147, 149, 154
Middle meatus (of nose)	130, 138, 163
Midpalatal suture	137
Mitochondria	23, 109
Mitosis	19, 26, 33, 175
Mitral valve	194
Monocytes	68, 317
Morula	35
Motor impulses	106
Motor nucleus (of trigeminal nerve)	114
Mouth (general)	159-162
Mucinogen	462

Mucogingival junction	416
Mucous cells	462
Multipolar neurons	109
Muscle (general)	27, 42, 73-78, 147
auricular	153
buccinator	125, 153, 159, 211, 212, 327, 505
cricothyroid	153
development	147
digastric	151, 153, 212, 507
genioglossus	155, 505
geniohyoid	507
hyoglossus	155
intrinsic (of tongue)	155
lateral pterygoid	151, 200, 223, 451, 454
levator palati	153, 506
masseter	151, 408, 454
medial pterygoid	151, 408
mylohyoid	151, 214, 505, 507
occipitofrontalis	125, 153
of face	153
of larynx	153
of mastication	147, 151
of pharynx	153
omohyoid	212
orbicularis oris	327
palatoglossus	153, 155, 157, 159, 505
palatopharyngeus	153, 159, 507
platysma	125, 153
smooth	73
stapedius	153
sternocleidomastoid	212
striated cardiac	73, 78
striated skeletal	73, 74
styloglossus	155, 505
stylohyoid	153, 507
stylopharyngeus	153, 507
temporalis	151, 454
tensor palati	151, 506
tensor tympani	151
trapezius	212
Myelencephalon	100
Myelin	103, 105, 112
Myoblasts	74, 147
Myocardium	42
Myoepithelial cells	465
Myofibrils	73, 75, 147
Myofilaments	74, 147
Myosin	147
Myotome	42
Myxedema	81

N

Nasal capsule	133, 135, 138
Nasal cavities	133, 163
Nasal fin	118
Nasal sac	121
Nasal septum (secondary)	130, 135, 138, 166
Nasolacrimal duct	125, 404
Nasopalatine canal	130
Nasopharynx	163
Natural selection	217
Nerve	
anterior superior alveolar	135
auriculotemporal	471
cervical	42, 103
chorda tympani	155, 157, 471
classification in dental pulp	438
classification of neurons	438
coccygeal	42, 103
cranial	105
facial	153, 157
glossopharyngeal	153, 157, 471
hypoglossal	155
impulse conduction	437
incisive	141
inferior alveolar	141, 408
infraorbital	135, 137
internal laryngeal	157
lingual	141, 157, 471
lumbar	42, 103
mandibular	112, 141, 151, 471
maxillary	112, 404, 471
membrane potential	437
mental	141
mylohyoid	408
olfactory	135
opthalmic	112
optic	97
posterior superior alveolar	404
recurrent laryngeal	153, 206
sacral	42, 103
spinal	42, 103, 105
superior laryngeal	153, 157
thoracic	42, 103
trigeminal	112, 157, 321
vagus	153, 157, 196
Nerve supply and sensitivity of teeth	435-441
Nervous system (general)	105-115
autonomic	97, 103, 106
central	97, 105
development	95-104

functions	105-115
ganglia	97, 100, 103, 106, 257
histology	109-112
parasympathetic	103, 106, 471
peripheral	97, 105
somatic	97, 106
sympathetic	103, 106, 471
Neural crest cells	97, 114, 257
Neural groove	97, 257
Neural plate	97
Neural tube	97, 257
Neurilemma	109
Neuroblasts	100, 130
Neurofibrils	109
Neurons	100, 103, 105, 109
Neuropore	97
Nissl substance	109
Nociceptors	437
Nodes of Ranvier	112
Notochord	30, 38, 42, 95, 127, 199, 220
Nuclear envelope (membrane)	25
Nuclear pore	25
Nucleolus	26
Nucleus	19, 24

O

Oblique fibers	424
Odontoblasts	240, 244, 254, 263, 264, 301, 320
Olfactory bulb	130
Olfactory epithelium	130
Olfactory (nasal) pits	118
Olfactory (nasal) placodes	118, 130
Olfactory sulcus	163
Olfactory tracts	130
Oligodendrocytes	100, 112
Omnivores	235
Ontogeny	29
Optic (lens) placodes	118
Optic vesicle	79
Oral mucosa (general)	379-388
classification	380
origin of epithelium	41, 379
pigmentation	382, 416, 418
Oropharynx	159, 163, 379, 506
Orthokeratin	49, 380
Osteichthyes	228
Osteoblasts	53, 58, 391, 415
Osteoclasts	392
Osteocytes	58, 393, 411

Osteogenesis (ossification)	58, 391-397
chemical aspects	391
endochondral	58, 391, 451
intramembranous	58, 141, 391
Osteoid	393
Otic capsule	133, 141
Otic ganglion	108, 471
Otic placode	154
Otic vesicle	79
Ovaries	84
Ovists	30
Ovulation	33
Ovum	29, 33, 36, 84, 173
Oxyhemoglobin	65
Oxytalan fibers	424

P

Palaeozoic Era	227
Palatal process	127
Palate	
closure (fusion)	130
development	127-132
hard	130, 159, 161, 383
histology	383
morphology	161
primary	121, 127, 130
rugae	161, 383
secondary	127, 130
soft	130, 159, 161, 163, 383
Palatine tonsils	154, 159
Pancreas	41, 81, 84, 166
Parafollicular (clear) cells (C cells)	82
Parakeratin	49, 382
Paranasal sinus	41, 163
Parasympathetic ganglion	108
Parathormone	82, 154, 390
Parathyroid gland	41, 81, 82, 154
Paravertebral ganglia (sympathetic chain)	108
Parotid (Stensen's) duct	159, 459
Parotid gland	108, 159
Parthenogenesis	30
Pericardial swelling, sac	118, 193
Perichondrium	56, 61, 395, 451
Perikaryon	109
Perimysium	78
Perineurium	112
Periodontal ligament	240, 423-426, 429-430, 432-433
blood supply	426

cell rests (of Malassez)	240, 302
development	42, 430
fibers	242, 245, 415, 424
metabolism	429
nerve supply	429
proprioception	106, 114
space	242, 423
width	242
Periodontium (general)	242, 415-434
Periosteum (see bone)	
Perissodactyla	220
Permeability of teeth	373-375
Peroxidase	66
Phagocytosis	20, 25, 66
Phagolysosome	25
Phagosome	25
Pharyngeal arches (general)	118, 149-158
arteries	149, 199
development	149
grooves	80, 149
hyoid (second)	125, 143, 149
mandibular (first)	118, 135, 141, 149
nerves	149, 151, 153
pouches	149, 154
Pharyngotympanic tube (Eustachian tube)	41, 154
Pharynx	163, 199
Phenocopy	176
Phenotype	91, 176, 217
Philtrum (of lip)	125
Pinocytosis	25
Pisces	220
Pituitary gland	81, 82, 90, 97, 117
Placenta	35, 38, 196, 220
Plasma cells	56, 211, 317
Plasma membrane (cell membrane)	19
Pleura	193
Plexus of Raschkow	321, 437
Polyphyodontism	229, 231
Pongidae	221
Pons	100, 105, 112
Posterior horn	108
Postganglionic fibers	106, 112
Post-trematic nerve	157
Precambrian Era	227
Prechordal plate	38, 95, 117
Preformation	29, 215
Preganglionic fibers	103, 106
Premaxilla	130, 135, 137
Pre-odontoblasts	263
Pre-trematic nerve	153, 157

Prévost	30
Primary brain vesicles	97
Primary epithelial thickening	247
Primary jaw joint	143, 147
Primary (primitive) nasal septum	127
Primary nodules	211
Primary palate	118, 121, 127, 130
Primates	219, 221
Primitive streak	38
Proboscidae	220
Process	
alveolar	130, 236, 239, 242, 404, 409
alveolar (of maxilla)	135, 411
condyloid (condylar)	405
coronoid	405
frontal	135, 404, 412
frontonasal	118, 121, 125, 127, 137
globular	121
mandibular	118, 125
maxillary	118, 121, 125, 127
palatal	127, 135, 404
zygomatic	135, 404, 412, 449
Progesterone	84
Prognathism	222
Prokariotic cells	19
Proline	54
Proprioceptive impulses	106, 114
Prosencephalon	97
Prosimii	221
Proteoglycan	45, 54, 285, 302
Prothrombin	188
Protoplasm	21
Pseudo-unipolar neuron	109
Pterygomaxillary fissure	200
Pterygopalatine fossa	200 404, 471
Pulmonary valve	194
Pulse	194

R

Rami communicantes	103
Rathke's pouch	117
Recessivity (genetic)	177
Reduced enamel epithelium	281, 286, 430
Reflex arc	108
Reptilia	220
Respiratory system	169-172
Reticular cells	56
Reticular fibers	53, 54, 56
Reticular lamina	45

Retromolar cushion	161
Rhesus (Rh) factor	183-185
Rhombencephalon	97
Ribonucleic acid (RNA)	21
Ribosomes	21, 109
Right lymphatic duct	209, 212
Rodentia	220
Root sheath (of Hertwig)	240, 252, 301

S

Saliva (general)	469-479
amylase activity	474
buffering action	476
composition	477
constituents	469
factors controlling flow	470
functions	474
immunoglobulins	474
modification of composition	477
nervous control	471
pH	469
volume	469
Salivary glands (general)	459-467
classification	460, 466
development	459
histology of secretory cells	462
major	466
minor	466
mixed glands	464
nerve supply	108, 471
secretory ducts	462, 465
structural components	460
Sarcolemma	73, 74
Sarcomere	74
Sarcopterygii	228
Schleiden	30
Schwann cell	30, 97, 112
Sclerotome	42
Sebaceous glands	41
Secondary dentition	235
Secondary nasal septum	127
Secondary nodule	211
Septomaxillary complex (general)	133-139
Septum transversum	38, 95
Serotonin	187
Serous cells	464
Sex chromosomes	26, 35, 175, 177
Sex determination	35, 177
Sex-linked inheritance	178
Sharks	229
Sharpey's fibers	302, 309, 310, 411, 415, 423
Shedding of primary teeth	314
Singer-Nicholson model	20
Somatic cells	35, 173
Somites	41, 42, 79
Spallanzani	30
Specialized mucosa	380, 385
Species	219
Spermatozoon	29, 33, 84, 173
Sphenoidal sinus	130
Sphenomalleolar ligament	143, 151
Sphenomandibular ligament	143, 151, 408, 452
Sphenopalatine ganglion	108, 471
Spinal cord	97, 103, 105
Spinal nerves (development)	103
Spinal nucleus	114
Spinothalamic tracts	108
Spleen	42
Spontaneous generation	29, 215
Squamotympanic fissure	449, 452
Stapes	143, 151
Stellate reticulum	239, 249
Stomach	165
Stomodeum	95, 117
Stratum corneum (keratin layer)	49, 380
Stratum germinativum (basal cell layer)	49, 380, 418
Stratum granulosum (granular cell layer)	49, 380, 418
Stratum intermedium (epithelium)	382
Stratum spinosum (prickle cell layer)	45, 49, 380, 418
Stratum superficiale	382
Striated ducts	465
Stylohyoid ligament	151
Styloid process	151
Sublingual gland	108, 159
Submandibular ganglion	108, 471
Submandibular gland	108, 159, 212
Submucosa	383
Subnucleus caudalis	114, 438
Subnucleus interpolaris	114
Subnucleus rostralis	114
Sulcus limitans	100
Superior cervical ganglion	108, 471
Superior concha (turbinate)	130, 135
Superior meatus	163
Superior parathyroid gland	154
Sweat glands	41

Sympathetic ganglia	108	permeability	373, 374
Sympathetic nervous system (see nervous system)		premolars	347-355
		primary (deciduous)	235, 323, 329-338
Symphyseal secondary cartilage	145	pulp cavity (see dental pulp)	
Synapse	108, 109	secondary (permanent)	235, 323, 339-363
		shedding	314
		surface terminology	327, 328

T

		Telencephalon	97
		Temporomandibular joint (general)	449-457
Tachycardia	197	articular disc (meniscus)	145, 454
Tail fold	95	articular eminence	147
Taste (general)	501-504	articular surfaces	449
adaptation	503	collateral ligaments	449, 452
and chemical structure	502	development	145-147
associated stimuli	503	fibrous capsule	147, 452
definition	501	joint cavities	145, 454
flavor	503	mandibular condyle	451
location of taste buds	501	mandibular (glenoid) fossa	449
preferences	502	movements	455, 493
primary tastes	501	proprioception	106
sensitivity of tongue	501, 502	Tendons	56
Taxon	219	Terminal sulcus	155, 385
Teeth (general)	323-365	Tertiary Period	227
age changes	375	Testes	33, 84, 90
apices	305, 307, 311	Testosterone	84, 90, 91
arches	326	Tetany	82
canines	329-331 (primary), 342, 345-347 (secondary)	*Tetrapoda*	230
chronology of development	338 (primary), 363 (secondary)	Thalamus	97, 114
		Thoracic duct	209, 212
cingulum	326	Thrombin	188
comparative anatomy	227-236	Thromboplastin	70, 178, 187
composition	367-371	Thymus gland	41, 81, 83, 154
contact areas (points)	326	Thyroglossal duct	155
crown morphology, determination	252, 281	Thyroid cartilage	81
development (general)	239-240, 247-255, 257-261	Thyroid diverticulum	155
		Thyroid follicles	81
differences between primary and secondary teeth	363	Thyroid gland	41, 81, 154, 155
		Thyroxin	81, 90
embrasures	326	Tongue (general)	41, 159, 385-388
eruption dates	338 (primary), 363 (secondary)	abnormalities	157
		development	154-157
FDI numbering system	324	lymphatic drainage	214
formulae of dentitions	324	morphology	385
incisors	329, 331 (primary), 339, 342 (secondary)	mucosa	385
		nerve supply	153, 157
		papillae	155, 385-387
molars	331, 333-336 (primary), 355-363 (secondary)	taste buds	153, 155, 385-388, 501
		Toothache (see dental pain)	
nerve supply	435	Tooth buds	239, 248
occlusal surface	326	Tooth eruption (general)	240, 427, 428, 430-433, 443-448
occlusion (primary dentition)	337		

active	432
biochemical changes	430
chronology	323, 338 (primary), 363 (secondary)
definition and phases	443
histology	427, 428, 430-433
passive	421
rodent experiments	445
role of collagen	447
role of fibroblasts	447
theories	444
Tooth germ	240, 254
Trachea	41, 48, 169
Tracheal cartilages	151
Tracheobronchial groove	155
Trans-septal fibers	421
Trigeminal ganglion	112
Trophoblast	35
Tropocollagen	54
Tuberculum impar	155
Tubotympanic recess	153, 154
Turner syndrome	88
Tympanic membrane	41, 153

U

Umbilical cord	38
Unipolar neurons	109
Unit membrane	20
Upper lip	125, 159
Uterine tube	33
Uvula	130, 161

V

Van Leeuwenhoek	30
Vasovagal episode	196
Veins	
coronary	193
external jugular	212
facial	211
hepatic	196
inferior vena cava	194
internal jugular	209, 212
posterior auricular	212
pulmonary	194
retromandibular	212
subclavian	209
superior vena cava	194
umbilical	196
vitelline	196
Ventricles (of brain)	97
Ventricles (of heart)	193
Vertebrata	220
Vestibule (of mouth)	159, 239, 247
Vestibule (of nose)	163
Vitamin C	432, 447
Vitamin D	390
Volkmann's canals	60, 244, 411, 423
Von Baer	30
Von Beneden	30
Von Korff, fibers of	263, 264, 320

W

White matter	100, 103
Wiener	184
Wolff	30

Y

Yolk sac	36, 95

Z

Zygomatic process	135
Zygote	19, 29, 33, 173